Functional Nanostructures for Sensors, Optoelectronic Devices and Drug Delivery

Functional Nanostructures for Sensors, Optoelectronic Devices and Drug Delivery

Editor

Maria Angela Castriciano

MDPI • Basel • Beijing • Wuhan • Barcelona • Belgrade • Manchester • Tokyo • Cluj • Tianjin

Editor
Maria Angela Castriciano
CNR-Ismn Istituto per lo Studio dei Materiali Nanostrutturati
Italy

Editorial Office
MDPI
St. Alban-Anlage 66
4052 Basel, Switzerland

This is a reprint of articles from the Special Issue published online in the open access journal *Nanomaterials* (ISSN 2079-4991) (available at: https://www.mdpi.com/journal/nanomaterials/special_issues/func_nano_sens_opt_devic).

For citation purposes, cite each article independently as indicated on the article page online and as indicated below:

LastName, A.A.; LastName, B.B.; LastName, C.C. Article Title. *Journal Name* **Year**, *Article Number*, Page Range.

ISBN 978-3-03943-216-5 (Hbk)
ISBN 978-3-03943-217-2 (PDF)

© 2020 by the authors. Articles in this book are Open Access and distributed under the Creative Commons Attribution (CC BY) license, which allows users to download, copy and build upon published articles, as long as the author and publisher are properly credited, which ensures maximum dissemination and a wider impact of our publications.
The book as a whole is distributed by MDPI under the terms and conditions of the Creative Commons license CC BY-NC-ND.

Contents

About the Editor vii

Maria Angela Castriciano
Functional Nanostructures for Sensors, Optoelectronic Devices, and Drug Delivery
Reprinted from: *Nanomaterials* 2020, *10*, 1195, doi:10.3390/nano10061195 1

Mario Prosa, Margherita Bolognesi, Lucia Fornasari, Gerardo Grasso, Laura Lopez-Sanchez, Franco Marabelli and Stefano Toffanin
Nanostructured Organic/Hybrid Materials and Components in Miniaturized Optical and Chemical Sensors
Reprinted from: *Nanomaterials* 2020, *10*, 480, doi:10.3390/nano10030480 7

Dongping Xue, Yan Wang, Jianliang Cao and Zhanying Zhang
Hydrothermal Synthesis of CeO_2-SnO_2 Nanoflowers for Improving Triethylamine Gas Sensing Property
Reprinted from: *Nanomaterials* 2018, *8*, 1025, doi:10.3390/nano8121025 39

Kaixiang Yang, Zhengguang Yan, Lin Ma, Yiping Du, Bo Peng and Jicun Feng
A Facile One-Step Synthesis of Cuprous Oxide/Silver Nanocomposites as Efficient Electrode-Modifying Materials for Nonenzyme Hydrogen Peroxide Sensor
Reprinted from: *Nanomaterials* 2019, *9*, 523, doi:10.3390/nano9040523 51

Edyta Turek, Bogumila Kumanek, Slawomir Boncel and Dawid Janas
Manufacture of Networks from Large Diameter Single-Walled Carbon Nanotubes of Particular Electrical Character
Reprinted from: *Nanomaterials* 2019, *9*, 614, doi:10.3390/nano9040614 65

Yuan-Chang Liang and Che-Wei Chang
Improvement of Ethanol Gas-Sensing Responses of ZnO–WO_3 Composite Nanorods through Annealing Induced Local Phase Transformation
Reprinted from: *Nanomaterials* 2019, *9*, 669, doi:10.3390/nano9050669 77

Amine Achour, Mohammad Islam, Sorin Vizireanu, Iftikhar Ahmad, Muhammad Aftab Akram, Khalid Saeed, Gheorghe Dinescu and Jean-Jacques Pireaux
Orange/Red Photoluminescence Enhancement Upon SF_6 Plasma Treatment of Vertically Aligned ZnO Nanorods
Reprinted from: *Nanomaterials* 2019, *9*, 794, doi:10.3390/nano9050794 89

André Luiz Tessaro, Aurore Fraix, Ana Claudia Pedrozo da Silva, Elena Gazzano, Chiara Riganti and Salvatore Sortino
"Three-Bullets" Loaded Mesoporous Silica Nanoparticles for Combined Photo/Chemotherapy
Reprinted from: *Nanomaterials* 2019, *9*, 823, doi:10.3390/nano9060823 105

Zhaotian Cai, Yabing Ye, Xuan Wan, Jun Liu, Shihui Yang, Yonghui Xia, Guangli Li and Quanguo He
Morphology–Dependent Electrochemical Sensing Properties of Iron Oxide–Graphene Oxide Nanohybrids for Dopamine and Uric Acid
Reprinted from: *Nanomaterials* 2019, *9*, 835, doi:10.3390/nano9060835 119

Ganesan Mohan Kumar, Pugazhendi Ilanchezhiyan, Hak Dong Cho, Shavkat Yuldashev, Hee Chang Jeon, Deuk Young Kim and Tae Won Kang
Effective Modulation of Optical and Photoelectrical Properties of SnS_2 Hexagonal Nanoflakes via Zn Incorporation
Reprinted from: *Nanomaterials* **2019**, *9*, 924, doi:10.3390/nano9070924 139

Mariachiara Trapani, Maria Angela Castriciano, Andrea Romeo, Giovanna De Luca, Nelson Machado, Barry D. Howes, Giulietta Smulevich and Luigi Monsù Scolaro
Nanohybrid Assemblies of Porphyrin and Au_{10} Cluster Nanoparticles
Reprinted from: *Nanomaterials* **2019**, *9*, 1026, doi:10.3390/nano9071026 151

Xiangyang Li, Ling Li, Huancheng Zhao, Shuangchen Ruan, Wenfei Zhang, Peiguang Yan, Zhenhua Sun, Huawei Liang and Keyu Tao
$SnSe_2$ Quantum Dots: Facile Fabrication and Application in Highly Responsive UV-Detectors
Reprinted from: *Nanomaterials* **2019**, *9*, 1324, doi:10.3390/nano9091324 165

Valentina Paolucci, Seyed Mahmoud Emamjomeh, Michele Nardone, Luca Ottaviano and Carlo Cantalini
Two-Step Exfoliation of WS_2 for NO_2, H_2 and Humidity Sensing Applications
Reprinted from: *Nanomaterials* **2019**, *9*, 1363, doi:10.3390/nano9101363 175

Karolina Kniec, Karolina Ledwa and Lukasz Marciniak
Enhancing the Relative Sensitivity of V^{5+}, V^{4+} and V^{3+} Based Luminescent Thermometer by the Optimization of the Stoichiometry of $Y_3Al_{5-x}Ga_xO_{12}$ Nanocrystals
Reprinted from: *Nanomaterials* **2019**, *9*, 1375, doi:10.3390/nano9101375 193

Marta Rubio-Camacho, Yolanda Alacid, Ricardo Mallavia, María José Martínez-Tomé and C. Reyes Mateo
Polyfluorene-Based Multicolor Fluorescent Nanoparticles Activated by Temperature for Bioimaging and Drug Delivery
Reprinted from: *Nanomaterials* **2019**, *9*, 1485, doi:10.3390/nano9101485 203

Ana Isabel Ruiz-Carmuega, Celia Garcia-Hernandez, Javier Ortiz, Cristina Garcia-Cabezon, Fernando Martin-Pedrosa, Ángela Sastre-Santos, Miguel Angel Rodríguez-Perez and Maria Luz Rodriguez-Mendez
Electrochemical Sensors Modified with Combinations of Sulfur Containing Phthalocyanines and Capped Gold Nanoparticles: A Study of the Influence of the Nature of the Interaction between Sensing Materials
Reprinted from: *Nanomaterials* **2019**, *9*, 1506, doi:10.3390/nano9111506 221

Massimiliano Magro and Fabio Vianello
Bare Iron Oxide Nanoparticles: Surface Tunability for Biomedical, Sensing and Environmental Applications
Reprinted from: *Nanomaterials* **2019**, *9*, 1608, doi:10.3390/nano9111608 235

Gerardo Grasso, Daniela Zane and Roberto Dragone
Microbial Nanotechnology: Challenges and Prospects for Green Biocatalytic Synthesis of Nanoscale Materials for Sensoristic and Biomedical Applications
Reprinted from: *Nanomaterials* **2020**, *10*, 11, doi:10.3390/nano10010011 255

Maria Angela Castriciano, Mariachiara Trapani, Andrea Romeo, Nicoletta Depalo, Federica Rizzi, Elisabetta Fanizza, Salvatore Patanè and Luigi Monsù Scolaro
Influence of Magnetic Micelles on Assembly and Deposition of Porphyrin J-Aggregates
Reprinted from: *Nanomaterials* **2020**, *10*, 187, doi:10.3390/nano10020187 275

About the Editor

Maria Angela Castriciano (Messina, Italy) is a researcher at the Institute for the Study of Nanostructured Material (ISMN), Department of Chemistry and Technology of Materials (DSCTM), National Research Council (CNR), Messina, Italy. She received her PhD in February 2002 from the Chemistry Department of the University of Messina, Italy. Her research interests span a broad range of topics at the interfaces of supramolecular, macromolecular and material chemistry. She is an expert in: (i) the synthesis, functionalization and characterization of nano and mesoscopic structures of organic, inorganic and hybrid systems for applications in the field of sensors, optoelectronics and drug release; (ii) aggregation phenomena in solution and supramolecular organization of chromophoric systems on polymeric and/or biological matrices and confined environments (with particular attention to the expression and transmission of chirality on a nano and mesoscopic scale); (iii) the design, synthesis and characterization of organic dyes and their derived metals for use as photosensitizers or composite polymeric membranes for applications in polymeric electrolyte fuel cell technology; and (iv) the characterization of chromophoric systems by static and time resolved spectroscopic and/or light scattering techniques. She has published more than 60 articles in SCIE journals and has been cited more than 1300 times. She has an h-index of 21 (May 2020).

Editorial

Functional Nanostructures for Sensors, Optoelectronic Devices, and Drug Delivery

Maria Angela Castriciano

Istituto Per Lo Studio Dei Materiali Nanostrutturati, c/o Dipartimento di Scienze Chimiche, Biologiche, Farmaceutich ed Ambientali, University of Messina, 31 98166 Messina, Italy; maria.castriciano@cnr.it

Received: 22 May 2020; Accepted: 16 June 2020; Published: 19 June 2020

Nanoparticles and nanostructured materials represent an active area of research for their impact in many application fields. The recent progress obtained in the synthesis of nanomaterials and the fundamental understanding of their properties led to significant advances for their technological applications. The scope of the Special Issue "Functional Nanostructures for Sensors, Optoelectronic Devices, and Drug Delivery" was to provide an overview of the current research activities in the field of nanostructured materials, with a particular emphasis on their potential applications for sensors [1–9], optoelectronic devices [10–15], and biomedical systems [16–18]. The Special Issue welcomed the submission of original research articles [1–6,9–17] and comprehensive reviews that demonstrated or summarized significant advances in the above-mentioned research fields. Next, the Special Issue collected fifteen selected original research papers and three comprehensive reviews [7,8,18] on various topics of nanostructured materials and relative characterization spanning from fundamental research to technological applications. More than 100 scientists from universities and research institutions participated with their research activities and expertise in the success of this Special Issue.

The scientific contributions are summarized here.

A facile one-step hydrothermal synthesis reaction for obtaining flower-like CeO_2-doped SnO_2 nanostructures was reported. The composite system showed improved gas-sensing performances, possibly due to the formation of n-n heterojunctions between CeO_2 and SnO_2 and due to the presence of Ce^{4+}/Ce^{3+} species in SnO_2 that facilitate the interaction of electrons [1].

An easy meta-ion mediated hydrothermal route was successfully used to prepare cubic, thorhombic, and discal Fe_2O_3 NPs with a uniform size and controllable structure that were coupled with graphene oxide (GO) nanosheets. The influence of Fe_2O_3/GO morphologies on electrochemical sensing performances was studied systematically. The synergistic effect from d-Fe_2O_3 NPs and GO nanosheets operated as sensing films for the simultaneous detection of dopamine and uric acid [2].

The potential to detect ethanol vapor in an open environment was reported for ZnO–WO_3 composite nanorods, which exhibited high selectivity for this alcohol among the various target gases of NH_3, H_2, and NO_2. These composite nanorods were obtained by a combination of hydrothermal growth, sputtering methodologies, and a thermal annealing procedure in a hydrogen-contained atmosphere to induce a microstructural modification of the material that allowed for the improvement of its sensing performance. The composite nanorods annealed at 400 °C exhibited a strong response of 16.2 at the gas concentration of 50 ppm, while the pristine ZnO–WO_3 could only reach 7.3 at the identical gas concentration [3].

Reproducible gas sensing responses to NO_2 and H_2 and humidity at 150 °C with detection limits of 200 ppb and 5 ppm to NO_2 and H_2, respectively, were reported for well-packed and interconnected WS_2 flakes with controlled and reproducible microstructure over large areas. The WS_2 flakes were obtained by a reproducible and high-yield exfoliation process followed by drop casting the centrifuged suspension, leading to the deposition of thin films, thus representing a fast, simple, and scalable method, compatible with standard microelectronic fabrication techniques [4].

Uniform and small-size Cu$_2$O/Ag nanocomposites showing electrochemical behavior and good electrocatalytic reduction performance towards H$_2$O$_2$ were successfully synthesized by an easy one-step procedure. The linear range of the Cu$_2$O/Ag/glass carbon electrode was estimated to be 0.2–4000 µM with a sensitivity of 87.0 µA mM^{-1} cm^{-2} and a low detection limit of 0.2 µM. The anti-interference capability experiment indicated that the Cu$_2$O/Ag nanocomposites have good selectivity toward H$_2$O$_2$. Furthermore, the H$_2$O$_2$ recovery test in the milk solution demonstrated the Cu$_2$O/Ag/GCE (glassy carbon electrode) potential application in routine H$_2$O$_2$ analysis [5].

New voltammetric sensors based on combinations of gold nanoparticles and sulfur-substituted zinc phthalocyanines, AuNP^{tOcBr}/ZnPc^{RS} and AuNP^{tOcBr}/ZnPc^R-S-ZnPc^R, have been developed and used as electrochemical sensors for the detection of catechol. The electron transfer process, as well as the existence of synergistic effects between both components in the absence and presence of covalent links, has been analyzed, showing that the investigated samples enhance the electron transfer rate of the catechol reduction [6].

J-aggregates of 5,10,15,20-tetrakis-(4-sulfonatophenyl)-porphyrin (TPPS) porphyrin are interesting nanomaterials, since by depending on the experimental conditions and templating agents, they showed a variety of different morphologies and physical-chemical properties. Easy and convenient methods to obtain, under mild acidic conditions, novel nanohybrid assembly of porphyrin J-aggregates with Au$_{10}$ cluster or nanoparticles, or superparamagnetic iron oxide nanoparticles (SPIONs) incorporated in micelles, have been reported [9,10]. In the first case, J-aggregates showed a chirality that is related to the configuration of the amino acid (D- and L-histidine) used in the metal clusters synthesis suggesting the important role of Au$_{10}$ clusters as chiral seeds in the growth of the porphyrin aggregates. Furthermore, the growth of the metallic nanostructures with the codeposition of TPPS J-aggregates on the substrates has been exploited as a test system for the potential use of these nanoparticles in SERS applications. This supramolecular system tends to spontaneously cover glass substrates with a co-deposit of gold nanoclusters and porphyrin nanoaggregates showing Surface-Enhanced Raman scattering (SERS) [9]. In the second system, the proper choice of experimental conditions and mixing protocol allows one to control the kinetics of growth of TPPS J-aggregates, leading to supramolecular structures in which the porphyrin nanoassemblies are embedded into the magnetic micelles. By applying an external magnetic field, a high level of alignment of the nanohybrids into the film has been achieved [10].

Hydrothermal methodology was used also for obtaining ZnO nanorods arrays with high surface area and well-aligned crystallographic orientation for fundamental studies and in optoelectronic applications, including visible-light-emitting devices and display systems. These nanostructures show, upon argon/SF$_6$ plasma treatment, enhancement of the PL intensity in the orange/red region of ZnO$_2$-fold, compared to the ZnO sample without plasma treatment. Moreover, the presence of hydroxyl group at the surface, more oxygen in the ZnO lattice (O_L), fluorine incorporation in terms of F–Zn and F–OH bonds, and passivation of the surface states as well as bulk defects have been reported [11].

Kumar at al. reported on the low temperature hydrothermal synthesis and structural and photoelectrical characterization of Sn$_{0.97}$Zn$_{0.03}$S$_2$ nanoflakes. They demonstrated that these nanostructured materials show higher visible-light absorption and significant improvement in conductivity and sensitivity to illuminations, compared to pristine SnS$_2$. Such an excellent performance of Sn$_{0.97}$Zn$_{0.03}$S$_2$ nanoflakes was reported for potential application in optoelectronic devices [12].

The impact of the host material composition on the temperature-dependent luminescent properties of vanadium-doped nanocrystalline garnets was investigated by Kniec et al. It was demonstrated that the incorporation of Ga^{3+} ions into the Y$_3$Al$_{5-x}$Ga$_x$O$_{12}$:V structure enables modification of the emission color of the phosphor by the stabilization of the vanadium ions on the V^{4+} oxidation. The abilities of the Y$_3$Al$_{5-x}$Ga$_x$O$_{12}$:V nanocrystals to noncontact temperature sensing in terms of spectral response, maximal relative sensitivity, and operating temperature range, by the Ga^{3+} doping, was also reported [13].

n-n heterostructures based on graphene monolayer and SnSe$_2$ quantum dots (QDs) showing good performance in the light absorption and the transportation of photocarriers were reported by Li et al. Uniformly distributed SnSe$_2$ quantum dots were synthesized at room temperature by a facile and environment-friendly methodology. The graphene monolayer and SnSe$_2$ quantum dots UV-detector showed fast photoresponse time of ~0.31 s, and its photoresponsivity was up to 7.5×10^6 mA W^{-1}, which are promising for optoelectronic applications [14].

Turek et al. reported on an aqueous two-phase extraction (ATPE), which allowed to differentiate among large diameter single-walled carbon nanotubes (CNTs) by electrical character. The introduction of hydration modulators (H$_2$O$_2$ and PEGme) significantly improved the resolution of the one-step system. Moreover, to isolate the separated CNTs from the matrices, the authors proposed a method based on precipitation and hydrolysis, which is easier than lengthy dialysis routines [15].

Blue, green, and red fluorescent nanoparticles composed of thermosensitive liposomes (TSLs) of phospholipid 1,2-dipalmitoyl-sn-glycero-3-phospho-rac-(1-glycerol) sodium salt, and three different conjugated polyfluorenes were investigated as fluorescent drug carriers for bioimaging applications. These systems showed stable fluorescence signals, good colloidal stability, spherical morphology, and ability to transport and control drug delivery. Preliminary experiments with mammalian cells showed the capability of the nanoparticles to mark and visualize cells in blue, green, and red colors, extending their applications as bioimaging probes [16].

"Three bullet" nanoconstructs based on mesoporous silica nanoparticles (MSNs) covalently integrating a nitric oxide (NO) photodonor (NOPD) and a singlet oxygen (^{1}O$_2$) photosensitizer (PS) and encapsulating the anticancer doxorubicin (DOX) in a noncovalent fashion were reported as good candidates for potential combined cancer photo-chemotherapy. Such a multifunctional nanoplatform is able to generate ^{1}O$_2$ and NO under selective excitation with green and blue light, respectively, and release the noncovalently entrapped anticancer DOX under physiological conditions. Preliminary biological results performed using A375 cancer cells showed a good tolerability of the functionalized MSNs in the dark and a potentiated activity of DOX upon irradiation, due to the effect of the NO photoreleased [17].

Beside the fascinating samples reported so far, the Special Issue includes interesting comprehensive reviews on nanostructured organic and hybrid compounds [7]; bare iron oxide nanoparticles (BIONs) [8]; and microbial biosynthesis of nanomaterials by bacteria, yeast, molds, and microalgae for the manufacturing of sensoristic devices and therapeutic/diagnostic applications [18]. In particular, Prosa et al. reported an interesting overview on the use of nanostructured organic and hybrid compounds in optoelectronic, electrochemical, and plasmonic components as constituting elements of miniaturized and easy-to-integrate biochemical sensors. They highlight that the new concept of having highly integrated architectures through a system-engineering approach may enable the full expression of the potential of the sensing systems in real-setting applications in terms of fast-response, high sensitivity, and multiplexity at low-cost and ease of portability [7]. Madro et al. highlighted the properties and advantages of BIONs with respect to pristine surface chemistry of iron oxide, providing ideas on the future expansion of these nanomaterials and emphasizing the opportunities achievable by tuning their pristine surfaces [8]. Biosynthesis of nanomaterials by microorganisms, attracting interest as a new, exciting approach, was herein reported. Grasso et al. reported on microbial nanotechnology as a fascinating and booming field for future breakthrough nanomaterial synthesis. The authors showed that, through a "green" and sustainable approach, microbial nanotechnology can really spur innovation in nanomanufacturing, with a potential strong impact in several fields, including sensoristics and biomedicine [18].

In conclusion, the papers collected in this Special Issue reflect the existing widespread interest in design, synthesis, characterization, and potential applications of functional nanostructured materials. Although the present Special Issue cannot fully reproduce the complete topic of functional nanostructured materials, I am confident that its contributions to fundamental research will open new perspectives in development and innovation, thus improving the application of these materials in technological fields.

Funding: This research received no external funding.

Acknowledgments: I am grateful to all the authors who contributed to this Special Issue. I also acknowledge the referees for reviewing the manuscripts. Due to their professionalism and expertise, reviewers helped to improve the quality and impact of all submitted manuscripts. Finally, I sincerely thank Mirabelle Wang and all the editorial staff of *Nanomaterials* for their support during the development and publication of this Special Issue.

Conflicts of Interest: The authors declare no conflict of interest.

References

1. Xue, D.; Wang, Y.; Cao, J.; Zhang, Z. Hydrothermal Synthesis of CeO_2–SnO_2 Nanoflowers for Improving Triethylamine Gas Sensing Property. *Nanomaterials* **2018**, *8*, 1025. [CrossRef] [PubMed]
2. Cai, Z.; Ye, Y.; Wan, X.; Liu, J.; Yang, S.; Xia, Y.; Li, G.; He, Q. Morphology-Dependent Electrochemical Sensing Properties of Iron Oxide–Graphene Oxide Nanohybrids for Dopamine and Uric Acid. *Nanomaterials* **2019**, *9*, 835. [CrossRef] [PubMed]
3. Liang, Y.-C.; Chang, C.-W. Improvement of Ethanol Gas-Sensing Responses of ZnO–WO_3 Composite Nanorods through Annealing Induced Local Phase Transformation. *Nanomaterials* **2019**, *9*, 669. [CrossRef]
4. Paolucci, V.; Emamjomeh, S.M.; Nardone, M.; Ottaviano, L.; Cantalini, C. Two-Step Exfoliation of WS_2 for NO_2, H_2 and Humidity Sensing Applications. *Nanomaterials* **2019**, *9*, 1363. [CrossRef]
5. Yang, K.; Yan, Z.; Ma, L.; Du, Y.; Peng, B.; Feng, J. A Facile One-Step Synthesis of Cuprous Oxide/Silver Nanocomposites as Efficient Electrode-Modifying Materials for Nonenzyme Hydrogen Peroxide Sensor. *Nanomaterials* **2019**, *9*, 523. [CrossRef] [PubMed]
6. Ruiz-Carmuega, A.I.; Garcia-Hernandez, C.; Ortiz, J.; Garcia-Cabezon, C.; Martin-Pedrosa, F.; Sastre-Santos, Á.; Rodríguez-Perez, M.A.; Rodriguez-Mendez, M.L. Electrochemical Sensors Modified with Combinations of Sulfur Containing Phthalocyanines and Capped Gold Nanoparticles: A Study of the Influence of the Nature of the Interaction between Sensing Materials. *Nanomaterials* **2019**, *9*, 1506. [CrossRef] [PubMed]
7. Prosa, M.; Bolognesi, M.; Fornasari, L.; Grasso, G.; Lopez-Sanchez, L.; Marabelli, F.; Toffanin, S. Nanostructured Organic/Hybrid Materials and Components in Miniaturized Optical and Chemical Sensors. *Nanomaterials* **2020**, *10*, 480. [CrossRef] [PubMed]
8. Magro, M.; Vianello, F. Bare Iron Oxide Nanoparticles: Surface Tunability for Biomedical, Sensing and Environmental Applications. *Nanomaterials* **2019**, *9*, 1608. [CrossRef] [PubMed]
9. Trapani, M.; Castriciano, M.A.; Romeo, A.; De Luca, G.; Machado, N.; Howes, B.D.; Smulevich, G.; Scolaro, L.M. Nanohybrid Assemblies of Porphyrin and Au_{10} Cluster Nanoparticles. *Nanomaterials* **2019**, *9*, 1026. [CrossRef] [PubMed]
10. Castriciano, M.A.; Trapani, M.; Romeo, A.; Depalo, N.; Rizzi, F.; Fanizza, E.; Patanè, S.; Monsù Scolaro, L. Influence of Magnetic Micelles on Assembly and Deposition of Porphyrin J-Aggregates. *Nanomaterials* **2020**, *10*, 187. [CrossRef] [PubMed]
11. Achour, A.; Islam, M.; Vizireanu, S.; Ahmad, I.; Akram, M.A.; Saeed, K.; Dinescu, G.; Pireaux, J.-J. Orange/Red Photoluminescence Enhancement Upon SF_6 Plasma Treatment of Vertically Aligned ZnO Nanorods. *Nanomaterials* **2019**, *9*, 794. [CrossRef] [PubMed]
12. Mohan Kumar, G.; Ilanchezhiyan, P.; Cho, H.D.; Yuldashev, S.; Jeon, H.C.; Kim, D.Y.; Kang, T.W. Effective Modulation of Optical and Photoelectrical Properties of SnS_2 Hexagonal Nanoflakes via Zn Incorporation. *Nanomaterials* **2019**, *9*, 924. [CrossRef] [PubMed]
13. Kniec, K.; Ledwa, K.; Marciniak, L. Enhancing the Relative Sensitivity of V^{5+}, V^{4+} and V^{3+} Based Luminescent Thermometer by the Optimization of the Stoichiometry of $Y_3Al_{5-x}Ga_xO_{12}$ Nanocrystals. *Nanomaterials* **2019**, *9*, 1375. [CrossRef] [PubMed]
14. Li, X.; Li, L.; Zhao, H.; Ruan, S.; Zhang, W.; Yan, P.; Sun, Z.; Liang, H.; Tao, K. $SnSe_2$ Quantum Dots: Facile Fabrication and Application in Highly Responsive UV-Detectors. *Nanomaterials* **2019**, *9*, 1324. [CrossRef] [PubMed]

15. Turek, E.; Kumanek, B.; Boncel, S.; Janas, D. Manufacture of Networks from Large Diameter Single-Walled Carbon Nanotubes of Particular Electrical Character. *Nanomaterials* **2019**, *9*, 614. [CrossRef] [PubMed]
16. Rubio-Camacho, M.; Alacid, Y.; Mallavia, R.; Martínez-Tomé, M.J.; Mateo, C.R. Polyfluorene-Based Multicolor Fluorescent Nanoparticles Activated by Temperature for Bioimaging and Drug Delivery. *Nanomaterials* **2019**, *9*, 1485. [CrossRef] [PubMed]
17. Tessaro, A.L.; Fraix, A.; Pedrozo da Silva, A.C.; Gazzano, E.; Riganti, C.; Sortino, S. "Three-Bullets" Loaded Mesoporous Silica Nanoparticles for Combined Photo/Chemotherapy. *Nanomaterials* **2019**, *9*, 823. [CrossRef] [PubMed]
18. Grasso, G.; Zane, D.; Dragone, R. Microbial Nanotechnology: Challenges and Prospects for Green Biocatalytic Synthesis of Nanoscale Materials for Sensoristic and Biomedical Applications. *Nanomaterials* **2020**, *10*, 11. [CrossRef] [PubMed]

© 2020 by the author. Licensee MDPI, Basel, Switzerland. This article is an open access article distributed under the terms and conditions of the Creative Commons Attribution (CC BY) license (http://creativecommons.org/licenses/by/4.0/).

Review

Nanostructured Organic/Hybrid Materials and Components in Miniaturized Optical and Chemical Sensors

Mario Prosa [1], Margherita Bolognesi [1], Lucia Fornasari [2], Gerardo Grasso [3], Laura Lopez-Sanchez [2], Franco Marabelli [4] and Stefano Toffanin [1],*

[1] Institute of Nanostructured Materials (ISMN), National Research Council (CNR), via P. Gobetti 101, 40129 Bologna, Italy; mario.prosa@cnr.it (M.P.); margherita.bolognesi@cnr.it (M.B.)
[2] Plasmore s.r.l., viale Vittorio Emanuele II 4, 27100 Pavia, Italy; lfornasari.plasmore@gmail.com (L.F.); llopez.plasmore@gmail.com (L.L.-S.)
[3] Institute of Nanostructured Materials (ISMN), National Research Council (CNR) c/o Department of Chemistry, 'Sapienza' University of Rome, Piazzale Aldo Moro 5, 00185 Rome, Italy; gerardo.grasso@ismn.cnr.it
[4] Physics Department, University of Pavia, via A. Bassi 6, 27100 Pavia, Italy; franco.marabelli@unipv.it
* Correspondence: stefano.toffanin@cnr.it; Tel.: +39-051-639-8514

Received: 8 February 2020; Accepted: 4 March 2020; Published: 7 March 2020

Abstract: In the last decade, biochemical sensors have brought a disruptive breakthrough in analytical chemistry and microbiology due the advent of technologically advanced systems conceived to respond to specific applications. From the design of a multitude of different detection modalities, several classes of sensor have been developed over the years. However, to date they have been hardly used in point-of-care or in-field applications, where cost and portability are of primary concern. In the present review we report on the use of nanostructured organic and hybrid compounds in optoelectronic, electrochemical and plasmonic components as constituting elements of miniaturized and easy-to-integrate biochemical sensors. We show how the targeted design, synthesis and nanostructuring of organic and hybrid materials have enabled enormous progress not only in terms of modulation and optimization of the sensor capabilities and performance when used as active materials, but also in the architecture of the detection schemes when used as structural/packing components. With a particular focus on optoelectronic, chemical and plasmonic components for sensing, we highlight that the new concept of having highly-integrated architectures through a system-engineering approach may enable the full expression of the potential of the sensing systems in real-setting applications in terms of fast-response, high sensitivity and multiplexity at low-cost and ease of portability.

Keywords: integration; smart-system; portable; sensors; biodiagnostics; optical; optoelectronics; electrochemical; analytics; organics; nanoplasmonics

---

1. Introduction

Sensors have become increasingly important in our daily lives. Home pregnancy tests, blood glucose meters, gas-leak sensors are some among the multitude of possible examples. However, the continuous growth of the global population in conjunction with the need for better standards of living are pushing for a rapid development of new technologies of sensing. The scientific interest is even more clear when considering the ca. 20,000 research articles on this topic that have been published last year (bibliographic sources: Web of Science and Scopus by searching for the word "sensor" in the title of articles and review papers).

In this context, biochemical sensors represent a fascinating class of sensing systems that has provided a change of paradigm in analytical chemistry and microbiology, passing from general analytical systems to dedicated systems [1]. The chemical information at the basis of the sensing effect is obtained in real time, possibly on site, as a result of the interaction between sensor and chemical and biological agents in a two-step process: recognition and signal processing. Because of a chemical or physical reaction/interaction with a sample, a change in a physical property occurs or is observed (remote sensing) in the sensor receptor. In cases where the chemical recognition in the receptor does not directly modify an electrical property (resistance, potential), but rather other properties like heat, mass, or light changes, some type of transduction is required to obtain an electrical signal compatible with the electronic circuits.

Depending on the signal-detection technique, sensors can be divided into different types such as resistive, catalytic, thermoconductive, electrochemical and optical [2,3]. Every class has its strengths and weaknesses. However, a lot of parameters must be considered in evaluating sensor performance: sensitivity, selectivity, stability, response time, accuracy, durability, maintenance, portability, cost, safety and lifetime [4]. As a remarkable example, we mention that electrochemical methods allow a signal that is proportional to the analyte concentration to be obtained, thus guaranteeing easy-to-calibrate sensors in a simple configuration. However, lowering the limit of detection is the major goal in the realization of electrochemical sensors, while rapid response-time, low-cost, ease of operation and potential for miniaturization have already been achieved [5].

Among others, selectivity—the ability to respond primarily to the analytes in the presence of other species—is another key issue with sensors, which can be achieved physically, by the selective interaction of the analyte with electrostatic or electromagnetic fields, or chemically, using an equilibrium-based or kinetically-based selective interaction with the layer containing the (bio)reagents [6]. By contrast with electrochemical sensors in which analyte specificity is an open issue [7], optical sensors allow for a highly selective analysis in real-time and automated operational configuration. Moreover, the suitable combination of label-free and multiplexing detection with the reduction of the susceptibility to environmental interference is the next expected breakthrough in the engineering of optical sensors.

Even though they are endowed with different characteristics, biosensors based on electrochemical and optical methods can aim at becoming powerful and robust analytical tools: nanostructuring the sensing active region for amplifying the collected signal and/or implementing functional nanomaterials for increasing analyte specificity are considered valuable strategies [3,5].

Finally, the concept of sensors can be exploited even more effectively if included in miniaturized instrumentation that makes it possible to convert the functions of an entire lab into user-friendly analytical instruments [8]. Highly integrated systems such as lab-on-a-chip (LOC) devices have shown themselves to be highly effective for laboratory-based research, where their superior analytical performance has established them as efficient tools for complex tasks in genetic sequencing, proteomics, and drug discovery applications [9]. Although the chips themselves are cheap and small, they must be generally used in conjunction with bulky detectors (especially in the case of optical sensors), which are needed to identify or quantify the analytes or reagents present. This feature prevents the use of these systems in point-of-care or in-field applications. Furthermore, most existing detectors are limited to analysis of a single analyte at a predetermined location on the chip.

The lack of an integrated and versatile detection scheme (one which is miniaturized, selective and able to monitor multiple locations on the chip) is a major obstacle to the deployment of diagnostic devices in the field and has prevented the development of more complex tests where rapid, kinetic or multipoint analysis is required.

In this scenario, we remark that an increasing trend among academics and in research and development (R&D) is towards monolithic-integrated sensor systems, which merge optical, electrochemical, optoelectronic and electronic elements (e.g., light sources and/or detectors) into one functional unity fabricated on one common substrate to build LOC systems. In particular, the great interest of the scientific community on organic/hybrid light-emitting and light-sensing devices that has characterized the last few decades has led to technologies mature enough to be exploited in real-setting applications.

Innovation in system engineering is, however, useless if not supported by advancements also in the synthesis and use of functional materials. The evolution of the technology requires indeed a combinatorial development of material science and engineering. Despite a multitude of review articles that have thoroughly described specific classes of organic and hybrid (multicomponent) materials and composites for sensing applications [10–13], here we focus on reporting about the advancements, the figures of merit and the effectiveness of different multipurpose integrated detection schemes in biosensors with the objective of inspiring readers towards a new way of visualizing devices as constituting elements of multimodal and miniaturized sensing platforms.

In particular, we report on the use of organic and hybrid nanostructured materials in optoelectronic, electrochemical and plasmonic elements of biosensors for contaminant detection and/or biodiagnostics. The cross-correlation between structural characteristics and functional properties of these classes of nanomaterials plays a major role in the definition and realization of next-generation portable and wearable sensors, if the architectures of the single-component devices are suitably engineered.

For this purpose, the review is organized as follows: in Section 2 we report notable examples of effective miniaturized and conformable light sources and detectors based on organic and hybrid devices; Section 3 describes the nanostructuring of organic and hybrid materials in 2D and 3D geometries for achieving smart-detection schemes in both optical and electrochemical biosensors; finally, Section 4 reports on the assembling of the constituting single-component devices into portable and flexible sensors for wearable contaminant detection and biodiagnostics.

2. Light-Sources and -Detectors Based on Organic/Hybrid Nanostructured Materials and Architectures

Organic and hybrid materials are widely applied in optoelectronic and photonic components and devices because of the fascinating possibility to tune their physical-chemical properties for an improved performance with respect to the corresponding bulk inorganic counterparts. In particular, thanks to the targeted design, synthesis and processing of new nanostructured materials and architectures, great progress in the realization of optoelectronic and photonic components and devices has been recently reported, such as the new generation of miniaturized, easy-to-integrate, highly performing photodetectors, light-emitting devices, light-sensors and imagers, optical fibers, plasmonic and photonic structures [14]. This progress has triggered great advances also in the engineering and development of new detection schemes as, for instance, in the fascinating class of optical sensors.

Optical sensors rely on the modification of an optical stimulus, in intensity, frequency and/or polarization, upon interaction with a target analyte. Optical modification (transduction) can occur through different mechanisms such as bioluminescence, photo- and electroluminescence, and surface plasmon resonance (SPR). In all these methods, light-emitting and light-sensing devices are fundamental parts of the sensor. In this section, a brief illustration on the use of nanostructured organic and hybrid materials for realizing light sources and light detectors is given.

Within photodetectors, the geometrical layout of the active layers and electrodes define some performance characteristics of the sensors. Photodetectors such as photoconductors and phototransistors are based on a lateral structure with (at least) two electrodes in a side-by-side geometry (Figure 1). These photodetectors, due to the micrometric electrode spacing, typically show a slow response at relatively high driving bias (but with switching and amplifying functionality, in the case of transistors). Photodetectors based on a vertical structure, where the active layer is sandwiched between vertically stacked electrodes, include photodiodes (PDs) and photomultipliers (PMs) (Figure 1). These photodetectors have smaller electrode vertical spacing with a short carrier transit length, which generally provides a fast response speed at relatively low driving bias, with small noise.

Within ultraviolet–visible–near-infrared (UV–vis–NIR) photodetectors, the simplest are PDs typically based on inorganic semiconductors such as silicon (Si), gallium arsenide phosphide (GaAsP), gallium phosphide (GaP) and InGaAs (indium gallium arsenide). Despite the good electrical properties, standard inorganic light sensing systems have a number of limitations that restrict their application. For example, silicon has poor mechanical flexibility [15]. In addition, high carrier mobility and long lifetimes can lead to crosstalk between neighboring pixels in image sensors [16]. Also, with broadband absorption, filter-less color discrimination is hard to be achieved [17]. These limitations can be partially overcome with organic and hybrid photodetectors.

Organic and hybrid photodetectors has seen outstanding progress thanks to the advance in the research and development on new nanostructured organic and hybrid materials for light-sensing and light-converting devices, such as third-generation photovoltaic (PV) devices which can be comprised by hybrid dye-sensitized solar cells (DSSC) [18], organic photovoltaic devices (OPVs) [19–21], quantum dot PVs (QD PVs) [22], perovskite photovoltaic devices (PeroPV) [23,24]. PDs based on new nanostructured semiconductors such as organic semiconductors, metal-halide perovskites and quantum dots, combine simple processing (i.e., solution-based), flexibility and conformability, programmable optoelectronic properties through chemical engineering, easy integration onto complementary metal oxide semiconductors (CMOS), and high performance [25]. Thanks to these properties, new PDs designs and architectures can be envisaged for applications outside traditional technologies, such as biomedical devices i.e., large-area and flexible UV–vis–NIR light and X-ray imagers [26], artificial retinas [27,28], machine vision and robotics, endoscope-based imaging [29] (Figure 1). In particular, organic semiconductors are ideally suited for interfacing with biological systems. Indeed, their mechanical and chemical nature offers better compatibility with tissues and suits the non-planar form factors often required for biomedical implantable sensors.

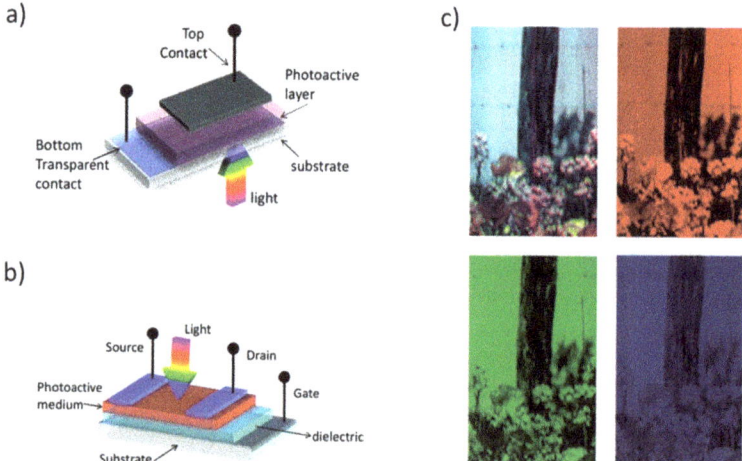

Figure 1. Typical device configurations of photodetectors: (**a**) a photodiode with vertical geometry illuminated from the bottom-side and (**b**) a phototransistor with the light window entrance on the top. Reproduced with permission from [30] © 2020 by WILEY-VCH Verlag GmbH & Co. KGaA, Weinheim. (**c**) full-color image (with the corresponding monocolor, R, G and B images) taken by an organic photodetector (OPD) array. Adapted with permission from [31] © 2020 by WILEY-VCH Verlag GmbH & Co. KGaA, Weinheim.

With reference to specific optical and electrical characteristics, organic semiconductors present an effective combination of higher extinction coefficient and lower dielectric constant [32] which must be considered when comparing them with respect to the inorganic counterparts. Thus, the amount of material used to absorb light, in comparison with silicon, can be reduced down to sub-micrometric thin films (in a compromise with sensitivity), enhancing conformability, flexibility and the capability to be miniaturized. On the other hand, to overcome the high exciton binding energy (due to the low dielectric constant) and the low charge-transport properties of organic semiconductors (1 cm^2 V^{-1} s^{-1}), organic and hybrid heterojunctions of two (electron donor and acceptor) different semiconductors are used into the same active region, thus allowing a high photocurrent, that is correlated to high dynamic range in luminous signal detection. This enables the tailoring of the absorption range of the optical sensors from the UV to the deep infrared.

Typically, highly conjugated molecular or polymeric semiconductors are used in heterojunction organic photodetectors (OPDs), thanks to their high conductivity and band-gap in the visible region. Within the different strategies for the molecular design of conjugated polymers and small molecules for optoelectronic applications, the push-pull approach is one of the most widely applied in literature. This strategy consists in combining, through covalent binding, electron-rich and electron-deficient moieties into the same structure: electron-rich units such as thiophene, bithiophene, fluorene, carbazole, dibenzosilole, benzodithiophene and their derivatives can be bound with electron-deficient units such as quinoxaline, benzothiadiazole, diketopyrrolopyrrole, isoindigo, etc. and their derivatives. Conjugation between the repeating units can be modulated through the design of side functional atoms and groups and through the nature of the covalent spacer between units, allowing the fine tuning of electronic properties (such as charge mobility and bandgap). This fine molecular design allows to obtain a plethora of medium-to-low band gap semiconductors [21].

Organometallic complexes are also effectively used as electron donor p-type semiconductors in heterojunction OPDs [33]. Indeed, metal ions can strongly alter the electronic and optical properties of organic conjugated molecules. For example, the presence of transition metal centers can promote the formation of triplet excited states, significantly enhancing exciton lifetime and diffusion, allowing in turn high OPD photocurrent. Indeed, metal (i.e., copper (II), Al (III), Zn (II)) phthalocyanine complexes exhibit excellent charge transport characteristics. Copper phthalocyanines are particularly well suited for red light, high-speed OPDs, while other metal-phtalocyanine-based OPDs are more suited for NIR OPDs, due to their typical absorption in the 600–700 nm range.

In turn, C_{60} or C_{70} fullerenes and their derivatives, perylene-diimmide and bisimmide polymers and derivatives can be used as electron accepting materials [34–36]. In particular, the strong absorption bands of C_{70} and perylene derivatives in the visible range can improve the spectral range and external quantum efficiency (EQE) of OPDs.

Thanks to the wide variety of organic and hybrid materials available, different choices can be done in relation with the target application, sensing method or analyte to be detected (for more examples please refer to Table 2 of Ref. [37]).

Regarding color selectivity, the benchmark for OPDs is a quasi-Gaussian spectral response with a full-width- at-half-maximum (FWHM) ≤ 100 nm. This can be achieved by appropriate design and coupling (with minimized spectral overlap) of semiconducting organic materials, and/or by the manipulation of the electro-optics of the device [38]. The use of thick active layers and of selective contacts (electron-/hole-blocking layers) in OPD can lead to dark currents as low as 10^{-9} A cm^{-2} and to a signal-to-noise ratio (S/N) competitive to that of silicon-based PDs. Also, the specific detectivity of OPDs, which can be defined as the S/N of a PD with an effective area of 1 cm^2 irradiated with an optical power of 1 W and detected at a bandwidth of 1 Hz, has reached and overcome that of silicon and GaP (10^{15}–10^{14} J) [32,39,40]. The highest operational light intensity range of an OPD, described by the linear dynamic range (LDR), is 9 orders of magnitude (180 dB) for visible, and 12 orders of magnitude for UV [41–43], which is comparable with that of GaP, GaAsP and Si photodiodes (220 dB) [41]. Finally, OPDs have been also demonstrated to reach temporal bandwidths as high as 400 MHz in the visible range, making these devices useful building blocks for organic photonic integrated circuits for imaging and sensing.

As for light detection, the research on innovative semiconductors such as organic materials, metal-halide perovskites and quantum dots and on the design of new device architectures has been the driving force for the progress of light-emitting devices for sensing applications.

Considering optical sensors for reliable in-field applications, lasers, inorganic light-emitting diodes (LEDs) and broad-spectrum lamps are the well-assessed light-sources used in literature. Lasers have the advantage of a high-power, collimated, narrow bandwidth and possibly frequency pulsed light emission. This allows efficient and selective excitation of targets to be achieved [44–46], but their application is unadaptable and restricted due to high costs, need for eye protection, bulky size and a limited selection of wavelengths. Inorganic LEDs and micro-LEDs as light sources are highly efficient, and arrays of LEDs can achieve large-area excitation on targets allowing even fluorescence imaging applications [47], although the emission uniformity of micro-LEDs arrays is limited, due to the point-like source nature of LEDs [48].

Broad-spectrum discharge lamps can be color-tuned with color filters (for selective excitation of absorbing or photosensitized targets), but undesirable side-effects due to heating which can interfere with measurements remain critical [49]. Therefore, for all applications where it is needed a light source with a sufficiently high irradiance (in the order of mW/cm^2), and a spatially controlled emission at a selected wavelength, new solutions have been proposed.

Organic light emitting diode (OLED)-based light sources are robust, portable, miniaturized, cost-effective and fast [50,51]. In addition, thanks to their soft-mechanical and chemical nature, and their low dimension, OLEDs can be easily interfaced with biological and environmental systems, while minimizing their perturbation, and they can be easily integrated into industrial value chains allowing for a real in-line and on-line process control.

To date, the OLEDs technology is already at an industrial stage thanks to the huge progress in OLED-based displays. The use of organometallic complexes based on noble metals such as platinum, iridium and osmium has guaranteed a wide color tunability (covering the whole visible and NIR range) and a high stability in operating conditions [52,53]. Red and green OLEDs easily reach lifetimes up to 10^6 h, while only recently blue OLEDs have shown improved device lifetimes up to 10^3 h [54]. OLEDs can reach incredibly high performance parameters such as a high-brightness at low driving voltages (10^6 cd m^{-2} at $\approx$ 8 V and 10^4 cd m^{-2} at $\approx$ 3 V) [55]. Blue OLEDs still cannot reach such high performance parameters, but they have been recently demonstrated to show a brightness >10^3 cd m^{-2} at the same low voltages and EQE as red and green OLEDs [56]. In addition, the use of phosphorescent organometallic complexes and thermally activated delayed fluorescent (TADF) emitters has enabled to obtain OLEDs showing an internal efficiency (IQE, electron-to-photon conversion) of nearly 100% [57,58]. Most recently, also the substitution of noble metals-based organometallic complexes as emitters with more abundant and convenient metals (i.e., copper or zinc) is object of many studies and will allow the increase in the volume of production and the reduction of costs both of raw materials and of processes [59,60].

If the class of light sources is broadened by considering multifunctional devices, organic light-emitting transistors (OLETs) are emerging as optoelectronic component capable of integrating the electrical properties of a transistor (electrical amplification, switching …) with the light-generating capacity and color tunability [61]. A clear advantage of OLETs, inherent in the structure of the device, is the higher quantum electroluminescence efficiency, compared to the corresponding OLEDs [62]. This feature is correlated to the lateral charge transport and in-plane radiative charge recombination which are intrinsic to the device architecture [63]. The as-realized planar μm-wide emission region in the channel of the transistor enables the integration of the light-emitting component with photonic planar structure [64], thus opening the way to the realization of highly integrated optical communications and optoelectronic systems. In the application field of miniaturized and integrated optical sensors, the use of intense nanoscale light sources with controlled light-emission pattern and tunable photonic integration in optoelectronic systems is a key enabling step for the development of effective low-cost, compact, portable and highly sensitive sensors (see Section 4). In this regard, in OLET devices not only can the light-emission zone be shifted across the transistor channel but also the lateral dimension can be tuned by acting on the gate voltage of (Figure 2) [65]. This peculiar feature with respect to OLEDs holds promise to be exploited for defining innovative excitation and detection schemes in next-generation optical sensors by fine-tuning the optical pig-tailing within the photonic components in a planar architecture [14].

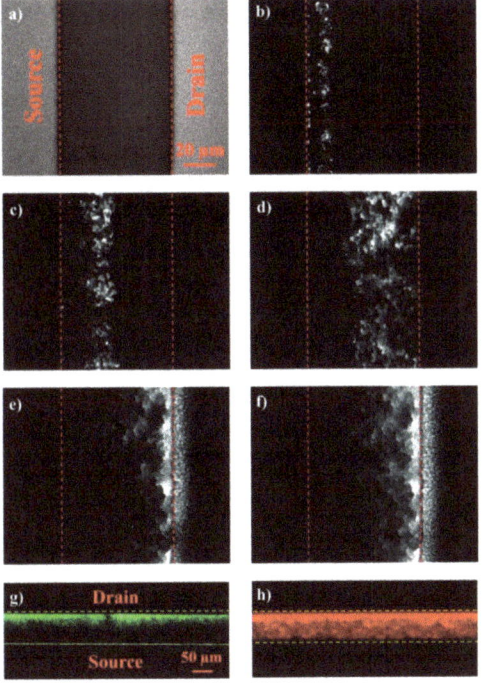

Figure 2. Images of the light-emitting area in the channel of an organic light-emitting transistor (OLET). A transmission optical microphotograph of the unbiased device is reported as the reference (**a**). The gate-source bias (V_{GS}) is varied at increasingly high voltages: −20 V (**b**), −40 V (**c**), −60 V (**d**), −80 V (**e**) and −100 V (**f**), while the drain-source bias (V_{DS}) is kept constant at −100 V. Illuminated channels of green- (**g**) and red- (**h**) emitting OLETs are reported, with recombination layers based on different organic active materials. Reproduced with permission from [65] © 2020 by WILEY-VCH Verlag GmbH & Co. KGaA, Weinheim.

OLEDs and OLETs often share a similar composition of the active layer, which follows the specific requirements of their sensor application (for more details, see also Section 4 of the present review). As a first example, label-free biosensors such as photoplethysmogram (PPG) sensors, extracting the oxygen saturation level from the difference in light absorbed by oxy-hemoglobin and deoxy-hemoglobin, work with at least two coupled light sources emitting in different spectral regions. To this aim, polyfluorene co-polymers can be effectively used as cheap and efficient building blocks for obtaining emissive layers with colour-tunable emission, depending on the co-monomer used (i.e., benzo-2,1,3-thiadiazole for green, hexylthiophene-benzothiadiazole for red [66]). As a further example, muscle contraction sensors, which rely on a different sensing mechanism, need for light-emitting sources operating in the red-NIR regions, for allowing the light to penetrate into tissue for centimeters and return to detectors. This has been achieved, for example, by using the efficient conjugated polymer Superyellow (SY) [67]. Finally, for time-resolved luminescence sensors based on frequency modulated OLEDs, the use of a spike- and tail-free emitter in the UV-blue spectral region, such as a carbazol-biphenyl derivative instead of the more classical molecular host-guest emissive layer such as coumarin-doped Alq_3, has resulted in more reliable sensors [68].

OLEDs and OLETs have been also demonstrated to reach a high-speed response, at a modulation frequency even higher than that of commercially available LEDs. In detail, a peak brightness of ≈10^6 cd m^{-2} at a 40 MHz modulation frequency under 10 ns pulse operation has been recently obtained in OLETs [69]. Thus, these devices are perfectly suited for the realization of photonic integrated circuits for imaging and sensing. Finally, high-frequency modulation has allowed also a large improvement in the electroluminescence properties of the most recent light-emitting transistors based on solution-processed hybrid perovskite emitters (PeLEFET) [70]. Alternated-current operation (AC) of PeLEFET at frequencies in the 10^5 Hz range has enabled an increase in their brightness by two orders of magnitude, enabling their operation at room temperature.

3. Nanostructured Components and Materials in Miniaturized Detection Schemes

In this section we will provide an insight into the detection methods that are enabled by the use of nanostructured materials and might then be integrated into the architecture of the sensor to guarantee high degree of miniaturization together with suitable sensitivity and selectivity for in-field and real-time applications. Given the very broad topic of this contribution, we focus our attention on the detection schemes based on optical probing (and specifically on the high-performing methods related to surface-plasmon resonance effects), but also on non-optical methods such as biochemical and electrochemical methods whose component constituents, however, belong to the same category of nanostructured organic and hybrid materials which has been presented in the previous section. In this way, the reader will have a more general and exhaustive view of the range of possible applications for this category of multifunctional materials.

3.1. Analyte Detection Based on Plasmonic Systems

Nanomaterials show a modified interaction pattern which has significant effects on the macroscopic features of the system, giving rise also to non-conventional behaviors [71]. Nanostructured materials are currently applied in various fields, including healthcare (in targeted drug delivery, regenerative medicine, and diagnostics) [72], energy harvesting [73,74], photovoltaics [75,76], cosmetics [77], gas sensing [78–80], electronics [81], environmental protection and food supplements [82]. From the point of view of sensing application, this evolution triggered an impressive development in the performance and miniaturization of sensing devices which are nowadays pervasively affecting our everyday life.

In many respects, this huge progress has been fostered by the evolution of optoelectronics and photonics, i.e., the possibility of manipulating light–matter interaction and light behaviour. Nowadays, the extremely large availability of miniaturized and well controlled light sources and detectors opened the way to very compact, ease of use, cheap and robust optical devices. Then, as observed in communication technologies, where an almost complete transition from electric-based to photon-based transmission occurred, optical sensing gained more and more importance. With respect to electrical signals, photons are less affected by electric- or magnetic- fields and self-interference effects. In general, the main detection mechanism can be described as follows: once a definite optical resonance has been identified (this can be related to an excited state of a molecule or a material or to a cavity mode), one is measuring the change induced in such a resonance by the presence of the target analyte. The change can be the direct switching on (or off) of the optical excitation; this is the case for fluorescence emission of some dyes in the presence, or otherwise, of the target molecule. In the more general case, the change occurs in the wavelength, the phase, the polarization or the angular dispersion of the optical signal. In this case, the effect is a combination of features of both device material(s) and nano-structuring. Nano structuring is also responsible of an enhancement effect of the material optical response via the localization of the photonic electromagnetic field. It is such an enhancement controlled by nanostructuring which is exploited to increase the sensitivity and allow the measurement of small analyte concentrations. Apart from SERS (surface-enhanced Raman spectroscopy), which is a real tool to identify molecules (but requires a spectral scan), the described scheme of detection is not specific for any given analyte without a recognition tool.

This can be the "labelling" of the target molecule e.g., by its binding with a known dye. Actually, the most part of photonic based sensors belongs to the so called "label-free" category, i.e., the detection is not depending on a previous selection of the analyte in the sample examined.

On the one hand, this is an advantage because the same detection scheme can be applied to detect a large panel of molecules. It is the bare presence of the molecule, with its mass and contribution to refractive index which is detected. On the other hand, a functionalization step is required to make the device able to select the specific target molecule. This is the very tricky point because one has to find a specific bio-chemical recognition mechanism for each analyte and provide its activation within the active region of the sensor.

Considering optical components and devices, nanostructured systems can be divided into two categories: those based on metallic behavior, concerning plasmonic excitations, and those built with dielectric components, paving the way to the realization of photonic devices.

The first class of devices exploit the properties of metals, mainly gold and silver, but also metallic alloys and heavily doped semiconductors or oxides [83]. The choice is driven by low losses criterium, but easy of fabrication and stability is playing a big role (e.g., silver shows better performance, but, due to oxidation affecting it, gold is often preferred). As for the photonic systems, a large variety of dielectric materials, from semiconductors to insulators, is considered. The most used is silicon and silicon-based compounds like silicon nitride. This is due to the compatibility with the microelectronics industry and the well assessed fabrication technologies. Nevertheless, besides silicon dioxide, also titania oxide structures are studied, which offer a good compatibility with e.g., biomolecules [84].

In this respect, interest in organic, polymeric materials and the possible related structures is relevant too. Besides the usual features characterizing polymers (low cost of material and fabrication, flexibility, lightness, ...), organic materials add in general a high compatibility with biological matter [85–87].

The features of the plasmonic and photonic systems are quite complementary and a new interest emerged in combining them into hybrid systems to take advantage of the best properties of both. Generally speaking, photonic systems have a much better-defined spectral response that can be finely tailored and is intrinsically stable and robust. On the other side, plasmonic systems offer a superior performance in terms of field confinement and enhancement, combined with the drawback of relatively large losses and a much broader spectral response [88]. Up to now, even considering that sensitivity is also depending on the chosen optical parameter (shift of spectrum, polarization, phase, intensity, in reflectance, transmittance and fluorescence mode), field enhancement remains the main factor affecting the final optical detection performance. Then, metallic based plasmonic systems offer the most suitable platform [89]. This is true even for fluorescence, provided that a thin dielectric layer is inserted between the metal and the dye to avoid fluorescence quenching [90,91]. Actually, fluorescent-based methods are, in general, the most sensitive tools used for detection of small quantities of substances in different matrices. On the other hand, they are heavily subjected to matrix effects and also need the use of fluorescent markers, which are provided by complex pretreatments of the sample to be analyzed. These often require laboratory-grade analytical techniques and are not suitable for integration "on chip". In other words, these are typical "label" techniques.

Among label-free methods, a relevant position is occupied by SPR.

SPR sensors exploit the excitation of charge waves: (i) in metallic nanoparticles (localized surface plasmon resonance (LSPR) [92], (ii) at the surface of metal layer (surface plasmon polariton (SPP)) or (iii) a combination of both. The detection mechanism is based on the change in the spectral- and/or angular-distribution of the optically excited mode, when the refractive index of the medium close to the surface is changing. Despite the fact that LSPRs can be easily optically excited, they exhibit an intrinsic spectral broadening which makes them less sensitive to variation in the medium refractive index.

By contrast, SPP are longitudinal waves characterized by a shorter wavelength than the electromagnetic wave having the same frequency. Consequently, some coupling strategies must be used to optically excite SPP. The largest part of SPP based SPR systems typically measure the attenuated total reflection (ATR) in the Kretschmann configuration, where a thin (less than 50 nm) layer of gold is deposited on a prism [93]. This implies a careful control of either angle of incidence and beam collimation, and then the use of costly and bulky optical setup and mechanics, not easy to miniaturize (Figure 3).

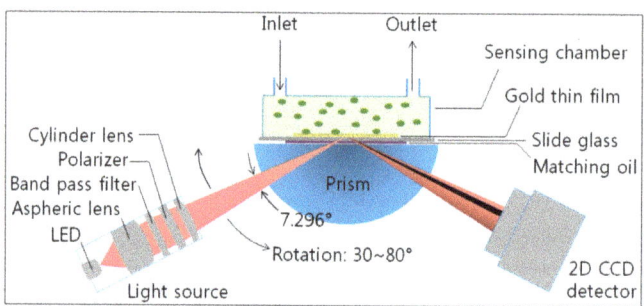

Figure 3. Scheme of a surface plasmon resonance (SPR) system in Kretschmann configuration. © 2020 MDPI, reproduced with permission from ref. [94].

In order to make the system simpler and, to some extent, more portable, an excitation through an optical fiber design has been developed. Drawbacks in this case are the need of stable laser sources and the cost of the disposable fiber sensors. A more extensive and complete review of these classes of instruments is given in ref. [95].

A different coupling strategy is offered by a periodic nanostructuring of the plasmonic surface. When the period of the nanostructure is comparable to the wavelength of the optical mode, diffraction effects (either of light and polaritons) allows the direct excitation of SPP at an arbitrary angle of incidence and even at normal incidence. This greatly simplifies the optical system even though it is at the cost of losing angular resolution, which is the most sensitive parameter used in commercial instruments for the detection of molecular interaction on the surface.

A further advantage of using nanostructured surfaces is the easy implementation of imaging capabilities. Image analysis implies that each portion of the active surface can be considered as an almost independent sensor, opening the way to multiplexing, i.e., the simultaneous detection of a whole panel of different analytes [96].

Some works have been published about systems based on grating of holes in a gold film [97]. The majority of them report on spectral/intensity light-modulation in a transmittance configuration in order to benefit of the extraordinary transmittance properties of such periodic surfaces (i.e., gratings). However, a great disadvantage of such configuration originates from the light path across the fluidic cell which is necessary for providing the sensitive surface with the analytes to detect: evidently, this detection scheme affects the measurement by reducing the sensitivity and introducing interference effects.

To disentangle the optical signal collection from the fluidic system, one should collect the light reflected by the grating from the (transparent) substrate side: however, this approach requires an effective cross-talking between plasmonic modes at the exposed (metal/fluid) surface and the back (metal/substrate) interface. Actually, this can be easily achieved considering the local excitations supported by nanostructuring. As a matter of fact, a hole in a metal behaves like a mirror-like metallic nanoparticle in a dielectric and exhibits a series of localized plasmonic excitations. In principle, these LSPRs are dispersionless: however, when they are active in the same spectral region where SPP occurs, hybridized modes between the two kind of excitations take place.

In order to better illustrate the interplay of all these effects, we can use as an example a specific nanostructured metal-dielectric grating, produced by Plasmore S.r.l, which might be implemented for a sensitive detection in a reflectance configuration. The system is constituted by a hexagonal lattice of polymeric pillars embedded in a relatively thick gold layer (about 150 nm). A scanning electron microscope (SEM) view of a typical surface so obtained is shown in Figure 4. The gratings are prepared through colloidal lithography and plasma etching techniques. The detailed fabrication protocol was presented by Giudicatti et al. [98]. In this context it is worth noting that the colloidal mask fixes the lattice pitch, while the etching procedure determines the pillar dimensions and shape, which also give a significant contribution to the optical response [99].

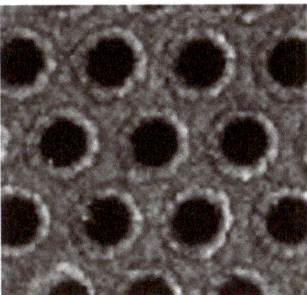

Figure 4. Scanning electron microscope (SEM) image of a typical nanostructured gold layer produced by Plasmore S.r.l., with a hexagonal plasmonic lattice (lattice pitch = 500 nm). Top of polymeric pillars corresponds to the black areas.

A good tool to explore the dispersion behavior of plasmonic excitation is to study the response as a function of the incident angle. Figure 5 shows the complicated interplay occurring among dispersionless localized modes and propagating polaritons in surfaces with a lattice pitch of 400 nm. Indeed, in the figure the theoretical dispersion of polaritons for the two main symmetry directions of a hexagonal lattice at the gold/glass interface is also shown (red and green lines), while the opening of a propagation gap at the folding point corresponding to zero incidence-angle is clearly visible. The crucial point is the spectral and modal superposition of polaritonic and localized modes by using suitably designed planar grating. The folding process of the polaritonic modes is controlled by the pitch of the grating, whereas the spectral position of the localized modes is mainly dictated by the size and the shape of the nanostructured elements forming the grating, namely, the diameter of the holes in the metal layer.

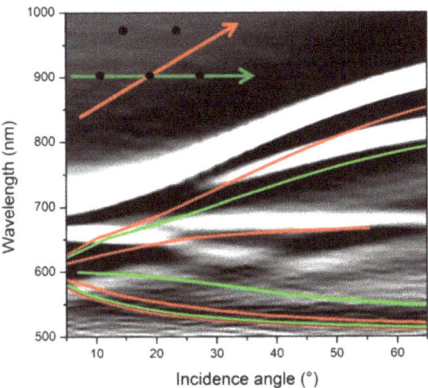

Figure 5. Second derivative of the spectral reflectance plotted as a function of the incident angle. The bright and dark signals correspond to minima and maxima in the reflectance spectrum, respectively. The superimposed red and green lines correspond to the calculated plasmonic dispersion for two orientations of the hexagonal lattice.

Figure 6a shows the reflectance measured from the back through the glass substrate of two gratings with different pitches. The hole dimension has been scaled to the pitch in order to have a similar relationship with the polariton frequency. The scaling effect of the whole response is evident, with the reflectance maximum corresponding to the folding gap shifting from 700 nm to 600 nm when passing from the 500 nm to 400 nm lattice pitches.

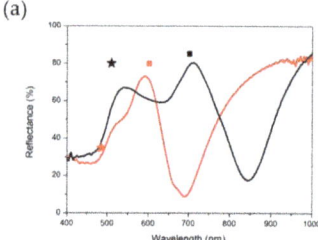
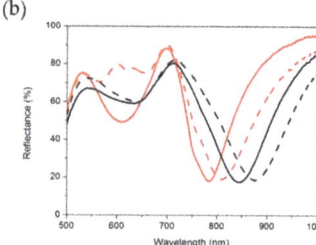

Figure 6. (a) Reflectance spectra for two gratings having lattice pitch of 500 nm (black line) and 400 nm (red line). The stars and the squares point out the calculated gold/air and gold/glass polariton wavelength, respectively. (b) Solid black and red lines correspond to the reflectance spectra recorded for two gratings having the same lattice pitch (500 nm) but different hole size. Dashed lines show the spectra acquired for the two samples when water is dispensed on the grating surface. These graphs are obtained from unpublished data from the authors.

As a general constraint upon the integration of the plasmonic-sensing surfaces into a working optical sensor, it is mandatory to tune the resonance wavelength of the grating to the emission specifications of the used light-source. In view of a fully effective system miniaturization of the complete sensor, the use of conformable easy-to-adapt planar or stripe-like light-emitting sources (as in the case of organic and hybrid light-emitting diodes and transistors, see Section 2) may play key-role in the engineering of portable and flexible plasmonics-based sensors.

The competitive advantage of these nanostructured plasmonic devices with respect to other standard detection schemes is the possibility to design and control the resonance localized modes by acting on the geometrical features of the gratings. Indeed, the hole size can be tuned by increasing or decreasing the etching time. Hole size also plays a role in defining the spectral response. As shown in Figure 6b (solid lines), by modulating the cavity diameter between 340 and 300 nm, it is possible to shift the reflectance peak wavelength of the grating from 850 nm to 770 nm.

It is worth noting that, despite reflectance spectra having been taken from the backside, through the substrate, they carry information about the top surface. Hence, by changing the medium on the top side from air to water, resonance minima in reflectance shift towards higher wavelengths (dashed spectra in the figure). It follows that spectra are sensitive to refractive index changes in the medium above the surface.

Indeed, this is the detection principle of this kind of detectors: any adhesion of molecules at the exposed surface changes the local refractive index close to surface and hence is inducing a spectral shift of the whole plasmonic resonance (the broad deep in reflectance spectra).

In the case of small refractive-index changes, and then small spectral shift, the change can be also detected as an intensity change in the reflectance at a given wavelength close to the resonance minimum. Thus, it is possible to build up sensors based on quasi-monochromatic light sources and a camera as light detectors [100]. By working in imaging modality, it is possible to be sensitive to the spatial positioning of the intensity change. In biosensing applications this finding allows the end user to monitor multiple interactions simultaneously.

The crucial point is related to the functionalization process. The functionalization of plasmonic nanostructures is much more challenging than that of continuous metal films. Nanostructured gratings often comprise multiple materials (e.g., glass and gold) and exhibit surface curvatures (e.g., edges, tips). On the other hand, the heterogeneous chemistry composition of nanostructures could be exploited to functionalize only the areas where the electromagnetic field is enhanced rather than the entire surface [101].

The simplest and straightforward functionalization method is the passive adsorption of receptors to the metal surface. It is also possible to covalently bind an array of different biological molecules through simple functionalization strategies as self-assembled monolayers (SAMs) which use thiol-containing compounds attached to the gold component of the plasmonic grating. However, these approaches often result in a uncontrollable release of receptors and high nonspecific adsorption of complex matrixes to the surface. The ideal functionalization method is expected to create a functional coating that: provides optimized concentration of receptors, preserves their biological activity and good accessibility and guarantees a low non-specific binding of non-target molecules.

In the development of diagnostic applications for the screening of complex matrixes of interest, as serum or milk, it is appropriate to follow a functionalization strategy that comprises the use of highly hydrophilic polymers (carboxymethyl dextran, polyacrylamide derivatives ...). In this way, the functionalization layer of the grating guarantees high stability during large number of regeneration cycles and low fouling properties [102]. As an example, we mention the commercially available copolymer MCP-2F (Lucidant Polymers) for the coating of a wide range of materials (glass, silicon oxide, silicon nitride, gold, PDMS, COC, and Teflon).

The MCP-2F is a poly(dimethylacrylamide) copolymer that incorporates a silane moiety and it is functionalized with nacryloyloxysuccinimide (NAS). The NAS moiety extends the succinimidyl ester from the surface. This group is highly reactive toward nucleophiles such as amines naturally present in most of the target molecules.

A film of the copolymer MCP-2F bearing active esters was successfully grafted by a combination of physical adsorption and covalent linking to the plasmonic nanostructure of Plasmore. It is worthwhile noticing that Plasmore plasmonic gratings have been used for example for the detection of long pentraxin 3 (PTX3), a biomarker for different human pathologies and infections [100] and for the detection of mycotoxins in barley and beers [103].

3.2. Analyte Detection Based on Electrochemical Bio/Chemosensoristic Devices

Selectivity and sensitivity are key parameters for the assessment of analytical performances of sensing devices, especially to evaluate their application for real matrix analyses and for multianalyte detection purposes. Optical sensors generally show a good sensitivity and selectivity against the analyte(s). However, a proper design of optical sensors based on the use of nanostructured materials and the adoption of suitable sample pretreatments (e.g., dilution and analyte extraction) can substantially improve the performance and sensitivity of various optical sensors. Similarly, the introduction of nanotechnologies into research and development in electrochemistry is nowadays addressing important issues in sensoristic field. Nanophase materials possess peculiar physical, electronic and chemical properties (compared to bulk materials) that can be exploited for the functionalization of electrode surfaces both as 'direct' active layers (e.g., in non-enzymatic electrochemical sensors) and interfacing layers between an electrode surface and biological recognition element (e.g., enzymes and antibodies). The use of nanomaterials to modify electrode surfaces can strongly affect the analytical performance of non-enzymatic electrochemical sensors through changes of operating parameters including working potential, surface morphology, signal amplification and catalytic efficiency.

In electrochemical sensors, both selectivity and sensitivity are strongly affected by the material of working electrode and modification of surface architecture. The chemical modification of the electrode can be achieved with nanostructured materials e.g., to increase electrode surface area, to enhanced electron transfer kinetics or to enhance selectivity and or sensitivity to the analyte(s) [3,104].

In this section, the introduction of the classes of nanomaterials used in the optimization of the performance of electrochemical devices is meant to broaden the overview of the status of progress of the engineering of sensors for real-setting applications. Furthermore, the cited nanomaterials could be also used in new and interesting applications based on the combination and the integration of electrochemical and optical techniques, as in spectroelectrochemistry approaches [105,106], in the fabrication of optically transparent electrodes and in bio/chemosensing applications [107] with enhanced selectivity, sensitivity and signal intensity. In this context, a recent analytical integrated approach for innovative monitoring and diagnostics of the environment and the agro-food supply chain is provided by the patented physicochemical sensing device called 'Snoop' [108]. The physico-chemical device 'Snoop' uses one or more tailor-made designable advanced chemical or biological 'sensitive materials' (SMs) (included nanomaterials) in the same sensor. The interaction with one or more target analytes (or substances belonging to the same chemical class) with selected SMs can induce specific or aspecific physicochemical (electric or optical) responses.

Referring to classes of innovative multifunctional nanomaterials, carbon allotrope nanomaterials (Figure 7) are widely used in the fabrication of electrochemical sensors because of interesting features like an increased electroactive surface area (large surface-to-volume ratio and specific surface area), a faster electron transfer kinetics and an enhanced interfacial adsorption properties (exploited for adsorption of molecules and to reduce surface fouling effects) [109].

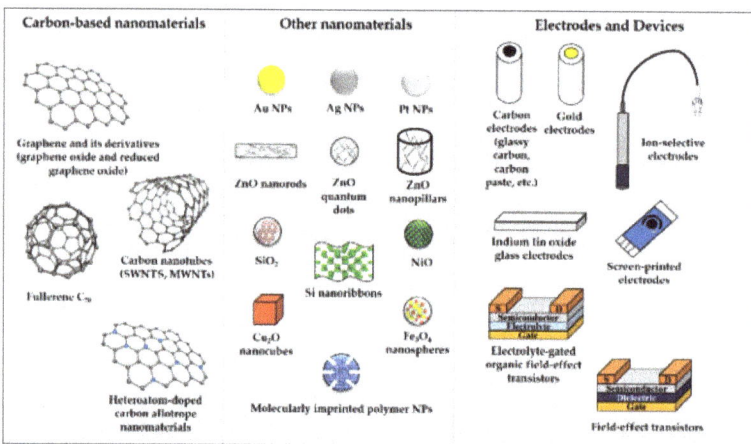

Figure 7. Schematic illustration of carbon allotrope nanomaterials and other nanomaterials together with electrochemical and electronic tools described in this review.

Graphene is the basic building block for other graphitic materials, which consists in a 2D single-layer sheet of sp^2-hybridized carbon atoms structured into a honeycomb-like hexagonal pattern. Beside high surface area, graphene exhibits remarkable electrochemical properties (e.g., such as high electric conductivity, a zero-bandgap semimetal behavior, a large potential window), low charge-transfer resistance, excellent electrochemical activity and fast electron transfer rate. Three-dimensional graphene (3D graphene) has been recently used as support and stabilizer of bimetallic electrocatalyst $NiCo_2O_4$ for non-enzymatic detection of urea in urine samples [110]. The sensor has shown an excellent analytical performance in both neutral and alkaline pH conditions, rapid response (approximately 1.0 s) and stability (no significant loss in activity after four months of storage at room temperature). Reduced graphene oxide (rGO) is a suitable alternative to the pristine graphene and graphene oxide for interesting features: a chemical reduction by removal of oxygen functional groups increases its conductivity (compared to graphene oxide). In addition, chemical reduction produces several chemically active lattice defects sites (compared to pristine graphene), making rGO a promising candidate for active layers in electrochemical (bio)sensors [111].

Carbon nanotubes (CNTs) (single-wall carbon-nanotubes or SWCNTs, double-wall carbon-nanotubes or DWCNTs, and multi-wall carbon-nanotubes or MWCNTs) are a graphene-derived class of carbon allotrope nanomaterials that display metallic or semi-conducting electron transport (depending on the sheet direction about which the graphite sheet is rolled) and they are extremely attractive carbon allotrope nanomaterials for a wide range of electrochemical sensing [109]. A potentiometric sensor for *Escherichia coli* O157:H7 detection in milk samples and apple juice samples was described in literature. In this work, SWCNTs were used as ion-to-electron transducers in potentiometric measures [112]. In the last 30 years, the integration of carbon nanotubes and graphene into field-effect transistor (FET)-type nanobiosensors has increasingly gained interest in sensing as a promising more sensitive, and portable label-free, analytical solution. Through the modification and functionalization of the gate electrode and the semiconducting channel of field-effect transistor, new possibilities have been opened for the development of new SWCNT and graphene FET-based biosensors [113]. A recent example is the functionalization of gold gate electrode in an organic electrochemical transistor with poly(diallyldimethyl-ammonium chloride) + MWCNTs and graphene nanocomposites for the determination of gallic acid in green and black tea samples. The electrocatalytic activity of the gate electrodes was enhanced by MWCNTs and graphene, with the best detection limit (as low as 10 nM) in the case of MWCNTs nanocomposite [114]. In recent years, CNTs have been also gradually exploited for the manufacturing of new electroactive nanomaterials for solid-contact ion-selective electrodes (SC-ISEs) for the potentiometric determination of ions K^+, Ca^{2+}, H_3O^+, Pb^{2+}, NH_4^+, NO_3^- and ClO_4^- also in real samples [115–118]. The application of nanomaterials in stack architectures is very promising for the development of high stable SC-ISEs with long operating lifetimes. [118] Several screen-printed electrodes (SPEs) modified with nanocomposites were also developed for electrochemical determinations of e.g., heavy metal and antibiotics in real environmental and food samples [119–122]. The use of SPEs modified by carbon black was also described for phenyl carbamate pesticides (carbofuran, isoprocarb, carbaryl and fenobucarb) detection in grain samples. The analytical performance of the sensor allowed a class-selective detection of several phenyl carbamates in food samples. The detection of carbaryl was possible for concentrations up to maximum residue limit levels (MRLs) [123]. Fullerenes and derived nanostructures [124] (e.g., nanorods) are very interesting nanomaterials in the electrochemical sensoristic field. A fullerene C_{70}/polyaniline (a conductive polymer) nanocomposite modified glassy carbon electrode was recently exploited for the detection of pyridine herbicide fungicide triclopyr in spiked tomatoes extracted by square-wave voltammetry [125]. Heteroatom-doped carbon allotrope nanomaterials (e.g., doping with phosphorus and nitrogen) is another interesting family of nanomaterials that possess improved physicochemical and structural properties that can be used in electrochemical sensor devices [126–129].

Metal and metal oxide NPs have been largely used for the modification of solid electrodes, especially in association with carbon allotrope nanomaterials. Indeed, NPs can be dispersed in inorganic–organic nanocomposites, by conferring new remarkable synergistic electronic properties that cannot be achieved by individual nanocomponents [130].

Noble metal nanoparticles (AuNPs, AgNPs and PtNPs) are extensively employed to form nanocomposite with unique electronic and catalytic properties to be exploited in electrochemical sensors (Figure 7). The presence of AuNPs in rGO-AuNPs modified glassy carbon electrode had showed a positive effect in the preconcentration step in stripping voltammetry analysis for the detection of As^{3+} in soil samples [131]. Other recent examples of noble metal NPs-modified electrodes used in electrochemical sensors with the limit of detection in the nanomolar range include the differential pulse voltammetric detection of the insecticide methyl parathion in spiked water samples [132], the differential pulse voltammetric detection of Ca^{2+} in pork meat [133], the detection of antibiotic neomycin in spiked milk and honey samples [134] and for the determination of H_2O_2 in human serum and saliva samples [135]. The determination of synthetic azo-colorant Sudan I in real food samples (chili powder, chili, tomato and strawberry sauce) have been obtained by a voltammetric sensor based on a Pt/CNTs nanocomposite modified ionic liquid carbon paste electrode. The sensor showed a very large current response (thanks to an enhanced conductivity), a good selectivity, a biocompatible interface and a low limit of detection (LOD, 3×10^{-9} mol L^{-1}) [136].

Zinc oxide (ZnO) is an inorganic semiconductor that forms several hierarchical nanostructures that have been recently exploited for electrochemical sensoristic purposes (Figure 2). ZnO nanopillars were recently used for coating gold electrode in direct chronoamperometric detection of Cd^{2+} in spiked river water samples (LOD = 3.6×10^{-8} mol L^{-1}) [137]. An amperometric glucose biosensor with glucose oxidase physically immobilized onto ZnO nanorods showed a large linear range (0.05 mM to 1 mM) and good sensitivity (48.75 µA/mM) [138]. A gold electrode modified with ZnO quantum dots (QDs) has been recently used for voltammetric detection of mercury in groundwater samples (LOD = 2.5×10^{-8} mol L^{-1}) [139]. Surface coating of electrode by ZnO helped electron migration between mercury and electrode surface and no sample pretreatments are required. Most QDs are inorganic semiconductor nanocrystal like ZnO nanocrystals or cadmium sulfide (CdS), whose electronic properties can be fine-tuned by varying the size of the nanostructures that can be used alone or as nanocomposite with CNTs e.g., for enzymes immobilization with higher direct electron transfer between the active site and electrode [140]. ZnO NPs co-doped with nickel and iron have been recently used for the fabrication of a FET sensor for the detection of hexahydropyridine in mineral water and tap water samples [141]. ZnO-CNTs- nanocomposites are peculiar nanomaterials that can strongly enhance the electroanalytical performance of modified electrode surfaces, especially in terms of charge transfer [142]. Some recent examples reported in literature include a carbon paste electrode modified by ZnO/CNTs nanocomposite for voltammetric determination of ascorbic acid in fresh vegetable juice, fruit juices and food supplement samples [138] and a screen-printed electrode modified with a gold NPs/graphene oxide nanocomposite for voltammetric detection of semi-synthetic β-lactam antibiotic cloxacillin in raw milk samples [143]. In the latter work, a pre-concentration step by addition of a cloxacillin-imprinted polymer to spiked milk samples was also described. Other recent electroanalytical application of pristine or metal oxide NPs nanocomposite include: NiO NPs-modified carbon paste electrode with the N-hexyl-3-methylimidazolium hexafluorophosphate ionic liquid for square-wave voltammetry measurement of p-nitrophenol in tap water, drinking water and river water samples [144], a glassy carbon electrode modified by Cu_2O nanocubes/Ag NPs nanocomposite for impedimetric measures of hydrogen peroxide (H_2O_2) in spiked commercial milk [145]. The combination with Ag NPs improves sensitivity and linear detection range of the electrochemical measure, compared to analytical performances of other previously reported in literature H_2O_2 sensors based on pristine Cu_2O. A carbon paste electrode modified with Fe_3O_4 nanospheres combined with a cobalt(II)-Schiff base complex was employed for square-wave voltammetric detection of NO_2^- ion in spiked spring water, mineral water and tap water samples [146]. The newly synthesized cobalt(II)-Schiff base complex was involved in preconcentration of NO_2^- ions at the electrode surface and no sample pretreatments were required.

Non-metal nanostructures and nonmetal oxide nanostructures provide effective surfaces and peculiar electronic characteristics that can enhance sensing performances. An electrode modified with nanosilica (nano-SiO_2) has been recently described for the detection of Pb^{2+} in black tea and in wastewater samples. The nano-SiO_2 characteristics of hydrophobic filler with a high specific surface area and a peculiar three-dimensional structure increase the number of Pb^{2+} binding sites. This can positively affect diffusion rates of target analyte(s), with possible involvement in the extraction step of ions at the electrode surface [147]. This modified electrode showed a good analytical performance (LOD = 7.3×10^{-8} mol L^{-1}) and a long lifetime (2 months). Silicon nanoribbons have been exploited to make the semiconducting channel of a pH FET device for H_3O^+ detection in milk samples (from condensed milk powder) and bovine blood plasma. The developed proton exchange lipid bilayer membrane proved to be a resistant and highly efficient proton-conductor antifouling coating suitable for good analytical performances of the pH FET device in milk and bovine blood plasma samples (pH sensitivity of 64% and 35% per unit pH, respectively) [148].

Molecularly imprinted polymers (MIPs) are chemical artificial polymeric receptors with a tailor-made designable selectivity and specificity for a target analyte [149,150]. Impedimetric detection of estradiol in spiked commercial milk with no pre-treated sample has been described using Au NPs and molecular imprinted polymer electrodeposited on the surface of a glassy carbon electrode [151]. Magnetic NiO NPs decorated by molecularly imprinted polymer were used to modify the surface of a glassy carbon electrode for chronoamperometric detection of phenylurea herbicide chlortoluron in irrigation water samples [152].

An improvement in the molecular imprinting polymerization method is the synthesis of molecularly imprinted polymeric NPs. Thanks to higher surface-to-volume ratio and a better accessibility of recognition sites for the analyte, improvements in binding kinetics and detection sensitivity in MIP NPs-based sensoristic devices can be achieved [153]. An example of uses of MIP NPs modified carbon paste electrodes for electroanalytical detection of target analytes in real samples is the modification of carbon paste electrode composition for square-wave voltammetric determination organophosphate insecticide diazinon in well water and apple fruit samples [154].

4. Towards Smart Integration of Nanostructured Components for the Realization of Miniaturized Optical Sensors

In this section, we will focus on the system engineering of optical bio- and chemosensors through the implementation the organic and hybrid device components that we previously introduced. Particular attention is devoted to specific structural characteristics of the sensors such as miniaturization, portability and wearability that have a huge impact on the effectiveness of the use of the sensors in real settings. We aim at highlighting that optical bio- and chemosensors based on innovative organic and hybrid materials are expected to be well-suited to point-of-care and in-field applications.

Integrated optical sensors are typically based on a light-emitting source and a light-detector, in conjunction with an element sensitive to the analyte of interest. In the working principle of the sensor, the light emitted by the source interacts with the sensitive element, where a signal variation occurs as a result of a bio/chemical event (Figure 8). The role of the photodetector is to monitor a spectral or intensity change of the light with respect to the original characteristics of the light emitted by the source. It is evident that, the three components must be endowed by compatible spectral characteristics.

Optical sensors are typically classified on the basis of the optical transduction modality (that is the working principle of the sensing element): (i) photoluminescence and (ii) absorption/transmission-reflection of light. In the first case the light interacts with the sensing element and hence is energetically modified (Figure 8a) while, in the latter case, the optical transduction involves a variation of the light intensity (Figure 8b).

One of the most common photoluminescence-based detection schemes relies on the dynamic luminescent quenching of the sensitive element, as a consequence of the presence of the analyte. That is, the emission of a dye is reduced in intensity by increasing the amount of the analyte that works as a quencher.

By using that detection scheme, Shinar et al. attempted the realization of an oxygen-sensor by integrating an OLED as light-source, an a-(Si, Ge):H p-i-n diode as photodiode and a sensor film based on a polystyrene matrix doped with a metalorganic complex (Platinum(II) 2,3,7,8,12,13,17,18-octaethyl-21H,23H-porphyrin, PtOEP) [155]. Oxygen indeed represents an ideal analyte because of the typically high sensitivity of dyes to oxygen quenching. In that work, the authors showed opportunities and challenges of that type of sensor. The principle of operation relies on the emission of light by the OLED via electroluminescence (see Section 2). The light is then absorbed by the sensing element, thus re-emitted at red-shifted wavelengths and eventually detected by the photodiode. The luminescent quantum yield of the sensing element is reduced by the presence of oxygen. It is evident that an effective overlap of the spectral characteristics of the components is fundamental to ensure high sensitivity and as low as possible LOD of the sensor. To solve this issues, Liu et al. integrated (i) OPDs with a more selective spectral response, (ii) a microcavity OLED with increased emission intensity due to a more efficient light outcoupling and (iii) a more effective sensing element [156] (Figure 9).

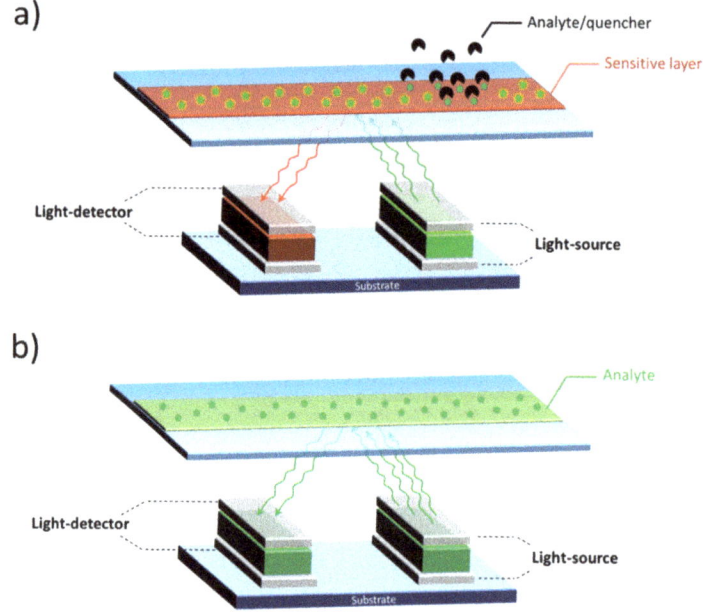

Figure 8. Schematic representation of optical sensors with a reflection-type architecture. The sensors comprise a light-emitting device, a light-sensitive device and a sensitive element, which acts as photoluminescent emitter sensitive to external quenchers (**a**) or absorber of the emitted light (**b**).

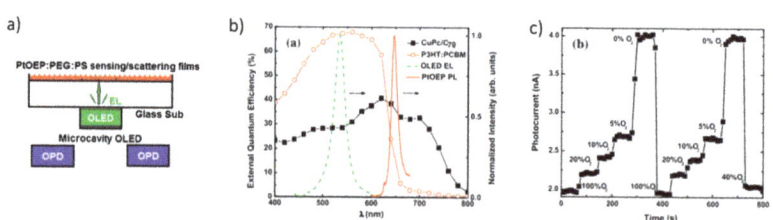

Figure 9. O_2 sensor based on a back-to-back integration of a microcavity organic light emitting diode (OLED) and two organic light photodiodes (OPDs) by using Platinum(II) 2,3,7,8,12,13,17,18-octaethyl-21H,23H-porphyrin (PtOEP) in a polymeric matrix of PEG-PS as sensing element (**a**). The external quantum efficiency (EQE) of CuPc/C$_{70}$ (black squares) and P3HT:PCBM-based OPDs (red dots) are reported with the EL (dashed green line) and PL (red line) characteristics of the OLED and the PtOEP sensing element, respectively (**b**). The signal of the O_2 sensor is reported at different analyte concentrations (**c**). Reprinted from [156], © 2020, with permission from Elsevier.

As shown in Figure 9a, two back-to-back glass slides respectively incorporating (i) a microcavity OLED and (ii) two back-detecting OPDs were attached. The sensing element was placed on the back of the OLED substrate and was improved in terms of efficiency of oxygen detection by optimizing the ratio between polyethylene glycol (PEG) and polystyrene (PS), in which the luminescent PtOEP is embedded. The spectral characteristic of the OLED (Figure 9b) is suitably narrow to favor a good level of sensitivity of the sensor. Moreover, the emission of the sensing element matches the absorptive spectral characteristics of the photodetector, that is based on a mixture of CuPc and C_{70}. For the sake of comparison, the EQE, i.e., the photon-to-charge conversion efficiency, of a spectrally mismatched OPD, based on poly(3-hexylthiophene) P3HT and 6,6-phenyl-C_{61}-butyric acid methyl ester (PCBM), is reported in Figure 9b. Remarkably, despite a high EQE of the OPD is generally desired, as in the case of the P3HT:PCBM-based OPD, the major contribution to the overall level of sensitivity arises from the spectral overlap with the OLED electroluminescence spectrum. Hence, CuPc:C_{70} is evidently preferred to P3HT:PCBM, as absorptive layer.

By following this approach, a multitude of integrated sensors based on fluorescence detection have been realized over the years [157]. Lefevre and coworkers demonstrated a portable sensor complete with microfluidic system that is based on OLEDs and OPDs in a sandwiched structure [158] (Figure 10a,b). The sensor is designed to detect algal fluorescence and additional excitation and emission filters are introduced in the detection scheme.

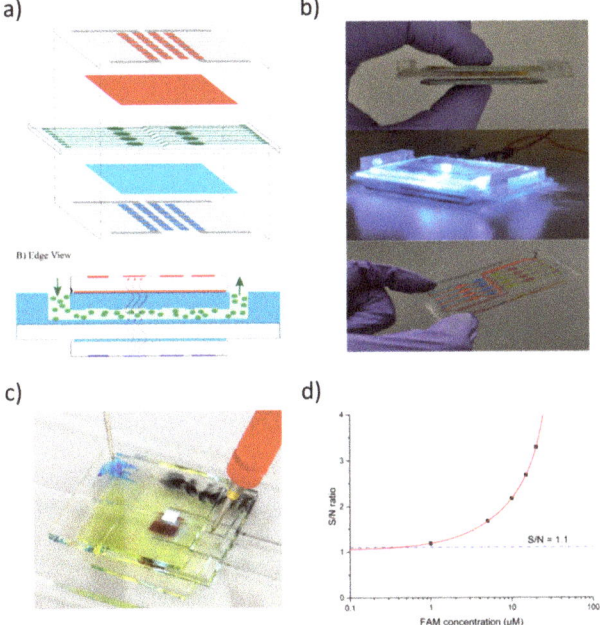

Figure 10. Schematic representation (**a**) and picture (**b**) of the algal fluorescence sensor based on a microfluidic chip comprising: (i) OLED, (ii) OPD, (iii) emission filter and (iv) excitation filter, adapted from Ref. [158] with permission from The Royal Society of Chemistry, © 2020. On-chip fluorescence sensor based on OLED and OPD (**c**) and the corresponding response to increasing concentration of fluorescein amidite (**d**), reproduced with permission from Ref [159], © 2020 The Royal Society of Chemistry.

Shu et al. reported on the first fluorescence light detector based on fully solution-processed organic electrochemical cells (OLECs) and an OPD for the detection of fluorescein amidite [159] (Figure 10c,d). Merfort and coworkers showed the monolithic integration of OLEDs and a-Si:H multispectral photodiodes for the detection of multiple dyes [160].

Miniaturized and disposable lab-on-a-chip devices have been realized within the European project "PHOTO-FET—Integrated photonic field-effect technology for bio-sensing functional components, Grant Agreement no. 248052" by using multifunctional field-effect transistors, in place of diodes, both as light source and detector. The aim of the project was the development of a miniaturized photonic device endowed with a microfluidic cartridge for quantitative bio-sensing for monitoring cardiovascular markers such as myoglobin and troponin-I.

Despite photoluminescent sensors comprising a transmission-type detection scheme having been successfully realized [161], the face-to-face architecture in which light source and light detector are vertically stacked on each other typically represents a limitation for the sensitivity of the sensor. Indeed, the direct illumination of photodetector by the light source represents the majority of the overall signal. Hence, excitation and emission filters are typically introduced in face-to-face photoluminescent architectures in order to improve the signal-to-noise ratio. Nevertheless, filters render the equipment relatively bulky as well as increasing the level of complexity of the system.

Over the years, the progressive engineering of the sensor architecture succeeded in avoiding the need of additional optical components. Indeed, independently from the modality of signal transduction, the relative positioning of the three sensor components (i.e., light source, light detector and sensitive element) was suitably modified to simplify the fabrication protocol and to improve the sensing performance. For instance, the monolithic, planar and concentric integration of an OLED and a ring-shaped OPD, allowed the production of a filter-less miniaturized oxygen sensor [162].

A fascinating approach to detection was developed by Ramuz et al. by evanescently coupling the light emitted by a polymeric LED (PLED) to a single-mode waveguide towards an array of polymer PDs (PPDs) [163]. The guided light interacts with the analyte, that is the mouse immunoglobulin G. While the waveguide's surface is functionalized with the mouse immunoglobulin G, the anti-mouse immunoglobulin G is injected via microfluidic system in conjunction with Cy5, that acts as luminescence quencher. The use of different detection schemes for the integration of optoelectronic components opened towards new sensing concepts such as the combination of the absorption/transmission signal transduction with a reflection-mode detection scheme. The working principle of the sensor is summarized well in the proximity sensor that Bürgi and coworkers developed in 2005 [164]. They monolithically integrated PLEDs and PPDs, in which the light reflected by an external object is then driven back to the PPD, where is absorbed and converted into a photocurrent. The proximity of the external object tunes the intensity of light reaching the PPD, and hence the generated photocurrent.

The use of that type of signal transduction and chip architecture found attractive applications in medicine. Bansal et al. recently combined OLEDs and OPDs to produce a wearable sensor for continuous health monitoring [67]. In particular, two different applications have been demonstrated. One is a muscle contraction sensor able to detect and distinguish between isotonic and isometric types of muscle contraction (Figure 11a). The second sensor consists of a bendable organic optoelectronic chip, that measures the oxygenation of human tissues. In the first case, an OLED, including Superyellow as emissive layer, has been integrated with 4 OPDs based on Poly [[4,8-bis[(2-ethylhexyl)oxy]benzo[1,2-b:4,5-b']dithiophene-2,6-diyl][3-fluoro-2-[(2-ethylhexyl)carbonyl] thieno[3,4-b]thiophenediyl]] (PTB7): 6,6-Phenyl-C_{71}-butyric acid methyl ester ($PC_{70}BM$) as absorbing compounds. In the latter application, the chip aims at monitoring the blood oxygenation by using hemoglobin and cytochrome aa_3 oxidase (Figure 11c). Compared to the first application, the emissive layer of the OLED has been replaced by OC_1C_{10}-PPV polymer (Figure 11b).

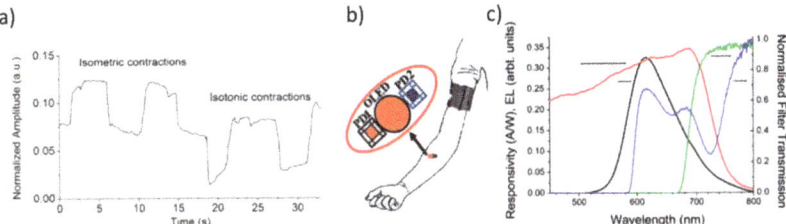

Figure 11. Response to isometric and isotonic muscle contractions reported by the wearable muscle contraction sensor based on the integration of one OLED and four OPDs (**a**). Representation of the bendable blood oxygenation sensor (**b**) and the corresponding electroluminescence (EL) spectra (left axis) of OC_1C_{10}-PPV-based OLED (black line), responsivity of $PTB7/PC_{70}BM$ photodiode (red line) together with their filter transmissions (right axis), i.e., 610 nm (blue line) and 700 nm (green line) (**c**). Adapted with permission from [67], © 2020 WILEY-VCH Verlag GmbH & Co. KGaA, Weinheim.

The integration of OLEDs and OPDs for medicine and biodiagnostics applications has been widely reported in the last years [165]. Several reflectance-based sensors have been fabricated for wireless monitoring of the photoplethysmogram (PPG) signal [166]. In particular, Arias' research group, one of the leaders of this research area [161,167,168], exhaustively showed a real-life application of a printed array of OLEDs and OPDs [167]. This was demonstrated through an accurate design of (i) single components, (ii) their relative distances/dimensions, (iii) their electrical connection and (iv) the collection and the analysis of the data. As a result, a flexible and printed integration of organic devices for measuring oxygen saturation with a high-quality biosignal was achieved (Figure 12). In this context, the development of the figures-of-merit of single components and the increase of the form-factor resulted in an all-day wearable pulse oxygen sensor operating at electric power as low as 24 µW [169].

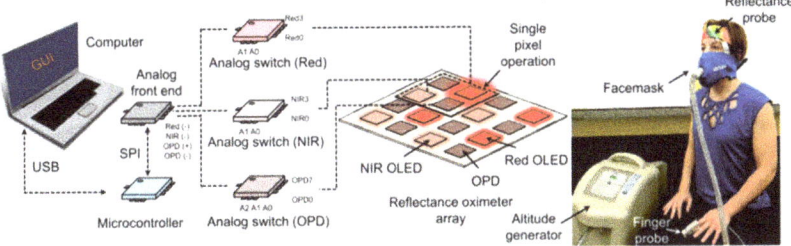

Figure 12. Design of the reflectance oximeter system, where each pixel comprising one red OLED, one near-infrared (NIR) OLED and two OPDs, is connected to an instrumet for driving the devices (**left** panel). Analog switches, Arduino microcontroller and universal serial bus (USB) connection are used. Setup of an altitude simulator that modifies the oxygen content of the air is showed during the operation on a volunteer breathing via a facemask (**right** panel). © 2020 by *Proceedings of the National Academy of Sciences* (PNAS), adapted with permission from [167].

The interest in portable and compact sensing systems for real-setting application opened towards the realization of a plethora of platforms with different functions, designs and concepts [170,171]. Most of the sensing systems rely on the integration of two or more optoelectronic devices. In this context, a breakthrough in the integration of components for sensing has recently been achieved by Shakoor et al., through the combination of photonic and optoelectronic devices [172]. The authors demonstrated the monolithic integration of a plasmonic grating with a CMOS, acting as photodiode (Figure 13). As reported in Section 3.1, the plasmonic nanostructures are widely used for sensing small variation of the refractive index of the medium taking place at their surface since it is correlated with variation of the LSPR wavelength. The integration of the plasmonic component onto the CSOM allowed detection of different concentrations of glycerol in aqueous solution (from 0 to 90% v/v) located at the grating surface, by illuminating with an external light source at 815 nm.

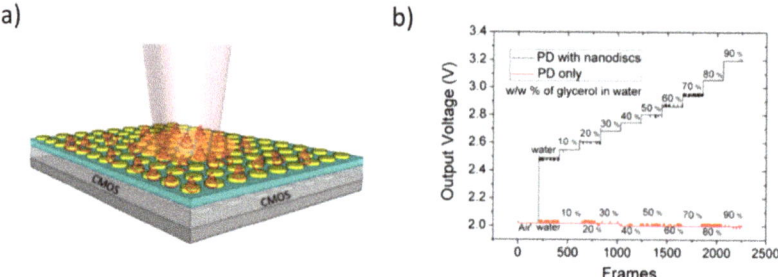

Figure 13. Schematic representation of a nanophotonic complementary metal oxide semiconductor (CMOS)-based biosensor (**a**) and the corresponding response at increasing glycerol concentration in water (**b**). The sensor is based on the monolithic integration of plasmonic nanostructures and a CMOS photodetector. © 2020 American Chemical Society, adapted with permission from [172].

Aiming at ultracompact and low-cost biosensors, the authors further developed the optoelectronic-plasmonic integration by structuring the photodiode electrical contact to obtain plasmonic nanoholes with sensitivities ≥1000 nm per refractive index unit [173].

In this context, it is worth highlighting a new detection scheme currently under development in the H2020 European project "MOLOKO, Multiplex photonic sensor for plasmonic-based online detection of contaminants in milk, Grant Agreement no. 780839". The project aims at developing a miniaturized and label-free sensor for the detection of analytes (i.e., proteins, toxins and antibiotics) in milk through the innovative combination of photonic and optoelectronic components, that is (i) a nanoplasmonic grating, (ii) an organic light-emitting transistor (OLET), used as light source, and iii) an organic photodiode, used as light detector, that is monolithically integrated onto the OLET. The three parts (that have been described in other sections of this review) are intended to work cooperatively in a reflection-like configuration for the detection of refractive index variations as small as 10^{-6} RIU upon biological stimulus. In detail, the light emitted by the OLET impinges the plasmonic grating, which modulates the luminous signal upon biochemical inputs occurring at its back-side, and then reflected back to the OPD, where is detected. By the implementation of both a suitable biofunctionalization of the sensing surface of the nanoplasmonic grating and ad hoc engineered microfluidic cartridge, the sensor is designed to detect up to 6 analytes of interest at the same time in a multiplexing configuration.

As a matter of fact, optical sensors are definitely a new scientifically and technologically challenging tool for improving the human lifespan as well as the quality of life when wearability, lightweight, and miniaturization factors are needed in the application. Nevertheless, the massive development of miniaturized and efficient single components based on nanostructured organic and hybrid materials, which has occurred over the last few decades, provides an enormous potential for the progress of new concepts and applications of sensors.

5. Conclusions

In this review we focused on the recent progress, main figures-of-merit and potential of different miniaturized sensing systems based on nanostructured materials. In particular, we reported on the use of organic and hybrid nanostructured materials into optoelectronic, electrochemical and plasmonic elements for highly integrated smart sensors.

The engineering of the functional properties of nanostructured materials through their chemical and structural design, together with the design of the architecture of the single-component devices, is demonstrated to be playing a major role in the realization of effective and integrated detection schemes. Indeed, for on-site bio-diagnostics and environmental/food monitoring purposes, the application of most standard and traditional analytical techniques is in contrast with the current need for rapid, cheap, easy-to-use and portable devices.

Above all, electrochemical and optical chemical sensors are promising tools with interesting analytical features. For electrochemical sensing, the major strategy that is implemented for enhancing both sensitivity and selectivity is the surface modification by nanophase-functional materials. Nanophase materials, such as carbon allotropes, metal- and metal oxide NPs, molecularly imprinted polymers, through different physical and chemical mechanisms, or acting as binders for biological recognition elements, can strongly enhance the limit of detection and analyte affinity of electrochemical sensors.

Optical sensors generally also show a good sensitivity and selectivity. Also, with respect to electronic devices, optoelectronic devices are less affected by external electric or magnetic fields and self-interference effects. The massive development of miniaturized and efficient single optoelectronic components based on nanostructured organic and hybrid materials, which has occurred over the last few decades, has provided enormous potential for the progress of new concepts and applications of optical and optical chemical sensors. In parallel, architecture engineering succeeded in avoiding the need for additional expensive optical components, thus allowing effective miniaturization and multiplexing.

On the other hand, plasmonic and photonic components obtained by nanostructuring allow the sensitivity of optical sensors to be increased. The features of the plasmonic and photonic systems are quite complementary and a new interest emerged in combining them into hybrid multicomponent systems to take advantage of the best properties of both. As one of the main advantages, the same plasmonic/photonic detection scheme can be applied to detect a large panel of molecules, often with a label-free approach (i.e., SPR). Furthermore, a specific functionalization step with biological recognition elements enable to obtain optoplasmonic or photonic chemical sensors with a high selectivity for specific target molecules.

Finally, we highlighted that integration is a key-factor to unravel the potentiality of optical-chemical sensors in terms of disposability, reliability, miniaturization and multiplexing while providing laboratory-quality analysis. In view of developing functional sensors for real-setting applications, a smart and effective system-engineering approach is necessary for realizing fast-responding, non-invasive, broadly adaptable, potentially highly sensitive and multiplexing sensors. Such systems could be employed to overcome existing limitations in measurements which are currently used in environmental and agri-food fields and advanced biodiagnostics.

Funding: This work received funding from the European Union's Horizon 2020 research and innovation program under grant agreement no. 780839 (MOLOKO project). We kindly acknowledge financial support from Programma Operativo POR-FESR 2014-2020 of Regione Emilia Romagna, Azione 1.2.2, through the project FORTRESS "Flexible, large-area patches for real-time detection of ionizing radiation", CUP I38D18000150009-PG/2018/629121.

Conflicts of Interest: The authors declare no conflict of interest.

References

1. Kozitsina, A.N.; Svalova, T.S.; Malysheva, N.N.; Okhokhonin, A.V.; Vidrevich, M.B.; Brainina, K.Z. Sensors based on bio and biomimetic receptors in medical diagnostic, environment, and food analysis. *Biosensors* **2018**, *8*, 35. [CrossRef] [PubMed]
2. Dincer, C.; Bruch, R.; Costa-Rama, E.; Fernández-Abedul, M.T.; Merkoçi, A.; Manz, A.; Urban, G.A.; Güder, F. Disposable Sensors in Diagnostics, Food, and Environmental Monitoring. *Adv. Mater.* **2019**, *31*, 1806739. [CrossRef] [PubMed]
3. Joshi, N.; Hayasaka, T.; Liu, Y.; Liu, H.; Oliveira, O.N.; Lin, L. A review on chemiresistive room temperature gas sensors based on metal oxide nanostructures, graphene and 2D transition metal dichalcogenides. *Microchim. Acta* **2018**, *185*, 213. [CrossRef] [PubMed]

4. Gauglitz, G. Direct optical sensors: Principles and selected applications. *Anal. Bioanal. Chem.* **2005**, *381*, 141–155. [CrossRef] [PubMed]
5. Majdinasab, M.; Mitsubayashi, K.; Marty, J.L. Optical and Electrochemical Sensors and Biosensors for the Detection of Quinolones. *Trends Biotechnol.* **2019**, *37*, 898–915. [CrossRef]
6. Janata, J. *Principles of Chemical Sensors*, 2nd ed.; Springer International Publishing: Berlin/Heidelberg, Germany, 2009; ISBN 978-0387699301.
7. Manikandan, V.S.; Adhikari, B.R.; Chen, A. Nanomaterial based electrochemical sensors for the safety and quality control of food and beverages. *Analyst* **2018**, *143*, 4537–4554. [CrossRef] [PubMed]
8. Yesilkoy, F. Optical interrogation techniques for nanophotonic biochemical sensors. *Sens. Switz.* **2019**, *19*, 4287. [CrossRef]
9. Janasek, D.; Franzke, J.; Manz, A. Scaling and the design of miniaturized chemical-analysis systems. *Nature* **2006**, *442*, 374–380. [CrossRef]
10. Tran-Thi, T.H.; Dagnelie, R.; Crunaire, S.; Nicole, L. Optical chemical sensors based on hybrid organic-inorganic sol-gel nanoreactors. *Chem. Soc. Rev.* **2011**, *40*, 621–639. [CrossRef]
11. Kaushik, A.; Kumar, R.; Arya, S.K.; Nair, M.; Malhotra, B.D.; Bhansali, S. Organic-Inorganic Hybrid Nanocomposite-Based Gas Sensors for Environmental Monitoring. *Chem. Rev.* **2015**, *115*, 4571–4606. [CrossRef]
12. Zhang, J.; Liu, X.; Neri, G.; Pinna, N. Nanostructured Materials for Room-Temperature Gas Sensors. *Adv. Mater.* **2016**, *28*, 795–831. [CrossRef] [PubMed]
13. Liu, X.; Ma, T.; Pinna, N.; Zhang, J. Two-Dimensional Nanostructured Materials for Gas Sensing. *Adv. Funct. Mater.* **2017**, *27*, 1–30. [CrossRef]
14. Muccini, M.; Toffanin, S. *Organic Light-emitting transistors: Towards the Next Generation Display Technology*; John Wiley & Sons: Hoboken, NJ, USA, 2016; pp. 158–160.
15. Konstantatos, G.; Clifford, J.; Levina, L.; Sargent, E.H. Sensitive solution-processed visible-wavelength photodetectors. *Nat. Photonics* **2007**, *1*, 531–534. [CrossRef]
16. Konstantatos, G.; Sargent, E.H. Nanostructured materials for photon detection. *Nat. Nanotechnol.* **2010**, *5*, 391–400. [CrossRef]
17. Lan, C.; Li, C.; Yin, Y.; Guo, H.; Wang, S. Synthesis of single-crystalline GeS nanoribbons for high sensitivity visible-light photodetectors. *J. Mater. Chem. C* **2015**, *3*, 8074–8079. [CrossRef]
18. Hagfeldt, A.; Boschloo, G.; Sun, L.; Kloo, L.; Pettersson, H. Dye-Sensitized Solar Cells. *Chem. Rev.* **2010**, *110*, 6595–6663. [CrossRef]
19. Xie, C.; Heumüller, T.; Gruber, W.; Tang, X.; Classen, A.; Schuldes, I.; Bidwell, M.; Späth, A.; Fink, R.H.; Unruh, T.; et al. Overcoming efficiency and stability limits in water-processing nanoparticular organic photovoltaics by minimizing microstructure defects. *Nat. Commun.* **2018**, *9*, 1–11. [CrossRef]
20. Meng, L.; Zhang, Y.; Wan, X.; Li, C.; Zhang, X.; Wang, Y.; Ke, X.; Xiao, Z.; Ding, L.; Xia, R.; et al. Organic and solution-processed tandem solar cells with 17.3% efficiency. *Science* **2018**, *361*, 1094–1098. [CrossRef]
21. Gedefaw, D.; Prosa, M.; Bolognesi, M.; Seri, M.; Andersson, M.R. Recent Development of Quinoxaline Based Polymers/Small Molecules for Organic Photovoltaics. *Adv. Energy Mater.* **2017**, *7*, 1700575. [CrossRef]
22. Hu, C.; Li, M.; Qiu, J.; Sun, Y.P. Design and fabrication of carbon dots for energy conversion and storage. *Chem. Soc. Rev.* **2019**, *48*, 2315–2337. [CrossRef]
23. Saliba, M.; Matsui, T.; Seo, J.Y.; Domanski, K.; Correa-Baena, J.P.; Nazeeruddin, M.K.; Zakeeruddin, S.M.; Tress, W.; Abate, A.; Hagfeldt, A.; et al. Cesium-containing triple cation perovskite solar cells: Improved stability, reproducibility and high efficiency. *Energy Environ. Sci.* **2016**, *9*, 1989–1997. [CrossRef] [PubMed]
24. Mosconi, E.; Quarti, C.; Ivanovska, T.; Ruani, G.; De Angelis, F. Structural and electronic properties of organo-halide lead perovskites: A combined IR-spectroscopy and ab initio molecular dynamics investigation. *Phys. Chem. Chem. Phys.* **2014**, *16*, 16137–16144. [CrossRef] [PubMed]
25. García De Arquer, F.P.; Armin, A.; Meredith, P.; Sargent, E.H. Solution-processed semiconductors for next-generation photodetectors. *Nat. Rev. Mater.* **2017**, *2*, 1–16. [CrossRef]
26. Chow, P.C.Y.; Someya, T. Organic Photodetectors for Next-Generation Wearable Electronics. *Adv. Mater.* **2019**, *1902045*, 1902045. [CrossRef] [PubMed]
27. Prosa, M.; Sagnella, A.; Posati, T.; Tessarolo, M.; Bolognesi, M.; Cavallini, S.; Toffanin, S.; Benfenati, V.; Seri, M.; Ruani, G.; et al. Integration of a silk fibroin based film as a luminescent down-shifting layer in ITO-free organic solar cells. *RSC Adv.* **2014**, *4*, 44815–44822. [CrossRef]

28. Manfredi, G.; Colombo, E.; Barsotti, J.; Benfenati, F.; Lanzani, G. Photochemistry of Organic Retinal Prostheses. *Annu. Rev. Phys. Chem.* **2019**, *70*, 99–121. [CrossRef]
29. Wu, Y.L.; Fukuda, K.; Yokota, T.; Someya, T. A Highly Responsive Organic Image Sensor Based on a Two-Terminal Organic Photodetector with Photomultiplication. *Adv. Mater.* **2019**, *31*, 1–7. [CrossRef]
30. Baeg, K.J.; Binda, M.; Natali, D.; Caironi, M.; Noh, Y.Y. Organic light detectors: Photodiodes and phototransistors. *Adv. Mater.* **2013**, *25*, 4267–4295. [CrossRef]
31. Yu, G.; Wang, J.; McElvain, J.; Heeger, A.J. Large-area, full-color image sensors made with semiconducting polymers. *Adv. Mater.* **1998**, *10*, 1431–1434. [CrossRef]
32. Jansen-van Vuuren, R.D.; Armin, A.; Pandey, A.K.; Burn, P.L.; Meredith, P. Organic Photodiodes: The Future of Full Color Detection and Image Sensing. *Adv. Mater.* **2016**, *28*, 4766–4802. [CrossRef]
33. Xu, H.; Chen, R.; Sun, Q.; Lai, W.; Su, Q.; Huang, W.; Liu, X. Recent progress in metal-organic complexes for optoelectronic applications. *Chem. Soc. Rev.* **2014**, *43*, 3259–3302. [CrossRef] [PubMed]
34. Yan, C.; Barlow, S.; Wang, Z.; Yan, H.; Jen, A.K.Y.; Marder, S.R.; Zhan, X. Non-fullerene acceptors for organic solar cells. *Nat. Rev. Mater.* **2018**, *3*, 1–19. [CrossRef]
35. Bolognesi, M.; Gedefaw, D.; Cavazzini, M.; Catellani, M.; Andersson, M.R.; Muccini, M.; Kozma, E.; Seri, M. Side chain modification on PDI-spirobifluorene-based molecular acceptors and its impact on organic solar cell performances. *N. J. Chem.* **2018**, *42*, 18633–18640. [CrossRef]
36. Bonetti, S.; Prosa, M.; Pistone, A.; Favaretto, L.; Sagnella, A.; Grisin, I.; Zambianchi, M.; Karges, S.; Lorenzoni, A.; Posati, T.; et al. A self-assembled lysinated perylene diimide film as a multifunctional material for neural interfacing. *J. Mater. Chem. B* **2016**, *4*, 2921–2932. [CrossRef]
37. Manna, E.; Xiao, T.; Shinar, J.; Shinar, R. Organic Photodetectors in Analytical Applications. *Electronics* **2015**, *4*, 688–722. [CrossRef]
38. Jansen Van Vuuren, R.; Johnstone, K.D.; Ratnasingam, S.; Barcena, H.; Deakin, P.C.; Pandey, A.K.; Burn, P.L.; Collins, S.; Samuel, I.D.W. Determining the absorption tolerance of single chromophore photodiodes for machine vision. *Appl. Phys. Lett.* **2010**, *96*, 2008–2011. [CrossRef]
39. Guo, F.; Yang, B.; Yuan, Y.; Xiao, Z.; Dong, Q.; Bi, Y.; Huang, J. A nanocomposite ultraviolet photodetector based on interfacial trap-controlled charge injection. *Nat. Nanotechnol.* **2012**, *7*, 798–802. [CrossRef]
40. Finlayson, G.D.; Hordley, S.D. Color constancy at a pixel. *J. Opt. Soc. Am. A* **2001**, *18*, 253. [CrossRef]
41. Armin, A.; Hambsch, M.; Kim, I.K.; Burn, P.L.; Meredith, P.; Namdas, E.B. Thick junction broadband organic photodiodes. *Laser Photonics Rev.* **2014**, *8*, 924–932. [CrossRef]
42. Guo, F.; Xiao, Z.; Huang, J. Fullerene Photodetectors with a Linear Dynamic Range of 90 dB Enabled by a Cross-Linkable Buffer Layer. *Adv. Opt. Mater.* **2013**, *1*, 289–294. [CrossRef]
43. Fang, Y.; Guo, F.; Xiao, Z.; Huang, J. Large gain, low noise nanocomposite ultraviolet photodetectors with a linear dynamic range of 120 dB. *Adv. Opt. Mater.* **2014**, *2*, 348–353. [CrossRef]
44. Redding, B.; Liew, S.F.; Sarma, R.; Cao, H. Compact spectrometer based on a disordered photonic chip. *Nat. Photonics* **2013**, *7*, 746–751. [CrossRef]
45. Jacques, S.L. Optical properties of biological tissues: A review. *Phys. Med. Biol.* **2013**, *58*, 5007–5008. [CrossRef]
46. Humar, M.; Yun, S.H. Intracellular microlasers. *Nat. Photonics* **2015**, *9*, 572–576. [CrossRef] [PubMed]
47. Poher, V.; Grossman, N.; Kennedy, G.T.; Nikolic, K.; Zhang, H.X.; Gong, Z.; Drakakis, E.M.; Gu, E.; Dawson, M.D.; French, P.M.W.; et al. Micro-LED arrays: A tool for two-dimensional neuron stimulation. *J. Phys. D Appl. Phys.* **2008**, *41*, 094014. [CrossRef]
48. Moseley, H.; Allen, J.W.; Ibbotson, S.; Lesar, A.; McNeill, A.; Camacho-Lopez, M.A.; Samuel, I.D.W.; Sibbett, W.; Ferguson, J. Ambulatory photodynamic therapy: A new concept in delivering photodynamic therapy. *Br. J. Dermatol.* **2006**, *154*, 747–750. [CrossRef] [PubMed]
49. Nammour, S.; Zeinoun, T.; Bogaerts, I.; Lamy, M.; Geerts, S.O.; Saba, S.B.; Lamard, L.; Peremans, A.; Limme, M. Evaluation of dental pulp temperature rise during photo-activated decontamination (PAD) of caries: An in vitro study. *Lasers Med. Sci.* **2010**, *25*, 651–654. [CrossRef]
50. Tang, C.W.; Vanslyke, S.A. Organic electroluminescent diodes. *Appl. Phys. Lett.* **1987**, *51*, 913–915. [CrossRef]
51. Fuhrman, J.; McCallum, K.; Davis, A. Flexible light-emitting diodes made from soluble conducting polymers. *Nature* **1992**, *359*, 710–713.
52. Fleetham, T.; Li, G.; Li, J. Phosphorescent Pt(II) and Pd(II) Complexes for Efficient, High-Color-Quality, and Stable OLEDs. *Adv. Mater.* **2017**, *29*, 1–16. [CrossRef]

53. Lo, Y.C.; Yeh, T.H.; Wang, C.K.; Peng, B.J.; Hsieh, J.L.; Lee, C.C.; Liu, S.W.; Wong, K.T. High-efficiency red and near-infrared organic light-emitting diodes enabled by pure organic fluorescent emitters and an exciplex-forming cohost. *ACS Appl. Mater. Interfaces* **2019**, *11*, 23417–23427. [CrossRef] [PubMed]
54. Kirlikovali, K.O.; Spokoyny, A.M. The Long-Lasting Blues: A New Record for Phosphorescent Organic Light-Emitting Diodes. *Chem* **2017**, *3*, 385–387. [CrossRef]
55. Ràfols-Ribé, J.; Will, P.A.; Hänisch, C.; Gonzalez-Silveira, M.; Lenk, S.; Rodríguez-Viejo, J.; Reineke, S. High-performance organic light-emitting diodes comprising ultrastable glass layers. *Sci. Adv.* **2018**, *4*, 1–10. [CrossRef] [PubMed]
56. Udagawa, K.; Sasabe, H.; Igarashi, F.; Kido, J. Simultaneous Realization of High EQE of 30%, Low Drive Voltage, and Low Efficiency Roll-Off at High Brightness in Blue Phosphorescent OLEDs. *Adv. Opt. Mater.* **2016**, *4*, 86–90. [CrossRef]
57. Kim, K.H.; Kim, J.J. Origin and Control of Orientation of Phosphorescent and TADF Dyes for High-Efficiency OLEDs. *Adv. Mater.* **2018**, *30*, 1–19. [CrossRef]
58. Komoda, T.; Sasabe, H.; Kido, J. Current Status of OLED Material and Process Technologies for Display and Lighting. *IEEE* **2018**, 1–4. [CrossRef]
59. Armaroli, N.; Bolink, H.J. *Photoluminescent Materials and Electroluminescent Devices*, 1st ed.; Springer International Publishing: Berlin/Heidelberg, Germany, 2017.
60. Costa, R.D.; Ortí, E.; Bolink, H.J.; Monti, F.; Accorsi, G.; Armaroli, N. Luminescent ionic transition-metal complexes for light-emitting electrochemical cells. *Angew. Chem. Int. Ed.* **2012**, *51*, 8178–8211. [CrossRef]
61. Prosa, M.; Benvenuti, E.; Pasini, M.; Giovanella, U.; Bolognesi, M.; Meazza, L.; Galeotti, F.; Muccini, M.; Toffanin, S. Organic Light-Emitting Transistors with Simultaneous Enhancement of Optical Power and External Quantum Efficiency via Conjugated Polar Polymer Interlayers. *ACS Appl. Mater. Interfaces* **2018**, *10*, 25580–25588. [CrossRef]
62. Capelli, R.; Toffanin, S.; Generali, G.; Usta, H.; Facchetti, A.; Muccini, M. Organic light-emitting transistors with an efficiency that outperforms the equivalent light-emitting diodes. *Nat. Mater.* **2010**, *9*, 496–503. [CrossRef]
63. Muccini, M.; Koopman, W.; Toffanin, S. The photonic perspective of organic light-emitting transistors. *Laser Photonics Rev.* **2012**, *6*, 258–275. [CrossRef]
64. Natali, M.; Quiroga, S.D.; Passoni, L.; Criante, L.; Benvenuti, E.; Bolognini, G.; Favaretto, L.; Melucci, M.; Muccini, M.; Scotognella, F.; et al. Simultaneous Tenfold Brightness Enhancement and Emitted-Light Spectral Tunability in Transparent Ambipolar Organic Light-Emitting Transistor by Integration of High-k Photonic Crystal. *Adv. Funct. Mater.* **2017**, *27*, 1–8. [CrossRef]
65. Toffanin, S.; Capelli, R.; Koopman, W.; Generali, G.; Cavallini, S.; Stefani, A.; Saguatti, D.; Ruani, G.; Muccini, M. Organic light-emitting transistors with voltage-tunable lit area and full channel illumination. *Laser Photonics Rev.* **2013**, *7*, 1011–1019. [CrossRef]
66. Lee, Y.H.; Kweon, O.Y.; Kim, H.; Yoo, J.H.; Han, S.G.; Oh, J.H. Recent advances in organic sensors for health self-monitoring systems. *J. Mater. Chem. C* **2018**, *6*, 8569–8612. [CrossRef]
67. Bansal, A.K.; Hou, S.; Kulyk, O.; Bowman, E.M.; Samuel, I.D.W. Wearable Organic Optoelectronic Sensors for Medicine. *Adv. Mater.* **2015**, *27*, 7638–7644. [CrossRef]
68. Liu, R.; Cai, Y.; Park, J.M.; Ho, K.M.; Shinar, J.; Shinar, R. Organic light-emitting diode sensing platform: Challenges and solutions. *Adv. Funct. Mater.* **2011**, *21*, 4744–4753. [CrossRef]
69. Ahmad, V.; Shukla, A.; Sobus, J.; Sharma, A.; Gedefaw, D.; Andersson, G.G.; Andersson, M.R.; Lo, S.C.; Namdas, E.B. High-Speed OLEDs and Area-Emitting Light-Emitting Transistors from a Tetracyclic Lactim Semiconducting Polymer. *Adv. Opt. Mater.* **2018**, *6*, 1–8. [CrossRef]
70. Maddalena, F.; Chin, X.Y.; Cortecchia, D.; Bruno, A.; Soci, C. Brightness Enhancement in Pulsed-Operated Perovskite Light-Emitting Transistors. *ACS Appl. Mater. Interfaces* **2018**, *10*, 37316–37325. [CrossRef]
71. Staude, I.; Schilling, J. Metamaterial-inspired silicon nanophotonics. *Nat. Photonics* **2017**, *11*, 274–284. [CrossRef]
72. Jeevanandam, J.; Barhoum, A.; Chan, Y.S.; Dufresne, A.; Danquah, M.K. Review on nanoparticles and nanostructured materials: History, sources, toxicity and regulations. *Beilstein J. Nanotechnol.* **2018**, *9*, 1050–1074. [CrossRef]
73. Goriparti, S.; Miele, E.; De Angelis, F.; Di Fabrizio, E.; Proietti Zaccaria, R.; Capiglia, C. Review on recent progress of nanostructured anode materials for Li-ion batteries. *J. Power Sources* **2014**, *257*, 421–443. [CrossRef]

74. Su, X.; Wu, Q.; Li, J.; Xiao, X.; Lott, A.; Lu, W.; Sheldon, B.W.; Wu, J. Silicon-Based nanomaterials for lithium-ion batteries: A review. *Adv. Energy Mater.* **2014**, *4*, 1–23. [CrossRef]
75. Yu, R.; Lin, Q.; Leung, S.F.; Fan, Z. Nanomaterials and nanostructures for efficient light absorption and photovoltaics. *Nano Energy* **2012**, *1*, 57–72. [CrossRef]
76. Prosa, M.; Li, N.; Gasparini, N.; Bolognesi, M.; Seri, M.; Muccini, M.; Brabec, C.J. Revealing Minor Electrical Losses in the Interconnecting Layers of Organic Tandem Solar Cells. *Adv. Mater. Interfaces* **2017**, *4*, 1700776. [CrossRef]
77. Raj, S.; Jose, S.; Sumod, U.S.; Sabitha, M. Nanotechnology in cosmetics: Opportunities and challenges. *J. Pharm. Bioallied Sci.* **2012**, *4*, 186–193. [CrossRef]
78. Sun, Y.F.; Liu, S.B.; Meng, F.L.; Liu, J.Y.; Jin, Z.; Kong, L.T.; Liu, J.H. Metal oxide nanostructures and their gas sensing properties: A review. *Sensors* **2012**, *12*, 2610–2631. [CrossRef]
79. Llobet, E. Gas sensors using carbon nanomaterials: A review. *Sens. Actuators B Chem.* **2013**, *179*, 32–45. [CrossRef]
80. Mirzaei, A.; Leonardi, S.G.; Neri, G. Detection of hazardous volatile organic compounds (VOCs) by metal oxide nanostructures-based gas sensors: A review. *Ceram. Int.* **2016**, *42*, 15119–15141. [CrossRef]
81. Theerthagiri, J.; Karuppasamy, K.; Durai, G.; Rana, A.U.H.S.; Arunachalam, P.; Sangeetha, K.; Kuppusami, P.; Kim, H.S. Recent advances in metal chalcogenides (MX; X = S, Se) nanostructures for electrochemical supercapacitor applications: A brief review. *Nanomaterials* **2018**, *8*, 256. [CrossRef]
82. Pathakoti, K.; Manubolu, M.; Hwang, H.M. Nanostructures: Current uses and future applications in food science. *J. Food Drug Anal.* **2017**, *25*, 245–253. [CrossRef]
83. West, P.R.; Ishii, S.; Naik, G.V.; Emani, N.K.; Shalaev, V.M.; Boltasseva, A. Searching for better plasmonic materials. *Laser Photonics Rev.* **2010**, *4*, 795–808. [CrossRef]
84. Ermolaev, G.A.; Kushnir, S.E.; Sapoletova, N.A.; Napolskii, K.S. Titania photonic crystals with precise photonic band gap position via anodizing with voltage versus optical path length modulation. *Nanomaterials* **2019**, *9*, 651. [CrossRef] [PubMed]
85. Missinne, J.; Teigell Benéitez, N.; Lamberti, A.; Chiesura, G.; Luyckx, G.; Mattelin, M.A.; Van Paepegem, W.; Van Steenberge, G. Thin and Flexible Polymer Photonic Sensor Foils for Monitoring Composite Structures. *Adv. Eng. Mater.* **2018**, *20*, 1–5. [CrossRef]
86. Wienhold, T.; Kraemmer, S.; Wondimu, S.F.; Siegle, T.; Bog, U.; Weinzierl, U.; Schmidt, S.; Becker, H.; Kalt, H.; Mappes, T.; et al. All-polymer photonic sensing platform based on whispering-gallery mode microgoblet lasers. *Lab Chip* **2015**, *15*, 3800–3806. [CrossRef] [PubMed]
87. Lova, P.; Manfredi, G.; Boarino, L.; Comite, A.; Laus, M.; Patrini, M.; Marabelli, F.; Soci, C.; Comoretto, D. Polymer distributed bragg reflectors for vapor sensing. *ACS Photonics* **2015**, *2*, 537–543. [CrossRef]
88. Bosio, N.; Šípová-Jungová, H.; Länk, N.O.; Antosiewicz, T.J.; Verre, R.; Käll, M. Plasmonic versus All-Dielectric Nanoantennas for Refractometric Sensing: A Direct Comparison. *ACS Photonics* **2019**, *6*, 1556–1564. [CrossRef]
89. Wang, A.X.; Kong, X. Review of recent progress of plasmonic materials and nano-structures for surface-enhanced raman scattering. *Mater. Basel* **2015**, *8*, 3024–3052. [CrossRef]
90. Bauch, M.; Toma, K.; Toma, M.; Zhang, Q.; Dostalek, J. Plasmon-Enhanced Fluorescence Biosensors: A Review. *Plasmonics* **2014**, *9*, 781–799. [CrossRef]
91. Suzuki, Y.; Yokoyama, K. Development of functional fluorescent molecular probes for the detection of biological substances. *Biosensors* **2015**, *5*, 337–363. [CrossRef]
92. Petryayeva, E.; Krull, U.J. Localized surface plasmon resonance: Nanostructures, bioassays and biosensing—A review. *Anal. Chim. Acta* **2011**, *706*, 8–24. [CrossRef]
93. Kretschmann, E.; Raether, H. Radiative Decay of Non Radiative Surface Plasmons Excited by Light. *Z. Naturforsch. Sect. A J. Phys. Sci.* **1968**, *23*, 2135–2136. [CrossRef]
94. Jang, D.; Chae, G.; Shin, S. Analysis of surface plasmon resonance curves with a novel sigmoid-asymmetric fitting algorithm. *Sens. Switz.* **2015**, *15*, 25385–25398. [CrossRef] [PubMed]
95. Schasfoort, R.B.M. *Handbook of Surface Plasmon Resonance*, 2nd ed.; Royal Society of Chemistry: London, UK, 2017; ISBN 9781788010283.
96. Wong, C.L.; Olivo, M. Surface Plasmon Resonance Imaging Sensors: A Review. *Plasmonics* **2014**, *9*, 809–824. [CrossRef]

97. Cetin, A.E.; Etezadi, D.; Galarreta, B.C.; Busson, M.P.; Eksioglu, Y.; Altug, H. Plasmonic Nanohole Arrays on a Robust Hybrid Substrate for Highly Sensitive Label-Free Biosensing. *ACS Photonics* **2015**, *2*, 1167–1174. [CrossRef]
98. Giudicatti, S.; Marabelli, F.; Valsesia, A.; Pellacani, P.; Colpo, P.; Rossi, F. Interaction among plasmonic resonances in a gold film embedding a two-dimensional array of polymeric nanopillars. *J. Opt. Soc. Am. B* **2012**, *29*, 1641. [CrossRef]
99. Giudicatti, S.; Marabelli, F.; Pellacani, P. Field Enhancement by Shaping Nanocavities in a Gold Film. *Plasmonics* **2013**, *8*, 975–981. [CrossRef]
100. Bottazzi, B.; Fornasari, L.; Frangolho, A.; Giudicatti, S.; Mantovani, A.; Marabelli, F.; Marchesini, G.; Pellacani, P.; Therisod, R.; Valsesia, A. Multiplexed label-free optical biosensor for medical diagnostics. *J. Biomed. Opt.* **2014**, *19*, 017006. [CrossRef]
101. Goerlitzer, E.S.A.; Speichermann, L.E.; Mirza, T.A.; Mohammadi, R.; Vogel, N. Addressing the plasmonic hotspot region by site-specific functionalization of nanostructures. *Nanoscale Adv.* **2020**, *2*, 394–400. [CrossRef]
102. Cretich, M.; Pirri, G.; Damin, F.; Solinas, I.; Chiari, M. A new polymeric coating for protein microarrays. *Anal. Biochem.* **2004**, *332*, 67–74. [CrossRef]
103. Joshi, S.; Segarra-Fas, A.; Peters, J.; Zuilhof, H.; Van Beek, T.A.; Nielen, M.W.F. Multiplex surface plasmon resonance biosensing and its transferability towards imaging nanoplasmonics for detection of mycotoxins in barley. *Analyst* **2016**, *141*, 1307–1318. [CrossRef]
104. Jadon, N.; Jain, R.; Sharma, S.; Singh, K. Recent trends in electrochemical sensors for multianalyte detection—A review. *Talanta* **2016**, *161*, 894–916. [CrossRef]
105. Hu, X.; Dong, S. Metal nanomaterials and carbon nanotubes—Synthesis, functionalization and potential applications towards electrochemistry. *J. Mater. Chem.* **2008**, *18*, 1279–1295. [CrossRef]
106. Plieth, W.; Wilson, G.S.; Gutiérrez De La Fe, C. Spectroelectrochemistry: A survey of in situ spectroscopic techniques (Technical Report). *Pure Appl. Chem.* **1998**, *70*, 1395–1414. [CrossRef]
107. Layani, M.; Kamyshny, A.; Magdassi, S. Transparent conductors composed of nanomaterials. *Nanoscale* **2014**, *6*, 5581–5591. [CrossRef] [PubMed]
108. Dragone, R.; Frazzoli, C.; Monacelli, F. Chemical-Physical Sensing Device for Chemical—Toxicological Diagnostics in Real Matrices 2012. *Front. Public Health* **2017**, *5*, 80. [CrossRef] [PubMed]
109. Yang, C.; Denno, M.E.; Pyakurel, P.; Venton, B.J. Recent trends in carbon nanomaterial-based electrochemical sensors for biomolecules: A review. *Anal. Chim. Acta* **2015**, *887*, 17–37. [CrossRef]
110. Nguyen, N.S.; Das, G.; Yoon, H.H. Nickel/cobalt oxide-decorated 3D graphene nanocomposite electrode for enhanced electrochemical detection of urea. *Biosens. Bioelectron.* **2016**, *77*, 372–377. [CrossRef]
111. Wu, S.; He, Q.; Tan, C.; Wang, Y.; Zhang, H. Graphene-based electrochemical sensors. *Small* **2013**, *9*, 1160–1172. [CrossRef]
112. Zelada-Guillén, G.A.; Bhosale, S.V.; Riu, J.; Rius, F.X. Real-time potentiometric detection of bacteria in complex samples. *Anal. Chem.* **2010**, *82*, 9254–9260. [CrossRef]
113. Tran, T.T.; Mulchandani, A. Carbon nanotubes and graphene nano field-effect transistor-based biosensors. *TrAC Trends Anal. Chem.* **2016**, *79*, 222–232. [CrossRef]
114. Xiong, C.; Wang, Y.; Qu, H.; Zhang, L.; Qiu, L.; Chen, W.; Yan, F.; Zheng, L. Highly sensitive detection of gallic acid based on organic electrochemical transistors with poly(diallyldimethylammonium chloride) and carbon nanomaterials nanocomposites functionalized gate electrodes. *Sens. Actuators B Chem.* **2017**, *246*, 235–242. [CrossRef]
115. Rius-Ruiz, F.X.; Crespo, G.A.; Bejarano-Nosas, D.; Blondeau, P.; Riu, J.; Rius, F.X. Potentiometric strip cell based on carbon nanotubes as transducer layer: Toward low-cost decentralized measurements. *Anal. Chem.* **2011**, *83*, 8810–8815. [CrossRef] [PubMed]
116. Yuan, D.; Anthis, A.H.C.; Ghahraman Afshar, M.; Pankratova, N.; Cuartero, M.; Crespo, G.A.; Bakker, E. All-Solid-State Potentiometric Sensors with a Multiwalled Carbon Nanotube Inner Transducing Layer for Anion Detection in Environmental Samples. *Anal. Chem.* **2015**, *87*, 8640–8645. [CrossRef] [PubMed]
117. Liu, Y.; Liu, Y.; Gao, Y.; Wang, P. A general approach to one-step fabrication of single-piece nanocomposite membrane based Pb^{2+}-selective electrodes. *Sens. Actuators B Chem.* **2019**, *281*, 705–712. [CrossRef]
118. Yin, T.; Qin, W. Applications of nanomaterials in potentiometric sensors. *TrAC Trends Anal. Chem.* **2013**, *51*, 79–86. [CrossRef]

119. Ping, J.; Wang, Y.; Ying, Y.; Wu, J. Application of electrochemically reduced graphene oxide on screen-printed ion-selective electrode. *Anal. Chem.* **2012**, *84*, 3473–3479. [CrossRef] [PubMed]
120. Trojanowicz, M. Impact of nanotechnology on design of advanced screen-printed electrodes for different analytical applications. *TrAC Trends Anal. Chem.* **2016**, *84*, 22–47. [CrossRef]
121. Yao, Y.; Wu, H.; Ping, J. Simultaneous determination of Cd(II) and Pb(II) ions in honey and milk samples using a single-walled carbon nanohorns modified screen-printed electrochemical sensor. *Food Chem.* **2019**, *274*, 8–15. [CrossRef]
122. Munteanu, F.D.; Titoiu, A.M.; Marty, J.L.; Vasilescu, A. Detection of antibiotics and evaluation of antibacterial activity with screen-printed electrodes. *Sens. Switz.* **2018**, *18*, 901. [CrossRef]
123. Della Pelle, F.; Angelini, C.; Sergi, M.; Del Carlo, M.; Pepe, A.; Compagnone, D. Nano carbon black-based screen printed sensor for carbofuran, isoprocarb, carbaryl and fenobucarb detection: Application to grain samples. *Talanta* **2018**, *186*, 389–396. [CrossRef]
124. Cristofani, M.; Menna, E.; Seri, M.; Muccini, M.; Prosa, M.; Antonello, S.; Mba, M.; Franco, L.; Maggini, M. Tuning the Electron-Acceptor Properties of [60]Fullerene by Tailored Functionalization for Application in Bulk Heterojunction Solar Cells. *Asian J. Org. Chem.* **2016**, *5*, 676–684. [CrossRef]
125. Pandey, A.; Sharma, S.; Jain, R. Voltammetric sensor for the monitoring of hazardous herbicide triclopyr (TCP). *J. Hazard. Mater.* **2019**, *367*, 246–255. [CrossRef] [PubMed]
126. Ikhsan, N.I.; Pandikumar, A. Doped-Graphene Modified Electrochemical Sensors. In *Graphene-Based Electrochemical Sensors for Biomolecules*; Elsevier Inc.: Amsterdam, The Netherlands, 2019; pp. 67–87. ISBN 9780128153949.
127. Cui, R.; Xu, D.; Xie, X.; Yi, Y.; Quan, Y.; Zhou, M.; Gong, J.; Han, Z.; Zhang, G. Phosphorus-doped helical carbon nanofibers as enhanced sensing platform for electrochemical detection of carbendazim. *Food Chem.* **2017**, *221*, 457–463. [CrossRef] [PubMed]
128. Tsierkezos, N.G.; Othman, S.H.; Ritter, U.; Hafermann, L.; Knauer, A.; Köhler, J.M.; Downing, C.; McCarthy, E.K. Electrochemical analysis of ascorbic acid, dopamine, and uric acid on nobel metal modified nitrogen-doped carbon nanotubes. *Sens. Actuators B Chem.* **2016**, *231*, 218–229. [CrossRef]
129. Li, L.; Liu, D.; Wang, K.; Mao, H.; You, T. Quantitative detection of nitrite with N-doped graphene quantum dots decorated N-doped carbon nanofibers composite-based electrochemical sensor. *Sens. Actuators B Chem.* **2017**, *252*, 17–23. [CrossRef]
130. Georgakilas, V.; Gournis, D.; Tzitzios, V.; Pasquato, L.; Guldi, D.M.; Prato, M. Decorating carbon nanotubes with metal or semiconductor nanoparticles. *J. Mater. Chem.* **2007**, *17*, 2679–2694. [CrossRef]
131. Zhao, G.; Liu, G. Electrochemical deposition of gold nanoparticles on reduced graphene oxide by fast scan cyclic voltammetry for the sensitive determination of As(III). *Nanomaterials* **2019**, *9*, 41. [CrossRef]
132. Hou, X.; Liu, X.; Li, Z.; Zhang, J.; Du, G.; Ran, X.; Yang, L. Electrochemical determination of methyl parathion based on pillar[5]arene@AuNPs@reduced graphene oxide hybrid nanomaterials. *N. J. Chem.* **2019**, *43*, 13048–13057. [CrossRef]
133. Fan, X.; Xing, L.; Ge, P.; Cong, L.; Hou, Q.; Ge, Q.; Liu, R.; Zhang, W.; Zhou, G. Electrochemical sensor using gold nanoparticles and plasma pretreated graphene based on the complexes of calcium and Troponin C to detect Ca2+ in meat. *Food Chem.* **2020**, *307*, 125645. [CrossRef]
134. Lian, W.; Liu, S.; Yu, J.; Li, J.; Cui, M.; Xu, W.; Huang, J. Electrochemical sensor using neomycin-imprinted film as recognition element based on chitosan-silver nanoparticles/graphene-multiwalled carbon nanotubes composites modified electrode. *Biosens. Bioelectron.* **2013**, *44*, 70–76. [CrossRef]
135. Thirumalraj, B.; Sakthivel, R.; Chen, S.M.; Rajkumar, C.; Yu, L.K.; Kubendhiran, S. A reliable electrochemical sensor for determination of H_2O_2 in biological samples using platinum nanoparticles supported graphite/gelatin hydrogel. *Microchem. J.* **2019**, *146*, 673–678. [CrossRef]
136. Elyasi, M.; Khalilzadeh, M.A.; Karimi-Maleh, H. High sensitive voltammetric sensor based on Pt/CNTs nanocomposite modified ionic liquid carbon paste electrode for determination of Sudan i in food samples. *Food Chem.* **2013**, *141*, 4311–4317. [CrossRef] [PubMed]
137. Bhanjana, G.; Dilbaghi, N.; Singhal, N.K.; Kim, K.H.; Kumar, S. Zinc oxide nanopillars as an electrocatalyst for direct redox sensing of cadmium. *J. Ind. Eng. Chem.* **2017**, *53*, 192–200. [CrossRef]
138. Ridhuan, N.S.; Abdul Razak, K.; Lockman, Z. Fabrication and Characterization of Glucose Biosensors by Using Hydrothermally Grown ZnO Nanorods. *Sci. Rep.* **2018**, *8*, 1–12. [CrossRef] [PubMed]

139. Bhanjana, G.; Dilbaghi, N.; Kumar, R.; Kumar, S. Zinc Oxide Quantum Dots as Efficient Electron Mediator for Ultrasensitive and Selective Electrochemical Sensing of Mercury. *Electrochim. Acta* **2015**, *178*, 361–367. [CrossRef]
140. Kumar, S.; Ahlawat, W.; Kumar, R.; Dilbaghi, N. Graphene, carbon nanotubes, zinc oxide and gold as elite nanomaterials for fabrication of biosensors for healthcare. *Biosens. Bioelectron.* **2015**, *70*, 498–503. [CrossRef]
141. Kim, E.B.; Ameen, S.; Akhtar, M.S.; Shin, H.S. Iron-nickel co-doped ZnO nanoparticles as scaffold for field effect transistor sensor: Application in electrochemical detection of hexahydropyridine chemical. *Sens. Actuators B Chem.* **2018**, *275*, 422–431. [CrossRef]
142. Barthwal, S.; Singh, N.B. ZnO-CNT Nanocomposite:A Device as Electrochemical Sensor. *Mater. Today Proc.* **2017**, *4*, 5552–5560. [CrossRef]
143. Bijad, M.; Karimi-Maleh, H.; Khalilzadeh, M.A. Application of ZnO/CNTs Nanocomposite Ionic Liquid Paste Electrode as a Sensitive Voltammetric Sensor for Determination of Ascorbic Acid in Food Samples. *Food Anal. Methods* **2013**, *6*, 1639–1647. [CrossRef]
144. Mulaba-Bafubiandi, A.F.; Karimi-Maleh, H.; Karimi, F.; Rezapour, M. A voltammetric carbon paste sensor modified with NiO nanoparticle and ionic liquid for fast analysis of p-nitrophenol in water samples. *J. Mol. Liq.* **2019**, *285*, 430–435. [CrossRef]
145. Yang, K.; Yan, Z.; Ma, L.; Du, Y.; Peng, B.; Feng, J. A facile one-step synthesis of cuprous oxide/silver nanocomposites as efficient electrode-modifying materials for nonenzyme hydrogen peroxide sensor. *Nanomaterials* **2019**, *9*, 523. [CrossRef]
146. Parsaei, M.; Asadi, Z.; Khodadoust, S. A sensitive electrochemical sensor for rapid and selective determination of nitrite ion in water samples using modified carbon paste electrode with a newly synthesized cobalt(II)-Schiff base complex and magnetite nanospheres. *Sens. Actuators B Chem.* **2015**, *220*, 1131–1138. [CrossRef]
147. Ganjali, M.R.; Motakef-Kazami, N.; Faridbod, F.; Khoee, S.; Norouzi, P. Determination of Pb^{2+} ions by a modified carbon paste electrode based on multi-walled carbon nanotubes (MWCNTs) and nanosilica. *J. Hazard. Mater.* **2010**, *173*, 415–419. [CrossRef] [PubMed]
148. Chen, X.; Zhang, H.; Tunuguntla, R.H.; Noy, A. Silicon Nanoribbon pH Sensors Protected by a Barrier Membrane with Carbon Nanotube Porins. *Nano Lett.* **2019**, *19*, 629–634. [CrossRef] [PubMed]
149. Vasapollo, G.; Sole, R.D.; Mergola, L.; Lazzoi, M.R.; Scardino, A.; Scorrano, S.; Mele, G. Molecularly imprinted polymers: Present and future prospective. *Int. J. Mol. Sci.* **2011**, *12*, 5908–5945. [CrossRef] [PubMed]
150. Dragone, R.; Grasso, G.; Muccini, M.; Toffanin, S. Portable bio/chemosensoristic devices: Innovative systems for environmental health and food safety diagnostics. *Front. Public Heal.* **2017**, *5*, 1–6. [CrossRef] [PubMed]
151. Zhang, X.; Peng, Y.; Bai, J.; Ning, B.; Sun, S.; Hong, X.; Liu, Y.; Liu, Y.; Gao, Z. A novel electrochemical sensor based on electropolymerized molecularly imprinted polymer and gold nanomaterials amplification for estradiol detection. *Sens. Actuators B Chem.* **2014**, *200*, 69–75. [CrossRef]
152. Li, X.; Zhang, L.; Wei, X.; Li, J. A Sensitive and renewable chlortoluron molecularly imprinted polymer sensor based on the gate-controlled catalytic electrooxidation of H_2O_2 on Magnetic Nano-NiO. *Electroanalysis* **2013**, *25*, 1286–1293. [CrossRef]
153. Wackerlig, J.; Lieberzeit, P.A. Molecularly imprinted polymer nanoparticles in chemical sensing—Synthesis, characterisation and application. *Sens. Actuators B Chem.* **2015**, *207*, 144–157. [CrossRef]
154. Motaharian, A.; Motaharian, F.; Abnous, K.; Hosseini, M.R.M.; Hassanzadeh-Khayyat, M. Molecularly imprinted polymer nanoparticles-based electrochemical sensor for determination of diazinon pesticide in well water and apple fruit samples. *Anal. Bioanal. Chem.* **2016**, *408*, 6769–6779. [CrossRef]
155. Shinar, R.; Ghosh, D.; Choudhury, B.; Noack, M.; Dalal, V.L.; Shinar, J. Luminescence-based oxygen sensor structurally integrated with an organic light-emitting device excitation source and an amorphous Si-based photodetector. *J. Non. Cryst. Solids* **2006**, *352*, 1995–1998. [CrossRef]
156. Liu, R.; Xiao, T.; Cui, W.; Shinar, J.; Shinar, R. Multiple approaches for enhancing all-organic electronics photoluminescent sensors: Simultaneous oxygen and pH monitoring. *Anal. Chim. Acta* **2013**, *778*, 70–78. [CrossRef] [PubMed]
157. Nalwa, K.S.; Cai, Y.; Thoeming, A.L.; Shinar, J.; Shinar, R.; Chaudhary, S. Polythiophene-fullerene based photodetectors: Tuning of spectral esponse and application in photoluminescence based (Bio)chemical sensors. *Adv. Mater.* **2010**, *22*, 4157–4161. [CrossRef] [PubMed]

158. Lefèvre, F.; Chalifour, A.; Yu, L.; Chodavarapu, V.; Juneau, P.; Izquierdo, R. Algal fluorescence sensor integrated into a microfluidic chip for water pollutant detection. *Lab Chip* **2012**, *12*, 787–793. [CrossRef] [PubMed]
159. Shu, Z.; Kemper, F.; Beckert, E.; Eberhardt, R.; Tünnermann, A. Highly sensitive on-chip fluorescence sensor with integrated fully solution processed organic light sources and detectors. *RSC Adv.* **2017**, *7*, 26384–26391. [CrossRef]
160. Merfort, C.; Seibel, K.; Watty, K.; Böhm, M. Monolithically integrated μ-capillary electrophoresis with organic light sources and tunable a-Si:H multispectral photodiodes for fluorescence detection. *Microelectron. Eng.* **2010**, *87*, 712–714. [CrossRef]
161. Lochner, C.M.; Khan, Y.; Pierre, A.; Arias, A.C. All-organic optoelectronic sensor for pulse oximetry. *Nat. Commun.* **2014**, *5*, 1–7. [CrossRef]
162. Lamprecht, B.; Abel, T.; Kraker, E.; Haase, A.; Konrad, C.; Tscherner, M.; Köstler, S.; Ditlbacher, H.; Mayr, T. Integrated fluorescence sensor based on ring-shaped organic photodiodes. *Phys. Status Solidi Rapid Res. Lett.* **2010**, *4*, 157–159. [CrossRef]
163. Ramuz, M.; Leuenberger, D.; Bürgi, L. Optical biosensors based on integrated polymer light source and polymer photodiode. *J. Polym. Sci. Part B Polym. Phys.* **2011**, *49*, 80–87. [CrossRef]
164. Bürgi, L.; Pfeiffer, R.; Mücklich, M.; Metzler, P.; Kiy, M.; Winnewisser, C. Optical proximity and touch sensors based on monolithically integrated polymer photodiodes and polymer LEDs. *Org. Electron.* **2006**, *7*, 114–120. [CrossRef]
165. Yokota, T.; Zalar, P.; Kaltenbrunner, M.; Jinno, H.; Matsuhisa, N.; Kitanosako, H.; Tachibana, Y.; Yukita, W.; Koizumi, M.; Someya, T. Ultraflexible organic photonic skin. *Sci. Adv.* **2016**, *2*, 1–9. [CrossRef]
166. Elsamnah, F.; Bilgaiyan, A.; Affiq, M.; Shim, C.H.; Ishidai, H.; Hattori, R. Reflectance-based organic pulse meter sensor for wireless monitoring of photoplethysmogram signal. *Biosensors* **2019**, *9*, 87. [CrossRef]
167. Khan, Y.; Han, D.; Pierre, A.; Ting, J.; Wang, X.; Lochner, C.M.; Bovo, G.; Yaacobi-Gross, N.; Newsome, C.; Wilson, R.; et al. A flexible organic reflectance oximeter array. *Proc. Natl. Acad. Sci. USA* **2018**, *115*, E11015–E11024. [CrossRef] [PubMed]
168. Khan, Y.; Han, D.; Ting, J.; Ahmed, M.; Nagisetty, R.; Arias, A.C. Organic Multi-Channel Optoelectronic Sensors for Wearable Health Monitoring. *IEEE Access* **2019**, *7*, 128114–128124. [CrossRef]
169. Lee, H.; Kim, E.; Lee, Y.; Kim, H.; Lee, J.; Kim, M.; Yoo, H.J.; Yoo, S. Toward all-day wearable health monitoring: An ultralow-power, reflective organic pulse oximetry sensing patch. *Sci. Adv.* **2018**, *4*, 1–9. [CrossRef]
170. Ratcliff, E.L.; Veneman, P.A.; Simmonds, A.; Zacher, B.; Huebner, D.; Saavedra, S.S.; Armstrong, N.R. A planar, chip-based, dual-beam refractometer using an integrated organic light-emitting diode (OLED) light source and organic photovoltaic (OPV) detectors. *Anal. Chem.* **2010**, *82*, 2734–2742. [CrossRef]
171. Tchernycheva, M.; Messanvi, A.; De Luna Bugallo, A.; Jacopin, G.; Lavenus, P.; Rigutti, L.; Zhang, H.; Halioua, Y.; Julien, F.H.; Eymery, J.; et al. Integrated photonic platform based on InGaN/GaN nanowire emitters and detectors. *Nano Lett.* **2014**, *14*, 3515–3520. [CrossRef]
172. Shakoor, A.; Cheah, B.C.; Hao, D.; Al-Rawhani, M.; Nagy, B.; Grant, J.; Dale, C.; Keegan, N.; McNeil, C.; Cumming, D.R.S. Plasmonic Sensor Monolithically Integrated with a CMOS Photodiode. *ACS Photonics* **2016**, *3*, 1926–1933. [CrossRef]
173. Augel, L.; Kawaguchi, Y.; Bechler, S.; Körner, R.; Schulze, J.; Uchida, H.; Fischer, I.A. Integrated Collinear Refractive Index Sensor with Ge PIN Photodiodes. *ACS Photonics* **2018**, *5*, 4586–4593. [CrossRef]

© 2020 by the authors. Licensee MDPI, Basel, Switzerland. This article is an open access article distributed under the terms and conditions of the Creative Commons Attribution (CC BY) license (http://creativecommons.org/licenses/by/4.0/).

Article

Hydrothermal Synthesis of CeO$_2$-SnO$_2$ Nanoflowers for Improving Triethylamine Gas Sensing Property

Dongping Xue [1,2], Yan Wang [1,*], Jianliang Cao [1] and Zhanying Zhang [1,2,*]

[1] The Collaboration Innovation Center of Coal Safety Production of Henan Province, Jiaozuo 454000, China; xdongping1231@126.com (D.X.); caojianliang@hpu.edu.cn (J.C.)
[2] School of Materials Science and Engineering, Henan Polytechnic University, Jiaozuo 454000, China
* Correspondence: yanwang@hpu.edu.cn (Y.W.); zhangzy@hpu.edu.cn (Z.Z.); Tel.: +86-391-398-7440 (Y.W. & Z.Z.)

Received: 17 November 2018; Accepted: 5 December 2018; Published: 8 December 2018

Abstract: Developing the triethylamine sensor with excellent sensitivity and selectivity is important for detecting the triethylamine concentration change in the environment. In this work, flower-like CeO$_2$-SnO$_2$ composites with different contents of CeO$_2$ were successfully synthesized by the one-step hydrothermal reaction. Some characterization methods were used to research the morphology and structure of the samples. Gas-sensing performance of the CeO$_2$-SnO$_2$ gas sensor was also studied and the results show that the flower-like CeO$_2$-SnO$_2$ composite showed an enhanced gas-sensing property to triethylamine compared to that of pure SnO$_2$. The response value of the 5 wt.% CeO$_2$ content composite based sensor to 200 ppm triethylamine under the optimum working temperature (310 °C) is approximately 3.8 times higher than pure SnO$_2$. In addition, CeO$_2$-SnO$_2$ composite is also significantly more selective for triethylamine than pure SnO$_2$ and has better linearity over a wide range of triethylamine concentrations. The improved gas-sensing mechanism of the composites toward triethylamine was also carefully discussed.

Keywords: CeO$_2$-SnO$_2$; nanostructure; hydrothermal; triethylamine; gas sensor

1. Introduction

Triethylamine (TEA) is a colorless, transparent oily liquid with strong ammonia odor and was widely used as an organic solvent, raw material, polymerization inhibitor, preservative, catalyst and synthetic dye [1,2]. However, TEA is also a flammable, explosive and toxic volatile organic gas that can harm human health, such as eye and skin irritation, dyspnea, headache, nausea and even death [3–6]. Therefore, it is very important to develop a method for detecting and monitoring the concentration of TEA. Up to now, gas/liquid/solid chromatography, gel chromatography, ion mobility spectrometry, electrochemical analysis and colorimetry and other methods have been explored to monitor TEA gas. However, these methods require expensive equipment and complex detection processes, which hamper their widespread use in real life [6–12]. Thus, it is very necessary to develop a device that is simple to manufacture and easy to detect TEA.

Metal oxide semiconductor (MOS) based gas sensors have been widely investigated in recent years because of their advantages such as simple detection, simple preparation, low cost, high sensitivity and good real-time performance [13–15]. SnO$_2$ is a wide band gap n-type MOS material with a band gap width of 3.62 eV at room temperature. It is widely used in various fields such as photocatalysts [16,17], solar cells [18,19], lithium-ion batteries [20,21] and gas sensors [22–24]. As a gas sensor, SnO$_2$ is one of the most widely considered gas sensitive materials due to its better gas sensitivity to various organic and toxic gases. However, we also know that the traditional SnO$_2$ based gas sensors has some obvious problems of low gas response, high optimum working temperature, poor selectivity and stability [25–27]. Hence, researchers use another MOS doped SnO$_2$ to improve its gas sensitivity. For example, Zhai et al. [28] prepared Au-loaded ZnO/SnO$_2$ heterostructure, this sensor not only reduces the optimum operating temperature of SnO$_2$ but also has a higher response to TEA than pure SnO$_2$. Yang et al. [24] synthesized porous SnO$_2$/Zn$_2$SnO$_4$ composites and their response to 100 ppm of TEA was about 2–3 times higher than that of pure SnO$_2$. Yan et al. [29] reported a kind of porous CeO$_2$-SnO$_2$ nanosheets, which exhibited excellent gas sensing properties toward ethanol compared with the pure SnO$_2$. These works have confirmed that another MOSs doped SnO$_2$ can indeed improve gas sensing properties. As a rare earth element, Ce not only has the highest abundance of elements but also has some special characteristics such as high oxygen storage capacity, rich oxygen vacancies and low redox potential [29–31], these characteristics make it an ideal candidate for gas sensing materials [32,33], such as Motaung et al. [34] synthesizing CeO$_2$-SnO$_2$ nanoparticles with a dramatic improvement in sensitive and selective to H$_2$, Dan et al. [35] prepared novel Ce-In$_2$O$_3$ porous nanospheres for enhancing methanol gas-sensing performance. Xu et al. [12] reported that a cataluminescence gas sensor based on LaF$_3$-CeO$_2$ has better sensitivity and selectivity for TEA. However, as far as we know, there are few reports about the materials of CeO$_2$-SnO$_2$ used to detect TEA.

Herein, we synthesized the flower-like CeO$_2$-SnO$_2$ composites via a facile hydrothermal method. The gas sensing performance of the flower-like CeO$_2$-SnO$_2$ composites based sensors were tested. The experimental results indicate that the TEA gas sensing performance of the CeO$_2$-SnO$_2$ composites based sensors are significantly improved by the modification of a small amount of CeO$_2$ compared with pure SnO$_2$, especially in terms of sensitivity and selectivity. The improvement of gas sensing properties of the composite materials is mainly due to the formation of n-n heterojunction between CeO$_2$ and SnO$_2$.

2. Results and Discussion

2.1. Sample Characterization

Figure 1 displays XRD patterns of the pure SnO$_2$ (SC-0), 3, 5 and 7 wt.% CeO$_2$ decorated SnO$_2$ (SC-3, SC-5, SC-7) samples. It can be seen from Figure 1 that the XRD peaks are sharp and coincide with those of the tetragonal rutile of SnO$_2$ in the space group P42/mnm with lattice constants of a = b = 4.738 Å and c = 3.187 Å (JCPDS file No. 41−1445). Apart from, all samples showing the same crystal planes of (110), (101), (200), (211), (220), (310), (301) and (321), respectively. However, no reflections characteristic of CeO$_2$ was observed even at the maximum CeO$_2$ content of 7 wt.%, which may be mainly due to less CeO$_2$ loading. Nevertheless, one can also see from the figure that all the diffraction peak becomes sharper as the content of Ce increases. According to Debye-Scherrer principle as shown in Equation (1), where K (K = 0.89) is the Scherrer constant, D is the crystallite size, λ (λ = 0.15406 nm) is the X-ray wavelength, B is the half-height width of the diffraction peak of the measured sample, θ is the diffraction angle. The calculated average particle sizes of SC-0, SC-3, SC-5 and SC-7 were 11.2350, 14.0154, 14.4425, 14.633, respectively. The combined diffraction peaks became sharper, indicating that CeO$_2$ was indeed loaded onto SnO$_2$, since the n-n heterojunction is formed between CeO$_2$ and SnO$_2$ and the average particle size of the composite becomes larger as the CeO$_2$ dopant increases.

$$D = \frac{K\lambda}{B\cos\theta} \quad (1)$$

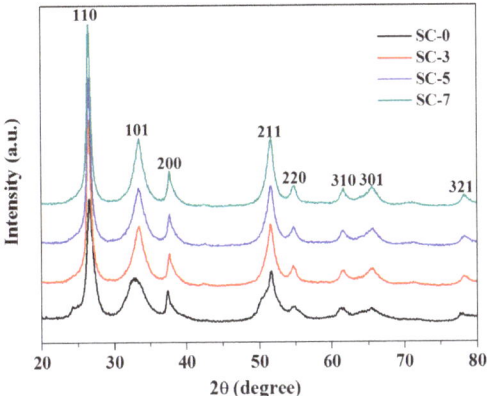

Figure 1. XRD patterns of the samples.

The morphology and structure of pure SnO$_2$ (Figure 2a,b) and SC-5 nanoflowers (Figure 2c,d) were observed by SEM and the presence of Ce dopant in the SC-5 nanoflowers (Figure 2e–g) was detected by EDS as illustrated in Figure 2. In Figure 2a,b, the SnO$_2$ sheets are gathered together to look like a spider web and have no flower-like structure. While, in Figure 2c,d, one can be clearly seen that SC-5 has a flower-like structure with an average diameter of about 1 μm and many nanoparticles are relatively evenly dispersed on the surface of the flower-like structure. By EDS detection as shown in Figure 2e–g, it was proved that only three elements of Sn, O and Ce were found in the composite, which proved that the sample had a relatively high purity.

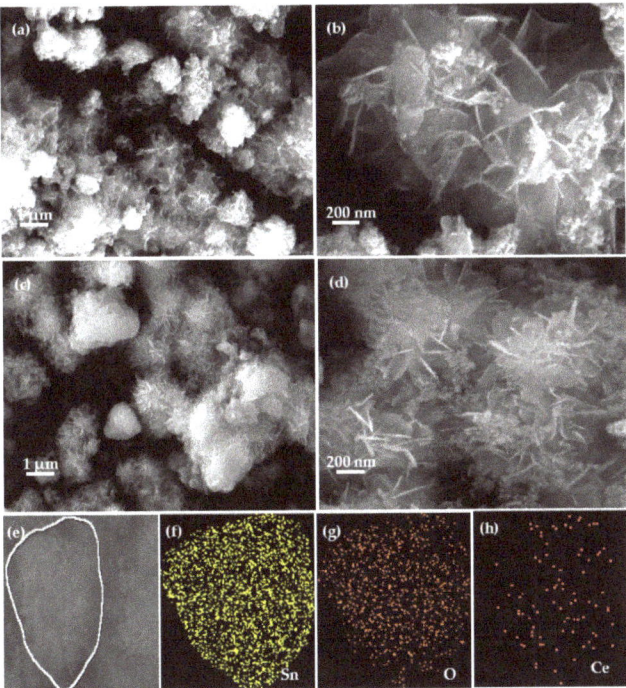

Figure 2. Field-emission scanning electron microscopy (FESEM) images of pure SnO$_2$ (**a,b**) and SC-5 nanocomposite (**c,d**) and the EDS images (**e-h**) of the SC-5 sample.

It is well known that the presence of a heterojunction affects the band gap width of a material. Hence, in order to further illustrate the formation of n-n heterojunctions, we investigated the band gap energies of the synthesized SnO$_2$ and SC-5 by UV-vis absorption spectra. As can be seen in Figure 3, the absorption peak of the red SC-5 sample moved significantly upward compared to the pure SnO$_2$ curve. The relationship diagram between $(\alpha h v)^2$ and photon energy hv (illustration in the upper right corner of Figure 3) is obtained according to the formula $(\alpha h v)^2 = A (hv - E_g)$, where α is the absorption index, h is the Planck constant, v is the light frequency, E_g is the semiconductor bandgap width, A is a constant associated with the material. It can be seen from the illustration that the band gap energy values of pure SnO$_2$ and SC-5 are 3.64 eV and 3.56 eV, respectively. The data show that the doping of CeO$_2$ narrows the band gap energy of the composite, which further proves that an n-n heterojunction is formed between composite nanomaterials.

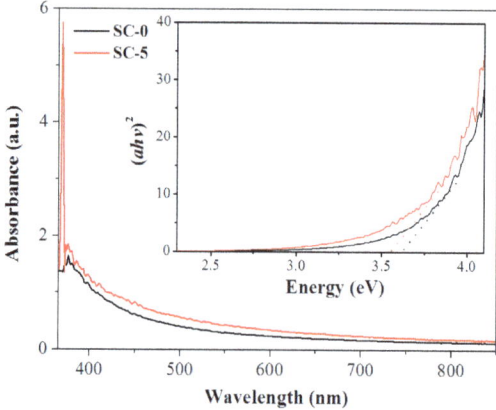

Figure 3. UV–vis absorption spectra of the synthesized SC-0 and SC-5 samples. The upper right corner inset is the relationship lines of $(\alpha h v)^2$ and hv.

2.2. Gas Sensing Performance

In order to study the gas sensing property of the as-synthesized samples to TEA, a series of examinations on the pure SnO$_2$ and CeO$_2$-SnO$_2$ composites were performed. Because the working temperature has great impact on the response of the gas sensor, the response of four sensors to 200 ppm TEA at different temperatures was first studied, as shown in Figure 4. One can see from Figure 4 that the four curves show similar variation tendency, which is first increase and then decrease as the working temperature increases. And the response of all the sensors reached their top value at the working temperature of 310 °C. This may be due to the amount of chemisorbed oxygen ions on the surface of the sensor has reached a sufficient amount to react with TEA and the effective reaction on the surface of MOS causes an eminent change in resistance [36]. Moreover, the response of the sensors based on SC-0, SC-3, SC-5 and SC-7 are 65.77, 218.12, 252.21, 156.38, respectively. It can also be clearly seen from the Figure 4 that all composite sensors have higher response to TEA than pure SnO$_2$ and the sensor based on SC-5 show higher response value than the response of the other two composite sensors. The TEA gas-sensitive properties of the SC-5 composite prepared in this work and the materials reported in other literatures [4,28,37–41] are shown in Table 1. Although the sensitivity of the CeO$_2$-SnO$_2$ sensor is higher than that of the reported results, the optimum operating temperature has not been improved very well. Therefore, we still need to do more in-depth research on reducing the working temperature, such as introducing photoexcitation equipment [42] or using p-type semiconductor doping modification [43].

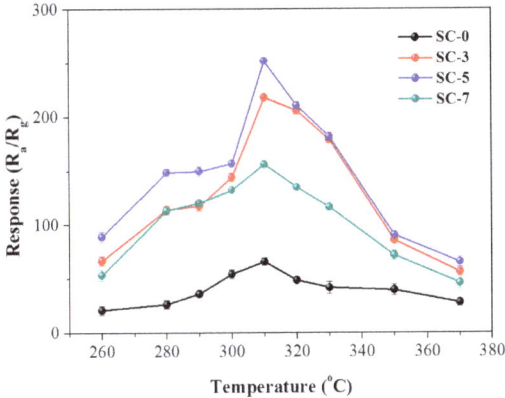

Figure 4. The response of the synthesized samples to 200 ppm TEA at different operating temperatures.

Table 1. TEA sensing performance comparison between this study and other reported results.

Materials	TEA Concentration (ppm)	Temperature(°C)	Response(R_a/R_g)	Ref.
SnO_2	200	350	5.9	[4]
Au@ZnO/SnO_2	200	300	160	[28]
Au@SnO_2/α-Fe_2O_3	200	300	63	[37]
$ZnFe_2O_4$/α-Fe_2O_3	200	305	65	[38]
$CoMoO_4$	200	600	110	[39]
CeO_2	100	Room temperature	4.67	[40]
Ce-doped In_2O_3	200	130	61.9	[41]
SC-5	200	310	252.2	this work

Sensitivity, selectivity and stability are also three important properties for evaluating sensor quality. Figure 5a exhibits the response of the four gas sensors to different TEA concentrations in the range of 20–2000 ppm at 310 °C. Obviously, the response of the four gas sensors increases with the increase of TEA concentration. It can also be seen that the response is almost linearly related to TEA concentration and the slope of the curve increase rapidly as TEA concentration increases in the concentration range of 20–200 ppm (inset of Figure 5a). Above 200 ppm, the responses increase slowly as the gas concentration increases, indicating that the adsorption of TEA by the sensor gradually becomes saturated. In addition, the SC-5 sensor exhibits higher response than that based on SC-0, SC-3 and SC-5 at different TEA concentrations. Figure 5b shows the dynamic response-recovery curves of the SC-0 and SC-5 sensors to TEA in the concentration range of 20–2000 ppm at 310 °C. As can be seen from Figure 5b that as the TEA concentration increases from 20 to 2000 ppm, the response amplitude of the two sensors gradually increases. The SC-5 composite sensor has a much higher response to the same TEA concentration than the pure SnO_2 sensor, indicating improved gas sensitivity of the composite. Moreover, after several times of gas injection, the output voltage of the sensor in air can still return to the original value, which means that the sensor has better repeatability. Figure 5c displays the results of selective testing of the SC-0 and SC-5 sensors for five different 200 ppm reducing gases, including formaldehyde, methanol, acetone, methane and triethylamine. As can be seen in Figure 5c, the flower-like SC-5 sensor has a significantly higher response to all detected reducing gases than the response of the SC-0 sensor, further demonstrating that the SC-5 sensor has better sensitivity to reducing gases. Moreover, The SC-5 sensor responds to 200 ppm of triethylamine up to 252.2, which is 10.9, 32, 40.6 and 117.2 times higher than formaldehyde, methanol, acetone and methane, respectively, which means that the SC-5 sensor has good selectivity for triethylamine. From the perspective of the practical application of the sensor, long-term stability is also an important factor in evaluating the property of the sensor. Figure 5d displays the durable response of the SC-5 sensor to 200 ppm TEA after storing for 30 days. As shown by the curve, the response remains almost constant and remains at around 250. It can be concluded that the SC-5 sensor had better stability and can be a promising candidate for TEA sensor.

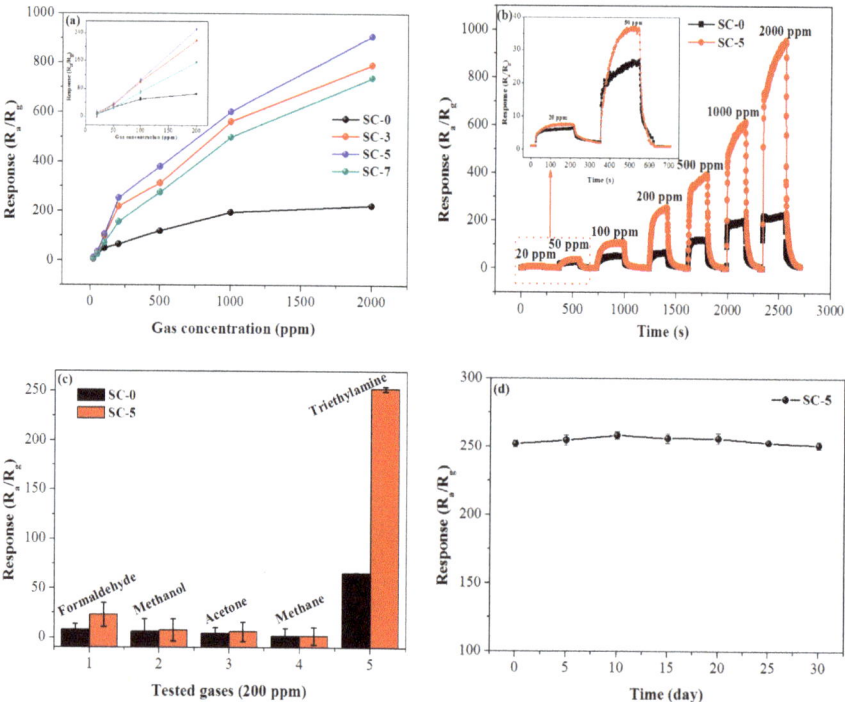

Figure 5. (a) Response of the sensors to different TEA concentrations at 310 °C (the inset shows the response curve in the range of 20–200 ppm), (b) dynamic response-recover curves of the sensors to different TEA concentrations at 310 °C, (c) responses of the SC-0 and SC-5 sensors to five gases of 200 ppm, (d) long-term stability measurements of the SC-5 sensor to 200 ppm TEA at 310 °C.

2.3. Gas Sensing Mechanism

As is known, the gas sensing mechanism of n-type MOS is based on the resistance change of the sensor by the adsorption and desorption reaction of oxygen molecules on the material surface with the gas to be detected [39]. When the undoped SnO_2 sensor is exposed to the air, as shown in Figure 6a, the oxygen molecule (O_2) are physically adsorbed on the surface of SnO_2 material and then convert from physisorption to chemisorption. Chemisorption oxygen molecules capture electrons from the SnO_2 conduction band to form oxygen anions O^{a-} (O^-_2, O^-, O^{2-}), which leads to the formation of electron depletion layer (EDL) on the surface of sensing material, the conduction channel in flower-like SnO_2 is narrowed, the carrier concentration and conductivity are lowered and the resistance rises (R_a). When TEA vapor is injected, the oxygen anion O^{a-} formed on the surface of the SnO_2 material reacts with the TEA and releases the trapped electrons back into the conduction band of the SnO_2 sensing material. Consequence, the EDL becomes narrow, the conduction channel becomes wider and the conductivity of the sensor is enhanced, thus reducing the resistance of the sensor (R_g).

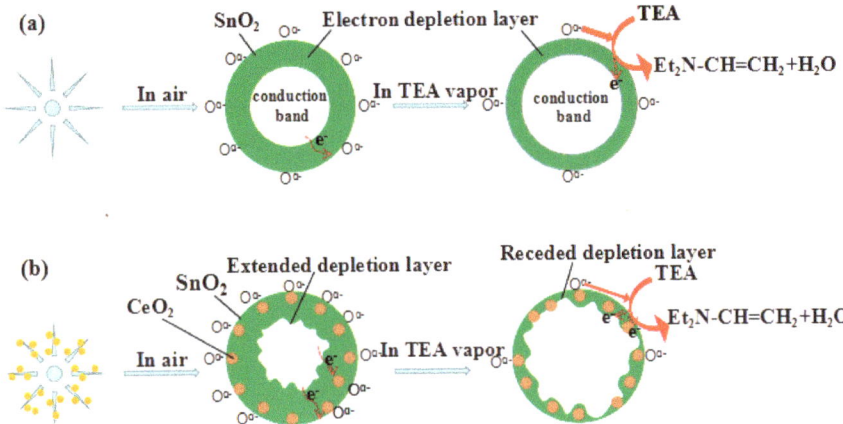

Figure 6. TEA sensing mechanisms diagram of (**a**) pure SnO$_2$ and (**b**) CeO$_2$/SnO$_2$ nanostructure.

The gas sensitivity of the CeO$_2$-SnO$_2$ sensor be improved may be mainly due to two factors. Firstly, when CeO$_2$ nanoparticles are supported on SnO$_2$ sheets, the electrons with a low work function flow from SnO$_2$ to CeO$_2$ with a higher work function due to different work functions of SnO$_2$ and CeO$_2$. Thus, an n-n heterojunction is formed between the junctions of SnO$_2$ and CeO$_2$, which further provides the surface of the SnO$_2$ sheets more active sites, so that the electron depletion layer of the composite undergoes a relatively large change depending on the atmosphere of the gas as shown in Figure 6b, thereby improving the gas sensitivity of SnO$_2$. When the composite sensor exposed to the air, more oxygen molecules will capture electrons from the conduction band of composite nanomaterials, which reduces the carrier concentration and the thickness of the EDL is further increased compared with the pure SnO$_2$ sensor, so the resistance of the CeO$_2$-SnO$_2$ sensor is greatly increased. When the composite sensor is exposed to TEA gas, more oxygen anions O^{a-} react with TEA and electrons are released back into the conduction band of the composite nanomaterials, concentration of free electrons in the conduction band increases and the electron depletion layer becomes thinner than the pure SnO$_2$, resulting in a significant drop in sensor resistance. Thereby, the gas sensitivity of the sensor was improved. On the other hand, the electronic properties of CeO$_2$ are also used to explain the sensitization mechanism [44]. Since CeO$_2$ is a strong acceptor of electrons, it induces an electron depletion layer at the interface with the host semiconductor SnO$_2$. By reacting with TEA, CeO$_2$ is induced to release electrons back into the semiconductor conduction band. Especially on the surface of CeO$_2$, the redox cycle of Ce^{4+}/Ce^{3+} can be realized quickly and repeatedly under certain conditions, which makes oxygen vacancies easy to produce and diffuse [31,33]. Therefore, the doping of CeO$_2$ can make the gas sensing material extract more oxygen more quickly, thereby enhancing the gas sensing performance. CeO$_2$-SnO$_2$ sensor has high selectivity for TEA in reducing gases such as methanol, formaldehyde, acetone and methane, which may be attributed to the electron-donating effect [4,45]. However, so far, there is no clear explanation for the sensitization mechanism, further research is still needed.

3. Materials and Methods

3.1. Sample Preparation

All of the chemical reagents was analytical grade (All of the chemical reagents was provided by Aladdin, Shanghai, China) and used without further purification in experiments, including Stannous chloride dihydrate (SnCl$_2$·2H$_2$O), Trisodium citrate dihydrate (Na$_3$C$_6$H$_5$O$_7$·2H$_2$O), Sodium hydroxide (NaOH), Cerium (III) nitrate hexahydrate (Ce(NO$_3$)$_3$·6H$_2$O) and absolute ethanol. Deionizer water was used throughout the experiments. In a typical process of synthesize CeO$_2$-SnO$_2$ composites, 3.6 g SnCl$_2$·2H$_2$O and 11.76 g Na$_3$C$_6$H$_5$O$_7$·2H$_2$O were dissolved into 40 mL of distilled water with stirring. Subsequently, 0.182 g (3 wt.%) Ce(NO$_3$)$_3$·6H$_2$O added into the above mixture with ultrasonic treatment for 20 min. Then 40 mL of NaOH solution was dropped into the above mixture under continuous magnetic stirring. Following, the above homogenous solution was transferred to a 100 mL of stainless autoclave lined with a Teflon vessel and heated to 180 °C for 12 h. The reaction system was then cooled naturally to room temperature after reaction. This pale yellow precipitate was collected by centrifugation and washed several times with distilled water and ethyl alcohol absolute and then dried at 60 °C for 12 h. Through varying the amount of Ce(NO$_3$)$_3$·6H$_2$O in the synthesis process, the CeO$_2$-SnO$_2$ composites with 0 wt.%, 3 wt.%, 5 wt.% and 7 wt.% CeO$_2$ decorated SnO$_2$ were prepared and denoted as SC-0, SC-3, SC-5, SC-7, respectively.

3.2. Characterizations

The phase and purity of the unloaded SnO$_2$ and CeO$_2$-doped SnO$_2$ samples was investigated by powder X-ray diffraction (XRD, Bruker-AXS D8, Bruker, Madison, WI, USA) with Cu Kα radiation at 40 kV and 150 mA in a scanning range of 20–80° (2θ) and in the continuous mode with step size of 0.02° (2θ) by scanning speed of 10°/min. The morphology and nanostructure were studied by field-emission scanning electron microscopy (FESEM, Quanta™ FEG 250) (FEI, Eindhoven, The Netherlands). Chemical composition analysis was tested by energy dispersive spectroscopy (EDS, INCA ENERGY 250) (FEI, Eindhoven, The Netherlands) integrated into the FESEM system.

3.3. Gas Sensor Fabrication and Analysis

The sensor is prepared similarly to the method we reported previously [46,47]. In brief, the sample was mixed with distilled water to form a uniform paste and coated onto a ceramic substrate (13.4 mm × 7 mm) with an Ag-Pd electrode to obtain a resistive sensor. The structure was shown in Figure 7. In order to improve the stability and repeatability of the sensor, the gas sensitivity test was performed after aging for 12 h at 60 °C. The gas sensitivity test was conducted on the intelligent gas sensing analysis system of CGS-4TPS (Beijing Elite Tech Co., Ltd., Beijing, China) under laboratory conditions (30 RH%, 30 °C). The sensitivity of the sensor was defined as S = R$_a$/R$_g$, where Ra and Rg represent the resistance of the sensor in air and target gas, respectively.

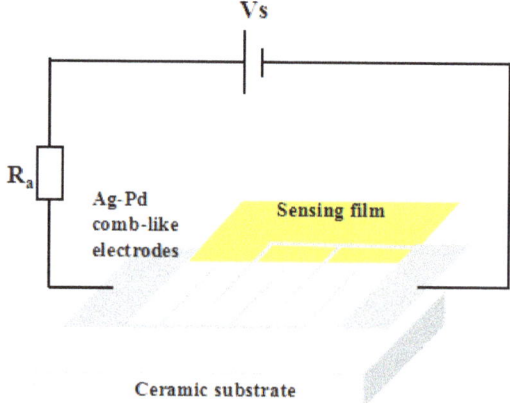

Figure 7. Structure diagram of gas sensor.

4. Conclusions

In summary, the flower-like CeO$_2$-doped SnO$_2$ nanostructures were successfully synthesized by a facile one-step hydrothermal synthesis reaction using Na$_3$C$_6$H$_5$O$_7$·2H$_2$O as stabilizer, SnCl$_2$·2H$_2$O and Ce(NO$_3$)$_3$·6H$_2$O as precursors. Various characterizations indicate that flower-like SnO$_2$ is a rutile structure with high crystallinity and CeO$_2$ well modified the surface of the flower-like SnO$_2$. The response of 5 wt.% CeO$_2$-SnO$_2$ sensor to 200 ppm TEA at the optimal working temperature of 310 °C is 252.2, which is about 3.8 times higher than that of undoped one. Moreover, SC-5 sensor displayed better selectivity for triethylamine. The improved gas-sensing performances of the composite were explained possibly due to the formation of n-n heterojunctions between CeO$_2$ and SnO$_2$ and the presence of Ce^{4+}/Ce^{3+} species in SnO$_2$ facilitates the interaction of electrons. Therefore, the CeO$_2$-doped SnO$_2$ sensors can be an ideal candidate for the detection of triethylamine gas.

Author Contributions: D.X. conceived and designed the experiments; J.C. performed the experiments; Y.W. and Z.Z. provided the concept of this research and managed the writing process as the corresponding authors.

Funding: This work was funded by the National Natural Science Foundation of China (U1704255), Program for Science & Technology Innovation Talents in Universities of Henan Province (19HASTIT042), the Fundamental Research Funds for the Universities of Henan Province (NSFRF170201, NSFRF1606, NSFRF1614), the Research Foundation for Youth Scholars of Higher Education of Henan Province (2017GGJS053) and Program for Innovative Research Team of Henan Polytechnic University (T2018-2).

Conflicts of Interest: The authors declare no conflict of interest.

References

1. Xu, H.Y.; Ju, D.X.; Li, W.R.; Zhang, J.; Wang, J.Q.; Cao, B.Q. Superior triethylamine sensing properties based on TiO$_2$/SnO$_2$ n-n heterojunction nanosheets directly grown on ceramic tubes. *Sens. Actuators B Chem.* **2016**, *228*, 634–642. [CrossRef]
2. Wu, Y.; Zhou, W.; Dong, W.; Zhao, J.; Qiao, X.Q.; Hou, D.F.; Li, D.S.; Zhang, Q.; Feng, P.Y. Temperature-Controlled Synthesis of Porous CuO Particles with Different Morphologies for Highly Sensitive Detection of Triethylamine. *Cryst. Growth Des.* **2017**, *17*, 2158–2165. [CrossRef]
3. Ju, D.; Xu, H.; Xu, Q.; Gong, H.; Qiu, Z.; Guo, J.; Zhang, J.; Cao, B. High triethylamine-sensing properties of NiO/SnO$_2$ hollow sphere P-N heterojunction sensors. *Sens. Actuators B Chem.* **2015**, *215*, 39–44. [CrossRef]
4. Liu, B.; Zhang, L.; Zhao, H.; Chen, Y.; Yang, H. Synthesis and sensing properties of spherical flowerlike architectures assembled with SnO$_2$ submicron rods. *Sens. Actuators B Chem.* **2012**, *173*, 643–651. [CrossRef]
5. Cai, T.; Chen, L.; Ren, Q.; Cai, S.; Zhang, J. The biodegradation pathway of triethylamine and its biodegradation by immobilized arthrobacter protophormiae cells. *J. Hazard. Mater.* **2011**, *186*, 59–66. [CrossRef] [PubMed]
6. Ju, D.; Xu, H.; Qiu, Z.; Zhang, Z.; Xu, Q.; Zhang, J.; Wang, J.; Cao, B. Near Room Temperature, Fast-Response, and Highly Sensitive Triethylamine Sensor Assembled with Au-Loaded ZnO/SnO$_2$ Core–Shell Nanorods on Flat Alumina Substrates, A.C.S. *Appl. Mater. Interfaces* **2015**, *7*, 19163–19171. [CrossRef] [PubMed]
7. Filippo, E.; Manno, D.; Buccolieri, A.; Serra, A. Green synthesis of sucralose-capped silver nanoparticles for fast colorimetric triethylamine detection. *Sens. Actuators B Chem.* **2013**, *178*, 1–9. [CrossRef]
8. Moore, W.M.; Edwards, R.J.; Bavda, L.T. An improved capillary gas chromatography method for triethylamine. Application to sarafloxacin hydrochloride and GnRH residual solvents testing. *Anal. Lett.* **1999**, *32*, 2603–2612. [CrossRef]
9. Haskin, J.F.; Warren, G.W.; Priestley, L.J.; Yarborough, V.A. Gas Chromatography. Determination of Constituents in Study of Azeotropes. *Anal. Chem.* **1958**, *30*, 217–219. [CrossRef]
10. Liu, C.H.J.; Lu, W.C. Optical amine sensor based on metallophthalocyanine. *J. Chin. Inst. Eng.* **2007**, *38*, 483–488. [CrossRef]
11. Li, Y.; Wang, H.C.; Cao, X.H.; Yuan, M.Y.; Yang, M.J. A composite of polyelectrolyte-Grafted multi-walled carbon nanotubes and in situ polymerized polyaniline for the detection of low concentration triethylamine vapor. *Nanotechnology* **2008**, *19*, 15503–15507. [CrossRef] [PubMed]
12. Xu, L.; Song, H.J.; Hu, J.; Lv, Y.; Xu, K.L. A cataluminescence gas sensor for triethylamine based on nanosized LaF$_3$-CeO$_2$. *Sens. Actuators B Chem.* **2012**, *169*, 261–266. [CrossRef]
13. Sun, P.; Cao, Y.; Liu, J.; Sun, Y.; Ma, J.; Lu, G. Dispersive SnO$_2$ nanosheets: Hydrothermal synthesis and gas-sensing properties. *Sens. Actuators B Chem.* **2011**, *156*, 779–783. [CrossRef]

14. Kim, S.J.; Hwang, I.S.; Choi, J.K.; Kang, Y.C.; Lee, J.H. Enhanced C_2H_5OH sensing characteristics of nano-porous In_2O_3 hollow spheres prepared by sucrose-mediated hydrothermal reaction. *Sens. Actuators B Chem.* **2011**, *155*, 512–518. [CrossRef]
15. Xu, S.; Wang, Z.L. One-dimensional ZnO nanostructures: Solution growth and functional properties. *Nano Res.* **2011**, *4*, 1013–1098. [CrossRef]
16. Niu, M.; Huang, F.; Cui, L.; Huang, P.; Yu, Y.; Wang, Y. Hydrothermal synthesis, structural characteristics, and enhanced photocatalysis of SnO_2/α-Fe_2O_3 semiconductor nanoheterostructures. *ACS Nano* **2010**, *4*, 681–688. [CrossRef] [PubMed]
17. Cheng, H.E.; Lin, C.Y.; Hsu, C.M. Fabrication of SnO_2-TiO_2 core-shell nanopillararray films for enhanced photocatalytic activity. *Appl. Surf. Sci.* **2017**, *396*, 393–399. [CrossRef]
18. Roose, B.; Johansen, C.M.; Dupraz, K.; Jaouen, T.; Aebi, P.; Steiner, U.; Abate, A. A Ga-doped SnO_2 mesoporous contact for UV stable highly efficient perovskite solar cells. *J. Mater. Chem. A* **2018**, *6*, 1850–1857. [CrossRef]
19. Ashok, A.; Vijayaraghavan, S.N.; Unni, G.E.; Nair, S.V.; Shanmugam, M. On the Physics of Dispersive Electron Transport Characteristics in SnO_2 Nanoparticle Based Dye Sensitized Solar Cells. *Nanotechnology* **2018**, *29*, 175401. [CrossRef]
20. Zhao, K.; Zhang, L.; Xia, R.; Dong, Y.; Xu, W.; Niu, C.; He, L.; Yan, M.; Qu, L.; Mai, L. SnO_2 quantum dots@graphene oxide as a high-rate and long-life anode material for lithium-ion batteries. *Small* **2016**, *12*, 588–594. [CrossRef]
21. Shi, Y.H.; Ma, D.Q.; Wang, W.J.; Zhang, L.F.; Xu, X.H. A supramolecular self-assembly hydrogel binder enables enhanced cycling of SnO_2-based anode for high-performance lithium-ion batteries. *J. Mater. Sci.* **2017**, *52*, 3545–3555. [CrossRef]
22. Narjinary, M.; Rana, P.; Sen, A.; Pal, M. Enhanced and selective acetone sensing properties of SnO_2-MWCNT nanocomposites: Promising materials for diabetes sensor. *Mater. Des.* **2017**, *115*, 158–164. [CrossRef]
23. Cao, J.L.; Qin, C.; Wang, Y. Synthesis of g-C_3N_4 nanosheets decorated flower-like tin oxide composites and their improved ethanol gas sensing properties. *J. Alloys Compd.* **2017**, *728*, 1101–1109. [CrossRef]
24. Yang, X.L.; Yu, Q.; Zhang, S.F.; Sun, P.; Lu, H.Y.; Yan, X.; Liu, F.M.; Zhou, X.; Liang, X.H.; Gao, Y.; et al. Highly sensitive and selective triethylamine gas sensor based on porous SnO_2/Zn_2SnO_4 composites. *Sens. Actuators B Chem.* **2018**, *266*, 213–220. [CrossRef]
25. Chen, W.; Zhou, Q.; Gao, T.; Su, X.; Wan, F. Pd-doped SnO_2-based sensor detecting characteristic fault hydrocarbon gases in transformer oil. *J. Nanomater.* **2013**, *2013*, 1–9. [CrossRef]
26. Kim, K.; Choi, K.; Jeong, H.; Kim, H.J.; Kim, H.P.; Lee, J. Highly sensitive and selective trimethylamine sensors using Ru-doped SnO_2 hollow spheres. *Sens. Actuators B Chem.* **2012**, *166*, 733–738. [CrossRef]
27. Wang, D.; Chu, X.; Gong, M. Gas-sensing properties of sensors based on single crystalline SnO_2 nanorods prepared by a simple molten-salt method. *Sens. Actuators B Chem.* **2006**, *117*, 183–187. [CrossRef]
28. Zhai, T.; Xu, H.Y.; Li, W.R.; Yu, H.Q.; Chen, Z.Q.; Wang, J.Q.; Cao, B.Q. Low-temperature in-situ growth of SnO_2 nanosheets and its high triethylamine sensing response by constructing Au-loaded ZnO/SnO_2 heterostructure. *J. Alloys Compd.* **2018**, *737*, 603–612. [CrossRef]
29. Yan, S.; Liang, X.; Song, H.; Ma, S.; Lu, Y. Synthesis of porous CeO_2-SnO_2 nanosheets gas sensors with enhanced sensitivity. *Ceram. Int.* **2018**, *44*, 358–363. [CrossRef]
30. Joy, N.A.; Nandasiri, M.I.; Rogers, P.H.; Jiang, W.; Varga, T.; Kuchibhatla, S.V.T.; Thevuthasan, S.; Carpenter, M.A. Selective plasmonic gas sensing: H_2, NO_2, and CO spectral discrimination by a single Au-CeO_2 nanocomposite film. *Anal. Chem.* **2012**, *84*, 5025–5034. [CrossRef]
31. Pourfayaz, F.; Khodadadi, A.; Mortazavi, Y.; Mohajerzadeh, S.S. CeO_2 doped SnO_2 sensor selective to ethanol in presence of CO, LPG and CH_4. *Sens. Actuators B Chem.* **2005**, *108*, 172–176. [CrossRef]
32. Jiang, Z.; Guo, Z.; Sun, B.; Jia, Y.; Li, M.Q.; Liu, J. Highly sensitive and selective bu-tanone sensors based on cerium-doped SnO_2 thin films. *Sens. Actuators B Chem.* **2010**, *145*, 667–673. [CrossRef]
33. Hamedani, N.F.; Mahjoub, A.R.; Khodadadi, A.A.; Mortazavi, Y. CeO_2 doped ZnO flower-like nanostructure sensor selective to ethanol in presence of CO and CH_4. *Sens. Actuators B Chem.* **2012**, *169*, 67–73. [CrossRef]
34. Motaung, D.E.; Mhlongo, G.H.; Makgwane, P.R.; Dhonge, B.P.; Cummings, F.R.; Swart, H.C.; Ray, S.S. Ultra-High Sensitive and Selective H_2, Gas Sensor Manifested by Interface of n-n Heterostructure of CeO_2-SnO_2 Nanoparticles. *Sens. Actuators B Chem.* **2018**, *254*, 984–995. [CrossRef]

35. Han, D.; Song, P.; Zhang, S.; Zhang, H.H.; Wang, Q.X. Enhanced methanol gas-sensing performance of Ce-doped In$_2$O$_3$ porous nanospheres prepared by hydrothermal method. *Sens. Actuators B Chem.* **2015**, *216*, 488–496. [CrossRef]
36. Xiao, Y.; Yang, Q.; Wang, Z.; Zhang, R.; Gao, Y.; Sun, P.; Sun, Y.; Lu, G.Y. Improvement of NO$_2$ gas sensing performance based on discoid tin oxide modified by reduced graphene oxide. *Sens. Actuators B Chem.* **2016**, *227*, 419–426. [CrossRef]
37. Yu, H.Q.; Li, W.R.; Han, R.; Zhai, T.; Xu, H.Y.; Chen, Z.Q.; Wu, X.W.; Wang, J.Q.; Cao, B.Q. Enhanced triethylamine sensing properties by fabricating Au@SnO$_2$/α-Fe$_2$O$_3$ core-shell nanoneedles directly on alumina tubes. *Sens. Actuators B Chem.* **2018**, *262*, 70–78.
38. Li, Y.; Luo, N.; Sun, G.; Zhang, B.; Ma, G.Z.; Jin, H.H.; Wang, Y.; Cao, J.L.; Zhang, Z.Y. Facile synthesis of ZnFe$_2$O$_4$/α-Fe$_2$O$_3$, porous microrods with enhanced TEA-sensing performance. *J. Alloys Compd.* **2018**, *737*, 255–262. [CrossRef]
39. Liu, F.; Yang, Z.; He, J.; You, R.; Wang, J.; Li, S.Q.; Lu, H.Y.; Liang, X.S.; Sun, P.; Yan, X.; Chuai, X.H.; Lu, G.Y. Ultrafast-response Stabilized Zirconia-Based Mixed Potential Type Triethylamine Sensor Utilizing CoMoO$_4$ Sensing Electrode. *Sens. Actuators B Chem.* **2018**, *272*, 433–440. [CrossRef]
40. Zito, C.A.; Perfecto, T.M.; Volanti, D.P. Porous CeO$_2$ nanospheres for a room temperature triethylamine sensor under high humidity conditions. *New J. Chem.* **2018**, *42*, 15954–15961. [CrossRef]
41. Wang, J.; Yu, L.; Wang, H.; Ruan, S.; Lj, J.J.; Wu, F. Preparation and Triethylamine Sensing Properties of Ce-Doped In$_2$O$_3$ Nanofibers. *Acta Phys. Chim. Sin.* **2010**, *26*, 3101–3105.
42. Liu, S.R.; Guan, M.Y.; Li, X.Z.; Guo, Y. Light irradiation enhanced triethylamine gas sensing materials based on ZnO/ZnFe$_2$O$_4$ composites. *Sens. Actuators B Chem.* **2016**, *236*, 350–357. [CrossRef]
43. Bai, S.; Liu, C.; Luo, R.; Chen, A. Metal organic frameworks-derived sensing material of SnO$_2$/NiO composites for detection of triethylamine. *Appl. Surf. Sci.* **2018**, *437*, 304–313. [CrossRef]
44. Franke, M.E.; Koplin, T.J.; Simon, U. Metal and metal oxide nanoparticles in chemiresistors: Does the nanoscale matter? *Small* **2006**, *2*, 36–50. [CrossRef] [PubMed]
45. Xie, Y.; Du, J.; Zhao, R.; Wang, H.; Yao, H. Facile synthesis of hexagonal brick-shaped SnO$_2$ and its gas sensing toward triethylamine. *J. Environ. Chem. Eng.* **2013**, *1*, 1380–1384. [CrossRef]
46. Gong, Y.; Wang, Y.; Sun, G.; Jia, T.; Jia, L.; Zhang, F.; Lin, L.; Zhang, B.; Cao, J.; Zhang, Z. Carbon nitride decorated ball-flower like Co$_3$O$_4$ hybrid composite: Hydrothermal synthesis and ethanol gas sensing application. *Nanomaterials* **2018**, *8*, 132. [CrossRef] [PubMed]
47. Cao, J.; Gong, Y.; Wang, Y.; Zhang, B.; Zhang, H.; Sun, G.; Hari, B.; Zhang, Z. Cocoon-like ZnO decorated graphitic carbon nitride nanocomposite: Hydrothermal synthesis and ethanol gas sensing application. *Mater. Lett.* **2017**, *198*, 76–80. [CrossRef]

© 2018 by the authors. Licensee MDPI, Basel, Switzerland. This article is an open access article distributed under the terms and conditions of the Creative Commons Attribution (CC BY) license (http://creativecommons.org/licenses/by/4.0/).

Article

A Facile One-Step Synthesis of Cuprous Oxide/Silver Nanocomposites as Efficient Electrode-Modifying Materials for Nonenzyme Hydrogen Peroxide Sensor

Kaixiang Yang [1,2], Zhengguang Yan [1,2,*], Lin Ma [1,2], Yiping Du [1,2], Bo Peng [1,2] and Jicun Feng [1,2]

[1] Institute of Microstructure and Property of Advanced Materials, Beijing University of Technology, Beijing 100124, China; ykx233210@gmail.com (K.Y.); hokingma@gmail.com (L.M.); duyp@emails.bjut.edu.cn (Y.D.); pengbo2017@emails.bjut.edu.cn (B.P.); fengjc0619@163.com (J.F.)
[2] Beijing Key Laboratory of Microstructure and Properties of Solids, Beijing University of Technology, Beijing 100124, China
* Correspondence: yanzg@bjut.edu.cn; Tel.: +86-10-67396143

Received: 6 March 2019; Accepted: 22 March 2019; Published: 3 April 2019

Abstract: Cuprous oxide/silver (Cu_2O/Ag) nanocomposites were prepared via a facile one-step method and used to construct an electrochemical sensor for hydrogen peroxide (H_2O_2) detection. In this method, $AgNO_3$ and $Cu(NO_3)_2$ were reduced to Cu_2O/Ag nanocomposites by glucose in the presence of hexadecyl trimethyl ammonium bromide (CTAB) at a low temperature. The optimum condition was the molar ratio of silver nitrate and copper nitrate of 1:10, the temperature of 50 °C. Under this condition, Cu_2O/Ag nanocomposites were obtained with uniformly distributed and tightly combined Cu_2O and Ag nanoparticles. The size of Cu_2O particles was less than 100 nm and that of Ag particles was less than 20 nm. Electrochemical experiments indicate that the Cu_2O/Ag nanocomposites-based sensor possesses an excellent performance toward H_2O_2, showing a linear range of 0.2 to 4000 µM, a high sensitivity of 87.0 µA mM^{-1} cm^{-2}, and a low detection limit of 0.2 µM. The anti-interference capability experiments indicate this sensor has good selectivity toward H_2O_2. Additionally, the H_2O_2 recovery tests of the sensor in diluted milk solution signify its potential application in routine H_2O_2 analysis.

Keywords: cuprous oxide/silver; nanocomposites; hydrogen peroxide; electrochemistry; sensor

1. Introduction

The rapid and sensitive detection of H_2O_2 has attracted a lot of attention because of the applications of H_2O_2 in food [1], medicine [2], chemical industry [3], and environmental protection [4] as a common intermediate and oxidant, as well as its involvement in many biological events and intracellular pathways [5]. Conventional techniques for H_2O_2 determination have been developed, such as titrimetry [6], colorimetry [7], chemiluminescence [8], fluorescence resonance energy transfer-based upconversion [9], chromatography [10], and electrochemical methods [11]. Among these techniques, the electrochemical method is considered to be a prospective approach for its good selectivity, high sensitivity, and simple manipulation [4]. Although enzyme-based H_2O_2 sensors exhibit prominent advantages of high selectivity, the complexity of the enzyme curing process and instability to toxic chemicals limit their practical applications [12]. Therefore, a growing interest in developing enzyme-free sensors for detecting H_2O_2 has been aroused in this field [13,14]. Catalytic active nanomaterials, including noble metals [15], transition metal oxides [16], and other transition metal compounds [17,18], thanks to their selectivity and high activity, have been widely used to construct nonenzyme H_2O_2 sensors.

In recent years, as a typical transition metal oxide, cuprous oxide (Cu_2O) has attracted increasing attention as a promising candidate for H_2O_2 sensors due to its proper redox potentials, easy production process, and low cost [19,20]. Unfortunately, pristine Cu_2O sensors demonstrate low sensitivity and narrow linear detection ranges [21,22]. Combination with other materials to prepare composites is one effective way to improve the performance of Cu_2O-based H_2O_2 sensors. The metal nanoparticles, thanks to their good conductivity and high electrocatalytic activity, could largely facilitate the electron transfer on the surface of transition-metal oxides and improve their electrocatalytic activity [23]. Up to now, different metal particles have been introduced to transition-metal oxides for H_2O_2 sensors, such as Au/MnO_2 [24], Au/Fe_3O_4 [25], $Ag/MnO_2/MWCNTs$ [26], Au/Cu_2O [27], and $Pt/Fe_3O_4/Graphene$ [28]. Particularly, Ag nanoparticles (AgNPs) exhibit higher conductivity and lower cost compared with Au and Pt, and could produce synergistic effects when combined with some metal oxides [26], thus they are a promising material for improving the catalytic performance of the transition-metal oxides. Therefore, it is promising to introduce Ag into Cu_2O-based composites to fabricate H_2O_2 sensors.

Although these transition-metal oxide/metal nanocomposites mentioned above do fairly well in H_2O_2 sensing, the preparation of these materials is usually complicated, multistep, and time-consuming. The conventional routes would synthesize metal oxides first, and then modify metal particle to the surface of metal oxides. Therefore, it makes sense to simplify the synthesis steps for material preparation.

In this work, we introduced a facile one-step procedure to combine Cu_2O with Ag to prepare Cu_2O/Ag nanocomposites. The effects of experimental conditions on composition and morphology of the nanocomposites were studied. The electrochemical measurements were applied to elucidate the sensing application of Cu_2O/Ag nanocomposites, and the anti-interference capability experiments and the H_2O_2 recovery tests indicate Cu_2O/Ag nanocomposites could be a promising material for H_2O_2 detection.

2. Materials and Methods

2.1. Reagents and Chemicals

All reagents were of analytical reagent grade and used without further purification. $Cu(NO_3)_2 \cdot 3H_2O$, $AgNO_3$, hexadecyl trimethyl ammonium bromide (CTAB), and ethanol were purchased from Beijing Chemical Reagents Company (Beijing, China). D-glucose, NaOH, urea, fructose, L-ascorbic acid, Na_2HPO_4, and H_2O_2 solution (30%) were purchased from Tianjin Fuchen Chemical Reagent Co, (Tianjin, China). Ltd. $K_3[Fe(CN)_6]$ and $NaH_2PO_4 \cdot 12H_2O$ were purchased from Aladdin Reagent Co (Shanghai, China). All aqueous solutions were prepared with double-distilled water.

2.2. Synthesis of Cu_2O/Ag Nanocomposites and Modification of Electrode

The preparation of Cu_2O/Ag nanocomposites was carried out in aqueous solution using glucose as reducing agent and CTAB as dispersing agent. A typical procedure is performed as illustrated in Figure 1. A 0.035 g portion of $AgNO_3$ (0.2 mmol) dissolved in 20 mL double-distilled water was marked as solution A. Next, 0.5 g $Cu(NO_3)_2 \cdot 3H_2O$ (2 mmol) and 0.5 g glucose (2.5 mmol) were dissolved in 50 mL double-distilled water, and then 10 mL aqueous solution of CTAB (0.014 mol L^{-1}) was added into the mixture under stirring. The solution was marked as solution B. The molar ratios of $AgNO_3$ and $Cu(NO_3)_2$ could be varied by changing the quantity of $AgNO_3$ according to the requirement. A 0.5 g portion of NaOH (12.5 mmol) dissolved in 20 mL double-distilled water was marked as solution C. The solutions A (20 mL), B (60 mL), and C (20 mL) were added into a flask under stirring at room temperature. The solution was stirred for another 10 min and a gray precipitate formed. Then the reaction suspension was heated under vigorous stirring (500 rpm) at a temperature of 50 °C for 30 min and the mixture turned brown-gray gradually. Finally, the product was separated by centrifugation and washed with water and ethanol for three times. The amount of ethanol and water used to wash the products was 20 mL per 100 mg each time, respectively. The products were dried at 70 °C overnight. Note, it is important to recover any organic solvent to reduce the environmental burden and improve the sustainability of the methodology [29]. The alcohol used to wash the products could be recovered by fractionation for secondary use.

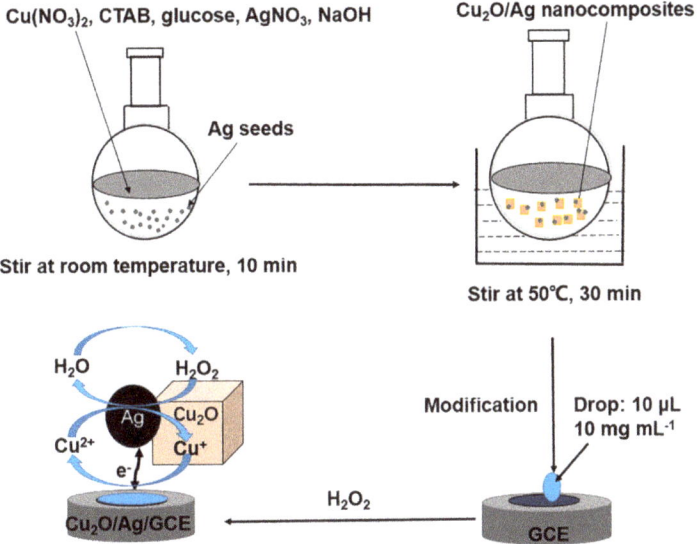

Figure 1. Schematic illustration for the facile method to prepare Cu$_2$O/Ag/GCE.

A glassy carbon electrode (GCE) was polished, cleaned, and dried for the fabrication of the sensor. Generally, 10 mg of Cu$_2$O/Ag nanocomposites were dispersed into 1 mL double-distilled water and sonicated for 15 min. A 10 µL portion of the suspension was dropped onto the GCE and then dried in air at room temperature. The modified electrode was marked as Cu$_2$O/Ag/GCE. The Cu$_2$O sample without Ag was used similarly to modify the electrode, which was marked as Cu$_2$O/GCE.

2.3. Electrochemical Experiments

Electrochemical measurements were carried out with a PARSTAT 2273 potentiostat galvanostat (Princeton Applied Research, Oak Ridge, TN, USA) in a three-electrode system, with the modified GCE (0.3 cm in diameter) as working electrode, Ag/AgCl/KCl (sat.) as reference electrode, and a platinum sheet as the counter electrode. The cyclic voltammetry profiles (CVs) and current–time profiles were measured in an N$_2$-saturated PBS solution (0.1 M, pH = 7.2) at room temperature. The electrochemical impedance spectroscopy (EIS) was tested in a 5 mM [Fe(CN)$_6$]$^{3-}$ solution containing 0.1 M KCl with a frequency range of 10^{-2}–10^5 Hz and an amplitude of 10 mV.

2.4. Material Characterization Techniques

The powder X-ray diffraction (XRD) patterns of the as-prepared materials were carried out on a D8 Advance X-ray diffractometer (Bruker AXS GmbH, Karlsruhe, Germany) with Cu Kα radiation (λ = 1.54178 Å). The scanning electron microscopy (SEM) images of the products were characterized using an FEI Quanta 600 field emission scanning electron microscope (FEI Company, Hillsboro, OR, USA). The transmission electron microscopy (TEM) images and electron diffraction (ED) patterns were obtained using an FEI T20 transmission electron microscope (FEI Company, Hillsboro, OR, USA) working at 180 kV. High resolution transmission electron microscopy (HRTEM) images and electron dispersive spectra mapping of the materials (EDS mapping) were obtained using an FEI Titan G2 spherical-aberration-corrected transmission electron microscope (FEI Company, Hillsboro, OR, USA) working at 200 kV. The X-ray photoelectron spectra (XPS) of materials were characterized by an

ESCALAB 250Xi X-ray Photoelectron Spectrometer (Thermo Fisher Scientific, Waltham, MA, USA) with a monochromatic Al Kα X-ray and a 500 µm nominal spot size, and the high-resolution scans were collected with a pass energy of 30 eV and a step size of 0.05 eV.

3. Results and Discussion

3.1. Effect of Experimental Conditions on Composition and Morphology

In this study, a simple one-step method was used to prepare Cu_2O/Ag nanocomposites successfully. The dose of $Cu(NO_3)_2$ 0.5 g (2 mmol) was kept unchanged, and the dose of $AgNO_3$ was changed. Different molar ratios of $AgNO_3$ and $Cu(NO_3)_2$ in the reactants (n_{AgNO3}:$n_{Cu(NO3)2}$ = 0, 1:20, 1:10, 1:5, respectively) were used to prepare nanomaterials with different compositions at the temperature of 50 °C. The XRD patterns of these nanocomposites prepared with different molar ratios of $AgNO_3$ and $Cu(NO_3)_2$ are shown in Figure 2a, from which we can easily find that all the nanocomposites show the strong diffraction peaks of the cubic crystal structure of the Cu_2O phase (space group: $Pn3m$, JCPDS 5-667 [30]) with fitted lattice parameter of a = 0.430 nm. The six peaks (square notations) with 2θ values of 29.68, 36.50, 42.40, 61.52, 73.70, and 77.57 were observed and could be assigned to diffraction from the (110), (111), (200), (220), (311), and (222) planes, respectively. In addition, the XRD pattern of products (n_{AgNO3}:$n_{Cu(NO3)2}$ = 1:5, 1:10, 1:20, respectively) showed extra peaks (round notations) because of the introduction of Ag, and the XRD peaks at 2θ degrees of 38.11, 44.28, 64.43, 77.47, and 81.54 can be attributed to the (111), (200), (220), (311), and (222) crystalline planes of the face-centered-cubic (fcc) crystalline structure of Ag, respectively (space group: $Fm-3m$, JCPDS 4-783 [31]) with fitted lattice parameter of a = 0.409 nm. In addition, with the molar ratio of n_{AgNO3}:$n_{Cu(NO3)2}$ decreased, the intensity of the Ag peaks decreased obviously, which indicated that the Ag content in the nanocomposites was positively correlated with the amount of $AgNO_3$ added.

The SEM was used to investigate the morphology of nanomaterials prepared with different molar ratios of $AgNO_3$ and $Cu(NO_3)_2$ under the temperature of 50 °C, as is shown in Figure 2b–e, from which we can easily find that the size of Cu_2O particles decreased obviously with the increase of molar ratio of $AgNO_3$:$Cu(NO_3)_2$. The average particle size of pure Cu_2O prepared without addition of $AgNO_3$ was between 400 nm and 1.2 µm (see the size distribution histograms shown in Figure S1a, SI). However, when n_{AgNO3}:$n_{Cu(NO3)2}$ = 1:20, Cu_2O particles of the nanocomposites became much smaller in size (50–300 nm) compared with the pure Cu_2O prepared; the size distribution histogram is in Figure S1b. The reason for the decrease in sizes for Cu_2O particles is that a lot of Ag nanoparticles were formed and acted as seeds before the Cu_2O nanoparticles appeared, which could be observed when the mixture quickly turned gray at room temperature in the process of synthesis. As shown in Figure S2, the size of Ag nanoparticles initially formed was smaller than 20 nm, and they would act as nucleation seeds for Cu_2O to nucleate on and grow. Therefore, the Cu_2O particles and Ag particles would form good contact in the step. Then, Cu_2O particles became small-sized because of these large numbers of Ag seeds. In addition, it can be seen from the SEM images in Figure 2d,e that the size of Cu_2O became very small (<100 nm) when n_{AgNO3}:$n_{Cu(NO3)2}$ = 1:10 and 1:5. However, when n_{AgNO3}:$n_{Cu(NO3)2}$ = 1:5, the nanoparticles tended to agglomerate. Considering the uniformity of particle size and the dispersion of nanocomposites, 1:10 is the appropriate dosage ratio to prepare Cu_2O/Ag nanocomposites.

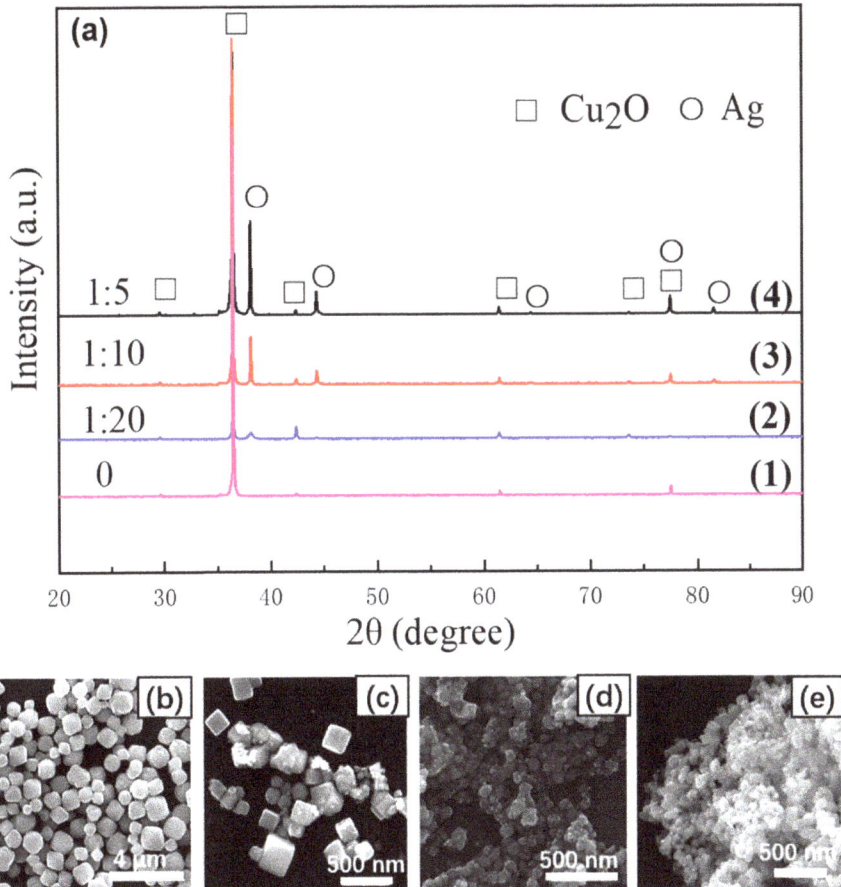

Figure 2. (a) XRD patterns of the as-synthesized nanomaterials prepared in different molar ratio of AgNO$_3$:Cu(NO$_3$)$_2$. n_{AgNO3}:$n_{Cu(NO3)2}$ = 0 (**1**), 1:20 (**2**), 1:10 (**3**), and 1:5 (**4**), respectively. SEM images of nanomaterials prepared at different molar ratio of AgNO$_3$:Cu(NO$_3$)$_2$. n_{AgNO3}:$n_{Cu(NO3)2}$ = 0 (**b**), 1:20 (**c**), 1:10 (**d**), 1:5 (**e**), respectively.

The formation of nanocomposites was also influenced by the reaction temperature. From the XRD patterns in Figure S3, we can easily find that the reaction temperature plays an important role in the formation of Cu$_2$O/Ag nanocomposites. At room temperature, only Ag was produced. In contrast, Cu(NO$_3$)$_2$ was partially reduced to Cu when the temperature was 70 °C, and a mixture of Cu and Ag was synthesized when the temperature raised to 100 °C. Only when the reaction temperature was around 50 °C were Cu$_2$O/Ag nanocomposites synthesized.

Additionally, the XPS measurement for the pure Cu$_2$O and Cu$_2$O/Ag nanocomposites (n_{AgNO3}:$n_{Cu(NO3)2}$ = 1:10) was further carried out to elucidate the valence states of the Cu and Ag element. Figure 3a shows the XPS survey spectra of pure Cu$_2$O and Cu$_2$O/Ag nanocomposites. The C, Cu, and O elements were detected for both samples [32,33], and the survey spectrum of Cu$_2$O/Ag nanocomposites (red line) shows extra peaks which can be assigned to the AgNPs [34]. Figure 3b shows the XPS spectra in Cu 2p regions of the Cu$_2$O/Ag nanocomposite, which indicate the existence of Cu$_2$O (932.3 eV: Cu(I) 2p$_{3/2}$, 952.1 eV: Cu(I) 2p$_{1/2}$ of Cu$_2$O) and the surface of Cu$_2$O nanoparticles was slightly oxidized (933.6 eV: Cu(II) 2p$_{3/2}$, 953.4 eV: Cu(II) 2p$_{1/2}$). Figure 3c shows the Ag 3d region of Cu$_2$O/Ag nanocomposites with doublet peaks at 374.5 eV and 368.3 eV, which were assigned to the Ag 3d$_{3/2}$ and Ag 3d$_{5/2}$ of Ag(0), respectively. Figure 3d shows the O 1s regions of the Cu$_2$O/Ag nanocomposites. The O 1s peak is around 529.7–532.4 eV, which is consistent with the O peak of Cu$_2$O reported [33]. We can see clearly from the XPS data above that the AgNPs was introduced to Cu$_2$O/Ag nanocomposites successfully.

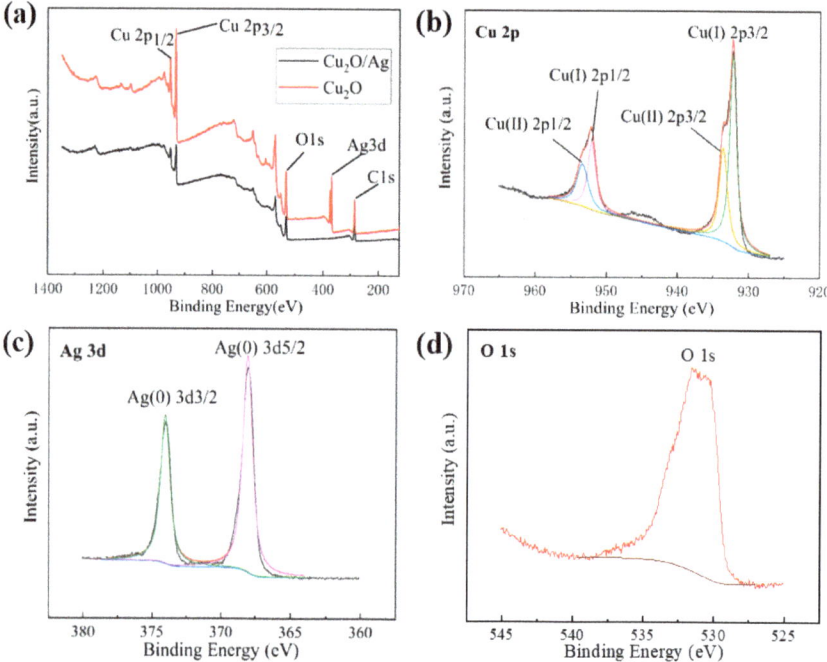

Figure 3. (a) XPS survey spectrum of the as-synthesized pure Cu$_2$O and Cu$_2$O/Ag nanocomposites obtained with $n_{AgNO3}:n_{Cu(NO3)2}$ = 1:10. (b) Cu 2p regions of Cu$_2$O/Ag nanocomposites. (c) Ag 3d regions of the Cu$_2$O/Ag nanocomposites. (d) O 1s regions of Cu$_2$O/Ag nanocomposites.

Figure 4a shows the TEM image and selected-area electron diffraction (SAED) image of pure Cu$_2$O particles. The SAED patterns were taken at the edge of the particle and demonstrate a typical fcc structure of Cu$_2$O crystals which are of highly crystalline nature [35]. Figure 4b shows the TEM image and SAED pattern of the Cu$_2$O/Ag nanocomposites. It can be seen clearly from the TEM image that the size of AgNPs in the nanocomposites is smaller than 20 nm. Meanwhile, the size of Cu$_2$O nanocubes is smaller than 100 nm, which is about less than 1/10 the size of the pure Cu$_2$O cubes prepared by the same way (Figure S1a). Figure 4c,d are HRTEM images of the Cu$_2$O/Ag nanocomposites. The lattice fringes in the particle in Figure 4d are separated by 0.236 nm, in good agreement with the (111) lattice spacing of Ag. In addition, it can be seen clearly that Ag particles are closely attached to Cu$_2$O cubes from the HRTEM images.

To further observe the combination of Ag and Cu$_2$O, EDS mapping was employed as shown in Figure 5. The EDS mapping images confirmed the coexistence of Ag, Cu, and O elements in the Cu$_2$O/Ag nanocomposites and further confirmed that the composite material is not a simple mixture of Ag particles and Cu$_2$O particles, but a nanoscale composite which is tightly bound together.

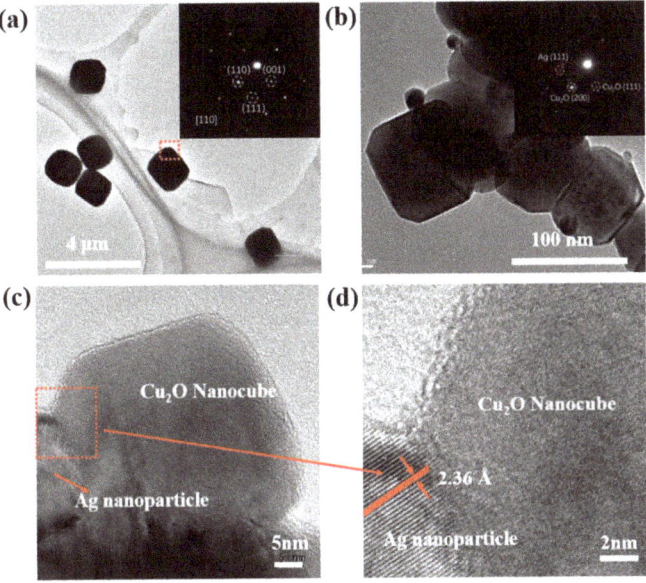

Figure 4. TEM images of the pure Cu$_2$O particles and the Cu$_2$O/Ag nanocomposites obtained with n_{AgNO3}:$n_{Cu(NO3)2}$ = 1:10. (**a**) The TEM image of Cu$_2$O particles (Inset: the SAED pattern of pure Cu$_2$O particles); (**b**) The TEM image of Cu$_2$O/Ag nanocomposites (Inset: the SAED pattern of Cu$_2$O/Ag nanocomposites); (**c**) HRTEM images of Cu$_2$O/Ag nanocomposites; (**d**) Enlarged HRTEM image of rectangular region of (**c**).

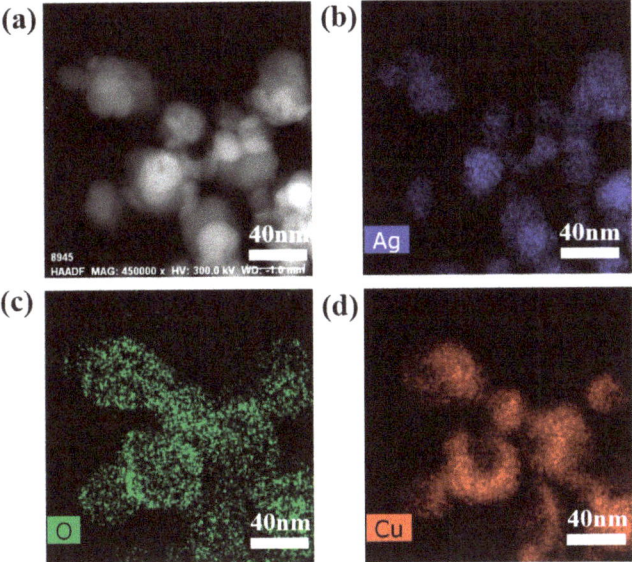

Figure 5. The images of the Cu$_2$O/Ag sample obtained with n_{AgNO3}:$n_{Cu(NO3)2}$ = 1:10. (**a**) A scanning transmission microscopy image, and (**b–d**) the corresponding EDS mapping images: (**b**) Ag element, (**c**) O element, (**d**) Cu element.

3.2. Electrochemical Sensing Performances of the Cu$_2$O/Ag/GCE for H$_2$O$_2$ Detection

The Cu$_2$O/Ag nanocomposites were successfully prepared with the molar ratios of n_{AgNO_3}:$n_{Cu(NO_3)_2}$ = 1:10 at 50 °C and used to fabricate a sensor (Cu$_2$O/Ag/GCE). In order to study the interfacial properties of the electrodes, electrochemical impedance spectroscopy (EIS) experiments were conducted. A typical Nyquist plot consists of a semicircle controlled by the electron transfer process in the high-frequency region and a straight line controlled by the diffusion process in the low-frequency region. The semicircle diameter of the curve reflects the electron transfer resistance (R$_{et}$) at the interface between the electrode material and the electrolyte [36]. Figure 6a shows the Nyquist plots of GCE, Cu$_2$O/GCE, Cu$_2$O/Ag/GCE in 0.1 M KCl solution containing 5 mM [Fe(CN)$_6$]$^{3-}$]. It is easy to find that the semicircular diameter of the Cu$_2$O/Ag/GCE Nyquist plots is smaller than that of the Cu$_2$O/GCE curves, which indicates that the introduction of Ag reduces the propagation resistance between the electrode material and the electrolyte improves the electron transfer rate and is beneficial to improving the electrocatalytic performance to some extent.

The electrochemical properties of the electrodes were studied by cyclic voltammetry (CV). Figure 6b shows CV response of the bare GCE, Cu$_2$O/GCE, and Cu$_2$O/Ag/GCE in the presence of 1 mM H$_2$O$_2$ in 0.1 M PBS (pH = 7.2) at scan rate of 100 mV/s. From Figure 6b, it can be seen that the responses of the bare GCE toward the reduction of H$_2$O$_2$ are quite weak. Cu$_2$O/GCE exhibits electrochemical response and the cathodic peak (−0.4~−0.17 V) and anodic peak (−0.17~0.1 V) can be ascribed to electrochemical reactions of conversion of Cu$_2$O to CuO (oxidation) and CuO to Cu$_2$O (reduction), respectively [22]. The electrode reactions involved in the reduction of H$_2$O$_2$ by the Cu$_2$O/Ag nanocomposites can be proposed as follows [37]:

$$Cu_2O + 2OH^- - 2e^- \rightarrow 2CuO + H_2O \tag{1}$$

$$2CuO + H_2O + 2e^- \rightarrow Cu_2O + 2OH^- \tag{2}$$

$$H_2O_2 + 2e^- \rightarrow 2OH^- \tag{3}$$

In comparison, Cu$_2$O/Ag/GCE showed much higher current response than Cu$_2$O/GCE and bare GCE, which proved the point that the introduction of silver improves the electrochemical properties towards H$_2$O$_2$ of nanocomposites. The enhanced electrocatalytic activity could be ascribed to the synergistic effect of Cu$_2$O and Ag. On the one hand, the appearance of a large number of silver seeds causes the Cu$_2$O nanocubes to have a small size of less than 100 nm in the process of synthesis. On the other hand, the introduction of silver could enhance the charge transport channels and accelerate the transfer rate of electrons in the reaction [38]. Meanwhile, the active area of reaction is increased by the combination of silver on the Cu$_2$O surface, which is beneficial to the adsorption and reaction of H$_2$O$_2$.

Figure 6c shows CV curves of Cu$_2$O/Ag/GCE in the presence of different concentrations of H$_2$O$_2$. It is obvious that the reduction currents gradually increased with the increase of the H$_2$O$_2$ concentrations, indicating the good electrocatalytic activity of Cu$_2$O/Ag/GCE toward H$_2$O$_2$ reduction. To investigate the possible kinetic mechanism, the effect of scan rate on the cathodic current was also investigated. As shown in Figure 6d, with the increasing scan rate from 50 to 150 mV s^{-1}, the reduction current increased linearly. Figure S4 shows that the linear relationship between cathodic peak current versus square root of scan rate can be obtained (R^2 = 0.9898), indicating this process was diffusion-controlled.

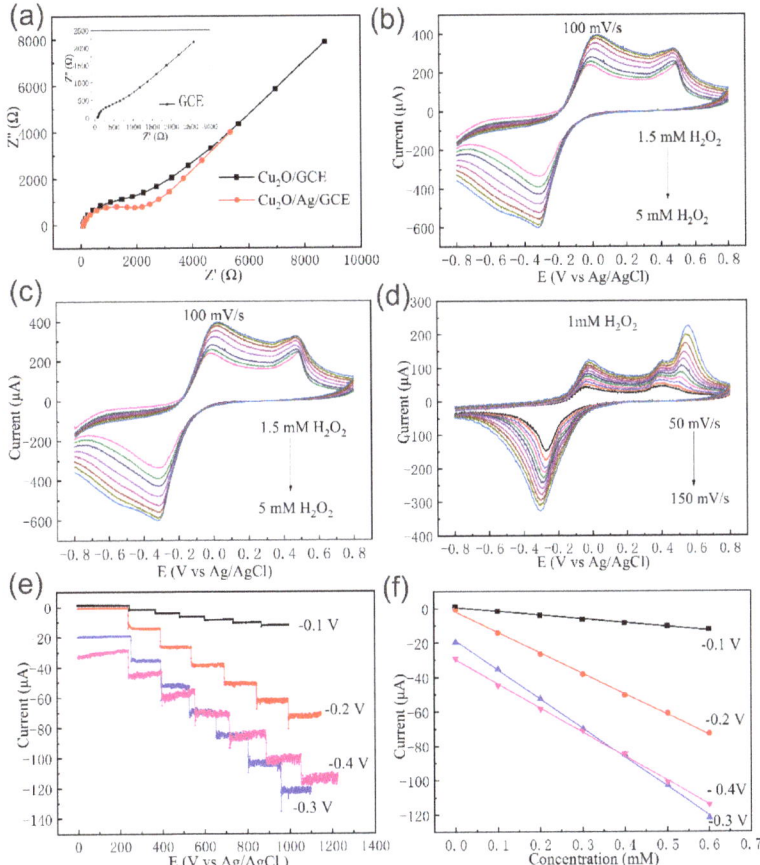

Figure 6. (a) Electrochemical impedance plots (Nyquist plots) of Cu$_2$O/GCE and Cu$_2$O/Ag/GCE in 5 mM [Fe(CN$_6$)$^{3-}$] containing 0.1 M KCl (Inset: Nyquist plots of bare GCE). (b) CVs of bare GCE, Cu$_2$O/GCE, and Cu$_2$O/Ag/GCE in N$_2$-saturated 0.1 M PBS (pH 7.2) in the presence of 1.0 mM H$_2$O$_2$ at a scan rate of 100 mV/s. (c) CVs of Cu$_2$O/Ag/GCE in N$_2$-saturated 0.1 M PBS (pH 7.2) at a scan rate of 100 mV s^{-1} in the presence of H$_2$O$_2$ with different concentrations of 1.5, 2.0, 2.5, 3.0, 3.5, 4.0, 4.5, and 5.0 mM. (d) CVs of Cu$_2$O/Ag/GCE in N$_2$-saturated 0.1 M PBS (pH 7.2) containing 1.0 mM H$_2$O$_2$ at different scan rates (50, 60, 70, 80, 90, 100, 110, 120, 130, 140, and 150 mV s^{-1}). (e) Current–time curves of the Cu$_2$O/Ag/GCE upon successive addition of 0.1 mM H$_2$O$_2$ into N$_2$-saturated 0.1 M PBS (pH = 7.2) under different applied potential of −0.10, −0.20, −0.30, and −0.40 V (vs. Ag/AgCl). (f) The corresponding calibration curves of currents vs. H$_2$O$_2$ concentrations under different potentials (−0.10, −0.20, −0.30, −0.40 V).

It is incontrovertible that the detection potential has much influence on the sensitivity of electrochemical sensors. When choosing the detection potential, the peak voltages in CV (−0.4~−0.2 V vs. Ag/AgCl) is preferred for the best reduction performance for H$_2$O$_2$, while the interference of possible impurities should be considered. The electroactive impurities such as ascorbic acid and uric acid can also be oxidized under high voltages, making it highly likely that their concurrent presences in real applications will interfere with the detection of H$_2$O$_2$ [39]. Figure 6e shows the current response at different detection potentials upon the successive addition of 0.1 mM H$_2$O$_2$. Figure 6f shows the corresponding calibration curves of currents vs. H$_2$O$_2$ concentrations under different potentials.

According to Figure 6e,f, though the sensitivity with −0.2 V is lower than that with −0.3 V and almost the same as that with −0.4 V, the profile is more stable and has less background noise. Therefore, the potential of −0.20 V was chosen as the working potential for the detection of H_2O_2.

3.3. Linear Range, Detection Limit, and Sensitivity of the $Cu_2O/Ag/GCE$ for H_2O_2 Detection

The Cu_2O/Ag nanocomposites-modified electrode was chosen as the sensor electrode for further investigation of H_2O_2 sensing for the outstanding electrochemical behavior and the good electrocatalytic reduction performance towards H_2O_2 detection. Figure 7a shows the current–time curves of the $Cu_2O/Ag/GCE$ to the successive addition of H_2O_2 into the stirred N_2-saturated PBS (pH = 7.2) solution at an applied potential of −0.20 V. It can be seen clearly from the enlargement of the current–time curve at low concentrations that the detection limit of $Cu_2O/Ag/GCE$ for hydrogen peroxide is as low as 0.2 µM (the signal-to-noise ratio of 3, S/N = 3). Figure 7b shows the calibration curve for the H_2O_2 sensor, and the linear regression equation was I (µA) = −0.0870 C (µM) −1.559 with a highly linear relationship (R^2 = 0.9972), in which I is the current and C is concentration of H_2O_2. Meanwhile, this sensor has a linear detection range from 0.2 to 4000 µM and a sensitivity of 87.0 µA mM^{-1} cm^{-2}. In summary, $Cu_2O/Ag/GCE$ exhibited excellent performance towards the reduction of H_2O_2.

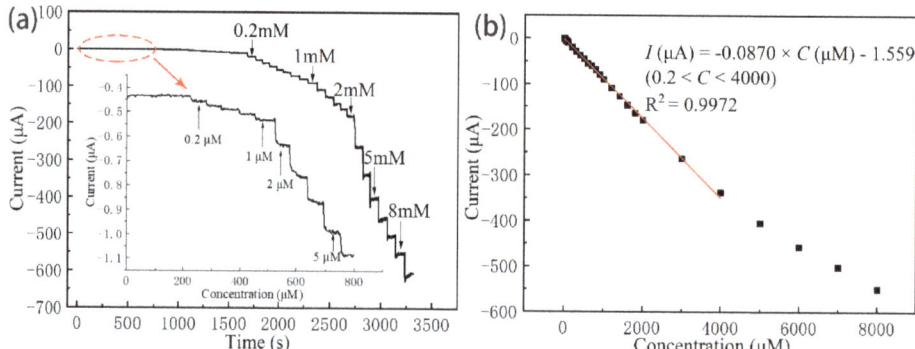

Figure 7. (a) Steady-state current–time responses of the $Cu_2O/Ag/GCE$ upon successive addition of H_2O_2 in N_2-saturated 0.1 M PBS (pH = 7.2) under an applied potential of −0.20 V (vs. Ag/AgCl). Insert: Enlarged image of circle region of (a). (b) The corresponding calibration curve of currents vs. H_2O_2 concentrations. Each dot in (b) shows the current value at the corresponding H_2O_2 concentration which was obtained in (a) and the line is a linear fitting for the experiment points with 0.2 < C < 4000 µM.

Table 1 demonstrates the comparison in the performances of the H_2O_2 nonenzyme sensors fabricated based on the use of similar materials as the electrodes in previous literature reports and in this work. It is shown that our Cu_2O/Ag sensor has a good performance in terms of a high sensitivity, a low detection limit, and a wide linear range. The enhanced electrocatalytic activity could be ascribed to the introduction of silver, which probably provides reaction sites and promotes the electron transfer on the surface of Cu_2O.

Table 1. The comparison of H_2O_2 determination with differently modified electrodes.

Electrode Materials	Detection Potential (V)	Sensitivity (μA mM^{-1} cm^{-2})	Limit of Detection (μM)	Linear Range (μM)	Reference
Porous Cu_2O	−0.2	50.6	1.5	1.5–1500	[40]
Mesocrystalline Cu_2O	−0.3	156.6	1.03	2–150	[21]
Graphene/Cu_2O	−0.4	285	3.3	300–3300	[41]
AgNPs			2.0		[15]
Ag-Au/Cu_2O	−0.2	4.16	1.3	1.3–1400	[23]
Pt-Cu_2O/Nafion	−0.25	20.32	10.3	10–6000	[42]
Cu_2O/Ag	−0.2	87.0	0.2	0.2–4000	This work

3.4. Interference Study

To explore the anti-interference ability of the synthesized Cu_2O/Ag/GCE (red line) and Cu_2O/GCE (black line) for H_2O_2 detection, we added interfering impurities into a continuous testing system. As shown in Figure 8, between the injections of 0.1 mM H_2O_2 solutions, 1 mM NaCl, 1 mM glucose, 1 mM ascorbic acid, and 1 mM urea solutions were added into the 0.1 M PBS solution (pH = 7.2) at −0.20 V in turn. Notably, compared with the Cu_2O/GCE, the Cu_2O/Ag/GCE was more sensitive to H_2O_2.

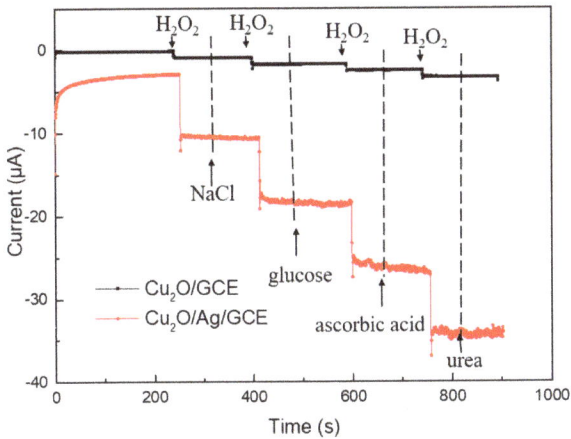

Figure 8. Amperometric response of the Cu_2O/Ag/GCE and Cu_2O/GCE successive addition of H_2O_2 (0.1 mM), NaCl (1 mM), glucose (1 mM), ascorbic acid (1 mM), and urea (1 mM).

The currents for the Cu_2O/Ag/GCE had obvious changes only when H_2O_2 was added. In contrast, the currents did not show any change when the interrupters mentioned above were added. The results indicate that these possible interfering substances do not yield a significant current response, which shows that Cu_2O/Ag/GCE has a good selectivity for H_2O_2.

3.5. Reliability and Recovery Test

The reliability test of the Cu_2O/Ag/GCE was performed by measuring the current response of the electrode upon 1 mM of H_2O_2 in 0.1 M PBS solution (pH = 7.2). The average relative standard deviation (RSD) was not more than 4.2%. In a series of eight sensors prepared in the same way, an RSD of 4.8% was obtained, indicating the reliability of this sensor.

To explore the application of the sensor in the practical environment, the recovery test was constructed by adding a certain amount of H_2O_2 into milk samples. Before the recovery test experiments were conducted, 5 mL milk purchased from a supermarket was diluted into 50 mL solution using 0.1 M PBS solution first. Then, H_2O_2 was added into the as-prepared milk sample with the amounts as shown in Table 2. The results indicate that $Cu_2O/Ag/GCE$ has the potential to be applied in practical environments.

Table 2. Determination of H_2O_2 in milk samples.

Sample	H_2O_2 Added (µM)	H_2O_2 Found (µM)	Recovery (%)	RSD (%)
1	50	48.4	96.8	1.3
2	100	104.2	104.2	6.1
3	150	142.6	95.1	3.0
4	200	192.4	96.2	1.9

What we need to be careful about is that the sensors would be better kept in a cool and dry environment to prevent the material from being oxidized in moisture. The service life of the sensor might be improved by using curing materials such as Nafion [36].

4. Conclusions

In summary, uniform and small-size Cu_2O/Ag nanocomposites (size of Cu_2O particle <100 nm, size of Ag particle <20 nm) were synthesized successfully via a facile one-step process, and successfully used to fabricate an H_2O_2 sensor. The electrochemical experiment results reveal that the $Cu_2O/Ag/GCE$ exhibits outstanding electrochemical behavior and good electrocatalytic reduction performance towards H_2O_2. The linear range of the $Cu_2O/Ag/GCE$ is estimated to be 0.2–4000 µM with a sensitivity of 87.0 µA mM^{-1} cm^{-2} and a low detection limit of 0.2 µM. The anti-interference capability experiment indicated that the Cu_2O/Ag nanocomposites have good selectivity toward H_2O_2. Additionally, the H_2O_2 recovery test in the milk solution demonstrates the potential application of $Cu_2O/Ag/GCE$ in routine H_2O_2 analysis.

Supplementary Materials: The Supplementary Materials are available online at http://www.mdpi.com/2079-4991/9/4/523/s1.

Author Contributions: Conceptualization and Methodology, K.Y. and Z.Y.; Formal Analysis and Investigation, K.Y., Z.Y., and L.M.; Data Curation, B.P. and Y.D.; Writing-Original Draft Preparation, K.Y.; Visualization, J.F.; Project Administration, K.Y.; Funding Acquisition, Z.Y.

Funding: This research was funded by Beijing Natural Science Foundation (2172005, Z180014), the National Natural Science Foundation of China (11674015, 91860202), and the "111" project under the grant of DB18015.

Conflicts of Interest: The authors declare no conflict of interest.

References

1. Xiao, Y.; Ju, H.X.; Chen, H.Y. Hydrogen peroxide sensor based on horseradish peroxidase-labeled Au colloids immobilized on gold electrode surface by cysteamine monolayer. *Anal. Chim. Acta* **1999**, *391*, 73–82. [CrossRef]
2. Sabahudin, H.; Yali, L.; Male, K.B.; Luong, J.H.T. Electrochemical biosensing platforms using platinum nanoparticles and carbon nanotubes. *Anal. Chem.* **2004**, *76*, 1083–1088.
3. Huang, Y.; Ferhan, A.R.; Dandapat, A.; Chong, S.Y.; Ji, E.S.; Cho, E.C.; Kim, D.H. A strategy for the formation of gold-palladium supra-nanoparticles from gold nanoparticles of various shapes and their application to high-performance H_2O_2 sensing. *J. Phys. Chem. C* **2015**, *119*, 26164–26170. [CrossRef]
4. Chen, S.H.; Yuan, R.; Chai, Y.Q.; Hu, F.X. Electrochemical sensing of hydrogen peroxide using metal nanoparticles: A review. *Microchim. Acta* **2013**, *180*, 15–32. [CrossRef]
5. Wei, C.; Shu, C.; Ren, Q.Q.; Wei, W.; Zhao, Y.D. Recent advances in electrochemical sensing for hydrogen peroxide: A review. *Analyst* **2011**, *137*, 49–58.
6. Hurdis, E.C.; Romeyn, H. Accuracy of determination of hydrogen peroxide by cerate oxidimetry. *Anal. Chem.* **1954**, *26*, 320–325. [CrossRef]

7. Kosman, J.; Juskowiak, B. Peroxidase-mimicking DNAzymes for biosensing applications: A review. *Anal. Chim. Acta* **2011**, *707*, 7–17. [CrossRef]
8. Greenway, G.M.; Leelasattarathkul, T.; Liawruangrath, S.; Wheatley, R.A.; Youngvises, N. Ultrasound-enhanced flow injection chemiluminescence for determination of hydrogen peroxide. *Analyst* **2006**, *131*, 501–508. [CrossRef] [PubMed]
9. Wang, H.; Li, Y.; Yang, M.; Wang, P.; Gu, Y. FRET-Based upconversion nanoprobe sensitized by Nd^{3+} for the ratiometric detection of hydrogen peroxide in vivo. *ACS Appl. Mater. Interfaces* **2019**, *11*, 7441–7449. [CrossRef] [PubMed]
10. Pinkernell, U.; Effkemann, S.; Karst, U. Simultaneous HPLC determination of peroxyacetic acid and hydrogen peroxide. *Anal. Chem.* **1997**, *69*, 3623–3627. [CrossRef] [PubMed]
11. Dai, H.; Chen, D.; Cao, P.; Li, Y.; Wang, N.; Sun, S.; Chen, T.; Ma, H.; Lin, M. Molybdenum sulfide/nitrogen-doped carbon nanowire-based electrochemical sensor for hydrogen peroxide in living cells. *Sens. Actuators B* **2018**, *276*, 65–71. [CrossRef]
12. Cheng, C.; Zhang, C.; Gao, X.; Zhuang, Z.; Du, C.; Chen, W. 3D network and 2D paper of reduced graphene oxide/Cu_2O composite for electrochemical sensing of hydrogen peroxide. *Anal. Chem.* **2018**, *90*, 1983–1991. [CrossRef]
13. Guan, H.; Zhang, J.; Liu, Y.; Zhao, Y.; Zhang, B. Rapid quantitative determination of hydrogen peroxide using an electrochemical sensor based on PtNi alloy/CeO_2 plates embedded in N-doped carbon nanofibers. *Electrochim. Acta* **2019**, *295*, 997–1005. [CrossRef]
14. Han, L.; Tang, L.; Deng, D.; He, H.; Zhou, M.; Luo, L. A novel hydrogen peroxide sensor based on electrodeposited copper/cuprous oxide nanocomposites. *Analyst* **2019**, *144*, 685–690. [CrossRef]
15. Welch, C.M.; Banks, C.E.; Simm, A.O.; Compton, R.G. Silver nanoparticle assemblies supported on glassy-carbon electrodes for the electro-analytical detection of hydrogen peroxide. *Anal. Bioanal. Chem.* **2005**, *382*, 12–21. [CrossRef]
16. Li, L.; Du, Z.; Liu, S.; Hao, Q.; Wang, Y.; Li, Q.; Wang, T. A novel nonenzymatic hydrogen peroxide sensor based on mno_2/graphene oxide nanocomposite. *Talanta* **2010**, *82*, 1637–1641. [CrossRef]
17. Benvidi, A.; Nafar, M.T.; Jahanbani, S.; Tezerjani, M.D.; Rezaeinasab, M.; Dalirnasab, S. Developing an electrochemical sensor based on a carbon paste electrode modified with nano-composite of reduced graphene oxide and $CuFe_2O_4$ nanoparticles for determination of hydrogen peroxide. *Mater. Sci. Eng. C* **2017**, *75*, 1435–1447. [CrossRef]
18. Sarkar, A.; Ghosh, A.B.; Saha, N.; Bhadu, G.R.; Adhikary, B. Newly designed amperometric biosensor for hydrogen peroxide and glucose based on vanadium sulfide nanoparticles. *ACS Appl. Nano Mater.* **2018**, *1*, 1339–1347. [CrossRef]
19. Zhong, Y.M.; Li, Y.; Li, S.; Feng, S.; Zhang, Y. Nonenzymatic hydrogen peroxide biosensor based on four different morphologies of cuprous oxide nanocrystals. *RSC Adv.* **2014**, *4*, 40638–40642. [CrossRef]
20. Li, Y.; Zhong, Y.; Zhang, Y.; Weng, W.; Li, S. Carbon quantum dots/octahedral Cu_2O nanocomposites for non-enzymatic glucose and hydrogen peroxide amperometric sensor. *Sens. Actuators B* **2015**, *206*, 735–743. [CrossRef]
21. Gao, Z.; Liu, J.; Chang, J.; Wu, D.; He, J.; Wang, K.; Fang, X.; Kai, J. Mesocrystalline Cu_2O hollow nanocubes: Synthesis and application in non-enzymatic amperometric detection of hydrogen peroxide and glucose. *CrystEngComm* **2012**, *14*, 6639–6646. [CrossRef]
22. Li, S.; Zheng, Y.; Qin, G.W.; Ren, Y.; Pei, W.; Zuo, L. Enzyme-free amperometric sensing of hydrogen peroxide and glucose at a hierarchical Cu_2O modified electrode. *Talanta* **2011**, *85*, 1260–1264. [CrossRef]
23. Li, D.; Meng, L.; Dang, S.; Jiang, D.; Shi, W. Hydrogen peroxide sensing using Cu_2O nanocubes decorated by Ag-Au alloy nanoparticles. *J. Alloys Compd.* **2017**, *690*, 1–7. [CrossRef]
24. Li, Y.; Zhang, J.; Zhu, H.; Yang, F.; Yang, X. Gold nanoparticles mediate the assembly of manganese dioxide nanoparticles for H_2O_2 amperometric sensing. *Electrochim. Acta* **2010**, *55*, 5123–5128. [CrossRef]
25. Youngmin, L.; Miguel Angel, G.; Huls, N.A.F.; Shouheng, S. Synthetic tuning of the catalytic properties of Au-Fe_3O_4 nanoparticles. *Angew. Chem. Int. Ed.* **2010**, *41*, 1271–1274.
26. Han, Y.; Zheng, J.; Dong, S. A novel nonenzymatic hydrogen peroxide sensor based on Ag–MnO_2–MWCNTs nanocomposites. *Electrochim. Acta* **2013**, *90*, 35–43. [CrossRef]
27. Chen, T.; Tian, L.; Chen, Y.; Liu, B.; Zhang, J. A facile one-pot synthesis of Au/Cu_2O nanocomposites for nonenzymatic detection of hydrogen peroxide. *Nanoscale Res. Lett.* **2015**, *10*, 252. [CrossRef] [PubMed]

28. Zhao, X.; Li, Z.; Cheng, C.; Wu, Y.; Zhu, Z.; Zhao, H.; Lan, M. A novel biomimetic hydrogen peroxide biosensor based on Pt flowers-decorated Fe_3O_4/graphene nanocomposite. *Electroanalysis* **2017**, *29*, 1518–1523. [CrossRef]
29. Schaepertoens, M.; Didaskalou, C.; Kim, J.F.; Livingston, A.G.; Szekely, G. Solvent recycle with imperfect membranes: A semi-continuous workaround for diafiltration. *J. Membr. Sci.* **2016**, *514*, 646–658. [CrossRef]
30. Feng, L.; Zhang, C.; Gao, G.; Cui, D. Facile synthesis of hollow Cu_2O octahedral and spherical nanocrystals and their morphology-dependent photocatalytic properties. *Nanoscale Res. Lett.* **2012**, *7*, 276. [CrossRef]
31. Zhang, N.; Sheng, Q.; Zhou, Y.; Dong, S.; Zheng, J. Synthesis of FeOOH@PDA-Ag nanocomposites and their application for electrochemical sensing of hydrogen peroxide. *J. Electroanal. Chem.* **2016**, *781*, 315–321. [CrossRef]
32. Lv, J.; Kong, C.; Xu, Y.; Yang, Z.; Zhang, X.; Yang, S.; Meng, G.; Bi, J.; Li, J.; Yang, S. Facile synthesis of novel CuO/Cu_2O nanosheets on copper foil for high sensitive nonenzymatic glucose biosensor. *Sens. Actuators B* **2017**, *248*, 630–638. [CrossRef]
33. Wang, Y.; Lü, Y.; Zhan, W.; Xie, Z.; Kuang, Q.; Zheng, L. Synthesis of porous Cu_2O/CuO cages using Cu-based metal–organic frameworks as templates and their gas-sensing properties. *J. Mater. Chem. A* **2015**, *3*, 12796–12803. [CrossRef]
34. Veisi, H.; Kazemi, S.; Mohammadi, P.; Safarimehr, P.; Hemmati, S. Catalytic reduction of 4-nitrophenol over Ag nanoparticles immobilized on Stachys lavandulifolia extract-modified multi walled carbon nanotubes. *Polyhedron* **2019**, *157*, 232–240. [CrossRef]
35. Tsai, Y.; Chanda, K.; Chu, Y.; Chiu, C.; Huang, M.H. Direct formation of small Cu_2O nanocubes, octahedra, and octapods for efficient synthesis of triazoles. *Nanoscale* **2014**, *6*, 8704–8709. [CrossRef]
36. Zhang, L.; Ni, Y.; Li, H. Addition of porous cuprous oxide to a Nafion film strongly improves the performance of a nonenzymatic glucose sensor. *Microchim. Acta* **2010**, *171*, 103–108. [CrossRef]
37. Kumar, J.S.; Murmu, N.C.; Samanta, P.; Banerjee, A.; Ganesh, R.S.; Inokawa, H.; Kuila, T. Novel synthesis of a Cu_2O–Graphene nanoplatelet composite through a two-step electrodeposition method for selective detection of hydrogen peroxide. *New J. Chem.* **2018**, *42*, 3574–3581. [CrossRef]
38. Qi, C.; Zheng, J. Novel nonenzymatic hydrogen peroxide sensor based on Ag/Cu_2O nanocomposites. *Electroanalysis* **2016**, *28*, 477–483. [CrossRef]
39. Chen, Y.; Hsu, J.; Hsu, Y. Branched silver nanowires on fluorine-doped tin oxide glass for simultaneous amperometric detection of H_2O_2 and of 4-aminothiophenol by SERS. *Microchim. Acta* **2018**, *185*, 106. [CrossRef]
40. Zhang, L.; Li, H.; Ni, Y.; Li, J.; Liao, K.; Zhao, G. Porous cuprous oxide microcubes for non-enzymatic amperometric hydrogen peroxide and glucose sensing. *Electrochem. Commun.* **2009**, *11*, 812–815. [CrossRef]
41. Liu, M.; Liu, R.; Chen, W. Graphene wrapped Cu_2O nanocubes: Non-enzymatic electrochemical sensors for the detection of glucose and hydrogen peroxide with enhanced stability. *Biosens. Bioelectron.* **2013**, *45*, 206–212. [CrossRef] [PubMed]
42. Lv, J.; Kong, C.; Liu, K.; Yin, L.; Ma, B.; Zhang, X.; Yang, S.; Yang, Z. Surfactant-free synthesis of Cu_2O yolk–shell cubes decorated with Pt nanoparticles for enhanced H_2O_2 detection. *Chem. Commun.* **2018**, *54*, 8458–8461. [CrossRef] [PubMed]

© 2019 by the authors. Licensee MDPI, Basel, Switzerland. This article is an open access article distributed under the terms and conditions of the Creative Commons Attribution (CC BY) license (http://creativecommons.org/licenses/by/4.0/).

Article

Manufacture of Networks from Large Diameter Single-Walled Carbon Nanotubes of Particular Electrical Character

Edyta Turek, Bogumila Kumanek, Slawomir Boncel and Dawid Janas *

Department of Organic Chemistry, Bioorganic Chemistry and Biotechnology, Silesian University of Technology, B. Krzywoustego 4, 44-100 Gliwice, Poland; ea.turek@o2.pl (E.T.); Bogumila.Kumanek@polsl.pl (B.K.); Slawomir.Boncel@polsl.pl (S.B.)
* Correspondence: Dawid.Janas@polsl.pl; Tel.: +48-32-237-10-82

Received: 18 March 2019; Accepted: 12 April 2019; Published: 14 April 2019

Abstract: We have demonstrated that the aqueous two-phase extraction (ATPE) can differentiate between large diameter single-walled carbon nanotubes (CNTs) by electrical character. Introduction of "hydration modulators" to the ATPE machinery has enabled us to isolate metallic and semiconducting CNTs with ease. We have also shown that often there is a trade-off between the purity of the obtained fractions and the ability to separate both metallic and semiconducting CNTs at the same time. To isolate the separated CNTs from the matrices, we have proposed a method based on precipitation and hydrolysis, which can eliminate the need to use lengthy dialysis routines. In the final step, we prepared thin free-standing films from the sorted material and probed how electrical charge is transported through such macroscopic ensembles.

Keywords: carbon nanotubes; electrical character; aqueous two-phase extraction

1. Introduction

The term carbon nanotubes (CNTs) may be somewhat misleading by implying that it is a plurality of exact copies of individual CNTs. However, the members of this population differ slightly from one another and these seemingly negligible discrepancies have a dramatic influence on the properties of the material. For instance, a diameter difference on the order of 0.1 Å is enough to make one CNT semiconducting and the other to have a metallic character [1,2]. The effect is valid not only for the electrical properties; their thermal [3,4], optical [5–7], and other attributes are also strongly affected. To tackle this problem, a spectrum of sorting methods has been established, which enable differentiation between CNTs of different types at the level of electrical character, chirality, or even handedness [8].

Large diameter single-walled carbon nanotubes (SWCNTs) are particularly attractive for various fields of science and technology, including photonics and microelectronics [9]. Their ability to emit light in the telecom range [10,11], as well as maintain high saturation current [12] at a reduced Schottky barrier [13], make them a very promising material. Unfortunately, with the increase in diameter, the differences between individual CNTs fade and sorting becomes challenging. Moreover, for some applications, macroscopic ensembles from defined CNTs such as films or fibers are necessary, but there are no straightforward techniques to reach this goal. The main obstacles are the lack of effective methods to obtain CNTs of selected electrical type at the large scale and, even if such sorted material is obtained, it can rarely be transformed into a CNT network free of contamination. A promising way to handle this problem was provided by He et al., who produced spontaneously aligned CNT films from a variety of parent materials [14].

In this contribution, we present our results of sorting large diameter SWCNTs by using a single-step adaptation [15] of the aqueous two-phase extraction (ATPE) method [16]. The key to reaching very high purity CNTs was to introduce a hydration modulator into the ATPE system, which enabled much more effective partitioning of semiconducting and metallic CNTs into top and bottom fractions, respectively. Large-diameter CNTs were conveniently divided into semiconducting and metallic rich fractions, and then they were transformed into thin free-standing films by using a method recently developed by us [17]. Characterization of the composition and electrical conductivity of such materials enabled us to validate whether the separation was successful. The developed methodology also demonstrated how to desorb the polymer species from the surface of the CNTs, which is a significant problem in the processing of sorted CNTs and which, to this day, is most commonly handled by lengthy dialysis.

2. Materials and Methods

Large-diameter single-walled carbon nanotubes (CNTs) of high purity with lengths exceeding 5 μm were purchased from OCSiAl (Tuball™, Leudelange, Luxembourg) and purified by air treatment and reflux in HCl, according to a published method [18]. Purified CNTs (1 mg/mL) were dispersed in water using 20 g/L of sodium cholate (SC). The mixture was processed by a tip sonicator for 2 h (Hielscher UP200St, Teltow, Germany; 200 W, 50% amplitude, 50 mL, 2 h). Due to the evolution of heat over time, ice baths were used continuously to ensure proper homogenization of the material. Next, the dispersion was centrifuged at 11,000 rpm (Eppendorf 5804R centrifuge, Hamburg, Germany; 15,314 × g) for 2 h to remove the bundled CNTs from the dispersion (the top 80% of the supernatant volume was used for the study; the rest was discarded).

One-step separation by ATPE was executed by modifying an approach published by us [15] and that of Gui et al. [16] (by the introduction of a range of hydration modulators). CNT dispersion, Dextran (DEX, 20%, aq.), poly(ethylene glycol) (PEG, 50%, aq.), sodium cholate (SC, 10%, aq.), sodium dodecyl sulfate (SDS, 10%, aq.), hydration modulator (alanine, β-cyclodextrin, diethanolamine, hydrogen peroxide, ethylenediaminetetraacetic acid (EDTA), ethylene glycol, N,N-dimethylformamide, imidazole, poly(ethylene glycol) methyl ether (PEGme) Mw = 5,000 g/mol, polyvinylpyrrolidone (PVP), potassium persulfate, potassium phthalimide, sodium borohydride, sodium hypochlorite, thioacetamide, thiourea, or urea) and H_2O were combined in an Eppendorf tube in the specified order (Figure 1) (exact experimental details are given in the Supplementary information file). The mixture was shaken vigorously until all the components were appropriately combined. Finally, to speed up the process of phase separation (which would otherwise take a few hours), the mixture was centrifuged for 3 min at 2,000 rpm (506 × g). After this time, the bottom DEX layer and the top PEG emerged spontaneously. The total volume was 1.53 mL each time.

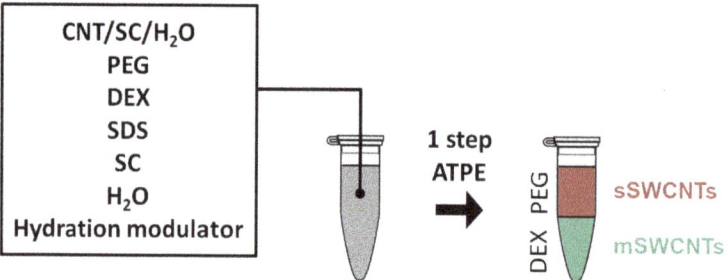

Figure 1. The method of separation used in the study—one-step aqueous two-phase extraction (ATPE), which, in this case, gives s-SWCNTs (semiconducting single-walled CNTs) and m-SWCNTs (metallic single-walled CNTs).

Absorbance spectra were obtained from 400 to 1000 nm using Hitachi U-2910 spectrophotometer (Tokyo, Japan). Bottom and top phases were separated from each other and introduced to quartz cuvettes for measurements. Intensities were normalized to the global absorbance minimum which can be often found within or near the 700–800 nm range. Presented data were offset to improve the clarity of the presentation.

Scanning electron microscopy (SEM, FEI Nova NanoSEM, Hillsboro, OR, USA; 10 keV acquisition voltage) was employed to analyze the microstructure of the material.

To remove PEG and DEX, the following routines have been developed. Acetone was added to the PEG phase to precipitate the CNTs and the mixture was filtered under reduced pressure. CNTs were washed using warm distilled water (this step was repeated three times). The obtained solid was dried in a desiccator at room temperature to reach a constant weight. With regard to the DEX phase, 0.7 M HNO_3 was used to precipitate the CNTs and the mixture was filtered under reduced pressure using warm distilled water, acetone, and methanol as the washing media. Then, Dextran was hydrolyzed by introducing the obtained solid into 1 M HCl solution, kept at 60 °C for 1 h under sonication. CNTs were separated by filtration under reduced pressure, washed with warm distilled water and subjected to another sonication step at 60 °C for 1 h, this time in distilled water. Finally, the obtained material was filtered under reduced pressure and washed with warm distilled water, acetone, and methanol. After the process, the product was dried in a desiccator at room temperature to reach a constant weight.

Thermograms (Linseis TA system, Selb, Germany) were acquired in the flow of air (50 mL/min) at a 10 °C/min heating rate. A total of 1.5 mg of material was used for each measurement.

To make the films, a previously reported routine was employed [17]. In brief, one equivalent of CNTs was combined with one equivalent of ethyl cellulose (both 1 wt% with respect to the solvent) in acetone/toluene mixture (1:1, V/V). Then, the mixture kept in an ice bath was sonicated to obtain a uniform dispersion. Subsequently, the CNT paint was deposited onto a Kapton® film (RS Components, Corby, UK), detached from its surface and annealed in air to remove the ethyl cellulose. CNT films produced this way (using unsorted CNTs or predominantly metallic/semiconducting) were further characterized.

Raman spectroscopy (inVia Renishaw Raman microscope, Wotton-under-Edge, UK; λ = 514, 633, or 780 nm, where indicated) was used to record inelastic scattering from the samples within a 100 to 3200 cm^{-1} range. To lower the effect of background, 50 accumulations were acquired for each sample.

Electrical resistivity was measured by using a four-probe method (Keithley 2450, Cleveland, OH, USA). The values were recorded at room temperature for neat and p-doped films (BF_3 solution in methanol was dripped onto the film and the solvent was allowed to evaporate).

3. Results and Discussion

3.1. Characterization of the Material

Sorting of CNTs is based on the principle that various CNT structures have slightly dissimilar interactions with the environment based on the minute differences in their structure. As a consequence, any type of unwanted functionalization can not only influence the course of separation, but after reaching a certain threshold, it makes the differentiation impossible. First, we confirmed that the selected material was suitable for the study by characterization of its microstructure and composition (Figure 2). The presence of non-CNT macroscopic adulterants was minimal, as observed under SEM. What is more, the CNTs were very much bundled-up as expected for single-walled CNTs.

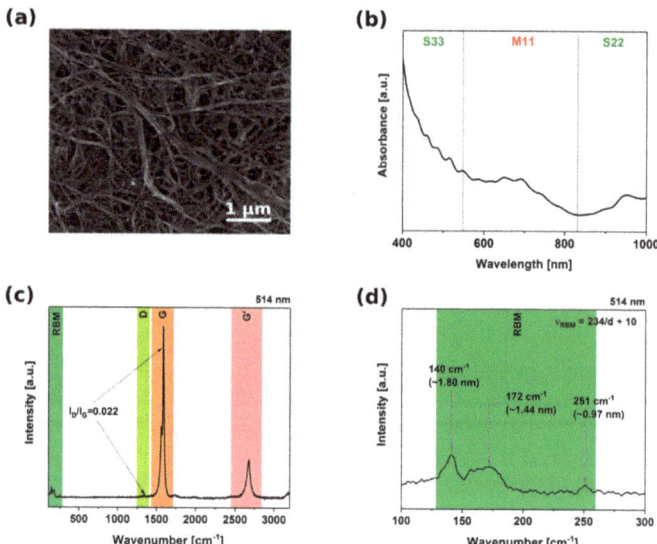

Figure 2. Characterization of the starting material. (**a**) Scanning Electron Microscopy (SEM) micrograph, (**b**) absorbance spectrum, (**c**) Raman spectrum, (**d**) close-up plot of Radial Breathing Mode (RBM) area from the Raman spectrum (514 nm excitation).

The absorbance spectra showed signals both in the M_{11} and S_{22}/S_{33} areas, indicating that the unsorted material was indeed composed of a mixture of CNTs of different electrical character. Furthermore, characterization of the surface by Raman spectroscopy confirmed high purity of the material. I_D/I_G, indicative of the level of crystal imperfection (D band—sp^3 impurities, G band—sp^2 graphitic lattice), was as low as 0.022. Additionally, splitting of the G band into G+ and G− components could be detected. It is important to note that no defects were introduced even during prolonged sonication of the material to make a CNT dispersion (Figure S1). Analysis of the Radial Breathing Mode (RBM) area confirmed the large-diameter character of the material. According to the manufacturer, the diameter distribution of this material is within the 1.8 ± 0.4 nm range. We detected CNTs with diameters from about 1 nm up to 1.87 nm by recalculating the wavenumbers to diameter [19–21] (please refer to Figure S2 to see Raman spectra acquired at 633 and 780 nm). Raman spectroscopy is a resonant method of characterization, so only those CNTs which are in tune with the wavelength of the lasers are detected (the rest remains invisible). Secondly, the peaks are Lorentzian in nature, so wavenumber of their maxima should be taken into consideration.

Therefore, in light of these results we can claim that the diameter of the material used for the study was within the 1–2 nm range.

3.2. Regular ATPE Separation

First, we wanted to find out how the ATPE works on large-diameter single-walled CNTs without introducing any hydration modulator (Figure 3).

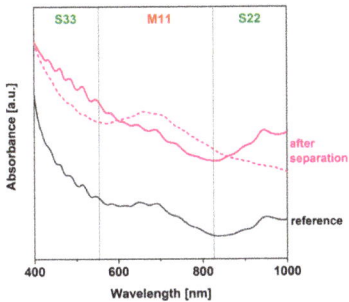

Figure 3. ATPE of large-diameter single-walled carbon nanotubes (CNTs) without addition of hydration modulators. Dextran-rich bottom phase (dashed line), PEG-rich top phase (solid line). Reference dispersion shown in black.

Preliminary results were encouraging, as we managed to separate a fraction significantly enriched with the metallic type of CNTs in the PEG phase. The peak of absorbance emerged in the M_{11} range and the intensity of the signal coming from the neighboring S_{33} and S_{22} areas [19] was significantly reduced as compared with the parent material. On the other hand, the bottom phase did not show an appropriate degree of isolation of the corresponding semiconducting fraction, as it resembled the reference. It was clear that the system had to be tuned to reach simultaneous preference towards these types of CNTs by the two phases and hence isolate them from each other. We decided to introduce another component into the ATPE system, which could influence how the surfactants encapsulating the CNTs arrange themselves on their surface.

3.3. Effect of H_2O_2 Addition

The first compound that we decided to exploit for this purpose was hydrogen peroxide because of its high compatibility with the ATPE components and well-documented electronic interactions with carbon nanostructures [22–24] (Figure 4). At a high CNT loading of 300 µL with added 80 µL of 30 wt% H_2O_2, the separation attempt was not successful. The excess of metallic CNTs was lower than in the reference sample and again we did not manage to separate the semiconducting ones from the mixture. We decided to lower the content of CNTs to 150 µL and tried introducing high (200 µL) or low (40 µL) content of H_2O_2.

To our delight, the differentiation of these two CNT types was achieved in the latter case. Not only was the peak of metallic CNTs much more defined than when no additive was employed, but also the semiconducting CNTs were obtained. Because of our previous experience, which suggests that, often, lowering the relative content of CNTs in the ATPE mixture improves the selectivity, we decided to lower the starting CNT amount by half, down to 75 µL, and also added 20 µL of H_2O_2. Unfortunately, as can be seen in the spectrum of the semiconducting-rich fraction, the intensity of interband metallic transitions increased slightly between 550 and 650 nm. Simultaneously, the peak of the corresponding metallic-rich fraction became less pronounced, which indicated that some of the metallic CNTs migrated to the top phase. Finally, and interestingly, by increasing the amount of added H_2O_2 to 100 µL, one could obtain the highly semiconducting fraction in the top phase, but a significant part of metallic CNTs would be lost from the bottom phase and shifted to the interface. This is completely opposite of the results from the neat ATPE, wherein only the metallic fraction was obtained. Closer investigation of the spectra clearly showed a trade-off between the resolution of the system toward a particular CNT type and a high yield of separation. When both metallic and semiconducting fractions were isolated to the respective phases, the purity could be high, but not necessarily the highest possible. Further enrichment in the metallic or semiconducting content was possible at the expense of purity of the complementary phase. We can conclude that the optimum parameters to obtain the highest amount of metallic and semiconducting species in one experiment are 150 μL CNT and 40 μL H$_2$O$_2$. Figure 5 depicts what these two phases look like after separation.

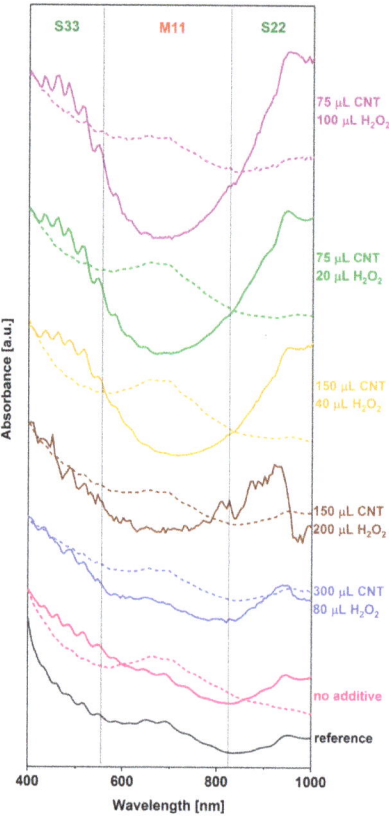

Figure 4. The influence of the volume of the CNT dispersion and hydration modulating agent (H$_2$O$_2$) in the ATPE system on the course of the separation. Dextran-rich bottom phase (dashed line), PEG-rich top phase (solid line).

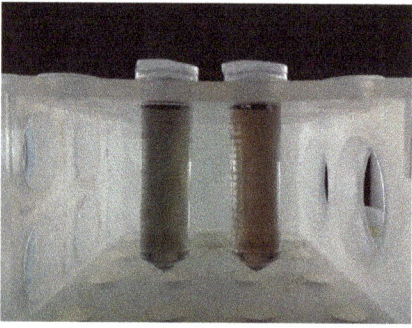

Figure 5. Photograph of the two phases after separation into CNT dispersions, composed predominantly of metallic (left) and semiconducting (right) species.

3.4. Effect of Poly(Ethylene Glycol) Methyl Ether Addition

More complicated structures than H_2O_2 can also fine tune the metallic–semiconducting separation ability of the ATPE. Poly(ethylene glycol) methyl ether (PEGme) also enabled us to significantly enrich the purity of the metallic and semiconducting CNT fractions (Figure 6).

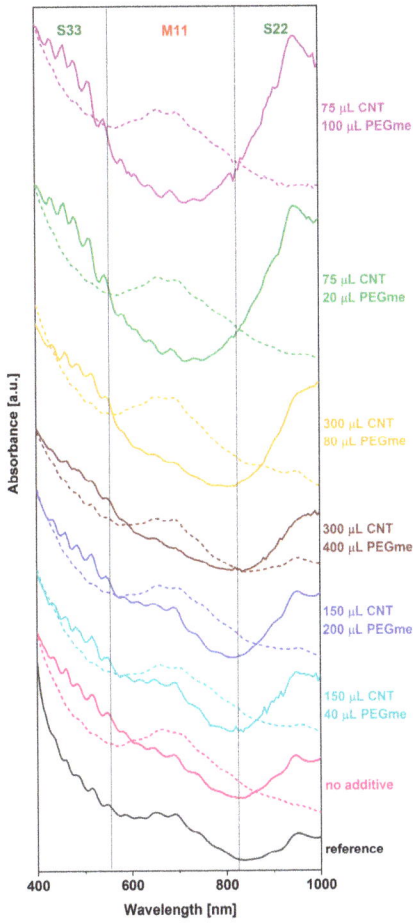

Figure 6. The influence of the volume of the CNT dispersion and hydration modulating agent (PEGme) in the ATPE system on the course of the separation. Dextran-rich bottom phase (dashed line), PEG-rich top phase (solid line).

Again, the best results were obtained when a relatively small amount of CNTs was introduced into the ATPE system (ca. 75 µL). In this case, however, the contamination of semiconducting fractions with the metallic CNTs was more obvious at this level. Only when the volume of the CNT dispersion and PEGme was increased four-fold to 300 and 80 µL, respectively, the purity was improved, but again at the expense of the yield of separation (the desired decrease in the M_{11} range of the semiconducting fraction was accompanied by a loss of intensity in the S_{22} zone). Other combinations of parameters involving high PEGme content were much less successful than the aforementioned conditions. It appears that PEGme, because of its hydrophobic methyl head, can somehow change the interaction of the CNTs with the ATPE matrix. Simple modification of the PEG content did not lead to the same results.

3.5. Effect of Addition of Other Hydration Modulators

We tried a wide range of compounds as hydration modulators (please refer to the Supplementary information file for all the conditions employed). Firstly, introducing redox active inorganic compounds, such as $NaBH_4$ (Figure S3), $K_2S_2O_8$ (Figure S4), or NaClO (Figure S5), did not lead to any separation at all, as most of the CNTs aggregated at the interface. Secondly, we explored organic compounds with and without heteroatoms in their structure (Figures S6–S17, ethylene glycol, alanine, diethanolamine, ethylenediaminetetraacetic acid (EDTA), urea, dimethylformatide (DMF), phthalimide, imidazole, thiourea, thioacetamide, β-cyclodextrin, and polyvinylpyrrolidone (PVP)). Either no improvement was observed, or the shape of the spectra did not resemble CNTs enriched with particular electrical character. In fact, addition of certain chemical species resulted in the collapse of the ATPE system or lack of clarity of the constituting phases. From our experience so far, the introduction of only the two aforementioned species (H_2O_2 and PEGme) had a positive influence on the course of separation.

3.6. Removal of ATPE Remnants

We carried out a large scale separation (100 fold increase—total volume of 153 mL) of CNTs to isolate both semiconducting and metallic fractions at the highest possible purity in one go (relative parameters: 150 μL CNT and 40 μL H_2O_2) for further examination. In the first step, we developed a process of isolation of these species from PEG and DEX matrices, respectively. Usually, this is accomplished by a lengthy dialysis, which requires high pressure and expensive membranes. Our chemical approach was designed to precipitate out the selected material and hydrolyze certain chemical compounds to enable a simple and convenient separation from the CNTs by filtration under reduced pressure. As shown in the thermograms, both metallic- and semiconducting-rich fractions were essentially purified from PEG and DEX (Figure 7). It should be noted, however, that the final samples contained a certain amount of SC, which was used for dispersion at the very beginning. Oxidation of the material took place at relatively high temperatures for single-walled CNTs (ca. 600 °C), but this can be justified by taking into the consideration their large diameter.

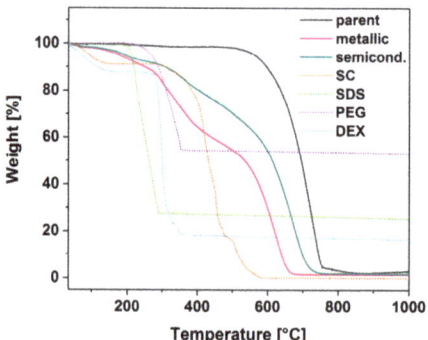

Figure 7. Thermograms of parent and sorted material as well as those of ATPE components.

3.7. Sorting Outcome Characterization by Raman Spectroscopy

To get a better insight into the separation, we studied the obtained samples by Raman spectroscopy (Figure 8). It can be seen that enrichment of the samples with particular electrical character was successful, although it led to complete differentiation between these two CNT families. The parent material was 60% semiconducting and 40% metallic. Separation by modified ATPE resulted in manufacture of the material having 60% metallic character or 85% semiconducting character of the bottom and top phases, respectively. It should be noted that the Raman technique is a resonant process, therefore the conclusions are valid only for the specified wavelength (633 nm). As a consequence, the observed change in RBM intensities should be considered as a qualitative measure of the observed phenomena, since there are more precise methods for appropriate quantification. According to the absorbance spectra presented earlier, the extent of the separation may be much larger because, under certain conditions, the metallic fraction does not seem to contain semiconducting CNTs and vice versa.

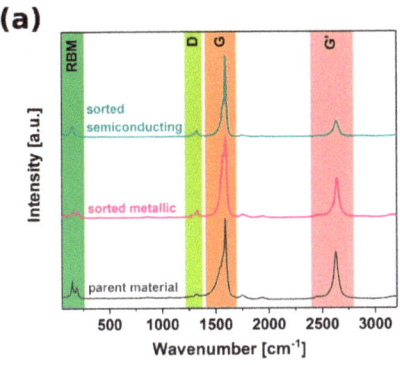

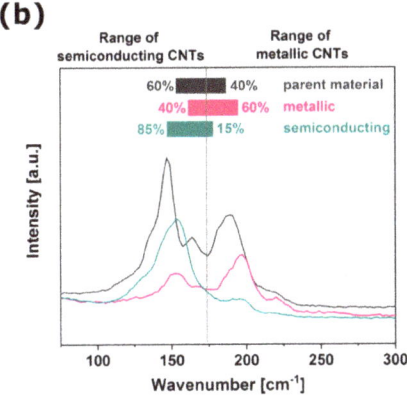

Figure 8. Raman spectra: (**a**) Overview of all the peaks, (**b**) magnification of RBM area.

3.8. Electrical Conductivity of Sorted CNT Films

Finally, we wanted to gauge the electrical performance of these materials (Figure 9). To accomplish this goal, we created thin free-standing films from them and characterized how their electrical conductivity (normalized to unity to eliminate uncertainties in the determination of samples dimensions) responds to doping. BF_3 was selected as a doping agent because of its very strong influence on the electrical conductivity of CNTs [25]. Addition of BF_3 in methanol to the film from as-made unsorted SWCNTs caused almost a five-fold increase in electrical conductivity. Very similar results were obtained in the case of the semiconducting-rich film, whereas the film composed predominantly of metallic CNTs experienced only a two-fold reduction in electrical resistance.

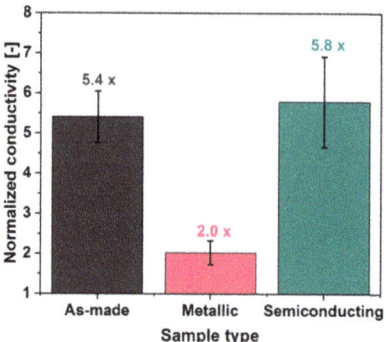

Figure 9. The influence of BF_3 on the electrical conductivity of unsorted and sorted CNT films.

This proves our earlier suspicion that the large-diameter SWCNT material Tuball™ is of predominantly semiconducting character [25]. BF_3 improves electrical conductivity most probably by introducing an impurity band to the density of states (DOS) similarly to the action of interhalogen compounds [1], which leads to lowering the bandgap of semiconducting CNTs. The reason why semiconducting-rich CNT films did not show an even larger increase in conductivity (as compared with unsorted samples) may be explained by the fact that the electronic transport in these materials is often limited by the contact resistance between individual CNTs, or caused by other extrinsic factors such as CNT misalignment or presence of impurities, as recently reported by Bulmer et al. [26].

On the other hand, electrical conductivity of metallic-rich CNT films doubled. As indicated by the Raman spectra, these films are adulterated with a minor amount of semiconducting CNTs, which can respond to BF_3 doping. This suggests that the percolation pathway in this case may not be constructed exclusively from metallic CNTs, but "bridges" exist, which are made of semiconducting CNTs. An increase in electrical conductivity of these semiconducting CNTs could then be responsible for the increase in the overall electrical conductivity of the network. One could also consider that the newly introduced impurity band increases the density of states near the Fermi energy, which is otherwise low for intrinsic metallic CNTs. Doping could then increase the Fermi energy beyond M_{11} and improve the electrical conductivity of the network composed predominantly of metallic CNTs.

It should also be considered that, when the doping agent is introduced to a CNT by using a volatile solvent as a vector, it often causes some densification of the material [1]. As a consequence, contact between the constituting CNTs is improved and the resistance is reduced. This could explain why the conductivity of metallic-rich CNT films improved upon exposure of the material to BF_3. In such a scenario, improvement to the electrical conductivity is caused by two factors: Reduction of intrinsic electrical resistance of semiconducting CNTs and optimization of geometrical factors, which alleviate the problem of contact resistance (the extrinsic contribution to the electrical transport).

4. Conclusions

We presented how large-diameter SWCNTs can be sorted according to the electrical character by using the ATPE method. Introduction of hydration modulators (H_2O_2 and PEGme) very much improved the resolution of the one-step system, which we developed. The obtained fractions were predominantly metallic or semiconducting, as observed by both absorbance and Raman spectroscopy. After isolation of these species from the corresponding matrices using our approach based on precipitation and hydrolysis, we manufactured thin free-standing films from them. Although surfactant was still detected on the surface of CNTs, this technique was more rapid for removal of Dextran or PEG than common hydrolysis. The doping tests indicated the way the electrical charge is transported through such (un)sorted CNT ensembles. In parallel, the electrical experiments manifested once again the problem of contact resistance—one of the most serious issues for CNT fibers and films, which must be taken care of to implement them in real life.

Supplementary Materials: The following are available online at http://www.mdpi.com/2079-4991/9/4/614/s1, Detailed description of parameters used for the preparation of the material (CNT dispersion and hydration modulator solutions), Figure S1: Raman spectra of the parent CNT material before and after sonication, Figure S2: Raman spectra of CNT dispersions acquired at 633 nm and 780 nm laser excitation wavelength, Table S1: ATPE parameters for the experiments with H_2O_2 as hydration modulator, Table S2: ATPE parameters for the experiments with PEG as hydration modulator, Figure S3: Absorbance spectra for the experiments with $NaBH_4$ as hydration modulator, Table S3: ATPE parameters for the experiments with $NaBH_4$ as hydration modulator, Figure S4: Absorbance spectra for the experiments with $K_2S_2O_8$ as hydration modulator, Table S4: ATPE parameters for the experiments with $K_2S_2O_8$ as hydration modulator, Figure S5: Absorbance spectra for the experiments with NaClO as hydration modulator, Table S5: ATPE parameters for the experiments with NaClO as hydration modulator, Figure S6: Absorbance spectra for the experiments with ethylene glycol as hydration modulator, Table S6: ATPE parameters for the experiments with ethylene glycol as hydration modulator, Figure S7: Absorbance spectra for the experiments with alanine as hydration modulator, Table S7: ATPE parameters for the experiments with alanine as hydration modulator, Figure S8: Absorbance spectra for the experiments with diethanolamine as hydration modulator, Table S8: ATPE parameters for the experiments with diethanolamine as hydration modulator, Figure S9: Absorbance spectra for the experiments with ethylenediaminetetraacetic acid as hydration modulator, Table S9: ATPE parameters for the experiments with ethylenediaminetetraacetic acid as hydration modulator, Figure S10: Absorbance spectra for the experiments with urea as hydration modulator, Table S10: ATPE parameters for the experiments with urea as hydration modulator, Figure S11: Absorbance spectra for the experiments with *N,N*-dimethylformamide as hydration modulator, Table S11: ATPE parameters for the experiments with *N,N*-dimethylformamide as hydration modulator, Figure S12: Absorbance spectra for the experiments with phthalimide as hydration modulator, Table S12: ATPE parameters for the experiments with phthalimide as hydration modulator, Figure S13: Absorbance spectra for the experiments with imidazole as hydration modulator, Table S13: ATPE parameters for the experiments with imidazole as hydration modulator, Figure S14: Absorbance spectra for the experiments with thiourea as hydration modulator, Table S14: ATPE parameters for the experiments with thiourea as hydration modulator, Figure S15: Absorbance spectra for the experiments with thioacetamide as hydration modulator, Table S15: ATPE parameters for the experiments with thioacetamide as hydration modulator, Figure S16: Absorbance spectra for the experiments with β-cyclodextrin as hydration modulator, Table S16: ATPE parameters for the experiments with β-cyclodextrin as hydration modulator, Figure S17: Absorbance spectra for the experiments with polyvinylpyrrolidone as hydration modulator, Table S17: ATPE parameters for the experiments with polyvinylpyrrolidone as hydration modulator.

Author Contributions: D.J. conceived and designed the experiments; E.T., B.K., and D.J. performed the experiments; E.T., B.K., S.B., and D.J. analyzed the data; S.B. and D.J. contributed reagents/materials/analysis tools; E.T., B.K., S.B., and D.J. wrote the paper.

Funding: E.T. and D.J. thank National Science Center, Poland (under the Polonez program, grant agreement UMO-2015/19/P/ST5/03799) and the European Union's Horizon 2020 research and innovation programme (Marie Skłodowska-Curie grant agreement 665778). D.J. would also like to acknowledge the Ministry for Science and Higher Education for the scholarship for outstanding young scientists (0388/E-367/STYP/12/2017). B.K. thanks National Center for Research and Development, Poland (under the Leader program, grant agreement LIDER/1/0001/L-8/16/NCBR/2017). S.B. greatly acknowledges the support from the Silesian University of Technology in the framework of Rector's Professorial Grant: RGP 04/020/RGP18/0072.

Conflicts of Interest: The authors declare no conflict of interest.

References

1. Janas, D.; Milowska, K.Z.; Bristowe, P.D.; Koziol, K.K. Improving the electrical properties of carbon nanotubes with interhalogen compounds. *Nanoscale* **2017**, *9*, 3212–3221. [CrossRef] [PubMed]
2. Zhang, M.; Li, J. Carbon nanotube in different shapes. *Mater. Today* **2009**, *12*, 12–18. [CrossRef]
3. Kumanek, B.; Janas, D. Thermal conductivity of carbon nanotube networks—Review. *J. Mater. Sci.* **2019**, *54*, 7397–7427. [CrossRef]
4. Mir, M.; Ebrahimnia-Bajestan, E.; Niazmand, H.; Mir, M. A novel approach for determining thermal properties of single-walled carbon nanotubes. *Comput. Mater. Sci.* **2012**, *63*, 52–57. [CrossRef]
5. Kataura, H.; Kumazawa, Y.; Maniwa, Y.; Umezu, I.; Suzuki, S.; Ohtsuka, Y.; Achiba, Y. Optical properties of single-wall carbon nanotubes. *Synth. Met.* **1999**, *103*, 2555–2558. [CrossRef]
6. Sfeir, M.Y.; Beetz, T.; Wang, F.; Huang, L.M.; Huang, X.M.H.; Huang, M.Y.; Hone, J.; O'Brien, S.; Misewich, J.A.; Heinz, T.F.; et al. Optical spectroscopy of individual single-walled carbon nanotubes of defined chiral structure. *Science* **2006**, *312*, 554–556. [CrossRef] [PubMed]

7. Janas, D.; Czechowski, N.; Krajnik, B.; Mackowski, S.; Koziol, K.K. Electroluminescence from carbon nanotube films resistively heated in air. *Appl. Phys. Lett.* **2013**, *102*, 181104. [CrossRef]
8. Janas, D. Towards monochiral carbon nanotubes: A review of progress in the sorting of single-walled carbon nanotubes. *Mater. Chem. Front.* **2018**, *2*, 36–63. [CrossRef]
9. Yang, D.H.; Hu, J.W.; Liu, H.P.; Li, S.L.; Su, W.; Li, Q.; Zhou, N.G.; Wang, Y.C.; Zhou, W.Y.; Xie, S.S.; et al. Structure Sorting of Large-Diameter Carbon Nanotubes by NaOH Tuning the Interactions between Nanotubes and Gel. *Adv. Funct. Mater.* **2017**, *27*, 1700278. [CrossRef]
10. He, X.W.; Htoon, H.; Doorn, S.K. Tunable Room-Temperature Single-Photon Emission at Telecom Wavelengths from sp3 Defects in Carbon Nanotubes. *Nat. Photonics* **2017**, *11*, 577. [CrossRef]
11. Graf, A.; Zakharko, Y.; Schiessl, S.P.; Backes, C.; Pfohl, M.; Flavel, B.S.; Zaumseil, J. Large scale, selective dispersion of long single-walled carbon nanotubes with high photoluminescence quantum yield by shear force mixing. *Carbon* **2016**, *105*, 593–599. [CrossRef]
12. Tulevski, G.S.; Franklin, A.D.; Frank, D.; Lobez, J.M.; Cao, Q.; Park, H.; Afzali, A.; Han, S.J.; Hannon, J.B.; Haensch, W. Toward High-Performance Digital Logic Technology with Carbon Nanotubes. *ACS Nano* **2014**, *8*, 8730–8745. [CrossRef] [PubMed]
13. Chen, Z.H.; Appenzeller, J.; Knoch, J.; Lin, Y.M.; Avouris, P. The role of metal-nanotube contact in the performance of carbon nanotube field-effect transistors. *Nano Lett.* **2005**, *5*, 1497–1502. [CrossRef]
14. He, X.; Gao, W.; Xie, L.; Li, B.; Zhang, Q.; Lei, S.; Robinson, J.M.; Haroz, E.H.; Doorn, S.K.; Wang, W.; et al. Wafer-scale monodomain films of spontaneously aligned single-walled carbon nanotubes. *Nat. Nanotechnol.* **2016**, *11*, 633–638. [CrossRef] [PubMed]
15. Turek, E.; Shiraki, T.; Shiraishi, T.; Shiga, T.; Fujigaya, T.; Janas, D. Single-step isolation of carbon nanotubes with narrow-band light emission characteristics. *Sci. Rep.* **2019**, *9*, 535. [CrossRef]
16. Gui, H.; Streit, J.K.; Fagan, J.A.; Walker, A.R.H.; Zhou, C.W.; Zheng, M. Redox Sorting of Carbon Nanotubes. *Nano Lett.* **2015**, *15*, 1642–1646. [CrossRef] [PubMed]
17. Janas, D.; Rdest, M.; Koziol, K.K.K. Free-standing films from chirality-controlled carbon nanotubes. *Mater. Des.* **2017**, *121*, 119–125. [CrossRef]
18. Clancy, A.J.; White, E.R.; Tay, H.H.; Yau, H.C.; Shaffer, M.S.P. Systematic comparison of conventional and reductive single-walled carbon nanotube purifications. *Carbon* **2016**, *108*, 423–432. [CrossRef]
19. Fagan, J.A.; Haroz, E.H.; Ihly, R.; Gui, H.; Blackburn, J.L.; Simpson, J.R.; Lam, S.; Walker, A.R.H.; Doorn, S.K.; Zheng, M. Isolation of > 1 nm Diameter Single-Wall Carbon Nanotube Species Using Aqueous Two-Phase Extraction. *ACS Nano* **2015**, *9*, 5377–5390. [CrossRef]
20. Souza, A.G.; Chou, S.G.; Samsonidze, G.G.; Dresselhaus, G.; Dresselhaus, M.S.; An, L.; Liu, J.; Swan, A.K.; Unlu, M.S.; Goldberg, B.B.; et al. Stokes and anti-Stokes Raman spectra of small-diameter isolated carbon nanotubes. *Phys. Rev. B* **2004**, *69*, 115428. [CrossRef]
21. Fantini, C.; Jorio, A.; Souza, M.; Strano, M.S.; Dresselhaus, M.S.; Pimenta, M.A. Optical transition energies for carbon nanotubes from resonant Raman spectroscopy: Environment and temperature effects. *Phys. Rev. Lett.* **2004**, *93*, 147406. [CrossRef]
22. Goran, J.M.; Phan, E.N.H.; Favela, C.A.; Stevenson, K.J. H_2O_2 Detection at Carbon Nanotubes and Nitrogen-Doped Carbon Nanotubes: Oxidation, Reduction, or Disproportionation? *Anal. Chem.* **2015**, *87*, 5989–5996. [CrossRef]
23. Miao, Z.Y.; Zhang, D.; Chen, Q. Non-enzymatic Hydrogen Peroxide Sensors Based on Multi-wall Carbon Nanotube/Pt Nanoparticle Nanohybrids. *Materials* **2014**, *7*, 2945–2955. [CrossRef] [PubMed]
24. Xie, A.J.; Liu, Q.X.; Ge, H.L.; Kong, Y. Novel H_2O_2 electrochemical sensor based on graphene-polyacrylamide composites. *Mater. Technol.* **2015**, *30*, 50–53. [CrossRef]
25. Janas, D. Powerful doping of chirality-sorted carbon nanotube films. *Vacuum* **2018**, *149*, 48–52. [CrossRef]
26. Bulmer, J.S.; Gspann, T.S.; Orozco, F.; Sparkes, M.; Koerner, H.; Di Bernardo, A.; Niemiec, A.; Robinson, J.W.A.; Koziol, K.K.; Elliott, J.A.; et al. Photonic Sorting of Aligned, Crystalline Carbon Nanotube Textiles. *Sci. Rep.* **2017**, *7*, 12977. [CrossRef] [PubMed]

© 2019 by the authors. Licensee MDPI, Basel, Switzerland. This article is an open access article distributed under the terms and conditions of the Creative Commons Attribution (CC BY) license (http://creativecommons.org/licenses/by/4.0/).

Article

Improvement of Ethanol Gas-Sensing Responses of ZnO–WO$_3$ Composite Nanorods through Annealing Induced Local Phase Transformation

Yuan-Chang Liang [1,*] and Che-Wei Chang [2]

1. Institute of Materials Engineering, National Taiwan Ocean University, Keelung 20224, Taiwan
2. Undergraduate Program in Optoelectronics and Materials Technology, National Taiwan Ocean University, Keelung 20224, Taiwan; jf860218@gmail.com
* Correspondence: yuanvictory@gmail.com

Received: 28 March 2019; Accepted: 22 April 2019; Published: 30 April 2019

Abstract: In this study, ZnO–WO$_3$ composite nanorods were synthesized through a combination of hydrothermal growth and sputtering method. The structural analysis results revealed that the as-synthesized composite nanorods had a homogeneous coverage of WO$_3$ crystallite layer. Moreover, the ZnO–WO$_3$ composite nanorods were in a good crystallinity. Further post-annealed the composite nanorods in a hydrogen-containing atmosphere at 400 °C induced the local phase transformation between the ZnO and WO$_3$. The ZnO–WO$_3$ composite nanorods after annealing engendered the coexistence of ZnWO$_4$ and WO$_3$ phase in the shell layer which increased the potential barrier number at the interfacial contact region with ZnO. This further enhanced the ethanol gas-sensing response of the pristine ZnO–WO$_3$ composite nanorods. The experimental results herein demonstrated a proper thermal annealing procedure of the binary composite nanorods is a promising approach to modulate the gas-sensing behavior the binary oxide composite nanorods.

Keywords: sputtering; composite nanorods; phase transformation; annealing

1. Introduction

Composite nanorod systems composed of various binary semiconductor oxides have been shown a promising approach to enhance the gas-sensing properties of the constituent compounds. For example, hydrothermally derived flower-like CeO$_2$–SnO$_2$ composites exhibit improved trimethylamine gas-sensing response than that of the pristine SnO$_2$ [1]. Moreover, hydrogen-sensing properties of ZnO nanofibers are significantly enhanced through NiO loading in a composite structure [2]. SnO$_2$/ZnO hetero-nanofibers demonstrate improved acetone gas-sensing responses in comparison with that of the pristine ZnO nanofibers [3]. By tuning the sputtering coated VO$_x$ morphology on the one-dimensional ZnO, ZnO–VO$_x$ composites demonstrate improved oxidizing gas-sensing responses than that of the pristine ZnO [4].

Among various binary oxides, ZnO is one of the most studied n-type semiconductor oxides which was widely used for gas-sensing material because of its low cost, high chemical stability, and versatile preparation methods. Furthermore, ZnO in a low-dimensional structure is of potential interest for gas sensor device applications because its high specific surface area enables efficient reaction between the oxide surface and target gas molecules [5,6]. However, developing high gas-sensing responses of the ZnO nanostructures toward various target gases is still highly desired and is technically challenging. Various ZnO-based composite systems incorporated with another binary oxide have been proposed to improve the ZnO gas-sensing properties based on the aforementioned demand [2,4,7]. By contrast, WO$_3$ is another promising gas-sensing binary oxide. It also has advantages of low cost, high chemical stability, and excellent process-dependent reproducibility. Recent progress has shown that WO$_3$ with various morphologies is promising in applications of gas sensors to detect toxic gases [8–10]. Moreover, WO_3 crystals can be synthesized through various physical and chemical methods and the crystalline quality and morphology can be easy controlled through varying the process conditions. [11,12]. Although various ZnO-based composite systems have been proposed to improve the gas-sensing responses to target gases, the reports on construction of one-dimensional ZnO–WO_3 composite system are still limited in number. Moreover, thermal annealing of solid materials is an efficient method to modulate their microstructures and to control their physical and chemical properties [13,14]. The past research works show that conducting proper thermal annealing procedures causes the possible solid-state reaction between the constituent oxides in a nanoscale in low-dimensional oxide systems [15,16]. The existence of phase transformation between the binary oxides in a low-dimensional composite system modifies the electric properties of the original composite systems without thermal annealing procedures. This broadens the design of the functionality and changes property performance of the pristine oxide composite system.

2. Materials and Methods

In this study, ZnO-based composite nanorods coated with the WO_3 and $ZnWO_4$ shell layers were synthesized through a combinational methodology of hydrothermal growth and sputtering. Hydrothermally synthesized high-density ZnO nanorods were used as templates for growing the ZnO-based composite nanorods. The hydrothermal growth reactions of the ZnO nanorods were conducted at 95 °C for 9 h. The detailed process procedures were reported elsewhere [14]. During sputtering growth of the WO_3 shell layer, the sputtering power of tungsten metallic target was fixed at 80 W. The thin-film growth temperature was maintained at 375 °C with an Ar/O_2 ratio of 3:2. Then, the as-synthesized ZnO–WO_3 composite nanorods were subsequently annealed in a 95% N_2/5% H_2 atmosphere for 20 min at the temperatures of 400~500 °C to induce a solid-state reaction between the ZnO surface and WO_3 ultra-thin layer.

The scanning electron microscopy (SEM; Hitachi S-4800, Tokyo, Japan) was used to investigate the surface morphology of nanorod samples. X-ray diffraction (XRD; Bruker D2 PHASER, Karlsruhe, Germany) was further used to investigate crystallographic structures of the samples. Moreover, the detailed microstructures of the composite nanorods with and without the thermal annealing were characterized by high-resolution transmission electron microscopy (HRTEM; Philips Tecnai F20 G2, Amsterdam, The Netherland). X-ray photoelectron spectroscope (XPS; PHI 5000 VersaProbe, Chigasaki, Japan) analysis was performed to determine the chemical binding states of the constituent elements of the composite nanorods. Silver contact electrodes were formed on the surface of the nanorod samples for gas-sensing measurements. The ethanol vapor concentrations of 25–500 ppm were used as target gas. The dry air was used as carrier gas herein. The variation of sensor resistance before and after introducing ethanol vapor was recorded. The gas-sensing response of the sensors to ethanol vapor herein is defined as the R_a/R_g, which is the electric resistance ratio of the resistance of the gas sensor in the absence of ethanol vapor to the resistance of the sensor in ethanol vapor.

3. Results

The crystallographic structures of the as-prepared ZnO nanorods, ZnO–WO_3 nanorods with and without thermal annealing at 400 °C were identified using XRD measurements (Figure 1). Figure 1a shows the XRD pattern of the ZnO nanorods which were used as a template for preparing various composite nanorods. The Bragg reflections in Figure 1a demonstrate that the ZnO nanorods has a hexagonal wurtzite structure and exhibits preferred (002) orientation (JCPDS no. 005-0664). The XRD pattern of the ZnO nanorods in situ sputtering coated with WO_3 thin films was exhibited in Figure 1b. The Bragg reflections centered approximately 24.5° and 34.2° are ascribed to (001) and (201) crystallographic planes of orthorhombic WO_3 (JCPDS no. 20-1324), respectively. Well crystalline ZnO–WO_3 composite nanorods were successfully formed via sputtering WO_3 thin films onto the surfaces of the ZnO nanorods. Figure 1c shows the XRD patterns of the ZnO–WO_3 nanorods annealed at

400 °C. Figure 1c shows that the WO$_3$ peaks are disappeared completely after the annealing procedure and several new diffraction peaks are observed at approximately 19.6°, 24.5°, 25.3° and 31.3° which are assigned to (100), (011), (110) and (111) of monoclinic ZnWO$_4$ (JCPDS no. 15-0774). The XRD result transformation of ultra-thin WO$_3$ thin film with the ZnO into the ternary ZnWO$_4$ phase occurred after the thermal annealing procedure. M. Bonanni et al. reported that the solid-state reaction between ZnO and WO$_3$ starts to develop at 350 °C [17]. The annealing temperature of 400 °C herein might have a sufficient thermal energy to activate the phase transformation between the WO$_3$ and ZnO of the ZnO–WO$_3$ composite nanorods. The XRD results revealed the ZnWO$_4$ layers with a polycrystalline feature were formed on the surfaces of residual ZnO nanorods after the annealing procedure.

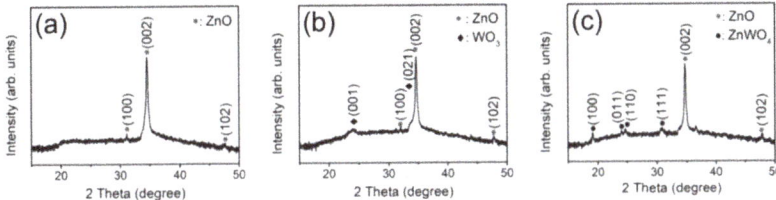

Figure 1. X-ray diffraction (XRD) patterns: (**a**) ZnO nanorods, (**b**) ZnO–WO$_3$ composite nanorods, (**c**) ZnO–WO$_3$ composite nanorods annealed at 400 °C.

Figure 2 shows SEM images of the ZnO, ZnO–WO$_3$ nanorods with and without annealing at 400 °C. Figure 2a shows a typical SEM image of as-synthesized ZnO nanorods, revealing a hexagonal crystal featured cross-section of the ZnO nanorods with a diameter in the range of 85–100 nm. The surface of the ZnO nanorods was smooth. Figure 2b presents the SEM image of ZnO nanorods after sputtering coated with ultra-thin WO$_3$ thin film. Compared to the bare ZnO nanorods, the surface feature of the ZnO–WO$_3$ composite nanorods became more rugged when the WO$_3$ crystallites were decorated onto the surfaces of the ZnO nanorods. Notably, the hexagonal cross-sectioned morphology of the ZnO nanorods was maintained after coating the WO$_3$ thin film, revealing the deposition of the ultra-thin WO$_3$ layer on the ZnO nanorods' surfaces. Figure 2c shows the ZnO–WO$_3$ composite nanorods annealed at 400 °C. The ZnO–WO$_3$ composite nanorods annealed at 400 °C did not exhibit substantial morphology change. The composite nanorods maintained a visible hexagonal cross-sectional crystal feature. By contrast, a high solid-state reaction temperature above 600 °C involves the marked surface roughening process of oxide composite nanorods; this was observed in phase transformation of other oxide nanocomposite systems [16].

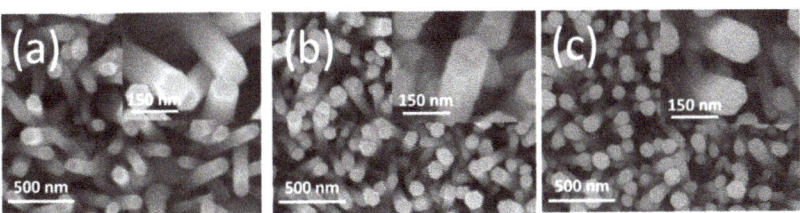

Figure 2. Scanning electron microscopy (SEM) micrographs: (**a**) ZnO nanorods, (**b**) ZnO–WO$_3$ composite nanorods, (**c**) ZnO–WO$_3$ composite nanorods annealed at 400 °C.

Figure 3a shows a low-magnification TEM image of a ZnO nanorod coated with a thin WO$_3$ layer. The surface of the composite nanorod exhibited an uneven feature. Figure 3b,c demonstrate high-resolution TEM (HRTEM) images of a ZnO–WO$_3$ composite nanorod taken from the different positions at the WO$_3$/ZnO interface. Figure 3b reveals an ultra-thin WO$_3$ layer covered on the surface of the nanorod and the interface between the ZnO nanorod and WO$_3$ layer is abrupt. By contrast in Figure 3c, tiny, nanoscaled surface bumps appeared on the surface of the composite nanorod. This might engender the uneven surface feature of the composite nanorod. The sputtering growth of binary oxides at an elevated temperature is likely to form island- or bump-like crystals on the hetero-substrates [4,7]. The in situ sputtering growth of WO_3 crystals onto the surfaces of the ZnO nanorods at 375 °C herein might cause locally inhomogeneous crystal growth and formed WO_3 bumps on the ZnO nanorods. The ordered lattice fringes in the outer region of the composite nanorod revealed the coverage of well-crystallized WO_3 crystals on the surface of the nanorod. The lattice fringe spacing of approximately 0.38 nm corresponds to the interplanar distance of orthorhombic WO_3 (001). Furthermore, the lattice fringe spacing of approximately 0.26 nm in the figures demonstrated the interplanar distance of hexagonal ZnO (002). The selected area electron diffraction (SAED) pattern taken from several ZnO–WO_3 composite nanorods revealed the crystalline feature and a composite structure of the hexagonal ZnO nanorods sputtering coated with the orthorhombic WO_3 thin film. Furthermore, elemental line-scan profiles of Zn, W, and O elements across the ZnO–WO_3 composite nanorod were displayed in Figure 3e. The line-scan profiles revealed that the W element was well distributed on the surface of the ZnO nanorod, revealing a formation of the compositionally defined composite structure of core-ZnO and shell-WO_3.

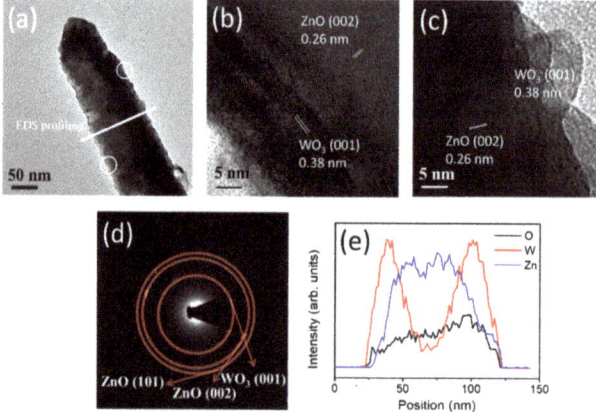

Figure 3. Transmission electron microscopy (TEM) analysis of the ZnO–WO_3 composite nanorod: (a) low-magnification image, (b,c) high-resolution images from local regions, (d) selected area electron diffraction (SAED) pattern, (e) line-scan profiling spectra across the composite nanorod.

Figure 4a shows a low-magnification TEM image of a ZnO–WO_3 composite nanorod with a thermal annealing at 400 °C. HRTEM images of the composite taken from various interfacial regions are demonstrated in Figure 4b,c. In Figure 4b, well-ordered and long-range arrangement of lattice fringes appear at the outer region of the composite nanorod. Moreover, the ordered lattice fringes arranged in the other orientation were found in the inner region of the composite nanorod. The lattice fringes with a spacing of approximately 0.36 nm in the outer region of the composite nanorod were attributed to the interplanar distance of monoclinic $ZnWO_4$ (110). By contrast, the lattice fringes with a spacing of 0.26 nm in the inner region of the composite nanorod were assigned to the interplanar distance of hexagonal ZnO (002). Figure 4b reveals that the WO_3 phase in the outer region of the ZnO–WO_3 composite nanorod transforms into a ternary phase of $ZnWO_4$ through a solid-state reaction process during the post-annealing procedure in this study. The $ZnWO_4$ crystals exhibited a good crystalline feature on the outer region of the composite nanorod and the interface of the $ZnWO_4$ and ZnO phase was sharp. However, the mixed lattice fringes arrangements were observed in the outer region of the composite nanorod in Figure 4c. In addition to the ordered lattice fringes which originated from the ZnWO4 (110) as indexed in the figure, some local region demonstrated that the lattice fringes were arranged in a slightly chaotic state. This revealed the presence of crystalline ZnWO$_4$ and deteriorated WO$_3$ crystals in the outer region of the composite nanorod. The observation of the TEM analysis demonstrated that most WO$_3$ crystals transformed into crystalline ZnWO$_4$ after annealing; whereas, partial WO$_3$ crystals did not yield the phase transformation with the ZnO due to the insufficient reaction condition. Moreover, the residual WO$_3$ phase region demonstrated the deteriorated crystallinity after annealing because of the presence of hydrogen in the annealing atmosphere. In conclusion, the composite nanorod is mainly composed of crystalline ZnWO$_4$ phase with a larger range and spatially distributed residual WO$_3$ phase in a smaller content in the outer layer. The SAED pattern taken from the several composite nanorods in Figure 4d indicate the various groups of diffraction rings, suggesting the presence of crystalline ZnO and ZnWO$_4$ in the composite nanorods. The elemental mapping images were further used to analyze the distribution of Zn, W, and O in a single ZnO–WO$_3$ composite nanorod treated with a thermal annealing at 400 °C (Figure 4e). The EDS mapping images clearly identified the spatial distributions of Zn, W, and O in the composite structure. The Zn and O elements existed the entire area of the nanorod. In particular, W is the main element distributed in the outer shell of the composite nanorod. TEM results indicated that a solid-state reaction occurred at the interface of WO$_3$/ZnO and formed a new shell layer phase of ZnWO$_4$ on the surface of the residual ZnO core in the composite nanorod. Similar solid-state reaction between the different binary oxides in low-dimensional systems to form a ternary phase in the outer region of the composite nanorods has also been reported in ZnO–SnO$_2$ and ZnO–TiO$_2$ [16,18].

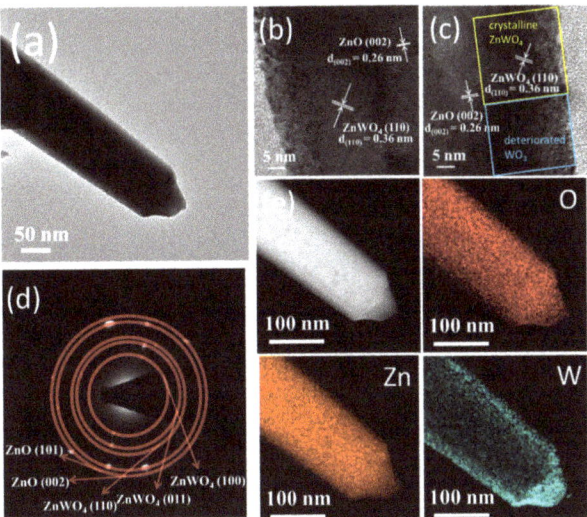

Figure 4. TEM analysis of the ZnO–WO$_3$ composite nanorod annealed at 400 °C: (**a**) low-magnification image, (**b**,**c**) high-resolution images from local regions, (**d**) SAED pattern, (**e**) elemental mapping images of the nanorod.

The elemental binding states of the ZnO–WO$_3$ composite with and without a thermal annealing at 400 °C were investigated by XPS. The narrow spectra of Zn 2p, W 4f, and O 1s for the ZnO–WO$_3$ composite nanorods are recorded in Figure 5a–c. Zn 2p spectrum of the composite nanorods in Figure 5a shows two peaks at approximately 1044.8 eV and 1021.8 eV which are respectively attributed to Zn 2p$_{1/2}$ and Zn 2p$_{3/2}$ and suggest the presence of Zn^{2+} ions in the oxide [16]. Moreover, the W4 f spectrum (Figure 5b) of pristine ZnO–WO$_3$ nanorods consisted of two spin-orbit doublets corresponding to the different valence states of tungsten. The bigger doublet located at 37.4 eV and 35.3 eV is assigned to W 4f$_{5/2}$ and W 4f$_{7/2}$ of W^{6+}, respectively and the smaller one is allocated to

W^{5+} [11]. No metallic W component was detected from the sample. The asymmetric O 1s spectrum of ZnO–WO$_3$ composite nanorods was displayed in Figure 5c and that spectrum was deconvoluted into three subpeaks at approximately 530.1 eV, 530.9 eV, and 531.9 eV, matching the oxygen coordination in lattice oxygen, vacancy oxygen, and surface chemisorbed oxygen, respectively [18,19]. By contrast, in Figure 5d, the Zn 2p spectrum of the ZnO–WO$_3$ composite nanorods with a thermal annealing at 400 °C exhibited a similar spectrum feature as exhibited in Figure 5a, revealing the divalent state of the zinc in the nanorods. Figure 5e shows the W4f spectrum of the composite nanorods annealed in a hydrogen-contained atmosphere. Notably, even annealed in oxygen deficient atmosphere, the metallic W component was not detected on the surfaces of the composite nanorods under the given annealing condition. Comparatively, the W 4f spectrum with the deconvoluted peaks in Figure 5e exhibited that the ZnO–WO$_3$ composite nanorods with the thermal annealing procedure demonstrated the area ratio of W^{5+} spin-orbit doublet becomes larger, resulting from the existence of the crystal deterioration region in the shell oxide layer. The O1s spectrum of the corresponding sample was shown in Figure 5f and was further used to explain the W4f XPS result. The O 1s spectrum from the composite nanorods with a thermal annealing process showed a marked intensity decrease in the lattice oxygen subpeak and a relative intensity rise in the subpeaks associated with oxygen vacancy and chemisorbed oxygen, compared with those from the ZnO–WO$_3$ composite nanorods without a thermal annealing. The concentration of oxygen vacancies has a direct relationship with the state of the oxide's crystallinity. A similar phenomenon of increased oxygen vacancies in oxides annealed in hydrogen-contained atmosphere has been proposed in previous works [20]. A higher degree of oxygen deficiency in the outer region of the composite nanorods engendered a larger content of W^{5+} in the tungsten-based oxides of the composite nanorods. The XPS results herein demonstrated that the annealing temperature of 400 °C for the ZnO–WO$_3$ composite nanorods engendered more oxygen vacancies in the surfaces; the tungsten was still in an oxide binding status without reducing to the metallic binding form.

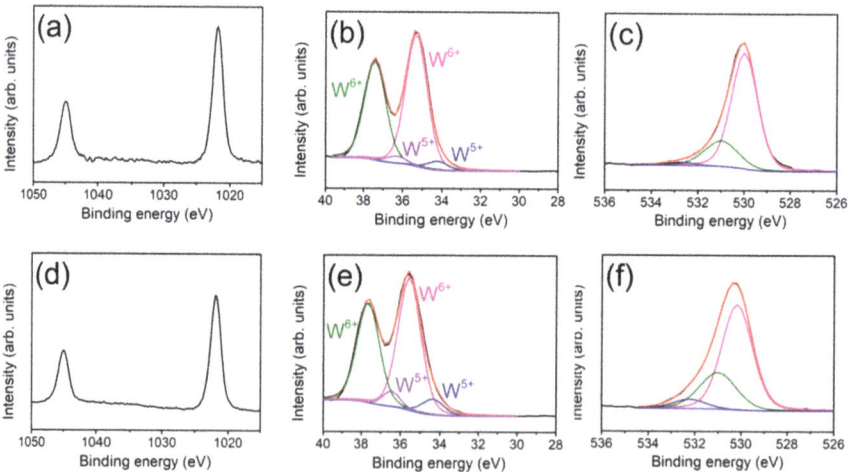

Figure 5. X-ray photoelectron spectroscope (XPS) analysis of the ZnO–WO$_3$ composite nanorods: (a) Zn 2p, (b) W 4f, (c) O1s. XPS analysis of the ZnO–WO$_3$ composite nanorods after annealing: (d) Zn 2p, (e) W 4f, (f) O1s.

The optimal operating temperature with the highest gas-sensing response for the various nanorods was determined. The gas-sensing responses of all samples on exposure to 50 ppm ethanol vapor were measured at the temperature range of 200–325 °C (Figure 6a). For the ZnO nanorods, the gas-sensing response to 50 ppm ethanol vapor varied from 1.1 to 2.8 corresponding with the temperature from 200 to 325 °C (black curve in Figure 6a). The gas-sensing responses of ZnO–WO$_3$ composite nanorods were from 1.3 to 7.3 (red curve in Figure 6a), which was substantially higher than that of the ZnO nanorods at all tested temperatures. Moreover, the gas-sensing responses of ZnO–WO$_3$ composite nanorods annealed at 400 °C ranged from 2.1 to 16.2 (blue curve in Figure 6a), which demonstrated the highest response among the samples at the tested temperatures. Significantly, the nanorod sensors herein exhibited the maximum gas-sensing responses to ethanol at 300 °C, suggesting that a resultant equilibrium between surface reaction with ethanol vapor molecules and the diffusion of ethanol vapor molecules to the nanorods' surfaces occurred at 300 °C [9]. Figure 6b–d show the dynamic response curves of ZnO nanorods, ZnO–WO$_3$ nanorods, and ZnO–WO$_3$ nanorods annealed at 400 °C, respectively, exposed to 25–500 ppm ethanol vapor at the operating temperature of 300 °C. For comparison, gas-sensing tests of the ZnO–WO$_3$ composite nanorods annealed at 500 °C were also conducted to evaluate whether the higher annealing temperature improves the gas-sensing performance of the initially-synthesized ZnO–WO$_3$ composite nanorods (Figure 6e). All nanorod samples exhibited reversible and stable response and recovery behaviors during gas-sensing tests. The nanorod samples herein showed a typical n-type sensing behavior because of the n-type conduction nature of the constituent oxides. The gas-sensing responses of the sensors made from various nanorods on exposure to various ethanol concentrations were summarized in Figure 6f. The WO$_3$-decorated ZnO nanorod sensor exhibited much higher responses than the pristine ZnO. Furthermore, the ZnO–WO$_3$ composite nanorods annealed at 400 °C showed the highest response in all ethanol vapor concentrations. Notably, the sensing ability of the ZnO–WO$_3$ composite nanorods annealed at the higher temperature of 500 °C was substantially weakened. To confirm the possible reason for the deterioration of the gas-sensing ability, XPS measurements were conducted. Figure 6g demonstrates that the W4f spectrum include not only W^{6+} and W^{5+} but also W^{4+} and W^0 in the ZnO–WO$_3$ composite nanorods annealed at 500 °C. The subpeaks located at 33.4 eV and 35.5 eV are ascribed to tetravalent bond of tungsten and that at 31.1 eV was associated with the contribution of metallic tungsten [21,22]. The existence of mixed binding states of tungsten implied that substantial deoxidization of the WO$_3$ shell layer occurred during the high-temperature annealing in the hydrogen-contained atmosphere. The appearance of metallic W component revealed that the WO$_3$ is not in a pure oxide phase and this might deteriorate the gas-sensing performance of the n-type WO$_3$ oxide. A similar deoxidization of metal oxides annealed in the hydrogen-contained atmosphere at a high temperature has been proposed and resultant deteriorated electric properties of the metal oxides are involved [20,23]. The reproducibility and stability of the sensor made from the ZnO–WO$_3$ nanorods annealed at 400 °C were further examined at its optimum operating temperature of 300 °C to 50 ppm ethanol vapor concentration (Figure 6h). The difference obtained in the gas response values of the sensor after cycling tests was very small and hence negligible. This suggests that the proposed composite nanorod sensor showed good reproducibility to detect ethanol vapor. Figure 6i shows the selectivity of the gas sensor based on the ZnO–WO$_3$ nanorods annealed at 400 °C. The sensor was exposed to ammonia gas, ethanol vapor, nitrogen dioxide gas, and hydrogen gas of the appropriate concentrations at 300 °C, respectively. It can be seen that the sensor exhibited the substantially highest response to ethanol vapor, revealing its suitability for detecting ethanol vapor in the test environment. Table 1 compares the gas-sensing responses of various ZnO-based composites exposed to appropriate ethanol vapor concentrations at 300 °C [24–26]. The ZnO–WO$_3$ composite nanorods annealed at 400 °C in this study presented superior ethanol vapor detecting ability among the various reference works.

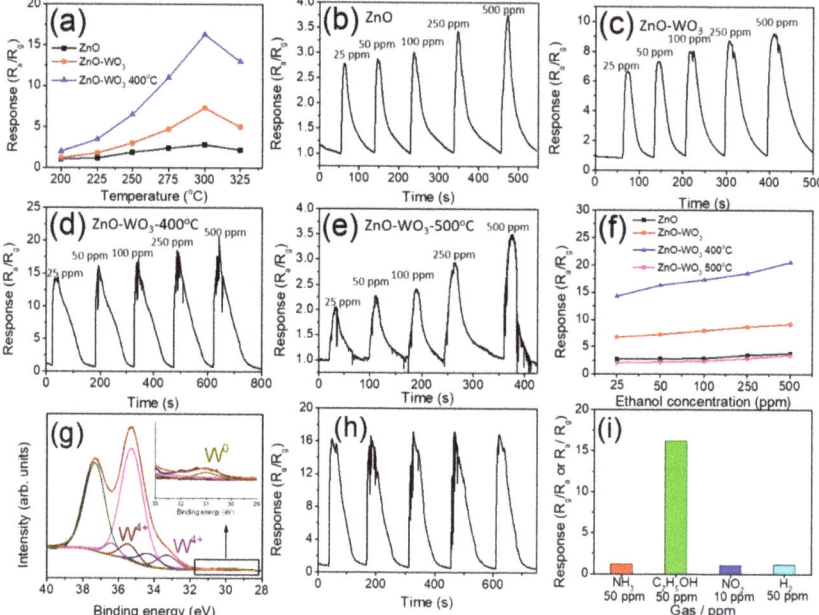

Figure 6. (a) Gas-sensing response vs. temperature curves of various nanorod sensors on exposure to 50 ppm ethanol vapor. Dynamic gas-sensing response curves of sensor on exposure to 25–500 ppm ethanol vapor: (b) ZnO nanorods, (c) ZnO–WO$_3$ composite nanorods, (d), ZnO–WO$_3$ composite nanorods annealed at 400 °C, (e) ZnO–WO$_3$ composite nanorods annealed 500 °C. (f) Gas-sensing response vs. ethanol vapor concentration curves. (g) XPS W 4f spectrum of the composite nanorods annealed at 500 °C. (h) Cycling testes of the composite nanorods annealed at 400 °C exposed to 50 ppm ethanol vapor. (i) Selectivity tests of the composite nanorods annealed at 400 °C.

Table 1. Comparison of ethanol gas-sensing responses of various ZnO-based heterogeneous sensors at 300 °C [24–26].

Materials	Preparation Method	Operating Temperature (°C)	Concentration (ppm)	Response
ZnO/NiO	Spray pyrolysis & Chemical bath	300	500	1.5
ZnO/SnO$_2$	Hydrothermal	300	50	11.2
ZnO/g-C$_3$N$_4$	Solvothermal treatment	300	100	9.2
ZnO/WO$_3$ annealed (this work)	Hydrothermal & sputtering	300	50	16.2

4. Discussion

The possible reasons caused various gas-sensing responses of the ZnO nanorods and various composite nanorods were further explained with the space-charge layer model [1,27]. In ambient air, oxygen molecules were absorbed on surfaces of the composite nanorods. These oxygen molecules become surface absorbed oxygen species (such as O$^-$$_{(ads)}$ and O^{2-}$_{(ads)}$) by capturing free electrons from the conducting bands of the oxides at the elevated sensor operating temperature of 300 °C. The reactions are described as follows:

$$O_2 \text{ (ambient)} \rightarrow O_{2(ads)} \text{ (n-type oxides)} \tag{1}$$

$$O_{2\,(ads)} + 2e^- \rightarrow 2O^-_{(ads)} \tag{2}$$

$$O^-_{(ads)} + e^- \rightarrow O^{2-}_{(ads)} \tag{3}$$

In this process, an electron depletion layer will be formed on the surfaces of the composite nanorods, resulting in a decrease of carrier concentration and an increase of sensor resistance. When the oxide nanorod sensor was exposed to ethanol vapor, the absorbed oxygen species will react with ethanol molecules according to the following possible reactions:

$$C_2H_5OH + 6O^- \rightarrow 2CO_2 + 3H_2O + 6e^- \tag{4}$$

$$C_2H_5OH + 6O^{2-} \rightarrow 2CO_2 + 3H_2O + 12e^- \tag{5}$$

As a result, the electrons trapped in the oxygen species are released back into the conduction band, leading to a decrease of the thickness of the depletion layer and the resistance of the oxides. In addition to the surface depletion layer, the contact of the different oxides in a one-dimensional heterostructure system engenders formation of interfacial depletion regions because of different work functions of the adjacent oxides. A proposed energy band structure diagram of the ZnO/WO$_3$ and ZnO/ZnWO$_4$ heterojunction herein are shown in Figure 7a [28,29]. The electrons will flow from ZnO nanorod to outer WO$_3$ (or ZnWO$_4$) crystals in the ZnO/WO$_3$ (or ZnO/ZnWO$_4$) heterostructures until their Fermi levels are equalized. Therefore, the exposure of the composite nanorods in air ambient result in formation of surface depletion regions in the surface WO$_3$ (or ZnWO$_4$) crystals and interfacial depletion regions inside the ZnO nanorod for the proposed various composite nanorods. This process creates an electron depletion layer on the surface of the ZnO core material and further bended the energy band and lead to a higher resistance of the composite nanorods with and without thermal annealing. The formation of additional interfacial depletion regions increased the potential barrier number in the composite nanorods than in the pristine ZnO nanorods; therefore, a larger resistance variation degree of the composite nanorods than that of the ZnO nanorods on exposure to the ethanol vapor is observed. This explained that the higher ethanol vapor sensing responses of the composite nanorods than those of the ZnO nanorods under the given gas-sensing tests herein. Similarly, a heterogeneous structure that improved the gas-sensing behavior of one-dimensional n-type oxide nanorods was demonstrated in ZnO–SnO$_2$, ZnO–Zn$_2$SnO$_4$, and TiO$_2$–CdO on exposure to test gases [16,19]. Comparatively, the ZnO–WO$_3$ nanorods with a thermal annealing at 400 °C substantially enhanced their ethanol gas-sensing responses at the given test conditions. This improvement of the gas-sensing response of the composite nanorods annealed at 400 °C might be attributed to the random thickness of the depletion layers of the surface and interface regions of the composite nanorods resulting from the local phase transformation of the ZnO–WO$_3$ composite system during the annealing process herein (Figure 7b). Compared to the chemically homogeneous shell layer of ZnO–WO$_3$ nanorods, the ZnO–WO$_3$ nanorods with a thermal annealing at 400 °C had a composite shell layer structure consisted of crystalline ZnWO$_4$ and deteriorated WO$_3$ as revealed in the structural analysis results. It is expected that the potential barrier number of the ZnO–WO$_3$ composite nanorods annealed at 400 °C was higher than that of the composite nanorods without a thermal annealing. Three types of depletion regions included surface depletions in the deteriorated WO$_3$ and crystalline ZnWO$_4$, depletion regions at WO$_3$/ZnWO$_4$ boundaries of the shell layer and interfacial depletion regions at the ZnO/WO$_3$ and ZnO/ZnWO$_4$ are expected to exist in the ZnO–WO$_3$ composite nanorods annealed at 400 °C. The local phase transformation of the ZnO/WO$_3$ after thermal annealing at 400 °C for the ZnO–WO$_3$ composite nanorod system in this study created a higher number of the potential barriers in the composite system as exhibited in Figure 7b. An increased potential barrier number in the composite systems has shown a substantial drop degree of the sensor resistance on exposure to the reducing gases and therefore this resulted in an enhanced gas-sensing response [3,9]. The cross-sectional potential barrier height variation alone the guided arrow red line in Figure 7b demonstrated that a more complex potential barrier height variation before and after introducing the ethanol vapor will be expected for the ZnO–WO$_3$ composite nanorods with an annealing procedure. Comparatively, further introducing the reducing vapor of ethanol into the test chamber, the injection of electrons from the adsorbed oxygen ions into the conduction bands of the constituent oxides of the composite nanorods caused a larger degree of resistance variation of the composite nanorods with thermal annealing because of their diverse microstructures in the composite system. The substantial microstructural differences in the ZnO–WO$_3$ composite nanorods with and without thermal annealing at 400 °C herein supported the different ethanol vapor sensing responses of various composite nanorods.

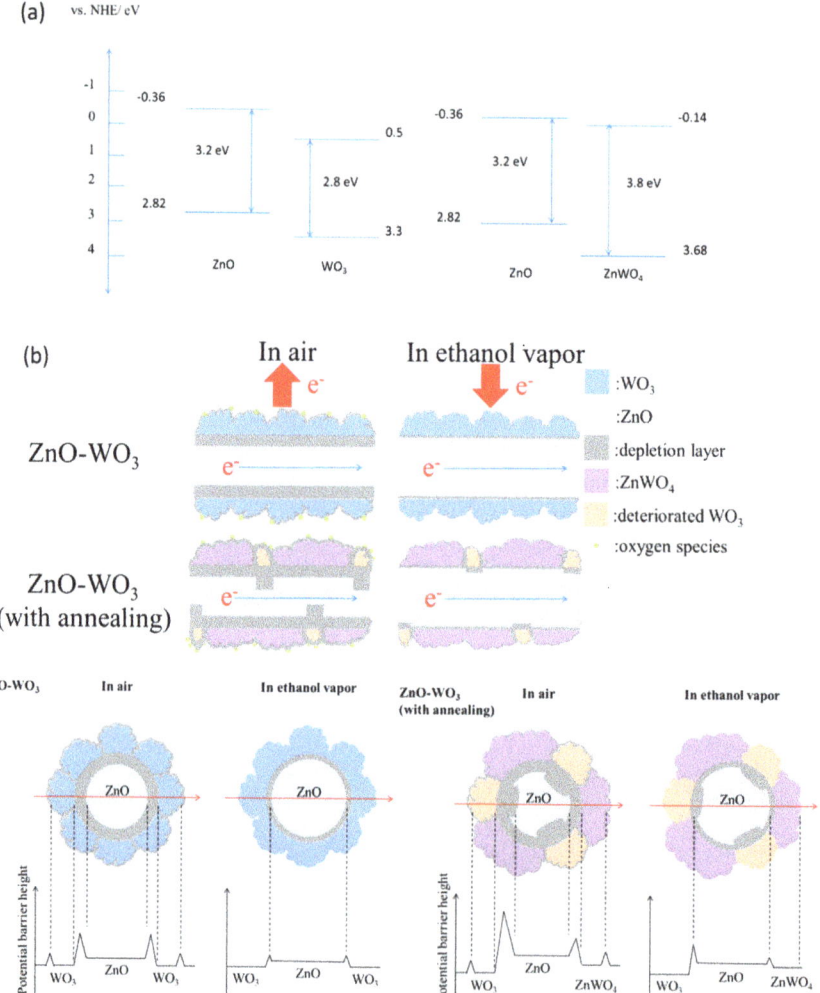

Figure 7. (a) Energy band alignments of ZnO/WO$_3$ and ZnO/ZnWO$_4$. (b) Possible ethanol gas-sensing mechanisms of the ZnO–WO$_3$ composite nanorods with and without a thermal annealing at 400 °C. The corresponding cross-sectional views for the possible potential barrier height variation was also shown in the bottom region of the plot.

5. Conclusions

A combinational methodology of hydrothermal growth and sputtering was used to synthesize ZnO–WO$_3$ composite nanorods. Furthermore, a thermal annealing procedure was conducted in a hydrogen-contained atmosphere to induce a microstructural modification of the composite nanorods. The structural analysis revealed that the ZnO nanorods sputtering coated with the ultra-thin WO$_3$

thin film formed well crystalline ZnO–WO$_3$ composite nanorods. The thermal annealing procedure at 400 °C further engendered the formation of ternary ZnWO$_4$ phase and deteriorated WO$_3$ phase on the surfaces of the ZnO nanorods. The ethanol gas-sensing test results demonstrated that the construction of the ZnO–WO$_3$ composite nanorods is advantageous for improving the gas-sensing response of the ZnO nanorods to ethanol vapor. The formation of the heterogeneous junction between the ZnO and WO$_3$ contributed to the enhanced ethanol gas-sensing responses. Moreover, an increase of potential barrier number in the ZnO–WO$_3$ composite nanorods annealed at 400 °C improved the gas-sensing responses of the composite nanorods without a thermal annealing. The composite nanorods annealed at 400 °C exhibited a strong response of 16.2 at the gas concentration of 50 ppm, while the pristine ZnO–WO$_3$ could only reach 7.3 at the identical gas concentration. Such intriguing ethanol gas-sensing response enhancement could be ascribed to the existence of heterogeneous junctions at interfaces of ZnO/ZnWO$_4$, ZnO/WO$_3$, and ZnWO$_4$/WO$_3$ in the composite nanorods after annealing at 400 °C. The local structural modification of the composite nanorods through a proper thermal annealing condition is feasible to control the gas-sensing behavior of the oxide composite nanorods. Moreover, the ZnO–WO$_3$ nanorods annealed at 400 °C exhibited high selectivity to ethanol vapor among the various target gases of NH$_3$, H$_2$, and NO$_2$. This composite nanorod system is of potential to effectively detect ethanol vapor in an open environment.

Author Contributions: Methodology, C.W.C.; formal analysis, C.W.C.; Writing—original draft preparation, Y.C.L. and C.W.C.; supervision, Y.C.L.; Writing—Review and Editing, Y.C.L.

Funding: This research was funded by Ministry of Science and Technology of Taiwan. Grant No. MOST 105-2628-E-019-001-MY3 & 107-2813-C-019-020-E.

Conflicts of Interest: The authors declare no conflict of interest.

References

1. Xue, D.; Wang, Y.; Cao, J.; Zhang, Z. Hydrothermal Synthesis of CeO$_2$-SnO$_2$ Nanoflowers for Improving Triethylamine Gas Sensing Property. *Nanomaterials* **2018**, *8*, 1025. [CrossRef] [PubMed]
2. Lee, J.-H.; Kim, J.-Y.; Mirzaei, A.; Kim, H.W.; Kim, S.S. Significant Enhancement of Hydrogen-Sensing Properties of ZnO Nanofibers through NiO Loading. *Nanomaterials* **2018**, *8*, 902. [CrossRef]
3. Du, H.; Li, X.; Yao, P.; Wang, J.; Sun, Y.; Dong, L. Zinc Oxide Coated Tin Oxide Nanofibers for Improved Selective Acetone Sensing. *Nanomaterials* **2018**, *8*, 509. [CrossRef]
4. Liang, Y.C.; Cheng, Y.R. Combinational physical synthesis methodology and crystal features correlated with oxidizing gas detection ability of one-dimensional ZnO–VO$_x$ crystalline hybrids. *CrystEngComm* **2015**, *17*, 5801–5807. [CrossRef]
5. Wang, P.P.; Qi, Q.; Zou, X.; Zhao, J.; Xuan, R.F.; Li, G.D. A precursor route to porous ZnO nanotubes with superior gas sensing properties. *RSC Adv.* **2013**, *3*, 19853–19856. [CrossRef]
6. Baratto, C. Growth and properties of ZnO nanorods by RF sputtering for detection of toxic gases. *RSC Adv.* **2018**, *8*, 32038–32043. [CrossRef]
7. Liang, Y.C.; Lin, T.Y.; Lee, C.M. Crystal growth and shell layer crystal-feature-dependent sensing and photoactivity performance of zinc oxide-indium oxide core-sehll nanorod heterostructures. *CrystEngComm* **2015**, *17*, 7948–7955. [CrossRef]
8. Xue, D.; Wang, J.; Wang, Y.; Sun, G.; Cao, J.; Bala, H.; Zhang, Z. Enhanced Methane Sensing Properties of WO$_3$ Nanosheets with Dominant Exposed (200) Facet via Loading of SnO$_2$ Nanoparticles. *Nanomaterials* **2019**, *9*, 351. [CrossRef]
9. Liang, Y.C.; Chao, Y. Crystal phase content-dependent functionality of dual phase SnO$_2$–WO$_3$ nanocomposite films via cosputtering crystal growth. *RSC Adv.* **2019**, *9*, 6482–6493. [CrossRef]
10. Fang, W.; Yang, Y.; Yu, H.; Dong, X.; Wang, T.; Wang, J.; Liu, Z.; Zhao, B.; Yang, M. One-step synthesis of flower-shaped WO$_3$ nanostructures for a high-sensitivity room-temperature NO$_x$ gas sensor. *RSC Adv.* **2016**, *6*, 106880–106886. [CrossRef]
11. Liang, Y.-C.; Chang, C.-W. Preparation of Orthorhombic WO$_3$ Thin Films and Their Crystal Quality-Dependent Dye Photodegradation Ability. *Coatings* **2019**, *9*, 90. [CrossRef]

12. Wang, J.M.; Sun, X.W.; Jiao, Z. Application of Nanostructures in Electrochromic Materials and Devices: Recent Progress. *Materials* **2010**, *3*, 5029–5053. [CrossRef]
13. Liang, Y.C.; Zhong, H. Materials synthesis and annealing-induced changes of microstructure and physical properties of one-dimensional perovskite-wurtzite oxide heterostructures. *Appl. Surf. Sci.* **2013**, *283*, 490–497. [CrossRef]
14. Liang, Y.C.; Liao, W.K.; Liu, S.L. Performance enhancement of humidity sensors made from oxide heterostructure nanorods via microstructural modifications. *RSC Adv.* **2014**, *4*, 50866–50872. [CrossRef]
15. Liang, Y.C.; Liao, W.K. Annealing induced solid-state structure dependent performance of ultraviolet photodetectors made from binary oxide-based nanocomposites. *RSC Adv.* **2014**, *4*, 19482–19487. [CrossRef]
16. Liang, Y.C.; Lo, Y.J. High-temperature solid-state reaction induced structure modifications and associated photoactivity and gas-sensing performance of binary oxide one-dimensional composite system. *RSC Adv.* **2017**, *7*, 29428–29439. [CrossRef]
17. Bonanni, M.; Spanhel, L.; Lerch, M.; Fuglein, E.; Muller, G. Conversion of Colloidal ZnO–WO$_3$ Heteroaggregates into Strongly Blue Luminescing ZnWO$_4$ Xerogels and Films. *Chem. Mater.* **1998**, *10*, 304–310. [CrossRef]
18. Liang, Y.C.; Hua, C.Y.; Liang, Y.C. Crystallographic phase evolution of ternary Zn-Ti-O nanomaterials during high-temperature annealing of ZnO-TiO$_2$ nanocomposites. *CrystEngComm* **2012**, *14*, 5579–5584. [CrossRef]
19. Liang, Y.C.; Xu, N.C.; Wang, C.C.; Wei, D.W. Fabrication of Nanosized Island-Like CdO Crystallites-Decorated TiO$_2$ Rod Nanocomposites via a Combinational Methodology and Their Low-Concentration NO$_2$ Gas-Sensing Behavior. *Materials* **2017**, *10*, 778. [CrossRef] [PubMed]
20. Liang, Y.C. Hydrogen-induced degradation in physical properties of dielectric-enhanced Ba$_{0.6}$Sr$_{0.4}$TiO$_3$/SrTiO$_3$ artificial superlattices. *Electrochem. Solid State Lett.* **2010**, *13*, 91–94. [CrossRef]
21. Hussain, T.; Al-Kuhaili, M.F.; Durrani, S.M.A.; Qurashi, A.; Qayyum, H.A. Enhancement in the solar light harvesting ability of tungsten oxide thin films by annealing in vacuum and hydrogen. *Int. J. Hydrogen Energy* **2017**, *42*, 28755–28765. [CrossRef]
22. Xie, F.Y.; Gong, L.; Liu, X.; Tao, Y.T.; Zhang, W.H.; Chen, S.H.; Meng, H.; Chen, J. XPS studies on surface reduction of tungsten oxide nanowire film by Ar$^+$ bombardment. *J Electron. Spectros. Relat. Phenomena* **2012**, *185*, 112–118. [CrossRef]
23. Liang, Y.C. Forming gas annealing induced degradation in nanoscale electrical homogeneity of bismuth ferrite thin films. *J. Electrochem. Soc.* **2011**, *158*, 137–140. [CrossRef]
24. Li, D.; Zhang, Y.; Liu, D.; Yao, S.; Liu, F.; Wang, B.; Sun, P.; Gao, Y.; Chuai, X.; Lu, G. Hierarchical core/shell ZnO/NiO nanoheterojunctions synthesized by ultrasonic spray pyrolysis and their gas-sensing performance. *CrystEngComm* **2016**, *18*, 8101–8107. [CrossRef]
25. Yang, X.; Zhang, S.; Yu, Q.; Zhao, L.; Sun, P.; Wang, T.; Liu, F.; Yan, X.; Gao, Y.; Liang, X.; et al. One step synthesis of branched SnO$_2$/ZnO heterostructures and their enhanced gas-sensing properties. *Sens. Actuator B Chem.* **2019**, *281*, 415–423. [CrossRef]
26. Wang, L.; Liu, H.; Fu, H.; Wang, Y. Polymer g-C$_3$N$_4$ wrapping bundle-like ZnO nanorod heterostructures with enhanced gas sensing properties. *J. Mater. Res.* **2018**, *10*, 1401–1410. [CrossRef]
27. Liang, Y.C.; Lee, C.M.; Lo, Y.J. Reducing gas-sensing performance of Ce-doped SnO$_2$ thin films through a cosputtering method. *RSC Adv.* **2017**, *7*, 4724–4734. [CrossRef]
28. Jiang, X.; Zhao, X.; Duan, L.; Shen, H.; Liu, H.; Hou, T.; Wang, F. Enhanced photoluminescence and photocatalytic activity of ZnO-ZnWO$_4$ nanocomposites synthesized by a precipitation method. *Ceram Int.* **2016**, *14*, 15160–15165. [CrossRef]
29. Lam, S.-M.; Sin, J.-C.; Abdullah, A.Z.; Mohamed, A.R. Transition metal oxide loaded ZnO nanorods: Preparation, characterization and their UV–vis photocatalytic activities. *Sep. Purif. Technol.* **2014**, *132*, 378–387. [CrossRef]

© 2019 by the authors. Licensee MDPI, Basel, Switzerland. This article is an open access article distributed under the terms and conditions of the Creative Commons Attribution (CC BY) license (http://creativecommons.org/licenses/by/4.0/).

Article

Orange/Red Photoluminescence Enhancement Upon SF$_6$ Plasma Treatment of Vertically Aligned ZnO Nanorods

Amine Achour [1], Mohammad Islam [2,*], Sorin Vizireanu [3], Iftikhar Ahmad [2], Muhammad Aftab Akram [4], Khalid Saeed [5], Gheorghe Dinescu [3] and Jean-Jacques Pireaux [1]

1. Laboratoire Interdisciplinaire de Spectroscopie Electronique (LISE), Namur Institute of Structured Matter (NISM), University of Namur, 61 Rue de Bruxelles, 5000 Namur, Belgium; a_aminph@yahoo.fr (A.A.); jean-jacques.pireaux@unamur.be (J.-J.P.)
2. Center of Excellence for Research in Engineering Materials, Deanship of Scientific Research, King Saud University, P.O. Box 800, Riyadh 11421, Saudi Arabia; ifahmad@ksu.edu.sa
3. National Institute for Laser, Plasma and Radiation Physics, Magurele, P.O. Box MG-16, 077125 Bucharest, Romania; s_vizi@infim.ro (S.V.); dinescug@infim.ro (G.D.)
4. School of Chemical and Materials Engineering, National University of Sciences and Technology, Sector H-12, Islamabad 44000, Pakistan; aftabakram@scme.nust.edu.pk
5. Department of Mechanical Engineering, College of Engineering, King Saud University, P.O. Box 800, Riyadh 11421, Saudi Arabia; khaliduetp@gmail.com
* Correspondence: mohammad.islam@gmail.com or miqureshi@ksu.edu.sa; Tel.: +966-544-523-909; Fax: +966-114-670-199

Received: 11 April 2019; Accepted: 20 May 2019; Published: 23 May 2019

Abstract: Although the origin and possible mechanisms for green and yellow emission from different zinc oxide (ZnO) forms have been extensively investigated, the same for red/orange PL emission from ZnO nanorods (nR) remains largely unaddressed. In this work, vertically aligned zinc oxide nanorods arrays (ZnO nR) were produced using hydrothermal process followed by plasma treatment in argon/sulfur hexafluoride (Ar/SF$_6$) gas mixture for different time. The annealed samples were highly crystalline with ~45 nm crystallite size, (002) preferred orientation, and a relatively low strain value of 1.45×10^{-3}, as determined from X-ray diffraction pattern. As compared to as-deposited ZnO nR, the plasma treatment under certain conditions demonstrated enhancement in the room temperature photoluminescence (PL) emission intensity, in the visible orange/red spectral regime, by a factor of 2. The PL intensity enhancement induced by SF$_6$ plasma treatment may be attributed to surface chemistry modification as confirmed by X-ray photoelectron spectroscopy (XPS) studies. Several factors including presence of hydroxyl group on the ZnO surface, increased oxygen level in the ZnO lattice (O_L), generation of F–OH and F–Zn bonds and passivation of surface states and bulk defects are considered to be active towards red/orange emission in the PL spectrum. The PL spectra were deconvoluted into component Gaussian sub-peaks representing transitions from conduction-band minimum (CBM) to oxygen interstitials (O_i) and CBM to oxygen vacancies (V_O) with corresponding photon energies of 2.21 and 1.90 eV, respectively. The optimum plasma treatment route for ZnO nanostructures with resulting enhancement in the PL emission offers strong potential for photonic applications such as visible wavelength phosphors.

Keywords: ZnO nanorods; photoluminescence; SF$_6$ plasma; visible emission; red emission

1. Introduction

With a direct band gap and high exciton binding energy values of 3.3 eV and 60 meV, respectively, the transparent n-type semiconductor zinc oxide (ZnO) has demonstrated use for numerous electronic, electrochemical, optoelectronic, and electromechanical devices [1–4], including light-emitting diodes [5], piezoelectric nanogenerators [6], field emission devices [7], high-performance nanosensors [8,9], solar cells [10,11], and ultraviolet (UV) lasers [12]. Due to their high surface-to-volume ratio and surface area, the ZnO nanostructures with zero- (quantum dots and ultrafine nanoparticles), one- (tubes, rods, belts, and wires), or two-dimensional morphology (sheets, flakes, and thin films) offer superior performance attributes than those of bulk ZnO structures.

The enhancement of visible emission from ZnO is also important besides any improvement in its UV emission characteristics. The presence of intrinsic defects in the ZnO lattice structure causes photoemission in the green or yellow/orange/red spectral regime. The visible photoluminescence (PL) emission in ZnO quantum dots (QD) was found to depend on the QD size as well as presence of singly ionized oxygen vacancies as defects and was ascribed to the transition of holes from valence band to preexisting deep donor energy levels [13]. Another study reported the effect of evaporation or chemical methods, as processing route, on the extent of green or yellow emission due to surface centers or defects in the bulk ZnO [14]. A decrease in aspect ratio owing to a change from nanoneedle to nanorod morphology was found to induce an increase in the PL intensity via nonradiative quenching by near-surface defects [15]. Various approaches explored for any enhancement in the visible PL emission include (i) control and induction of abundant intrinsic point defects such as zinc interstitials (Zn_i) and oxygen vacancies (V_O) in the ZnO nanostructures [16], (ii) doping with rare-earth ions [17,18], (iii) ZnO annealing and exposure to charge scavengers [19], and (iv) surface plasmon effect due to coupling with Au or Ag nanoparticles [20,21].

The solution processing routes offer an outstanding merit in terms of low equipment infrastructure and manufacturing costs. Most of these processes are also amenable to optoelectronic device fabrication that require vertical stacking of heterostructures. In this context, hydrothermal synthesis carries certain advantages, namely low processing temperatures, and inherent flexibility to in situ fabrication and integration. For intrinsic, doped, and core/shell ZnO nanorods arrays, hydrothermal process is a promising technique for tunable composition and desirable optical properties. In addition to other parameters, the shape and morphology of the resulting nanorods strongly depend on the ZnO source molarity in the solution, with hexagonal, faceted growth at high concentrations [18].

An investigation into the film thickness (70–220 nm) and annealing temperature (200–900 °C) effects on the atomic layer deposition ZnO films revealed green, orange, or red emission due to transition of the photoexcited electrons from the *CB* (conduction band) (2.17 eV) to V_O, Zn_i to O_i (interstitial oxygen) (2.09 eV), or from the *CB* to the mid-gap states (1.79–1.99 eV), respectively [22]. The lattice disorder or structural imperfections such as grain boundaries and surface defects in the ZnO crystals give rise to O_i and O_{Zn} oxygen antisites, thus causing orange/red emission [23]. Various mechanisms have been proposed for green emission based on V_{Zn}, V_O, Zn_i, and O_i, or O_{Zn} and O_i antisites [23–27]. Other mechanisms that may be active involve substituted Cu atoms at significantly high concentration, shallow donor to deep acceptor electronic transitions [25].

The PL visible emission from ZnO nanostructures can be tailored through surface modification treatments such as hydrogenation, polymer covering, argon ion milling, and annealing in different ambiences [28,29]. Polydorou et al. [30] reported an increase in the near band edge (NBE) intensity and reduction in the broad band intensity in the visible range upon ZnO thin film treatment in pure and mixed SF_6 plasma with subsequent improvements in device efficiency and lifetime of inverted polymer solar cells. On the other hand, Prucnal et al. [31] noticed an increase in the PL intensities for both UV (near-band-edge or NBE) and visible range (deep level emission or DLE) upon SF_6 plasma treatment of the ZnO films for 125 s. Treatment of the ZnO nanorods with hydrogen or oxygen plasma has been found to reduce V_O concentration and increase O_i concentration, respectively, with subsequent intensity changes in the PL spectra [32]. Yet, another study of the Ar plasma treatment of the ZnO nanowires revealed intensity enhancement of the NBE emission over DLE [33]. Although there are several reports on processing or plasma-induced surface modification of ZnO thin films and nanostructures, the relative enhancement in the visible emission (DLE) still needs to be further explored.

In this paper, we performed in-depth PL and XPS investigations of the as-produced and after argon/sulfur hexafluoride (Ar/SF$_6$) plasma treatment of ZnO nanorods (ZnO nR) produced by hydrothermal technique. The ZnO nR plasma treatment was accomplished by means of low-pressure plasmas generated by electrical discharges in Ar/SF$_6$ gas mixtures. The plasma treatments under controlled conditions preserved the nano/microfeatures of the ZnO nR surface morphology, while incorporation of chemical groups was achieved [34]. The surface chemistry of the samples was assessed by X-ray photoelectron spectroscopy which is a powerful surface characterization technique allowing detailed investigation of the chemical bonding. As compared to the initial ZnO nR, the photoluminescence (PL) emission in the visible region was doubled after the plasma treatment.

2. Experimental Procedure

2.1. Synthesis of ZnO Nanorods

10 mM zinc acetate dihydrate (ZAD, Zn(CH$_3$COO)$_2$·2H$_2$O) (98% purity, Sigma Aldrich, St. Louis, MO, USA) was dissolved in ethanol and the solution was spin-coated onto indium tin oxide (ITO) coated glass substrates at a 2000 rpm for 30 s. The spin casting cycle was repeated 3–5 times, and after each deposition cycle the samples were air dried at 130 °C for 5 min. Finally, a seed layer of ZnO nanoparticles was produced by annealing the samples in air at 340 °C for 10 min. Over the seeded substrates, ZnO nanorods were grown through immersion in 25 mM zinc nitrate hexahydrate (Zn(NO$_3$)$_2$·6H$_2$O) and 25 mM hexamethylenetetramine ((CH$_2$)$_6$N$_4$, HMTA) aqueous solution mixture at 90 °C for 6 h in a sealed autoclave. At high temperature in the oven, HMTA not only provides the basic environment needed for Zn(OH)$_2$ formation, it also performs coordination with and stabilization of Zn^{2+} ions, and subsequent dehydration into ZnO [35]. After that, the samples were extracted from the autoclave, washed thoroughly several times with deionized water to remove any residual reactants, and dried in air. As a final setup, the samples were annealed in air at 350 °C for 10 min to get rid of any trace of the organic precursor, as described elsewhere [36].

2.2. Ar/SF$_6$ Plasma Treatment

The as-produced ZnO nanorods were treated with a mixture of argon (Ar) and sulfur hexafluoride (SF$_6$) plasma using a glass bell jar reactor [37]. A capacitively coupled radio-frequency discharge (RF, 13.56 MHz) was generated between two parallel planar electrodes that were 4 cm apart. The upper electrode was connected to the RF power supply whereas the lower, grounded electrode was used as the substrate holder. The Ar/SF$_6$ gas mixture with volumetric flow rates of 10 and 20 sccm for the Ar and SF$_6$ gases, respectively, was introduced into the chamber. The plasma treatment was performed at 25 W applied RF power and 15 Pa working pressure for different time durations. Throughout this work, the as-prepared and plasma-treated zinc oxide nanorods arrays (after 5, 10, and 20 min of Ar/SF$_6$ plasma treatment) are referred to as Z-nR, Z-nR5, Z-nR10, and Z-nR20, respectively.

2.3. Nanostructures Characterization

For characterizing their morphology and composition, the as-deposited ZnO nanorods samples were investigated using scanning electron microscope (SEM) and the X-ray diffraction (XRD) machine, respectively. The microstructural examination was carried out on a JSM7600 model (JEOL, Tokyo, Japan) operating at 5 kV. The XRD studies were performed on a D500 MOXTEK apparatus (MOXTEK, Inc., Orem, UT, USA) in the θ–2θ Bragg Brentano configuration with monochromatic Cu-Kα radiation (λ = 1.5404 Å). The photoluminescence (PL) measurements were recorded from all samples at room temperature by means of a Jobin-Yvon Fluorolog®-3 spectrometer (Horiba Scientific, Piscataway, NJ, USA) with a 500 W Xenon lamp. To probe the surface composition of the nanostructures arrays, X-ray photoelectron spectroscopy (XPS) measurements were made using XPS measurements were carried out on K-Alpha spectrometer (Thermo Fisher Scientific, Waltham, MA, USA) using a monochromatic Al Kα radiation (1486.68 eV), with a spot size of 250 × 250 µm. A flood gun was used for charge compensation and the C1s line of 284.8 eV was used as a reference to correct the binding energies for charge energy shift.

3. Results and Discussion

The SEM microstructures of the ZnO nanorods assembly at low and high magnification are presented in Figure 1. From the top view of the as-deposited sample, the nanorods appear to exhibit growth in a direction that is almost perpendicular to the substrate surface. The electron microscopy also revealed well-faceted hexagonal growth with an area density of ~15/μm^2. The diameter and length of the nanorods was in the range of 100 to 150 nm and ~1 µm, respectively. The characteristic features of the nanorods assembly such as aspect ratio, area density, and degree of alignment depend on the crystalline quality of the seed as well as the processing conditions [38,39]. The faceted, hexagonal morphology becomes more evident at high magnification (Figure 1b). The deviation from ideal vertical growth of the ZnO nanorods and the not-so-perfect c-axis orientation (noticed in the XRD pattern) are caused by the polycrystalline nature of the ZnO seeds [40].

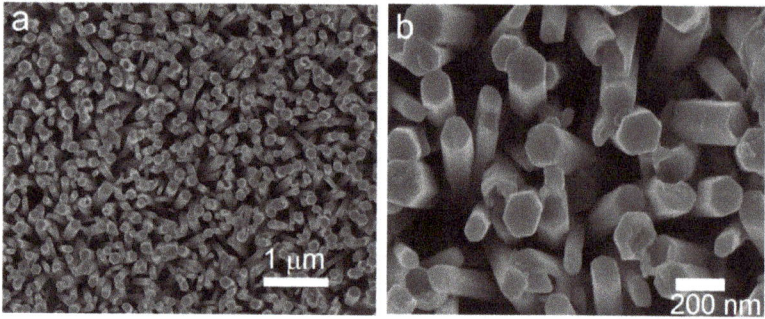

Figure 1. SEM microstructures of the as-prepared vertically aligned ZnO array (sample Z-nR) at (**a**) low magnification and (**b**) high magnification.

The X-ray diffraction pattern of the as-produced ZnO nanorods is presented in Figure 2. All the diffraction peaks were indexed to the ZnO hexagonal phase (space group: P63mc; JCPDS No. 36-1451). The most intense peak is located at 34.4° and can be assigned to the (002) crystallographic plane. From the presence of other peaks besides the preferred orientation, it can be deduced that the ZnO NR may either be polycrystalline with an overall preferred orientation or single crystals with most of them having the same preferred orientation [41]. The degree of vertical growth and the c-axis preferred orientation can be further enhanced through improvement in crystal quality of the ZnO seeds [42].

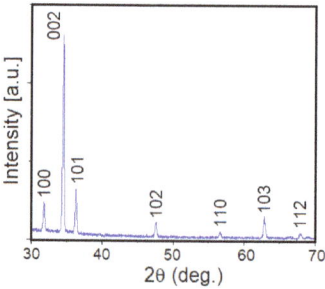

Figure 2. X-ray diffraction pattern of the as-made ZnO nanorods sample (Z-nR).

The values of the texture coefficient (T_C) for the (100), (002), (101), and (103) diffraction peaks and the crystallite size (D) corresponding to the most intense (002) peak were calculated through Equations (1)–(3). The strain induced in the film due to lattice distortions, as determined using the Williamson–Hall method, is a function of line breadth (β_{hkl}) and diffraction angle (θ) as given by Equation (4) [43–45],

$$T_c(hkl) = \frac{I(hkl)/I_0(hkl)}{\frac{1}{n}\sum_n I(hkl)/I_0(hkl)} \quad (1)$$

$$D = \frac{k\lambda}{\beta_{hkl}\cos\theta} \quad (2)$$

$$\beta_{hkl} = [(\beta_{hkl})^2_{measured} - (\beta)^2_{instrumental}]^{1/2} \quad (3)$$

$$\varepsilon = \frac{\beta_{hkl}}{4\tan\theta} \quad (4)$$

I and I_0 are the measured relative intensities and the JCPDS standard intensities for the plane with hkl indices. On the other hand, λ, k, β_{hkl}, $\beta_{intrument}$, and θ are monochromatic wavelength (1.5405 Å), shape-factor (0.94), full-width-half-maximum (FWHM) of the diffraction peak with hkl planar indices, instrumental corrected integral breadth of the diffraction at 2θ, and the diffraction angle, respectively. The XRD data clearly demonstrate preferred crystallographic orientation along (002), as indicated by a higher value of 2.79 for T_C (002) as compared to other diffraction peaks due its lowest surface free energy [46,47]. From the Scherrer Equation (2), the crystallite size was found to be 45.1 nm. Additionally, the annealing treatment induced high crystalline quality among the nanorods with strong (002) orientation. Assuming homogeneous isotropic nature of the ZnO crystals, the magnitude of strain in the annealed film was determined to be 1.45×10^{-3}, which is quite low for the annealed ZnO film. It is noteworthy that XRD patterns and SEM microstructures of the ZnO NR specimens after plasma treatment were similar to those for the as-made ZnO NR (without plasma treatment). The XPS analysis, however, revealed that the surface chemistry of the nanorods was altered, as discussed later.

The room temperature PL spectra of the ZnO NR samples before and after plasma treatment are shown in Figure 3. The band at 380 nm (Figure 3a) corresponds to 3.26 eV energy and is associated with the ZnO band gap emission [48]. The defect emissions due to structural defects in ZnO, such as V_O and Zn_i, are usually located in the spectral range of 450 to 650 nm [49]. Since the defect-related intensity is significantly higher than the band gap emission, the presence of crystallographic defects can be deduced in the as-deposited ZnO nanorods sample. It is noteworthy that the visible emission in all the samples was centered at 619 nm (2.0 eV) in the orange/red emission. The orange PL band with the maximum at 2.02–2.10 eV was earlier attributed either to oxygen interstitial atoms (O_i) due to excess oxygen on the ZnO surface (2.02 eV) or to the hydroxyl group (OH) (2.09 eV) [50]. The orange emission is caused by transition from the Zn interstitial (Zn_i) to the oxygen interstitial (O_i) states. Such phenomena are generally observed in oxygen rich systems [51]. In the Z-nR sample (Figure 3b), the highest PL peak is centered at ~620 nm, which indicates that oxygen interstitials are the dominant defects even before plasma treatment. The treatment of the Z-nR with Ar/SF$_6$ plasma led to a significant abatement of the intensity of the band gap emission in comparison with that of the Z-nR, particularly in case of Z-nR5 and Z-nR10 arrays. These reductions in band gap emission intensity may be ascribed to an increase in the defects density. These defects, in turn, trap the excited photons before the NBE recombination. In case of visible emission, the Z-nR5 and Z-nR20 samples demonstrate the maximum defect emission enhancement in the red/orange region. Thus, the degree of enhancement in the visible emission band for the three plasma-treated nanostructure arrays, in ascending order is Z-nR10 < Z-nR20 < Z-nR5. The I_{Vis}/I_{UV} ratio for the four samples, in descending order: Z-nR5 (36.4) > Z-nR10 (18.0) > Z-nR20 (8.16) > Z-nR (3.65). It seems that for the intermediate plasma treatment times of 5 and 10 min, the NBE peak is almost entirely quenched, followed by peak resurgence upon 20 min treatment, presumably due to generation of oxygen related defects. The effect of Ar/SF$_6$ plasma treatment for ~20 min should be further investigated for shorter time intervals in order to establish any concrete surface chemistry–PL properties correlation. There are conflicting reports on the relative decrease, increase or no change in the relative intensities of the NBE (UV) or DLE (visible range) peaks, with the reported findings depending on several factors including ZnO morphology, plasma composition, plasma treatment time, that in turn, influence the defect types and concentration on ZnO surface besides attachment of hydroxyl groups [30–33,52,53].

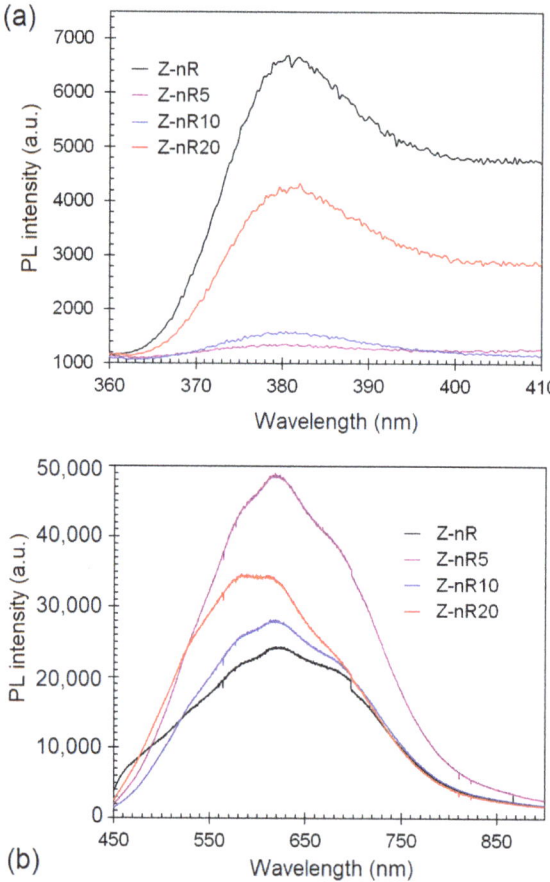

Figure 3. Photoluminescence spectra of the different Z-nR samples before and after SF_6 plasma treatment in the (**a**) ultraviolet (UV) spectral regime and (**b**) visible spectrum.

There are many reports about quenching and suppression of green emission band in fluorine-doped ZnO (ZnO:F) films due to oxygen vacancies (V_O) [52–54]. Upon F-doping, the V_O defect density is decreased, causing a reduction in the green emission intensity since the F atoms effectively fill V_O sites. In our case, an improvement in the orange/red emission of the visible spectral regime was noticed. Although the V_O defects are responsible for the green emission (centered at ~500 nm) [49,50], the peak intensity is very weak in the as-made Z-nR sample and almost absent in the plasma-treated ZnO NR, as revealed by the PL spectra in Figure 3b. On the contrary, there are also reports on the appearance or band intensity enhancement of the red/orange emission from ZnO:F owing to an increased oxygen defect density [55] or replacement of oxygen by fluorine through the occupation of interstitial sites [56]. The orange/red emission enhancement upon plasma treatment of the ZnO nanorods samples, as noticed in our case, will be further discussed in the next section.

The PL spectra of the as-produced and plasma-treated ZnO nR samples were deconvoluted into component peaks, as demonstrated in Figure 4. In all the samples, the total PL visible emission had contributions from red, orange and green emissions. Although the red band intensity was much greater than those from orange or green bands, it is interesting to compare the relative spectral evolution of each band. Upon Ar/SF_6 plasma treatment, a red shift in the green band position (2.44 eV) was observed with a maximum change by ~0.08 eV for 5 min treatment, in agreement with reports about annealed or O_2 plasma-treated ZnO nanorods. On the other hand, there was no change in the respective positions of red (1.89 eV) and orange (2.13 eV) emission bands. A comparison of the red and orange band intensities with respect to that of green emission in terms of the I_R/I_G and I_O/I_G ratio for all the samples pointed towards a gradual decrease in the green emission intensity, presumably due to reduction in oxygen vacancies (V_O) upon plasma treatment. The density of oxygen defects (O_i), however, increased due to interaction between plasma species and the surface layers, thus giving rise to electron transitions responsible for red/orange emissions. The XPS data presented and discussed later in this section further confirm this observation.

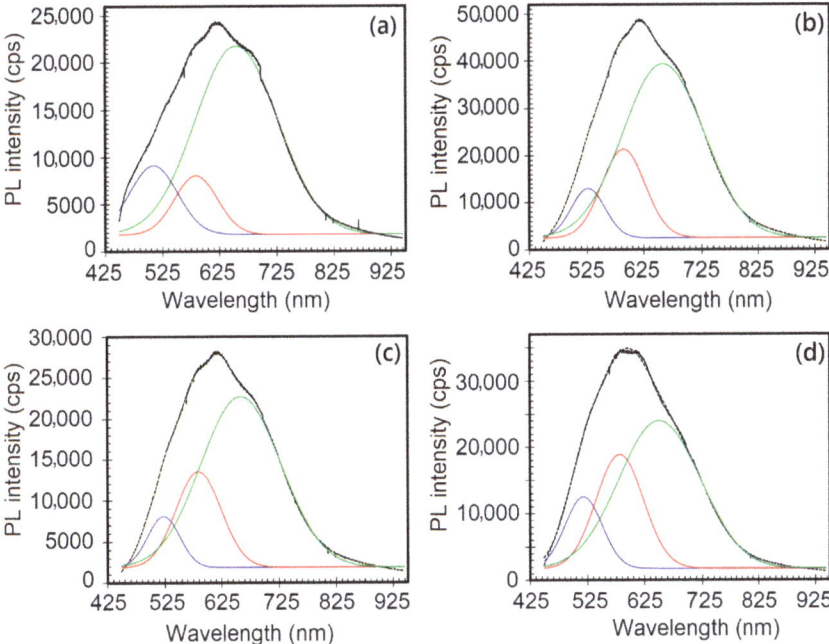

Figure 4. Deconvoluted PL spectra of the ZnO nanorods: (**a**) As-deposited nanorods (Z-nR), and after Ar/SF_6 plasma treatment for (**b**) 5 min (Z-nR5), (**c**) 10 min (Z-nR10), and (**d**) 20 min (Z-nR20).

The change in intensities of the orange (I_O), red (I_R), and green (I_G) emissions relative to the green or total spectral emissions, in terms of the I_O/I_G, I_R/I_G, and I_R/I_T ratios, are depicted for different ZnO nanorods samples in Figure 5. Upon Ar/SF_6 plasma treatment, the intensity of red emission decreased from 2.72 (for the as-deposited sample) to the minimum value of 1.27 (for 20 min treatment) with an associated increase in the orange emission intensity. The sample Z-nR5 exhibited the greatest extent of reduction in the NBE intensity as well as the maximum enhancement in the broad emission in the visible range with the latter arising mainly from deep level emissions due to zinc vacancy related defects. It is understandable SF_6 plasma generates SF_6^+ radicals that effectively passivate the ZnO surface through saturation of dangling bonds and diminishing the V_O concentration [31].

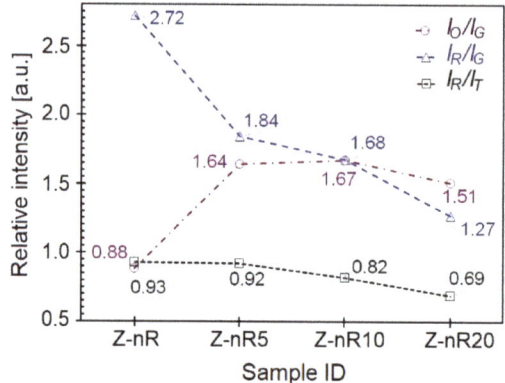

Figure 5. Change in orange (I_O) and red (I_R) emission intensities with respect to the green (I_G) or total intensity (I_T) from the PL spectra of the as-deposited and Ar/SF$_6$ plasma-treated ZnO nanorods.

In order to elucidate the PL emission behavior, XPS analyses were performed on the different ZnO nanorods samples before and after plasma treatment. The elemental composition (in atomic percent) of the different surfaces before and after plasma treatments are shown in Figure 6. While the oxygen content was found to decrease with increasing duration of the mixed Ar/SF$_6$ plasma treatment, presumably due to incorporation of F at the surface region, the latter increases as the treatment time increases. There was no sulfur detected on any of the Z-nR5, Z-nR10, and Z-nR20 surfaces (after Ar/SF$_6$ plasma treatment), implying that there are more F ions in the plasma (SF$_6$) than S. Also, the reactivity of F ions is greater and stronger than that of S ions. The carbon element was also present in all the samples, it is attributed to the adsorbed carbon due to contamination.

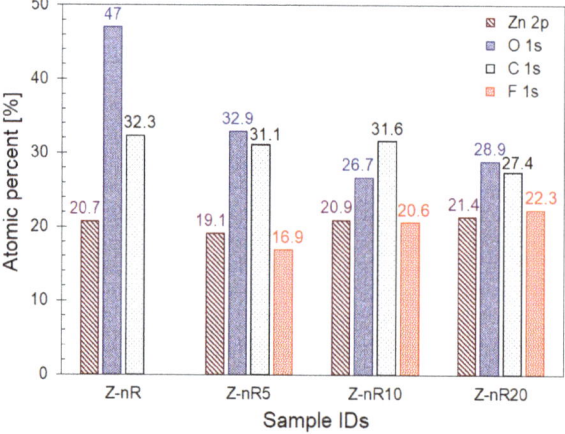

Figure 6. Bar chart showing atomic percentage of different elements as determined from the X-ray photoelectron spectroscopy (XPS) data.

The high-resolution XPS spectra of the as-deposited and plasma functionalized Z-nR samples of Zn 2p peaks are presented in Figure 7. The Zn 2p core lines of the Z-nR sample (no plasma treatment) comprise two distinct peaks that may be ascribed to the Zn 2p$_{3/2}$ (1021.4 eV) and Zn 2p$_{1/2}$ (1044.5 eV) with a spin–orbital splitting (Δ metal) of 23.1 eV. This is a direct confirmation of the Zn atoms to be in 2+ oxidation state [54]. From Figure 7, a slight shift of 0.2 V in the peak position towards higher binding energy values was noticed upon Ar/SF$_6$ plasma treatment, possibly due to F$^-$ ion incorporation into the ZnO lattice in the near-surface region thus causing the ZnO electronic band structure to be altered. Since electronegativity value for the F$^-$ (3.98) is greater than that of O^{2-} (3.44), the net charge transfer from Zn (1.65) to F will be more dominant than that from Zn to O. Nevertheless, the value of Δ (metal) remained constant for all of the samples, indicating that the subatomic structure and the core-level of Zn in the ZnO host lattice is independent of the F from the plasma treatment [57,58].

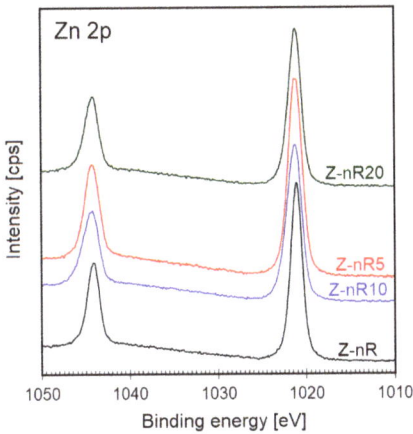

Figure 7. XPS Zn 2p core level spectra for the ZnO nanorods before and after Ar/SF$_6$ plasma treatment.

The F$^-$ incorporation into ZnO lattice can proceed in three possible ways: (i) formation of dative hydrogen bonds (OH–F) with surface hydroxyl groups, (ii) occupation of oxygen vacancies, and (iii) monosubstitution into oxygen lattice sites in the ZnO matrix. In order to quantify which one is dominant, the F 1s and O 1s XPS core level spectra were also carefully evaluated. The F 1s high-resolution spectra of the Z-nR samples at different plasma treatment times are shown in Figure 8. All the samples revealed the characteristic peak at 684.3 eV (F 1s peak) related to the F–Zn bond, which indicates replacement of the O by F upon fluorination, thus modifying the surface chemical composition of the ZnO. Upon SF$_6$ plasma treatment for 5 and 10 min (Z-nR5 and Z-nR10), a slight shift of ~0.1–0.2 eV in the BE to higher values was observed, which may be associated with the presence of dangling bonds on the ZnO surface such as F–OH bonds with more F surrounded by oxygen in the ZnO.

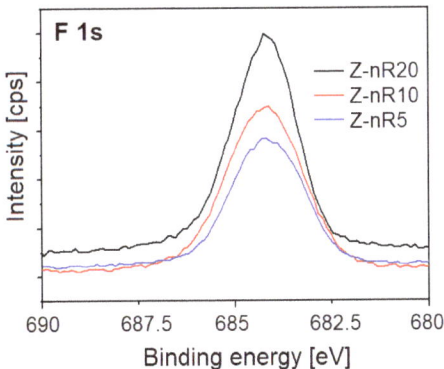

Figure 8. XPS F 1s core level spectra for the ZnO nanorods surface after Ar/SF$_6$ plasma treatment.

The O 1s deconvoluted core level spectra of the Z-nR before and after SF_6 plasma treatment are presented in Figure 9. The O 1s peak of ZnO is fitted by three nearly Gaussian components mainly centered at about 530, 531.5, and 532.5 eV with these peaks associated with O^{2-} species in the lattice (O_L), oxygen vacancies or defects (O_i), and chemisorbed or dissociated (O_C) oxygen species such as hydroxyl group, H_2O, O_2, H, and $-CO_3$, respectively [59–61]. During hydrothermal synthesis, oxygen adsorption from O_2, CO_2, H_2O, etc. promotes appearance of the PL component peak for O_C. The peak due to O_L represents O^{2-} in the lattice structure or presence of Zn–O bonds. Additionally, the component peak characteristic of O_i indicates generation of deep level point defects 0.7–1.0 eV below the bottom of conduction band with enhanced visible PL emission. The respective central positions for the three Gaussian components O_L, O_i, and O_C that fit the O 1s peak and their percentages for the as-deposited and SF_6 plasma-treated samples are listed in Table 1. It is evident from Table 1 that the relative fraction of the O_C component decreases from approximately 23% to 2% with increasing plasma treatment time, implying greater extent of dissociation/conversion of the O_C species initially present due to excess oxygen. Such intensity decrease in the O_C component peak maybe attributed to the removal of loosely bound contaminants by the energetic species in low-pressure plasma. Also, fluorine may substitute oxygen vacancies (dangling zinc bonds) and/or oxygen atoms in the ZnO lattice, or even passivation of dangling oxygen bonds through formation of some intermolecular hydrogen bonds [30]. Comparison of the changes in the O_C peak intensities with PL intensity evolution reveals that the PL intensity change is not linearly related with the O_C surface content (by percent). A correlation between the PL intensity enhancement and the O_C percentage on the nanostructures surface cannot be established. The presence of O_i on the samples' surfaces can be characterized by the equation, O_i (%) = $O_i/(O_i + O_C + O_L)$; this value was found to increase from ~1.26% for the as-prepared Z-nR to a maximum value of 3.2% for the Z-nR20 sample, probably due to fact that the Ar ions bombardment from the Ar/SF_6 mixture causes generation of new oxygen defects. Among all the SF_6 plasma-treated ZnO nanorods samples, the longer plasma treatment times of 10 and 20 min are considered to provide passivation to surface states and bulk defects along with maximum degree of F incorporation. This is in agreement with a recent study that reported formation of high-quality, highly-doped ZnO films upon flash lamp annealing in the millisecond range [62]. On the other hand, SF_6 plasma treatment of the ZnO nanrorods (Z-nR5) for only 5 min induces a maximum enhancement in the ZnO defect band intensity, presumably due to more favorable surface conditions in terms of OH bond physisorption, oxygen incorporation into the ZnO lattice (O_L peak in the XPS spectrum) and generation of F–OH and F–Zn bonds.

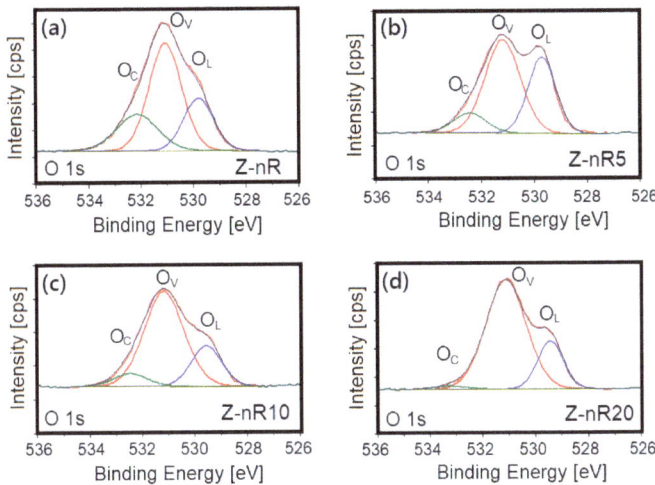

Figure 9. XPS O 1s deconvoluted core level spectra of (**a**) as-prepared ZnO NRs and ZnO NRs treated in Ar/SF$_6$ plasma: (**b**) Z-nR5, (**c**) Z-nR10, and (**d**) Z-nR20.

Table 1. Deconvolution of the O 1s XPS peak in the different ZnO nanorods in terms of binding energy values and atomic percentages for the three O 1s component peaks.

Sample	O 1s Binding Energy (eV)	Percentage (%)
Z-nR	O_L: 529.8	23.8
	O_i: 531.0	55.1
	O_C: 532.1	21.1
Z-nR5	O_L: 529.7	35.1
	O_i: 531.2	54.2
	O_C: 532.4	10.7
Z-nR10	O_L: 529.6	22.6
	O_i: 531.2	69.3
	O_C: 532.5	8.1
Z-nR20	O_L: 529.5	21.7
	O_i: 531.1	76.2
	O_C: 533.4	2.1

This work demonstrates that plasma modification of the nanostructures and thin films can lead to fine tuning of the functional properties without causing any structural changes. When compared with other recent works on different nanoscale materials involving hybrid nanostructures or surface chemistry modification [63–66], novel functional properties can be carefully tuned for subsequent incorporation into microelectronic and/or optoelectronic devices.

4. Conclusions

ZnO nanorods arrays with high surface area and well-aligned crystallographic orientation were produced via hydrothermal method. The annealed arrays exhibit a crystallite size of 45 nm with (002) preferred orientation and a relatively low lattice strain. Upon argon/SF$_6$ plasma treatment for up to 20 min, the PL intensity in the orange/red region of ZnO is enhanced by 2-fold compared to the ZnO sample without plasma treatment. Moreover, the PL intensity in the blue spectral regime is almost suppressed due to increased defect density. For 5 min Ar/SF$_6$ plasma treatment, the PL broad band intensity maximizes due to maximum oxygen content at the film surface. This finding implies presence of hydroxyl group at the surface, more oxygen in the ZnO lattice (O_L), fluorine incorporation in terms of F–Zn and F–OH bonds, and passivation of the surface states as well as bulk defects. These findings have implications for fundamental studies and industrial application of the plasma modified ZnO nanostructures in optoelectronic applications including visible light emitting devices and display systems.

Author Contributions: A.A. and J.-J.P. designed the scheme of experiments; A.A., I.A., S.V. and K.S. performed the experimental work; M.I., J.-J.P. and G.D. carried out the data analysis; M.A.A. and S.V. accomplished data validation; M.I., A.A. and J.-J.P. prepared the original draft; G.D., M.A.A., I.A., S.V. and K.S. did the necessary writing review and editing; J.-J.P. and M.I. administered the project and acquired funding for this research.

Funding: The authors would like to extend their sincere appreciation to the Deanship of Scientific Research at King Saud University for its funding of this research through the Research Group Project no. RGP-283. The authors also express their gratitude to Wallonia Region for financial support (Project Cleanair; convention 1510618, compl. Feder films).

Conflicts of Interest: The authors declare no conflict of interest.

References

1. Zang, Z.; Zeng, X.; Du, J.; Wang, M.; Tang, X. Femtosecond laser direct writing of microholes on roughened ZnO for output power enhancement of InGaN light-emitting diodes. *Opt. Lett.* **2016**, *41*, 3463–3466. [CrossRef]
2. Zang, Z. Efficiency enhancement of ZnO/Cu$_2$O solar cells with well oriented and micrometer grain sized Cu2O films. *Appl. Phys. Lett.* **2018**, *112*, 042106. [CrossRef]
3. Achour, A.; Soussou, M.A.; Ait Aissa, K.; Islam, M.; Barreau, N.; Faulques, E.; Le Brizoual, L.; Djouadi, M.A.; Boujtita, M. Nanostructuration and band gap emission enhancement of ZnO film via electrochemical anodization. *Thin Solid Films* **2014**, *571*, 168–174. [CrossRef]
4. Li, C.; Zang, Z.; Han, C.; Hu, Z.; Tang, X.; Du, J.; Leng, Y.; Sun, K. Highly compact CsPbBr$_3$ perovskite thin films decorated by ZnO nanoparticles for enhanced random lasing. *Nano Energy* **2017**, *40*, 195–202. [CrossRef]
5. Xu, Y.; Li, Y.; Zhang, H.; Jin, L.; Fang, X.; Shi, L.; Xu, L.; Ma, X.; Zou, Y.; Yin, J. Ultraviolet-enhanced electroluminescence from individual ZnO microwire/p-Si light-emitting diode by reverse tunneling effect. *J. Mater. Chem. C* **2017**, *5*, 6640–6646. [CrossRef]
6. Wang, Z.L.; Song, J. Piezoelectric nanogenerators based on zinc oxide nanowire arrays. *Science* **2006**, *312*, 242–246. [CrossRef] [PubMed]
7. Young, S.-J.; Liu, Y.-H. Field emission properties of Al-doped ZnO nanosheet based on field emitter device with UV exposure. *RSC Adv.* **2017**, *7*, 14219–14223. [CrossRef]
8. Tian, H.; Fan, H.; Ma, J.; Liu, Z.; Ma, L.; Lei, S.; Fang, J.; Long, C. Pt-decorated zinc oxide nanorod arrays with graphitic carbon nitride nanosheets for highly efficient dual-functional gas sensing. *J. Hazard. Mater.* **2018**, *341*, 102–111. [CrossRef]
9. Hatamie, A.; Khan, A.; Golabi, M.; Turner, A.P.F.; Beni, V.; Mak, W.C.; Sadollahkhani, A.; Alnoor, H.; Zargar, B.; Bano, S.; et al. Zinc oxide nanostructure-modified textile and its application to biosensing, photocatalysis, and as antibacterial material. *Langmuir* **2015**, *31*, 10913–10921. [CrossRef] [PubMed]
10. Manzoor, U.; Kim, D.K.; Islam, M.; Bhatti, A.S. Removal of micrometer size morphological defects and enhancement of UV emission by thermal treatment of Ga-doped ZnO nanostructures. *PLoS ONE* **2014**, *9*, e86418. [CrossRef] [PubMed]
11. Alam, M.; Islam, M.; Achour, A.; Hayat, A.; Ahsan, B.; Rasheed, H.; Salam, S.; Mujahid, M. Solution processing of cadmium sulfide buffer layer and aluminum-doped zinc oxide window layer for thin films solar cells. *Surf. Rev. Lett.* **2014**, *21*, 1450059. [CrossRef]
12. Wang, Y.; Zhu, G.; Mei, J.; Tian, C.; Liu, H.; Wang, F.; Zhao, D. Surface plasmon-enhanced two-photon excited whispering-gallery modes ultraviolet laser from Zno microwire. *AIP Adv.* **2017**, *7*, 115302. [CrossRef]
13. Zhang, L.; Yin, L.; Wang, C.; Lun, N.; Qi, Y.; Xiang, D. Origin of visible photoluminescence of ZnO quantum dots: Defect-dependent and size-dependent. *J. Phys. Chem. C* **2010**, *114*, 9651–9658. [CrossRef]

14. Li, D.; Leung, Y.H.; Djurišić, A.B.; Liu, Z.T.; Xie, M.H.; Shi, S.L.; Xu, S.J. Different origins of visible luminescence in ZnO nanostructures fabricated by the chemical and evaporation methods. *Appl. Phys. Lett.* **2004**, *85*, 1601–1603. [CrossRef]
15. Baral, A.; Khanuja, M.; Islam, S.S.; Sharma, R.; Mehta, B.R. Identification and origin of visible transitions in one dimensional (1D) ZnO nanostructures: Excitation wavelength and morphology dependence study. *J. Lumin.* **2017**, *183*, 383–390. [CrossRef]
16. Wang, Z.; Wang, F.; Cui, Y.; Li, X.; Toufiq, A.M.; Lu, Y.; Li, Q. Novel method to enhance the visible emission of ZnO nanostructures. *Chem. Phys. Lett.* **2014**, *614*, 53–56. [CrossRef]
17. Jadwisienczak, W.M.; Lozykowski, H.J.; Xu, A.; Pate, B. Visible emission from ZnO doped with rare-earth ions. *J. Electron. Mater.* **2002**, *31*, 776–784. [CrossRef]
18. Jung, Y.-I.; Noh, B.-Y.; Lee, Y.-S.; Baek, S.-H.; Kim, J.H.; Park, I.-K. Visible emission from Ce-doped ZnO nanorods grown by hydrothermal method without a post thermal annealing process. *Nanoscale Res. Lett.* **2012**, *7*, 43. [CrossRef]
19. Reish, M.E.; Zhang, Z.; Ma, S.; Harrison, I.; Everitt, H.O. How annealing and charge scavengers affect visible emission from ZnO nanocrystals. *J. Phys. Chem. C* **2016**, *120*, 5108–5113. [CrossRef]
20. Gwon, M.; Lee, E.; Kim, D.-W.; Yee, K.-J.; Lee, M.J.; Kim, Y.S. Surface-plasmon-enhanced visible-light emission of ZnO/Ag grating structures. *Opt. Express* **2011**, *19*, 5895–5900. [CrossRef] [PubMed]
21. Gogurla, N.; Bayan, S.; Chakrabarty, P.; Ray, S.K. Plasmon mediated enhancement of visible light emission of Au-ZnO nanocomposites. *J. Lumin.* **2018**, *194*, 15–21. [CrossRef]
22. Tian, J.-L.; Zhang, H.-Y.; Wang, G.-G.; Wang, X.-Z.; Sun, R.; Jin, L.; Han, J.-C. Influence of film thickness and annealing temperature on the structural and optical properties of ZnO thin films on Si (100) substrates grown by atomic layer deposition. *Superlattices Microstruct.* **2015**, *83*, 719–729. [CrossRef]
23. Manzano, C.V.; Alegre, D.; Caballero-Calero, O.; Alén, B.; Martín-González, M.S. Synthesis and luminescence properties of electrodeposited ZnO films. *J. Appl. Phys.* **2011**, *110*, 043538. [CrossRef]
24. Ton-That, C.; Weston, L.; Phillips, M.R. Characteristics of point defects in the green luminescence from Zn- and O-rich ZnO. *Phys. Rev. B* **2012**, *86*, 115205. [CrossRef]
25. Brahma, S.; Khatei, J.; Sunkara, S.; Lo, K.Y.; Shivashankar, S.A. Self-assembled ZnO nanoparticles on ZnO microsheet: Ultrafast synthesis and tunable photoluminescence properties. *J. Phys. D Appl. Phys.* **2015**, *48*, 225305. [CrossRef]
26. Cao, B.; Cai, W.; Zeng, H. Temperature-dependent shifts of three emission bands for ZnO nanoneedle arrays. *Appl. Phys. Lett.* **2006**, *88*, 161101. [CrossRef]
27. Liu, M.; Kitai, A.H.; Mascher, P. Point defects and luminescence centres in zinc oxide and zinc oxide doped with manganese. *J. Lumin.* **1992**, *54*, 35–42. [CrossRef]
28. Chua, S.J.; Loh, K.P.; Fitzgerald, E. The effect of post-annealing treatment on photoluminescence of ZnO nanorods prepared by hydrothermal synthesis. *J. Cryst. Growth* **2006**, *287*, 157–161.
29. Chen, C.; He, H.; Lu, Y.; Wu, K.; Ye, Z. Surface Passivation Effect on the photoluminescence of ZnO nanorods. *ACS Appl. Mater. Interfaces* **2013**, *5*, 6354–6359. [CrossRef] [PubMed]
30. Polydorou, E.; Zeniou, A.; Tsikritzis, D.; Soultati, A.; Sakellis, I.; Gardelis, S.; Papadopoulos, T.A.; Briscoe, J.; Palilis, L.C.; Kennou, S.; et al. Surface passivation effect by fluorine plasma treatment on ZnO for efficiency and lifetime improvement of inverted polymer solar cells. *J. Mater. Chem. A* **2016**, *4*, 11844–11858. [CrossRef]
31. Prucnal, S.; Gao, K.; Zhou, S.; Wu, J.; Cai, H.; Gordan, O.D.; Zahn, D.R.T.; Larkin, G.; Helm, M.; Skorupa, W. Optoelectronic properties of ZnO film on silicon after SF_6 plasma treatment and milliseconds annealing. *Appl. Phys. Lett.* **2014**, *105*, 221903. [CrossRef]
32. Lee, S.; Peng, J.W.; Liu, C.S. Photoluminescence and SERS investigation of plasma-treated ZnO nanorods. *Appl. Surf. Sci.* **2013**, *285*, 748–754. [CrossRef]
33. Baratto, C.; Comini, E.; Ferroni, M.; Faglia, G.; Sberveglieri, G. Plasma-induced enhancement of UV photoluminescence in ZnO nanowires. *CrystEngComm* **2013**, *15*, 7981–7986. [CrossRef]
34. Achour, H.; Achour, A.; Solaymani, S.; Islam, M.; Vizireanu, S.; Arman, A.; Ahmadpourian, A.; Dinescu, G. Plasma surface functionalization of boron nitride nano-sheets. *Diam. Relat. Mater.* **2017**, *77*, 110–115. [CrossRef]
35. Xu, S.; Wang, Z.L. One-dimensional ZnO nanostructures: Solution growth and functional properties. *Nano Res.* **2011**, *4*, 1013–1098. [CrossRef]

36. Law, M.; Greene, L.E.; Johnson, J.C.; Saykally, R.; Yang, P. Nanowire dye-sensitized solar cells. *Nat. Mater.* **2005**, *4*, 455–459. [CrossRef] [PubMed]
37. Vizireanu, S.; Ionita, M.D.; Dinescu, G.; Enculescu, I.; Baibarac, M.; Baltog, I. Post-synthesis carbon nanowalls transformation under hydrogen, oxygen, nitrogen, tetrafluoroethane and sulfur hexafluoride plasma treatments. *Plasma Process. Polym.* **2012**, *9*, 363–370. [CrossRef]
38. Lee, Y.; Zhang, Y.; Ng, S.L.G.; Kartawidjaja, F.C.; Wang, J. Hydrothermal growth of vertical ZnO nanorods. *J. Am. Ceram. Soc.* **2009**, *92*, 1940–1945. [CrossRef]
39. Riaz, A.; Ashraf, A.; Taimoor, H.; Javed, S.; Akram, M.A.; Islam, M.; Mujahid, M.; Ahmad, I.; Saeed, K. Photocatalytic and Photostability Behavior of Ag- and/or Al-Doped ZnO Films in Methylene Blue and Rhodamine B under UV-C Irradiation. *Coatings* **2019**, *9*, 202. [CrossRef]
40. Cao, X.L.; Zeng, H.B.; Wang, M.; Xu, X.J.; Fang, M.; Ji, S.L.; Zhang, L.D. Large scale fabrication of quasi-aligned ZnO stacking nanoplates. *J. Phys. Chem. C* **2008**, *112*, 5267–5270. [CrossRef]
41. Lepot, N.; van Bael, M.K.; van den Rul, H.; D'Haen, J.; Peeters, R.; Franco, D.; Mullens, J. Synthesis of ZnO nanorods from aqueous solution. *Mater. Lett.* **2007**, *61*, 2624–2627. [CrossRef]
42. Greene, L.E.; Law, M.; Tan, D.H.; Montano, M.; Goldberger, J.; Somorja, G.; Yang, P. General route to vertical ZnO nanowire arrays using textured ZnO seeds. *Nano Lett.* **2005**, *5*, 1231–1236. [CrossRef]
43. Zad, A.K.; Majid, W.H.A.; Abrishami, M.E.; Yousefi, R. X-ray analysis of ZnO nanoparticles by Williamson–Hall and size–strain plot methods. *Solid State Sci.* **2011**, *13*, 251–256.
44. Ilican, S.; Caglary, Y.; Caglar, M.; Demirci, B. Polycrystalline indium-doped ZnO thin films: Preparation and characterization. *J. Optoelectron. Adv. Mater.* **2008**, *10*, 2592–2598.
45. Yogamalar, R.; Srinivasan, R.; Vinu, A.; Ariga, K.; Bose, A.C. X-ray peak broadening analysis in ZnO nanoparticles. *Solid State Commun.* **2009**, *149*, 1919–1923. [CrossRef]
46. Vayssieres, L. Growth of arrayed nanorods and nanowires of ZnO from aqueous solutions. *Adv. Mater.* **2003**, *15*, 464–466. [CrossRef]
47. Pomar, F.S.; Martínez, E.; Meléndrez, M.F.; Tijerina, E.P. Growth of vertically aligned ZnO nanorods using textured ZnO films. *Nanoscale Res. Lett.* **2011**, *6*, 524. [CrossRef]
48. Macaluso, R.; Mosca, M.; Calì, C.; di Franco, F.; Santamaria, M.; di Quarto, F.; Reverchon, J.-L. Erroneous p-type assignment by Hall effect measurements in annealed ZnO films grown on InP substrate. *J. Appl. Phys.* **2013**, *113*, 164508. [CrossRef]
49. Børseth, T.M.; Svensson, B.G.; Kuznetsov, A.Y.; Klason, P.; Zhao, Q.X.; Willander, M. Identification of oxygen and zinc vacancy optical signals in ZnO. *Appl. Phys. Lett.* **2006**, *89*, 262112. [CrossRef]
50. Lozada, E.V.; González, G.M.C.; Torchynska, T. Photoluminescence emission and structure diversity in ZnO:Ag nanorods. *J. Phys. Conf. Ser.* **2015**, *582*, 012031. [CrossRef]
51. Kumar, V.; Swartn, H.C.; Ntwaeaborwa, O.M.; Kroon, R.E.; Terblans, J.J.; Shaat, S.K.K.; Yousif, A.; Duvenhage, M.M. Origin of the red emission in zinc oxide nanophosphors. *Mater. Lett.* **2013**, *101*, 57–60. [CrossRef]
52. Jiang, S.; Ren, Z.; Gong, S.; Yin, S.; Yu, Y.; Li, X.; Xu, G.; Shen, G.; Han, G. Tunable photoluminescence properties of well-aligned ZnO nanorod array by oxygen plasma post-treatment. *Appl. Surf. Sci.* **2014**, *289*, 252–256. [CrossRef]
53. Kim, Y.; Kim, M.; Leem, J.-Y. Optical and electrical properties of F-doped ZnO thin films grown on muscovite mica substrates and their optical constants. *J. Nanosci. Nanotechnol.* **2017**, *17*, 5693–5696. [CrossRef]
54. Papari, G.P.; Silvestri, B.; Vitiello, G.; de Stefano, L.; Rea, I.; Luciani, G.; Aronne, A.; Andreone, A. Morphological, structural, and charge transfer properties of F-doped ZnO: A spectroscopic investigation. *J. Phys. Chem. C* **2017**, *121*, 16012–16020. [CrossRef]
55. Xu, H.Y.; Liu, Y.C.; Ma, J.G.; Luo, Y.M.; Lu, Y.M.; Shen, D.Z.; Zhang, J.Y.; Fan, X.W.; Mu, R. Photoluminescence of F-passivated ZnO nanocrystalline films made from thermally oxidized ZnF_2 films. *J. Phys. Condens. Matter* **2004**, *16*, 5143–5150. [CrossRef]
56. Choi, Y.-J.; Kang, K.-M.; Park, H.H. Anion-controlled passivation effect of the atomic layer deposited ZnO films by F substitution to O-related defects on the electronic band structure for transparent contact layer of solar cell applications. *Sol. Energy Mater. Sol. Cells* **2015**, *132*, 403–409. [CrossRef]
57. El Hichou, A.; Bougrine, A.; Bubendorff, J.L.; Ebothe, J.; Addou, M.; Troyon, M. Structural, optical and cathodoluminescence characteristics of sprayed undoped and fluorine-doped ZnO thin films. *Semicond. Sci. Technol.* **2002**, *17*, 607–613. [CrossRef]

58. Kumar, P.M.R.; Kartha, C.S.; Vijayakumar, K.P.; Singh, F.; Avasthi, D.K. Effect of fluorine doping on structural, electrical and optical properties of ZnO thin films. *Mater. Sci. Eng. B* **2005**, *117*, 307–312. [CrossRef]
59. Achour, A.; Aissa, K.A.; Mbarek, M.; el Hadj, K.; Ouldhamadouche, N.; Barreau, N.; le Brizoual, L.; Djouadi, M.A. Enhancement of near-band edge photoluminescence of ZnO film buffered with TiN. *Thin Solid Films* **2013**, *583*, 71–77. [CrossRef]
60. Xu, H.Y.; Liu, Y.C.; Mu, R.; Shao, C.L.; Lu, Y.M.; Shen, D.Z.; Fan, X.W. F-doping effects on electrical and optical properties of ZnO nanocrystalline films. *Appl. Phys. Lett.* **2005**, *86*, 123107. [CrossRef]
61. Huang, T.-H.; Yang, P.-K.; Lien, D.-H.; Kang, C.-F.; Tsai, M.-L.; Chueh, Y.-L.; He, J.-H. Resistive memory for harsh electronics: Immunity to surface effect and high corrosion resistance via surface modification. *Sci. Rep.* **2014**, *4*, 4402. [CrossRef]
62. Hsieh, P.T.; Chen, Y.C.; Kao, K.S.; Wang, C.M. Luminescence mechanism of ZnO thin film investigated by XPS measurement. *Appl. Phys. A* **2008**, *90*, 317–321. [CrossRef]
63. Li, X.; Wang, Y.; Liu, W.; Jiang, G.; Zhu, C. Study of oxygen vacancies influence on the lattice parameter in ZnO thin film. *Mater. Lett.* **2012**, *85*, 25–28. [CrossRef]
64. Chen, Y.; Nayak, J.; Ko, H.U.; Kim, J. Effect of annealing temperature on the characteristics of ZnO thin films. *J. Phys. Chem. Solids* **2012**, *73*, 1259–1263. [CrossRef]
65. Achour, A.; Islam, M.; Solaymani, S.; Vizireanu, S.; Saeed, K.; Dinescu, G. Influence of plasma functionalization treatment and gold nanoparticles on surface chemistry and wettability of reactive-sputtered TiO$_2$ thin films. *Appl. Surf. Sci.* **2018**, *458*, 678–685. [CrossRef]
66. Javed, F.; Javed, S.; Akram, M.A.; Mujahid, M.; Islam, M.; Bhatti, A.S. Surface plasmon mediated optical properties of ZnO/Au/TiO$_2$ nanoheterostructure rod arrays. *Mater. Sci. Eng. B* **2018**, *231*, 32–39. [CrossRef]

© 2019 by the authors. Licensee MDPI, Basel, Switzerland. This article is an open access article distributed under the terms and conditions of the Creative Commons Attribution (CC BY) license (http://creativecommons.org/licenses/by/4.0/).

Article

"Three-Bullets" Loaded Mesoporous Silica Nanoparticles for Combined Photo/Chemotherapy †

André Luiz Tessaro [1,2,*], Aurore Fraix [1], Ana Claudia Pedrozo da Silva [3], Elena Gazzano [4], Chiara Riganti [4] and Salvatore Sortino [1,*]

1. Laboratory of Photochemistry, Department of Drug Sciences, University of Catania, 95125 Catania, Italy; fraix@unict.it
2. Department of Chemistry, Federal University of Technology, Paraná, R. Marcílio Dias, 635, Jardim Paraíso, Apucarana 86812-460, Paraná, Brazil
3. Department of Chemistry, Universidade Estadual de Maringá, Av. Colombo, 5.790, Maringá 87.020-900, Paraná, Brazil; anapedrozo1@gmail.com
4. Department of Oncology, University of Torino, Via Santena 5/bis, I-10126 Torino, Italy; elena.gazzano@unito.it (E.G.); chiara.riganti@unito.it (C.R.)
* Correspondence: andretessaro@utfpr.edu.br (A.L.T.); ssortino@unict.it (S.S.)
† Dedicated to the memory of Franca Perina.

Received: 26 April 2019; Accepted: 28 May 2019; Published: 31 May 2019

Abstract: This contribution reports the design, preparation, photophysical and photochemical characterization, as well as a preliminary biological evaluation of mesoporous silica nanoparticles (MSNs) covalently integrating a nitric oxide (NO) photodonor (NOPD) and a singlet oxygen ($^1O_2$) photosensitizer (PS) and encapsulating the anticancer doxorubicin (DOX) in a noncovalent fashion. These MSNs bind the NOPD mainly in their inner part and the PS in their outer part in order to judiciously exploit the different diffusion radius of the cytotoxic NO and $^1O_2$. Furthermore this silica nanoconstruct has been devised in such a way to permit the selective excitation of the NOPD and the PS with light sources of different energy in the visible window. We demonstrate that the individual photochemical performances of the photoactive components of the MSNs are not mutually affected, and remain unaltered even in the presence of DOX. As a result, the complete nanoconstruct is able to deliver NO and $^1O_2$ under blue and green light, respectively, and to release DOX under physiological conditions. Preliminary biological results performed using A375 cancer cells show a good tolerability of the functionalized MSNs in the dark and a potentiated activity of DOX upon irradiation, due to the effect of the NO photoreleased.

Keywords: multimodal therapy; singlet oxygen; doxorubicin; nitric oxide

1. Introduction

Multimodal cancer therapy involves the use of two or more treatment modalities with the aim to attack tumors on different sides by acting either on a single oncogenic pathway through different mechanisms or across parallel pathways without amplification of side effects [1–4]. In this frame, the combination of conventional chemotherapeutics with light-activated therapeutic treatments is a very appealing approach to potentiate the therapeutic outcome through anticancer synergistic/additive effects [5]. Light represents in fact as suitable and minimally invasive tool that permits the introduction of therapeutic species in biological environments with superb spatiotemporal control [6]. Among the light-activatable therapeutic treatments, photodynamic therapy (PDT) is so far one of the most promising to combat cancer diseases [7]. This treatment modality mainly exploits the cytotoxic effects of the highly reactive singlet oxygen ($^1O_2$), catalytically generated by energy transfer between the long-lived excited triplet state of a photosensitizer (PS) and the nearby molecular oxygen [8].

Another emerging light-activated approach is based on the photoregulated release of nitric oxide (NO) through the use of NO photodonors (NOPD) [9–11]. Although still confined to the research area, this NO-based PDT, namely NOPDT, has come to the limelight in recent years and holds very promising features in cancer treatment [12]. In fact, apart from playing multiple roles in the bioregulation of a broad array of physiological processes [13], NO has also proven to be an effective anticancer agent. If generated in an appropriate concentration range [14,15] this diatomic free radical can act not only as a cytotoxic agent [16], but also as inhibitor of the ATP binding cassette (ABC) transporters responsible of the efflux of chemotherapeutics and of the induction of multidrug resistance (MDR) in cancer cells [17,18]. In contrast to the photocatalytic mechanism leading to $^1O_2$ in PDT, the working principle of NOPDT exploits the excitation light to uncage NO leading, of course, to a consumption of the NOPD [9–11]. Noteworthy, both $^1O_2$ and NO are multitarget species that have not a reduced efficacy in MDR cells due to their short lifetimes, and confine their action to short distances from the production site inside the cells (<20 nm for $^1O_2$ and <200 μm for NO), reducing systemic toxicity issues common to many conventional drugs. Besides, since NO photorelease is independent of O_2 availability, it may successfully complement PDT at the onset of hypoxic conditions, typical for some tumors, where PDT may fail [19].

The significant breakthroughs in nanomedicine permit nowadays the development of multifunctional platforms in which more therapeutic agents are entrapped in a single nanocarrier. This offers an unprecedented opportunity to devise a better scheme for precise and controlled delivery of multiple therapeutics to the same area of the body at a predefined extra/intracellular level [20–23]. In this context, nanoplatforms combining PDT and NOPDT are opening new horizons towards more effective and less invasive cancer treatments entirely based on "nonconventional" chemotherapeutics [24]. Besides, an increasing large number of nanoconstructs in which PDT has been combined with chemotherapy in the management of several cancer types has been reported in literature [5]. In contrast, only recently NOPDT has been successfully used in combination with chemotherapeutics [25]. Based on this scenario, the achievement of a trimodal nanoplatform able to generate $^1O_2$ and NO and, simultaneously, to deliver a conventional chemotherapeutic is a very challenging objective to pursue. Mesoporous silica nanoparticles (MSNs) are very suitable scaffolds to this end due to their high loading capacity, ease of surface functionalization, good biocompatibility, and satisfactory light transparency [26–30]. Moreover these inorganic scaffolds have been extensively used as excellent carrier for (i) conventional chemotherapeutics to reverse MDR [31], (ii) PDT agents alone and in combinations with chemotherapeutics [32–34], and (iii) spontaneous NO donors [35] and NOPD [36].

In this manuscript, we report the first example of engineered MSNs able to deliver "three bullets" for potential anticancer therapy. These MSNs covalently integrate in their molecular scaffold erythrosine and a nitroaniline derivative as suitable PS and NOPD, respectively, and encapsulate the chemotherapeutic doxorubicin (DOX) in a noncovalent fashion. We demonstrate that the photochemical properties of the PS and NOPD are preserved in an excellent manner in the MSNs containing all active components. As a result, the spontaneous release of DOX under physiological conditions can be combined with the photoregulated generation of $^1O_2$, NO, or both by using light stimuli of appropriate energy in the visible light range.

2. Materials and Methods

2.1. Materials

All reagents (Sigma-Aldrich, Saint Louis, MO, USA) were of high commercial grade and were used without further purification. All solvents used (from Carlo Erba, Val de Reuil, France) were spectrophotometric grade. The free NOPD was synthesized according to our previously published procedure [37].

2.2. Synthetic Procedures

2.2.1. Synthesis of Amino-Modified MSNs (MSNs-NH$_2$)

MSNs-NH$_2$ were prepared according to literature procedures using cetyltrimethylammonium bromide (CTAB) as template [38]. Briefly, CTAB (1.9 mmoL) was dissolved in 340 mL of water followed by the addition of 2.45 mL of NaOH (2.0 mol L^{-1}). The temperature was adjusted at 80 °C and tetraethoxy silane (TEOS), 3.5 mL, 18.1 mmoL, and 3-aminopropyl triethoxy silane (APTES), (0.43 mL, 2.04 mmol) were simultaneously added dropwise to the solution during 4 min. The mixture was stirred at 80 °C for 2 h to give rise to white precipitate. The solid product was filtered, washed with deionized water and ethanol, and dried at 60 °C for 24 h.

2.2.2. Synthesis of the PS-Modified MSNs (PS-MSNs)

500 mg of MSNs-NH$_2$ were added in 15 mL of DMF and sonicated (90 s, 42 KHz). Then PS erythrosine (27 mg, 1 eq), O-(7-Azabenzotriazol-1-yl)-1,1,3,3-tetramethyl uronium hexafluorophosphate (11.1 mg, 1 eq) and N,N-di-isopropiletilamine (10 µL, 2 eq) were added to the suspension. The mixture was stirred at room temperature for 24 h under dark conditions. The solid product was filtered, washed with DMF (4 portions of 5 mL) and dried at 60 °C for 24 h. PS-MSNs were prepared with the nominal concentration of ca. 0.03% (w/w) of PS. The real concentration of the PS was indirectly determined by UV–Vis spectrum of the eluate (post washing) using ε_{543} = 112.000 M^{-1} cm^{-1} in DMF.

2.2.3. Synthesis of NOPD-Modified PS-MSNs (PS-MSNs-NOPD)

One-hundred milligrams of PS-MSNs was added in 10 mL of DMSO and sonicated (90 s, 50 KHz). Then cesium carbonate (10 mg, 0.2 eq) and 4-chloro-1-nitro-2(trifluoromethyl) benzene (30 µL, 2 eq) were added and refluxed overnight. The solid product was filtered and washed with ethyl acetate (to remove the unbounded 4-chloro-1-nitro-2(trifluoromethyl) benzene) and water (to remove the excess of cesium carbonate) and dried under vacuum to obtain PS-MSNs-NOPD. PS-MSNs-NOPD were prepared with the nominal concentration of ca. 0.4% (w/w) of NOPD. The real concentration of the NOPD was indirectly determined by UV–Vis spectrum of the eluate (post washing) using ε_{400} = 10.000 M^{-1} cm^{-1} in water.

2.2.4. Stability of the PS-MSNs-NOPD

Representative samples of PS-MSNs-NOPD (1 mg mL^{-1}) were either sonicated for 1 h or stirred at 70 °C for 2 h in the dark. Afterwards the samples were centrifuged for 25 min and the supernatant was collected and analyzed by UV-VIS absorption. The absence of any relevant absorption between 250 and 800 nm, suggested that no leaching of the chromogenic units from the MSN scaffold occurs.

2.3. Instrumentation

Transmission electron microscopy (TEM) experiments were performed with a TEM JEOL JEM-2010 (Pleasanton, CA, USA) using the bright field in conventional parallel beam (CTEM) mode (BF).

The X-ray diffraction analysis were done at the Complexo de Apoio à Pesquisa (COMCAP/UEM) in Bruke diffractometer model D8 Advance with radiation Cu-Kα (λ = 1.54062 Å). The scanning was done from 1 to 10° of 2θ with 0.45°/min.

UV-Vis spectra absorption and fluorescence emission spectra were recorded with a Jasco V-560 spectrophotometer (Easton, MD, USA) and a Spex Fluorolog-2 (mod. F-111) spectrofluorimeter, respectively, in air-equilibrated solutions, using either quartz cells with a path length of 1 cm. Fluorescence lifetimes were recorded with the same fluorimeter equipped with a TCSPC Triple Illuminator (Kyoto, Japan). The samples were irradiated by a pulsed diode excitation source (Nanoled) at 455 nm and the decays were monitored at 500 nm. The system allowed measurement of fluorescence lifetimes from 200 ps. The multiexponential fit of the fluorescence decay was obtained using the following equation.

$$I(t) = \sum \alpha_I \exp(-t/\tau_i)$$

Absorption spectral changes were monitored by irradiating the sample in a thermostated quartz cell (1 cm path length, 3 mL capacity) under gentle stirring, using a continuum laser with $\lambda_{exc} = 405$ nm and $\lambda_{exc} = 532$ nm (ca. 100 mW) and having a beam diameter of ca. 1.5 mm for the excitation of the NOPD and the PS, respectively.

Direct monitoring of NO release in solution was performed by amperometric detection (World Precision Instruments, Sarasota, FL, USA), with a ISO-NO meter, equipped with a data acquisition system, and based on direct amperometric detection of NO with short response time (<5 s) and sensitivity range 1 nM to 20 µM. The analogue signal was digitalized with a four-channel recording system and transferred to a PC. The sensor was accurately calibrated by mixing standard solutions of $NaNO_2$ with 0.1 M H_2SO_4 and 0.1 M KI according to the following reaction.

$$4H^+ + 2I^- + 2NO_2^- \rightarrow 2H_2O + 2NO + I_2$$

Irradiation was performed in a thermostated quartz cell (1 cm path length, 3 mL capacity) using the continuum laser with $\lambda_{exc} = 405$ nm. NO measurements were carried out under stirring with the electrode positioned outside the light path in order to avoid NO signal artifacts due to photoelectric interference on the ISO-NO electrode.

2.4. Chemical Detection of NO

NO release was also measured indirectly by means of the well-known, highly sensitive (detection limit on the order of the picomoles) fluorimetric bioassay of Misko et al. [39] based on the ring closure of the nonfluorescent 2,3-diaminonaphthalene (DAN) with nitrite to form the highly fluorescent product 2,3-diaminonaphthotriazole (DANT). Aliquots of 2 mL of aqueous suspension of the desired MSNs were irradiated or kept in the dark. Afterwards, the samples were centrifuged by 25 min and 1 mL of the supernatant was collected. One-hundred microliters of DAN solution (DAN 0.30 M in 0.62 M HCl) was added to 1 mL of the supernatant and solutions were stirred for 20 min at room temperature. Three-hundred microliters of NaOH 3 M was added in previous solution and stirred for 20 min at room temperature. The resultant solution was put into the fluorescent cuvette and the fluorescence was recorded at $\lambda_{exc} = 360$ nm.

2.5. Chemical Detection of Singlet Oxygen

The release of $^1O_2$ by the MSNs was determined by indirect measurements using the well-established traps of $^1O_2$ such as 1,3-diphenylisobenzofuran (DPBF) [40] and p-nitrosodimethylaniline/histidine (RNO/His) [41] in ethanol and phosphate buffer (pH = 7.4), respectively. The reaction between the scavengers and $^1O_2$ can be followed by the UV-Vis absorption by monitoring the bleaching of the traps at 400 nm and 440 nm in the case of DPBF and RNO/His, respectively. The free PS in ethanol or in PBS was used as standard and the $^1O_2$ delivery efficiency (η_Δ) can be calculated by the equation the following equation [42].

$$\eta_{\Delta particle} = \Phi_{ERY} \frac{t_{ERY}}{t_{particle}}$$

Briefly, the scavenger is transferred into a suspension of the nanoparticle (1 mg mL^{-1}) or a solution containing the PS and, under constantly stirring, irradiated at time intervals. In order to avoid intense scattering the spectrum was registered after the deposition of the particles on the bottom of cuvette. The absorptions of both solutions were normalized at 532 nm assuming that the real absorption of the PS bounded in the MSNs can be calculated by measuring the absorption spectrum of a suspension of PS-MSNs and subtracting the baseline scattering by multiple point level baseline correction.

2.6. Loading and Release of DOX

Four-and-a-half micrograms of PS-MSNs-NOPD was mixed with 4.0 mL of PBS (pH = 7.4) containing DOX (220 μM). After stirring for 24 h, in the dark and at room temperature, the DOX-loaded PS-MSNs-NOPD (PS-MSNs-NOPD/DOX) were collected by centrifugation. The samples were washed with PBS and the content of the DOX was determined by comparison between the absorbance of the supernatant and the original solution at 482 nm (ε_{482} = 10.000 M^{-1} cm^{-1}). A dispersion containing 1 mg of PS-MSNs-NOPD/DOX in 2.0 mL of PBS (pH 7.4) was incubated at 37 °C, in the dark under continuous stirring. At specified time, 100 μL of the supernatant was taken out after centrifugation (15,000× g rpm for 2 min) and an equal volume of fresh PBS was added to keep the sink condition. The concentration was determined by reading the fluorescence emission of DOX at 594 nm.

2.7. Biological Assays

2.7.1. Cells Characterization

Human melanoma A375 cells (ATCC, Manassas, VA, USA) were maintained in DMEM medium supplemented with 10% v/v fetal bovine serum, 1% v/v penicillin-streptomycin, 1% v/v L-glutamine. Cells were maintained in a humidified atmosphere at 37 °C, 5% CO_2. To measure the expression of ABC transporters, 1 × 10^6 cells were rinsed and fixed with 2% w/v paraformaldehyde (PFA) for 2 min, washed three times with PBS and stained with anti-P-glycoprotein (Pgp/ABCB1) (Kamiya, Hamamatsu City, Japan), anti-MDR-related protein 1 (MRP1/ABCC1) (Abcam, Cambridge, UK) and anti-breast cancer resistance protein (BCRP/ABCG2) (SantaCruz Biotechnology Inc., Santa Cruz, CA, USA) antibodies for 1 h on ice, followed by an AlexaFluor 488-conjugated secondary antibody (Millipore, Billerica, MA, USA) for 30 min. One-hundred-thousand cells were analyzed with EasyCyte Guava™ flow cytometer (Millipore), equipped with the InCyte software (Millipore). Control experiments included incubation with non-immune isotype antibody.

2.7.2. Viability Assays

Cells were incubated for 4 h with the different MSNs samples (1 mg mL^{-1}). Cells were maintained in the dark or irradiated at room temperature for 30 min with a blue LED at 400 nm with an irradiance of 1.75 mW/cm^2; irradiance was measured with a Delta Ohm HD2302.0 lightmeter equipped with a Delta Ohm LP4771RAD light probe. Twenty-four hours after the irradiation, cells were stained with 5% w/v crystal violet aqueous solution in 66% v/v methanol, washed twice with water, solubilized with 10% v/v acetic acid. The absorbance was read at 570 nm, using a Packard EL340 microplate reader (Bio-Tek Instruments, Winooski, VT, USA). The absorbance of untreated cells was considered as 100% viability; the results were expressed as percentage of viable cells versus untreated cells.

2.7.3. Statistical Analysis

Data are provided as means ± SD. The results were analyzed by a one-way ANOVA and Tukey's test. $p < 0.05$ was considered significant.

3. Results and Discussion

Scheme 1 illustrates the "three-bullets" MSNs with their working principle, and the rationale behind their design is described in the following.

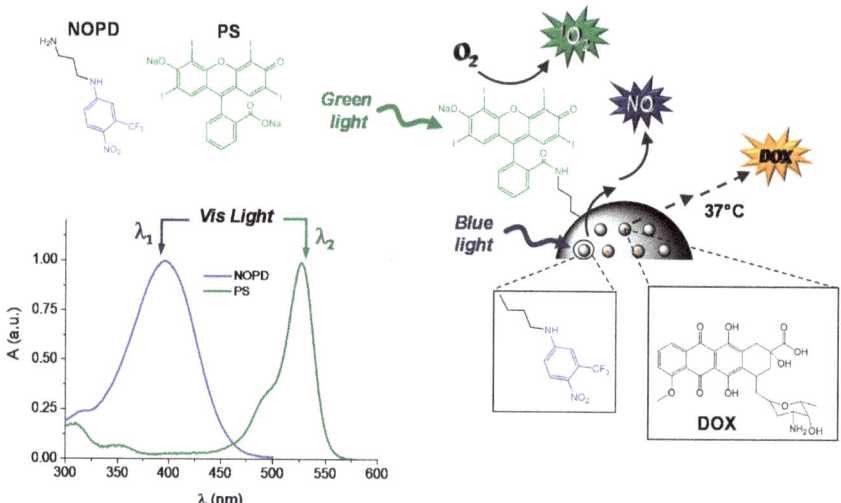

Scheme 1. Schematic for the "three-bullets" PS-MSNs-NOPD/DOX and their working principle. The molecular structures of the free photosensitizer (PS) and the nitric oxide photodonor (NOPD) and their normalized absorption spectra in PBS (10 mM, pH 7.4) are also shown.

As outlined in the introduction, $^1O_2$ and NO have diffusion radius of <20 nm and <200 μm, respectively, due to their short lifetimes: ca. 3 μs for $^1O_2$ [43] and ca. 5 s for NO [12]. Taking this into account we designed the MSNs in order to integrate mainly the PS in the outer part of the scaffold and the NOPD in the inner pores. This strategy avoids that $^1O_2$ decays before to get out from the scaffold. On the other hand, despite being photogenerated in the core of the scaffold NO can diffuse out due to its longer lifetime. To this end, initially we synthesized the PS-MSNs before to eliminate the CTAB template (see experimental) from the amino terminated MSNs-NH$_2$. In such a way, most of the condensation reaction between MSNs-NH$_2$ and the free PS takes place at the scaffold surface. Afterwards, most of the CTAB was eliminated by the pores allowing the covalent integration of the NOPD mainly therein, according with the procedure described in the experimental.

An important issue to be addressed in the fabrication of phototherapeutic systems aimed at exploiting the effects of $^1O_2$ and NO, regards the relative concentration of these species generated upon light excitation. As outlined above, $^1O_2$ is produced through a photocatalytic process that, in principle, does not consume the PS, whereas photogeneration of NO occurs through a neat photochemical reaction with consequent consumption of the NOPD. As a consequence, the regulation of the reservoir of NO with respect to $^1O_2$ is a critical point to be considered in view of an effective bimodal photodynamic action. This can be accomplished by (i) using a larger amount of the NOPD with respect to the PS, (ii) using PS and NOPD preferentially absorbing in different spectral regions, or (iii) combining both conditions. The appropriate synthetic protocol for the anchoring of the PS and NOPD permits to fulfill the first condition due to the regulation of the relative concentrations of distinct subsets of chromogenic units integrated within the very same scaffold. For such a reason, the amount of NOPD bounded to the MSNs was much larger than the PS (0.4% vs. 0.03%). Besides, the appropriate selection of the excitation wavelengths fulfills the second condition. On this basis, one can realize that the choice of erythrosine as PS and the nitroaniline derivative as NOPD is, of course, not casual. In fact, as shown in the spectra reported in Scheme 1, these two photoprecursors absorb in distinct regions of the visible spectral range and therefore can be selectively or simultaneously excited by using blue light, green light or both.

PS-MSNs were characterized by X-ray powder diffraction (XRD), TEM, and energy-dispersive X-ray spectroscopy (EDX) analysis in order to verify their structural parameters and particle size.

Figure 1A shows the diffractograms for the original material (MSNs) as well for the functionalized PS-MSNs. The typical reflections at 100, 110, 200, and 210 (not well resolved) confirm the production of an ordered hexagonal network material. The covalent addition of the PS did not affect the structural integrity of the material. The shift to higher values of 2θ can indicate a contraction of the porous due to the condensation of the sylanol groups. TEM microscopy (Figure 1B) shows particles with reasonable homogeneity regarding size and shape having an average size around 100 nm. EDX analysis (Figure 1C) confirms the anchoring of the PS by the signal of iodide in the external surface.

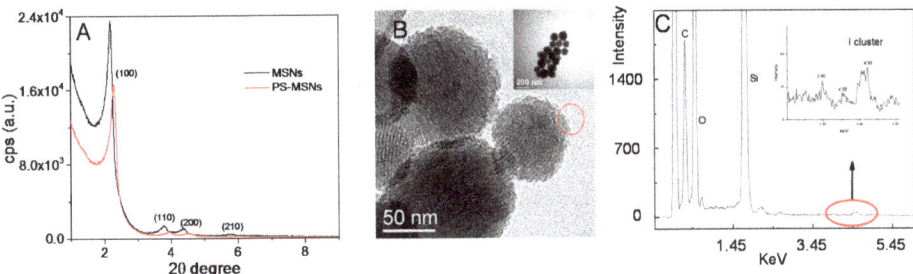

Figure 1. (**A**) X-ray powder diffraction (XRD) patterns of mesoporous silica nanoparticles (MSNs) (black line) and PS-MSNs (red line). (**B**) Representative TEM images and (**C**) EDX analysis of the PS-MSNs. The inset B shows a homogeneous distribution of the nanoparticles and the inset of C shows a zoom in the region of the iodide cluster.

The binding of the PS to the MSNs was further confirmed by the absorption and fluorescence emission spectra of the PS-MSNs (Figure 2). The aqueous suspension of PS-MSNs shows the characteristic absorption of the PS at 537 nm, slightly red-shifted with respect to that of unbounded PS (see absorption spectrum in Scheme 1), and its typical fluorescence emission in the green region with maximum at ca. 550 nm.

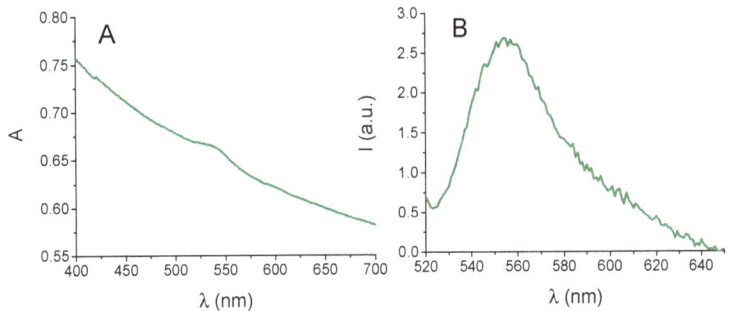

Figure 2. (**A**) Absorption and (**B**) fluorescence emission spectra (λ_{exc} = 510 nm) of PS-MSNs suspension (1 mg mL^{-1}) in PBS (pH 7.4, 10 mM).

The $^1O_2$ photogeneration of the PS-MSNs was preliminary demonstrated by using DPBF, a well-known trap for this reactive oxygen species (see experimental). Figure 3 shows that the bleaching of the typical absorption band of the DPBF at 411 nm is observed only in the case of the MSNs functionalized with the PS. Furthermore, a control experiment performed with an optically matched solution of the free PS clearly show that the bleaching kinetic of the PS is not affected upon its covalent binding with the silica scaffold, fully supporting its localization at the MSNs surface.

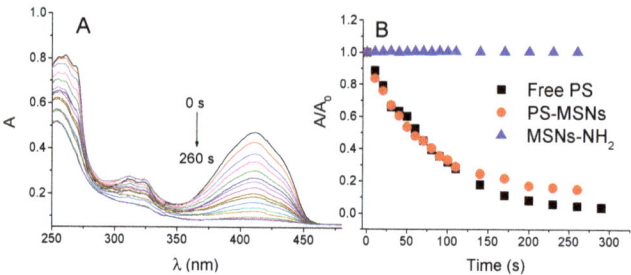

Figure 3. (**A**) Absorption spectral changes of DPBF (25 µM) observed upon irradiation at 532 nm (50 mW) in the presence of PS-MSNs suspension (1 mg mL^{-1}) in ethanol. (**B**) A/A0 ratio at 411 nm of DPBF (25 µM) observed upon irradiation at 532 nm (50 mW) in the presence of free PS and suspensions of PS-MSNs (1 mg mL^{-1}) and MSN-NH$_2$ (1 mg mL^{-1}) in ethanol.

The nitroaniline NOPD previously developed in our group [44] was covalently integrated in one-step into the PS-MSNs to give PS-MSNs-NOPD by reaction with the colorless chloro-nitroderivative (see experimental). Figure 4A shows unambiguous evidence for the successful synthesis of the PS-MSNs-NOPD. The absorption is in fact dominated by the typical charge transfer band of the nitroaniline chromophore in the blue region [44]. Note that, this absorption is ca. 10 nm blue-shifted if compared to that observed for the free NOPD in aqueous medium (see spectrum in Scheme 1, for sake of comparison), probably as a result of the different environment experienced by this chromogenic unit into the pore of the silica scaffold. This finding is in excellent agreement to what we recently observed for the same NOPD anchored on similar MSNs [36]. Note that, the introduction of the NOPD functionality into the nanoscaffold did not affect the particles size whose diameter was still of ~100 nm.

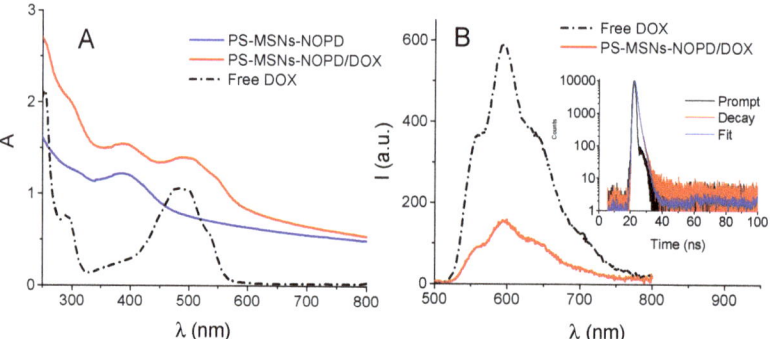

Figure 4. (**A**) Absorption spectra of free doxorubicin (DOX) and PS-MSNs-NOPD and PS-MSNs-NOPD/DOX suspensions (1 mg mL^{-1}) in PBS (pH 7.4, 10 mM). (**B**) Fluorescence emission spectra (λ_{exc} = 490 nm) of free DOX and PS-MSNs-NOPD/DOX suspension (1 mg mL^{-1}) in PBS (pH 7.4, 10 mM). The two samples are optically matched at the excitation wavelength. The inset shows the fluorescence decay and the related biexponential fitting for the PS-MSNs-NOPD/DOX suspension (λ_{exc} = 455 nm, λ_{em} = 590 nm).

The covalent binding of the NOPD mainly in the inner part of the MSNs does not preclude their capability to encapsulate in a noncovalent fashion additional therapeutic compounds. As a proof of principle, we selected Doxorubicin (DOX) as a suitable chemotherapeutic. This drug has a wide spectrum of activity in clinical application but it has to face the development of MDR in cancer cells [45]. The ABC transporters Pgp/ABCB1, MRP1/ABCC1 and BCRP/ABCG2 are the main transporters effluxing DOX from tumor cells [46]. The spectrum shown in Figure 4A clearly shows the effective encapsulation of DOX within the PS-MSNs-NOPD. In fact, the absorption band of the NOPD is accompanied by the characteristic band of the DOX with maximum at ca. 488 nm (see spectrum of the free DOX in Figure 4, for sake of comparison). DOX encapsulation was further confirmed by the appearance of its typical fluorescence emission in the red region (Figure 4B). Due to the scattering of the silica scaffold, the emission efficiency of the encapsulated DOX can be only estimated, resulting in ~70% smaller than the free drug. However, the biexponential decay observed for the PS-MSNs-NOPD/DOX (see inset Figure 4B) gave lifetimes of 0.54 and 2.1 ns, which were basically the same to those observed for the free DOX. These findings probably suggest the presence of some either non-emissive or scarcely emissive aggregates of DOX (these latter with lifetime below our time-resolution window) within the pores of the silica scaffold. The percentage of the DOX incorporated was ca. 85% corresponding to 90 µg of DOX per mg of MSNs

In order to demonstrate the validity of the "three bullets" MSNs, we performed $^1O_2$ and NO measurements under green and blue light stimuli, respectively, and DOX release experiments at 37 °C. In this case, we used *p*-nitrosodimethylaniline/histidine (RNO/His), a well-known $^1O_2$ scavenger in aqueous PBS. In fact, in contrast to the previously used DPBF, RNO absorption maximum is at ca. 440 nm and thus is less affected by the absorption of the NOPD unit. Figure 5A shows the evolution of the absorbance bleaching for the unfunctionalized MSNs, those integrating only PS, both PS and NOPD, and the complete nanoplatform also loading DOX. The bleaching monitored at 440 nm upon 532 nm light excitation was observed in all samples containing the PS and provides a clear cut evidence for the delivery of $^1O_2$, in contrast to the result observed for the unfunctionalized MSNs. Interestingly, the efficiency for the $^1O_2$ observed in the case of the complete system PS-MSNs-NOPD/DOX was basically comparable with the samples not loaded with DOX and not containing the NOPD, suggesting that the copresence of the NOPD and DOX does not influence the photophysical properties of the bound PS.

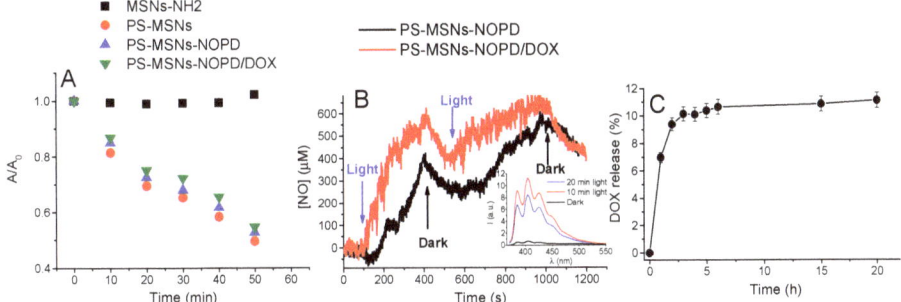

Figure 5. (**A**) A/A0 ratio at 440 nm of *p*-nitrosodimethylaniline/histidine (RNO) (17 µM) observed upon irradiation at 532 nm (50 mW) in the presence of different suspensions (1 mg mL^{-1}) of MSNs in PBS (pH 7.4, 10 mM). (**B**) NO release profile observed upon irradiation at 405 nm (100 mW) of different suspensions (1 mg mL^{-1}) of MSNs in PBS (pH 7.4, 10 mM). The inset shows the indirect e NO fluorimetric detection (λ_{exc} = 365 nm) (see experimental) observed for PS-MSNs-NOPD suspension (1 mg ml^{-1}) in the dark and after 10 and 20 min irradiation at 405 nm. (**C**) DOX release observed for PS-MSNs-NOPD/DOX suspension (1 mg mL^{-1}) in PBS (pH 7.4, 10 mM) at 37 °C.

The NO photorelease properties of PS-MSNs-NOPD/DOX were unambiguously demonstrated by direct real-time monitoring using an ultrasensitive NO electrode, which directly detects NO with nM concentration sensitivity employing an amperometric technique. The results illustrated in Figure 5B provide evidence that the nanoconstruct is stable in the dark, whereas it releases NO upon blue light excitation at 405 nm, stops as the light is turned off, and restarts again upon illumination. Also in this case, a comparison with the sample not loaded with DOX, confirmed that the efficiency of the NO photorelease is basically not affected by the presence of the chemodrug. NO photorelease was further proved by the indirect DAN assay (see experimental), one of the most sensitive and selective fluorescence-based methods for NO detection as nitrite (see experimental). As evident from the inset of Figure 5B, the significant fluorescence for of the nitrite probe was observed only in the case of the irradiated samples.

DOX release from the nanoparticles was performed under sink conditions at 37 °C in PBS at pH 7.4. Figure 5C illustrates the releasing profile of the DOX monitored for 24 h. After 3 h, ca. 10% of DOX was released suggesting that the presence of the NOPD and the PS in the silica scaffold do not change the typical release profile of DOX. Note that the amount of released chemodrug is comparable with that observed for DOX in similar MSN scaffolds [34].

As outlined in the introductory part, NO may play a double role when combined with chemotherapeutics. It can increase the therapeutic effectiveness of a chemodrug by acting as cytotoxic agent itself, if produced at significant concentrations, but also by inhibiting the ABC transporters responsible for MDR, when generated in non-toxic amount. Therefore, we considered it useful to perform some very preliminary biological experiments by using A375 melanoma cell line, overexpressing Pgp, MRP1, and BCRP (Figure 6), i.e., the main pumps effluxing DOX.

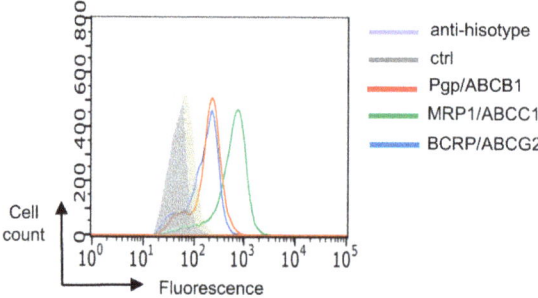

Figure 6. Representative histograms of Pgp/ABCB1, MRP1/ABCC1, and BCRP/ABCG2 surface levels in A375 cells, measured by flow cytometry in duplicate. The figure is representative of 1 out of 3 experiments with similar results.

Thanks to the logical design of the MSNs, irradiation with blue light permits the selective excitation of the NOPD component, ruling out any participation of $^1O_2$ (which would require green light excitation of the PS) in the cell viability. Figure 7 shows that the different samples of MSNs are well tolerated in the dark and that only a slight toxicity was observed upon blue light irradiation. Moreover, the very moderate phototoxicity observed in the sample with NOPD (PS-MSNs-NOPD) is basically the same to that of the sample, which does not integrate the NOPD (PS-MSNs/DOX), suggesting that under these experimental conditions NO is produced at not toxic doses.

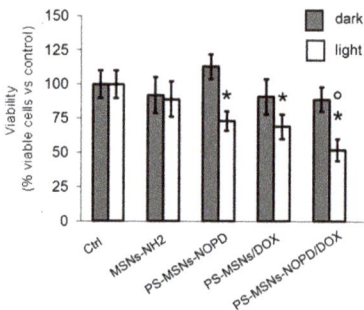

Figure 7. Cell viability of A375 cells, treated a reported in the Materials and Methods Section. Data are presented as means ± SD (n = 3 independent experiments). * $p < 0.02$: vs. Ctrl, irradiated cells (light); ° $p < 0.05$: PS-MSNs-NOPD/DOX vs. PS-MSNs-NOPD and PS-MSNs/DOX.

Interestingly, irradiation of the whole nanoconstruct (PS-MSNs-NOPD/DOX) increases cytotoxicity. PS-MSNs-NOPD/DOX reduced cell viability more than PS-MSNs-NOPD and PS-MSNs/DOX, suggesting that the presence of NOPD enhances the efficacy of DOX. We believe that such effect can be tentatively attributed to the inhibition of the ABC transporters present on A375 cells by NO photogenerated, an event already observed in melanoma cells treated with free NOPD/DOX hybrids [25].

4. Conclusions

We have designed, synthesized and characterized a novel "three bullet" nanoconstruct based on mesoporous silica nanoparticles for potential combined cancer photo-chemotherapy. Such a multifunctional nanoplatform is able to generate $^1O_2$ and NO under selective excitation with green and blue light, respectively, and release the noncovalently entrapped anticancer DOX under physiological conditions. A remarkable advantage of this system is the excellent preservation of the photochemical properties of the active components once simultaneously integrated in the same scaffold and despite the copresence of the DOX. This is the result of the strategy adopted which permitted to integrate the PS and the NOPD in the outer and the inner sites of the nanoscaffold, avoiding potential and undesired intermolecular photoprocesses (i.e., energy/electron transfer) which would have otherwise precluded the final goal. This allows the individual photoactivable components to be operated either singularly or in tandem upon the appropriate choice of the excitation wavelengths. At this regard, preliminary biological results performed on A375 cells expressing ABC transporters show a potentiated activity of DOX incorporated in our "three bullet" nanoconstruct upon irradiation with blue light, due to the effect of the NO photoreleased. Note that the reduction of the cell viability observed under light excitation (ca. 50%) is considered effective in overcoming DOX-resistance in cells expressing ABC transporters [47,48], as those used in the present work. Besides its efficacy, a plus of our approach is the absence of toxicity exerted by PS-MSNs/NOPD/DOX in the absence of irradiation. Since light can be easily controlled and directed towards tumor tissue, in a translational perspective, our approach can be considered a tumor-selective strategy that avoids undesired toxicity on nontransformed tissues and overcomes a severe limitation of currently used chemotherapy. Detailed photobiological investigations addressed to elucidate the single and the combined effects of all the active species produced by this novel multifunctional drug delivery system are currently under investigation in our laboratories.

Author Contributions: Conceptualization, A.L.T. and S.S.; Investigation, A.L.T., A.F., A.C.P.d.S., E.G., and C.R.; Methodology, C.R.; Supervision, S.S.; Writing—Original Draft, S.S.

Funding: This research was funded by Italian Association for Cancer Research (AIRC) IG-19859.

Acknowledgments: We thank S. Petralia (STMicroelectronics, Catania) for the help in TEM analysis. A.L.T. thanks UTFPR and CAPES for the financial support process no. 88881.119200/2016-01.

Conflicts of Interest: The authors declare no conflicts of interest.

References

1. Kemp, J.A.; Shim, M.S.; Heo, C.Y.; Kwon, Y.J. "Combo" nanomedicine: Co-delivery of multi-modal therapeutics for efficient, targeted, and safe cancer therapy. *Adv. Drug Deliv. Rev.* **2016**, *98*, 3–18. [CrossRef] [PubMed]
2. Li, Y.; Atkinson, K.; Zhang, T. Combination of chemotherapy and cancer stem cell targeting agents: Preclinical and clinical studies. *Cancer Lett.* **2017**, *396*, 103–109. [CrossRef] [PubMed]
3. Dawson, M.A. The cancer epigenome: Concepts, challenges, and therapeutic opportunities. *Science* **2017**, *355*, 1147–1152. [CrossRef] [PubMed]
4. Lehar, J.; Krueger, A.S.; Avery, W.; Heilbut, A.M.; Johansen, L.M.; Price, E.R.; Rickles, R.J.; Short, G.F.; Staunton, J.E.; Jin, X.; et al. Synergistic drug combinations tend to improve therapeutically relevant selectivity. *Nat. Biotechnol.* **2009**, *27*, 659–666. [CrossRef]

5. Quaglia, F.; Sortino, S. Polymer Nanoparticles for Cancer Photodynamic Therapy Combined with Nitric Oxide Photorelease and Chemotherapy. In *Applied Photochemistry. Lecture Notes in Chemistry*; Bergamini, S.G., Silvo, S., Eds.; Springer: Basel, Switzerland, 2016; Volume 92, pp. 397–426. ISBN 978-3-319-31669-7.
6. Sortino, S. Photoactivated nanomaterials for biomedical release applications. *J. Mater. Chem.* **2012**, *22*, 301–318. [CrossRef]
7. Castano, A.P.; Mroz, P.; Hamblin, M.R. Photodynamic therapy and anti-tumour immunity. *Nat. Rev. Cancer* **2006**, *6*, 535–545. [CrossRef]
8. Celli, J.P.; Spring, B.Q.; Rizvi, I.; Evans, C.L.; Samkoe, K.S.; Verma, S.; Pogue, B.W.; Hasan, T. Imaging and photodynamic therapy: Mechanisms, monitoring, and optimization. *Chem. Rev.* **2010**, *12*, 2795–2838. [CrossRef]
9. Sortino, S. Light-controlled nitric oxide delivering molecular assemblies. *Chem. Soc. Rev.* **2010**, *39*, 2903–2913. [CrossRef] [PubMed]
10. Ford, P.C. Photochemical delivery of nitric oxide. *Nitric Oxide* **2013**, *34*, 56–65. [CrossRef]
11. Fry, N.L.; Mascharak, P.K. Photoactive ruthenium nitrosyls as NO donors: How to sensitize them toward visible light. *Acc. Chem. Res.* **2011**, *44*, 289–298. [CrossRef]
12. Ostrowski, A.D.; Ford, P.C. Metal complexes as photochemical nitric oxide precursors: Potential applications in the treatment of tumors. *Dalton Trans.* **2009**, *48*, 10660–10669. [CrossRef] [PubMed]
13. Ignarro, L.J. (Ed.) *Nitric Oxide: Biology and Pathobiology*; Elsevier Inc.: Amsterdam, The Netherlands, 2009; ISBN 9780123738660.
14. Fukumura, D.; Kashiwagi, S.; Jain, R.K. The role of nitric oxide in tumour progression. *Nat. Rev. Cancer* **2006**, *6*, 521–534. [CrossRef]
15. Carpenter, A.W.; Schoenfisch, M.H. Nitric oxide release: Part II. Therapeutic applications. *Chem. Soc. Rev.* **2012**, *41*, 3742–3752. [CrossRef]
16. Wink, D.A.; Mitchell, J.R. Chemical biology of nitric oxide: Insights into regulatory, cytotoxic, and cytoprotective mechanisms of nitric oxide. *Free Radic. Biol. Med.* **1998**, *25*, 434–456. [CrossRef]
17. Riganti, C.; Miraglia, E.; Viarisio, D.; Costamagna, C.; Pescarmona, G.; Ghigo, D.; Bosia, A. Nitric oxide reverts the resistance to doxorubicin in human colon cancer cells by inhibiting the drug efflux. *Cancer Res.* **2005**, *65*, 516–525. [PubMed]
18. De Boo, S.; Kopecka, J.; Brusa, D.; Gazzano, E.; Matera, L.; Ghigo, D.; Bosia, A.; Riganti, C. iNOS activity is necessary for the cytotoxic and immunogenic effects of doxorubicin in human colon cancer cells. *Mol. Cancer* **2009**, *8*, 108. [CrossRef]
19. Fowley, C.; McHale, A.P.; McCaughan, B.; Fraix, A.; Sortino, S.; Callan, J.F. Carbon quantum dot–NO photoreleaser nanohybrids for two-photon phototherapy of hypoxic tumors. *Chem. Commun.* **2015**, *51*, 81–84. [CrossRef] [PubMed]
20. Couvreur, P. Nanoparticles in drug delivery: Past, present and future. *Adv. Drug Deliv. Rev.* **2013**, *65*, 21–23. [CrossRef] [PubMed]
21. Hu, C.M.; Fang, R.H.; Luk, B.T.; Zhang, L. Polymeric nanotherapeutics: Clinical development and advances in stealth functionalization strategies. *Nanoscale* **2014**, *6*, 65–75. [CrossRef]
22. Jain, R.K.; Stylianopoulos, T. Delivering nanomedicine to solid tumors. *Nat. Rev. Clin. Oncol.* **2010**, *7*, 653–664. [CrossRef] [PubMed]
23. Nazir, S.; Hussain, T.; Ayub, A.; Rashid, U.; MacRobert, A.J. Nanomaterials in combating cancer: Therapeutic applications and developments. *Nanomedicine* **2014**, *10*, 19–34. [CrossRef]
24. Fraix, A.; Sortino, S. Combination of PDT photosensitizers with NO photodononors. *Photochem. Photobiol. Sci.* **2018**, *17*, 1709–1727. [CrossRef]
25. Chegaev, K.; Fraix, A.; Gazzano, E.; Abd-Ellatef, G.E.F.; Blangetti, M.; Rolando, B.; Conoci, S.; Riganti, C.; Fruttero, R.; Gasco, A.; et al. Light-Regulated NO Release as a Novel Strategy to Overcome Doxorubicin MultiDrug Resistance. *ACS Med. Chem. Lett.* **2017**, *8*, 361–365. [CrossRef]
26. Coti, K.K.; Belowich, M.E.; Liong, M.; Ambrogio, M.W.; Lau, Y.A.; Khatib, H.A.; Zink, J.I.; Khashab, N.M.; Stoddart, J.F. Mechanised nanoparticles for drug delivery. *Nanoscale* **2009**, *1*, 16–39. [CrossRef]
27. Ambrogio, M.W.; Thomas, C.R.; Zhao, Y.L.; Zink, J.I.; Stoddart, J.F. Mechanized silica nanoparticles: A new frontier in theranostic nanomedicine. *Acc. Chem. Res.* **2011**, *44*, 903–913. [CrossRef] [PubMed]
28. Xia, X.; Zhou, C.; Ballell, L.; Garcia-Bennett, A.E. In vivo enhancement in bioavailability of atazanavir in the presence of proton-pump inhibitors using mesoporous materials. *ChemMedChem* **2012**, *7*, 43–48. [CrossRef]

29. Valetti, S.; Xin, X.; Costa-Gouveia, J.; Brodin, P.; Bernet-Camard, M.F.; Andersson, M.; Feiler, A. Clofazimine encapsulation in nanoporous silica particles for the oral treatment of antibiotic-resistant Mycobacterium tuberculosis infections. *Nanomedicine* **2017**, *8*, 831–844. [CrossRef]
30. Knezevic, N.Z.; Durand, J.O. Targeted Treatment of Cancer with Nanotherapeutics Based on Mesoporous Silica Nanoparticles. *ChemPlusChem* **2015**, *80*, 26–36. [CrossRef]
31. Shen, J.; He, Q.; Gao, Y.; Shi, J.; Li, Y. Mesoporous silica nanoparticles loading doxorubicin reverse multidrug resistance: Performance and mechanism. *Nanoscale* **2011**, *3*, 4314–4322. [CrossRef]
32. Qian, H.S.; Guo, H.C.; Ho, P.C.; Mahendran, R.; Zhang, Y. Mesoporous-silica-coated up-conversion fluorescent nanoparticles for photodynamic therapy. *Small* **2009**, *5*, 2285–2290. [CrossRef]
33. Brevet, D.; Gary-Bobo, M.; Raehm, L.; Richeter, S.; Hocine, O.; Amro, K.; Loock, B.; Couleaud, P.; Frochot, C.; MorHre, A.; et al. Mannose-targeted mesoporous silica nanoparticles for photodynamic therapy. *Chem. Commun.* **2009**, 1475–1477. [CrossRef] [PubMed]
34. Wong, R.C.H.; Ng, D.K.P.; Fong, W.-P.; Lo, P.-C. Encapsulating pH-Responsive Doxorubicin–Phthalocyanine Conjugates in Mesoporous Silica Nanoparticles for Combined Photodynamic Therapy and Controlled Chemotherapy. *Chem. Eur. J.* **2017**, *23*, 16505–16515. [CrossRef] [PubMed]
35. Soto, R.J.; Yang, L.; Schoenfisch, M.H. Functionalized Mesoporous Silica via an Aminosilane Surfactant Ion Exchange Reaction: Controlled Scaffold Design and Nitric Oxide Release. *ACS Appl. Mater. Interface Sci.* **2016**, *8*, 2220–2231. [CrossRef]
36. Afonso, D.; Valetti, S.; Fraix, A.; Bascetta, C.; Petralia, S.; Conoci, S.; Feiler, A.; Sortino, S. Multivalent mesoporous silica nanoparticles photo-delivering nitric oxide with carbon dots as fluorescence reporters. *Nanoscale* **2017**, *9*, 13404–13408. [CrossRef] [PubMed]
37. Callari, F.L.; Sortino, S. Amplified nitric oxide photorelease in DNA proximity. *Chem. Commun.* **2008**, *17*, 1971–1973. [CrossRef]
38. Wada, A.; Tamaru, S.; Ikeda, M.; Hamachi, I. MCM–Enzyme–Supramolecular Hydrogel Hybrid as a Fluorescence Sensing Material for Polyanions of Biological Significance. *J. Am. Chem. Soc.* **2009**, *131*, 5321–5330. [CrossRef] [PubMed]
39. Misko, T.P.; Schilling, R.J.; Salvemini, D.; Moore, W.M.; Currie, M.G. A fluorometric assay for the measurement of nitrite in biological samples. *Anal. Biochem.* **1993**, *214*, 11–16. [CrossRef]
40. Carloni, P.; Damiani, E.; Greci, L.; Stipa, P.; Tanfani, F.; Tartaglini, E.; Wozniak, M. On the use of 1,3-diphenylisobenzofuran (DPBF). Reactions with carbon and oxygen centered radicals in model and natural systems. *Res. Chem. Intermed.* **1993**, *19*, 395–405. [CrossRef]
41. Hartman, P.E.; Hartman, Z.; Ault, K.T. Scavenging of singlet molecular oxygen by imidazole compounds: High and sustained activities of carboxy terminal histidine dipeptides and exceptional activity of imidazole-4-acetic acid. *Photochem. Photobiol.* **1990**, *51*, 59–66. [CrossRef]
42. Silva, P.R.; Vono, L.L.R.; Espósito, B.P.; Baptista, M.S.; Rossi, L.M. Enhancement of hematoporphyrin IX potential for photodynamic therapy by entrapment in silica nanospheres. *Phys. Chem. Chem. Phys.* **2011**, *13*, 14946–14952. [CrossRef]
43. Ogilby, P.R. Singlet oxygen: There is indeed something new under the sun. *Chem. Soc. Rev.* **2010**, *39*, 3181–3209. [CrossRef] [PubMed]
44. Caruso, E.B.; Petralia, S.; Conoci, S.; Giuffrida, S.; Sortino, S. Photodelivery of nitric oxide from water-soluble platinum nanoparticles. *J. Am. Chem. Soc.* **2007**, *129*, 480–481. [CrossRef]
45. Krishna, R.; Mayer, L.D. Multidrug resistance (MDR) in cancer. Mechanisms, reversal using modulators of MDR and the role of MDR modulators in influencing the pharmacokinetics of anticancer drugs. *Eur. J. Pharm. Sci.* **2000**, *11*, 265–283. [CrossRef]
46. Gottesman, M.M.; Fojo, T.; Bates, S.E. Multidrug resistance in cancer: Role of ATP-dependent transporters. *Nat. Rev. Cancer* **2002**, *2*, 48–58. [CrossRef] [PubMed]
47. Gazzano, E.; Rolando, B.; Chegaev, K.; Salaroglio, I.C.; Kopecka, J.; Pedrini, I.; Saponara, S.; Sorge, M.; Buondonno, I.; Stella, B.; et al. Folate-targeted liposomal nitrooxy-doxorubicin: An effective tool against P-glycoprotein-positive and folate receptor-positive tumors. *J. Control. Release* **2018**, *270*, 37–52. [CrossRef] [PubMed]

48. Gazzano, E.; Buondonno, I.; Marengo, A.; Rolando, B.; Chegaev, K.; Kopecka, J.; Saponara, S.; Sorge, M.; Hattinger, C.M.; Gasco, A.; et al. Hyaluronated liposomes containing H_2S-releasing doxorubicin are effective against P-glycoprotein-positive/doxorubicin-resistant osteosarcoma cells and xenografts. *Cancer Lett.* **2019**, *456*, 29–39. [CrossRef] [PubMed]

© 2019 by the authors. Licensee MDPI, Basel, Switzerland. This article is an open access article distributed under the terms and conditions of the Creative Commons Attribution (CC BY) license (http://creativecommons.org/licenses/by/4.0/).

Article

Morphology–Dependent Electrochemical Sensing Properties of Iron Oxide–Graphene Oxide Nanohybrids for Dopamine and Uric Acid

Zhaotian Cai [1,†], Yabing Ye [1,†], Xuan Wan [1], Jun Liu [1], Shihui Yang [1], Yonghui Xia [2], Guangli Li [1,*] and Quanguo He [1,*]

1. Hunan Key Laboratory of Biomedical Nanomaterials and Devices, College of Life Sciences and Chemistry, Hunan University of Technology, Zhuzhou 412007, China; caizhaotian1998@163.com (Z.C.); yyb980501@163.com (Y.Y.); wanxuan1111@163.com (X.W.); liu.jun.1015@163.com (J.L.); yangshihui0522@163.com (S.Y.)
2. Zhuzhou Institute for Food and Drug Control, Zhuzhou 412000, China; Sunnyxia0710@163.com
* Correspondence: guangli010@hut.edu.cn (G.L.); hequanguo@hut.edu.cn (Q.H.); Tel.: +86-0731-2218-3382 (G.L. & Q.H.)
† These authors contributed equally to this work.

Received: 4 May 2019; Accepted: 20 May 2019; Published: 1 June 2019

Abstract: Various morphologies of iron oxide nanoparticles (Fe_2O_3 NPs), including cubic, thorhombic and discal shapes were synthesized by a facile meta-ion mediated hydrothermal route. To further improve the electrochemical sensing properties, discal Fe_2O_3 NPs with the highest electrocatalytic activity were coupled with graphene oxide (GO) nanosheets. The surface morphology, microstructures and electrochemical properties of the obtained Fe_2O_3 NPs and Fe_2O_3/GO nanohybrids were characterized by scanning electron microscopy (SEM), X-ray diffraction (XRD), cyclic voltammetry (CV) and electrochemical impedance spectroscopy (EIS) techniques. As expected, the electrochemical performances were found to be highly related to morphology. The discal Fe_2O_3 NPs coupled with GO showed remarkable electrocatalytic activity toward the oxidation of dopamine (DA) and uric acid (UA), due to their excellent synergistic effect. The electrochemical responses of both DA and UA were linear to their concentrations in the ranges of 0.02–10 μM and 10–100 μM, with very low limits of detection (LOD) of 3.2 nM and 2.5 nM for DA and UA, respectively. Moreover, the d-Fe_2O_3/GO nanohybrids showed good selectivity and reproducibility. The proposed d-Fe_2O_3/GO/GCE realized the simultaneous detection of DA and UA in human serum and urine samples with satisfactory recoveries.

Keywords: morphology-dependent; α-Fe_2O_3 nanoparticles; GO; uric acid; dopamine; voltammetric detection

1. Introduction

Dopamine (DA) and uric acid (UA) usually coexist in the serum and extracellular fluids of the central nervous system and play a significant role in regulating human metabolism activity [1]. As an indispensable catecholamine neurotransmitter, DA plays critical roles in regulating the function of the cardiovascular and central nervous systems, as well as maintaining emotional control and hormonal balance [2]. Abnormal levels of DA can lead to various neurological disorders such as schizophrenia, Parkinson's and Alzheimer's disease [3–5]. Therefore, the precise detection of the DA level in physiological fluids is essential for the early diagnosis of these neurological disorders. However, the rapid and reliable detection of DA in physiological samples remains critical and challenging due to the low DA concentration in the extracellular matrix (usually in the range of 0.01–1 μM) and its susceptibility to interference from endogenous substances such as UA and ascorbic acid (AA). Uric acid is another crucial biomolecule in physiological fluids and is often regarded as the end-product of purine metabolism in the human body [6]. For a heathy individual, the UA concentration is generally in the range of 4.1 ± 8.8 mg/100 mL [7]. The dysfunction of UA in bodily fluids likely causes several diseases, including gout, pneumonia and hyperuricemia [8]. As stated, both DA and UA are regarded as important biomolecules for the regulation of human metabolic activity. Thus, determining the concentration of DA and UA in biological matrices (i.e., human urine, serum) can provide valuable clues for healthcare and disease diagnosis. Since DA and UA usually coexist in physiological fluids, it is of the utmost importance to propose a highly efficient technique for the simultaneous determination of DA and UA.

To date, several analytical techniques have been proposed for the detection of DA and UA, including, but not limited to, high performance liquid chromatography [9,10], fluorescent [11], spectrophotometry [12], electrogenerated chemiluminescent [13] and surface plasmon resonance [14] etc. Although quite reliable, these techniques usually involve tedious and time-consuming analytical procedures that require expensive equipment, well-trained technical personnel or a large quantity of toxic solvents [15]. Compared with other techniques, the electroanalytical techniques are more suitable for sensing DA and UA because of their low price, fast response, facile operation and excellent anti-interfering ability [16–18]. Owing to the considerable superiorities including cost-effectiveness, rapidness, convenience and high efficiency, electroanalytical methods have drawn increasing attention for the detection of small biomolecules, food additives and contaminants [19–24]. DA and UA are electroactive species whose redox processes can be quantitatively detected by electroanalytical techniques. However, it becomes a great challenge to simultaneously detect DA and UA on bare electrodes due to the fouling effect that occurs during the oxidation [7] and cross-interferences as a result of similar oxidation potentials [25]. To resolve this problem, nanostructured materials were employed to achieve high sensitivity and prevent overlapping of the oxidation peaks [25]. In recent years, much effort has been devoted to developing promising alternatives as sensing materials for the simultaneous detection of DA and UA, including noble metal nanoparticles [25], metal oxides nanocomposites [26], alloyed nanoparticles [27], polymer films [28,29], and nanocarbon materials [30–32] etc.

Iron oxide (Fe_2O_3) has become one of the most versatile transition metal oxides not only due to low cost, more abundant, good biocompatibility and excellent electrochemical performances, but also for its widespread applications [33–38]. In particular, α-Fe_2O_3 nanoparticles (α-Fe_2O_3 NPs) have been considered as the most promising modifying material, because of the variable valence state of iron oxides that can be recovered in situ via electrochemical reducing or oxidizing during the sensing process, thus triggering the heterogeneous redox of the target analysts [39]. As far as we know, various morphologies of α-Fe_2O_3 NPs have been made available in previous reports, including wire [40], rod [41,42], tube [41], sphere [43,44], flower [45], spindle [44], cubic [44,46–49], thorhombic [46–48], discal [47], and shuttle [50]. Many studies demonstrate that the morphologies of nanostructured α-Fe_2O_3 have a significant impact on optical, magnetic, photocatalytic and electrochemical properties [46,47,51,52]. However, the morphology-dependent electrochemical sensing performances with respect to small biomolecules have rarely been investigated. Hence, it is essential to explore the morphology–dependent sensing properties of different α-Fe_2O_3 NPs. Unfortunately, the electrochemical sensing performances of pure α-Fe_2O_3 NPs modified electrodes are relatively poor probably because of their poor electrical conductivity and dispersibility [37,53,54]. To address this issue, iron oxides were often used in a composite with graphene for the detection of DA. For example, nitrogen and sulfur dual doped graphene-supported Fe_2O_3 (NSG-Fe_2O_3) has been utilized for the electrochemical detection of DA in the presence of AA, with a wide linear response range (0.3–210 µM) and low LOD (0.035 µM) [55]. However, the procedure for the doping of nitrogen and sulfur is very complicated. Moreover, the simultaneous detection of DA and UA is not clear for this nanocomposite. As an important derivant of graphene, graphene oxide (GO) usually works as a conductive component to enhance the electron transfer between electrode surface and target analysts [18,19]. Indeed, there are abundant hydrophilic oxygen-containing functional groups (OxFGs) presented on the hydrophobic basal planes of GO, which can behave like an amphiphilic surfactant to improve the dispersion of α-Fe_2O_3 NPs [56]. As far as we know, α-Fe$_2$O$_3$/GO nanocomposites have rarely reported for the simultaneous detection of DA and UA.

Herein, α-Fe$_2$O$_3$ NPs with various morphologies including cubic, thorhombic and discal shapes were synthesized by a facile meta-ion mediated hydrothermal route then composite with GO nanosheets. The electrocatalytic activities of DA and UA at Fe$_2$O$_3$/GO nanohybrids decorated glassy carbon electrodes (Fe$_2$O$_3$/GO/GCE) were measured in this work. The electrochemical measurements exhibited that the discal Fe$_2$O$_3$ NPs had the most remarkable electrochemical response toward the simultaneous detection of DA and UA, attributing to the more surface defects and rougher surface. After further coupled with GO, the discal α-Fe$_2$O$_3$ NPs/GO nanohybrids (d-Fe$_2$O$_3$/GO) showed superior electrochemical sensing performances toward DA and UA, due to the notable synergistic effect from d-Fe$_2$O$_3$ and GO. Therefore, a novel and ultrasensitive electrochemical sensor based on d-Fe$_2$O$_3$/GO nanohybrids was proposed for the simultaneous detection of DA and UA.

2. Materials and Methods

2.1. Chemicas and Solutions

Dopamine (DA), uric acid (UA), Iron (III) nitrate nonahydrate (Fe(NO$_3$)$_3$·9H$_2$O), cupric acetate anhydrous (Cu(Ac)$_2$), Zinc acetate (Zn(Ac)$_2$·2H$_2$O), aluminum acetate (Al(Ac)$_3$), disodium hydrogen phosphate dodecahydrate (Na$_2$HPO$_4$·12H$_2$O) and Sodium dihydrogen phosphate dihydrate (NaH$_2$PO$_4$·2H$_2$O) were purchased from Aladdin Reagents Co., Ltd. (Shanghai, China). Graphite powder, potassium permanganate (KMnO$_4$), hydrogen peroxide (H$_2$O$_2$, 30%), sodium nitrate (NaNO$_3$), potassium ferricyanide(K$_3$[Fe(CN)$_6$]), potassium ferrocyanide (K$_4$[Fe(CN)$_6$]), ammonia (NH$_3$·H$_2$O), sodium hydrate (NaOH), concentrated hydrochloric acid (HCl, 37%), concentrated sulfuric acid (H$_2$SO$_4$, 98%) and absolute ethanol (CH$_3$CH$_2$OH) were supplied by Sinopharm Chemical Reagent Co. Ltd. (Shanghai, China). All chemicals were analytically pure and directly used as received. Human serum samples were provided by Zhuzhou People's Hospital (Zhuzhou, China). The stock solutions of DA and UA (1 mM) were prepared by dissolving appropriate amount of DA and UA in 500 mL 0.1 M PBS. Then lower concentration series of the standard solution were obtained by appropriately diluting the stock solution with 0.1 M PBS. Deionized water (DI water, 18.2 MΩ) was used in all the experiments.

2.2. Apparatus

All the electrochemical measurements were performed on a CHI 760E electrochemical workstation (Chenhua Instrument Inc., Shanghai, China). Surface morphologies and crystalline structure of α-Fe$_2$O$_3$ NPs, and Fe$_2$O$_3$/GO were investigated by scanning electron microscopy (SEM) and powder X-ray diffractometry (XRD), respectively. SEM images were recorded on a cold field-emission SEM (Hitachi S-4800, Tokyo, Japan) at an accelerating voltage of 5.0 kV. The XRD patterns of α-Fe$_2$O$_3$ NPs was recorded using a powder X-ray diffractometer system (PANalytical, Holland) with monochromatized Cu Kα radiation (λ = 0.1542 nm) operating at 40 kV and 40 mA. The patterns were collected in a 2θ range from 10° to 80° with a scanning step of 0.02°2θ s^{-1}. A digital precision pH meter (Leici instrument Inc., Shanghai, China) was used to measure solution pH.

2.3. Synthesis of Discal, Thorhombic, and Cubic α-Fe$_2$O$_3$ NPs

Referring to previous reports [46–48], three types of α-Fe$_2$O$_3$ NPs were synthesized by a meta-ion mediated hydrothermal route (Scheme 1). In briefly, 0.808 g of Fe(NO$_3$)$_3$·9H$_2$O was adequately dissolved into 10 mL of DI water under magnetic stirring, and then 1 mM of acetate precursor was added to the above solution. After 15 min of stirring, 10 mL NH$_3$·H$_2$O was added and continuously stirring for another 15 min. Next, the mixture was poured into a 50 mL Teflon-lined stainless-steel autoclave and reacted at 160 °C for 16 h. Subsequently, the autoclave was naturally cooled down to room temperature. Finally, the resultant α-Fe$_2$O$_3$ product was alternately washed three times with absolute alcohol and DI water, and allowed to dry under vacuum at 60 °C for 12 h. Herein, cupric acetate, zinc acetate, and aluminum acetate were separately used as acetate precursor to obtain thorhombic, cubic and discal α-Fe$_2$O$_3$ NPs.

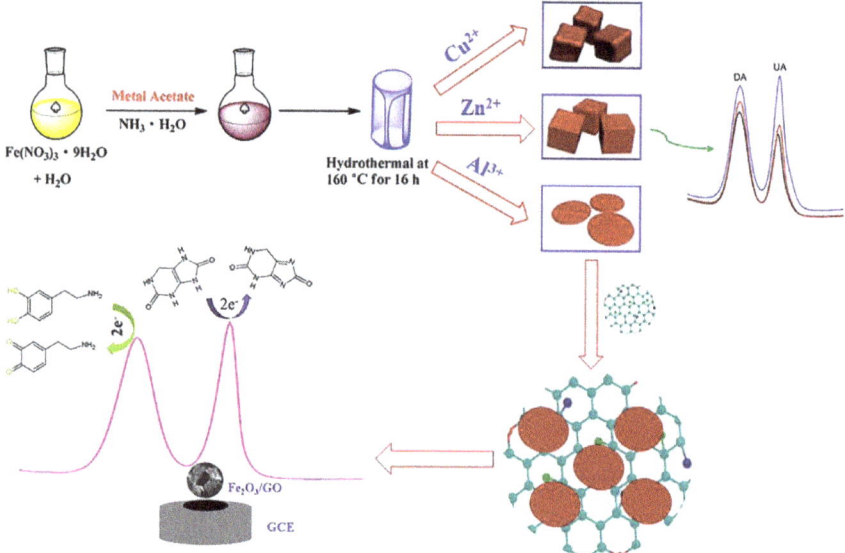

Scheme 1. The schematic illustration of meta-ion mediated hydrothermal route for the synthesis of discal, thorhombic, and cubic α-Fe$_2$O$_3$ NPs and possible mechanism for the electrochemical sensing of DA and UA at the d-Fe$_2$O$_3$/GO/GCE.

2.4. Preparation of Fe$_2$O$_3$/GO Composite

GO was synthesized by a slightly modified Hummer's method according to our earlier reports [20,23,24]. 10 mg GO was delaminated and dispersed in the 10 mL DI water under ultrasonication for 2 h to form 1 mg/mL GO dispersion. Then one aliquot of as-obtained α-Fe$_2$O$_3$ products was added into five aliquots of GO dispersion (1 mg/mL), and then ultrasonicated for 0.5 h to form uniform α-Fe$_2$O$_3$/GO dispersion.

2.5. Fabrication of Fe$_2$O$_3$/GO/GCE

At first, bare GCE (ca. 0.0707 cm^2) was carefully polished to mirror-like with 0.3 μm and 0.05 μm fine alumina slurry. Then the polished GCE was alternately washed by absolute ethanol and DI water several times and allowed to dry under an infrared lamp. The Fe$_2$O$_3$/GO/GCE was prepared by a simple drop-casting method. More specifically, 5 μL Fe$_2$O$_3$/GO dispersion was carefully dropped and coated on the GCE surface with a micropipette, then dried with an infrared lamp to form sensing film. For comparison, the discal, thorhombic, and cubic α-Fe$_2$O$_3$ NPs decorated GCEs (d-Fe$_2$O$_3$/GCE, t-Fe$_2$O$_3$/GCE, c-Fe$_2$O$_3$/GCE) and GO decorated GCE (GO/GCE) were also fabricated by similar procedure.

2.6. Electrochemical Measurements

The electrochemical behaviors of various decorated electrodes were measured by cyclic voltammetry (CV) and electrochemical impedance spectroscopy (EIS) in the 0.1 M PBS (pH 7.0) containing 0.5 mM [Fe(CN)$_6$]$^{3-/4-}$ as redox probe solution. EIS was measured at the frequency range from 100 kHz to 0.1 Hz at open circuit potential with 5 mV amplitude, using a redox couple of 0.5 mM [Fe(CN)$_6$]$^{3-/4-}$ (1:1) in 0.1 M KCl as an electrochemical probe. The electrochemical responses of 10 μM DA and UA (1:1) mixture standard solution at different electrodes were recorded by CV.

Differential pulse voltammetry (DPV) was used for the individual and simultaneous detection of DA and UA. A conventional three-electrode assembly was immersed into a 10 mL electrochemical cell for electrochemical measurements and detections, consisting of a bare or modified GCE as the working electrode, a saturated calomel electrode (SCE) as the reference electrode, and a platinum wire as auxiliary electrode, respectively. To increase response peak current, a suitable accumulation period was applied before CV and DPV measurements. DPV was performed from 0 V to 0.8 V, with step potential of 4 mV, the amplitude of 50 mV and pulse width of 0.06 s and pulse period of 0.5 s, respectively. 0.1 M PBS was used as the supporting electrolytes unless otherwise specified. All the electrochemical measurements were performed at 25 °C.

3. Results and Discussions

3.1. Morphology–Dependent Electrochemical Sensing of α-Fe$_2$O$_3$ NPs

The schematic diagram of the preparation of the α-Fe$_2$O$_3$ NPs is shown in Scheme 1. The α-Fe$_2$O$_3$ NPs with cubic (c-Fe$_2$O$_3$ NPs), discal (d-Fe$_2$O$_3$ NPs), and thorhombic (t-Fe$_2$O$_3$ NPs) shapes were prepared via different metal-ion mediated hydrothermal treatment [46–48]. The α-Fe$_2$O$_3$ NPs are produced as cubic, discal and thorhombic shapes, due to the addition of Zn^{2+}, Al^{3+}, and Cu^{2+} ions, respectively. Figure 1A–C shows the SEM images of as-obtained α-Fe$_2$O$_3$ NPs. Clearly, three α-Fe$_2$O$_3$ NPs with uniform cubic, thorhombic and discal shapes are observed, suggesting uniform cubic, thorhombic and discal shapes were successfully prepared. The crystalline structures of α-Fe$_2$O$_3$ NPs with three distinct morphologies were further investigated by XRD. As presented in Figure 2, the diffraction peaks in all the patterns can be well assigned to the rhombohedral phase of hematite (JCPDS PDF# 33-0664) [46]. Moreover, no apparent diffraction peaks relating to CuFe$_2$O$_4$, ZnFe$_2$O$_4$ and AlFeO$_3$ can be found, demonstrating the crystalline structures of the α-Fe$_2$O$_3$ NPs are not altered by the addition of Cu^{2+}, Zn^{2+} and Al^{3+} ions.

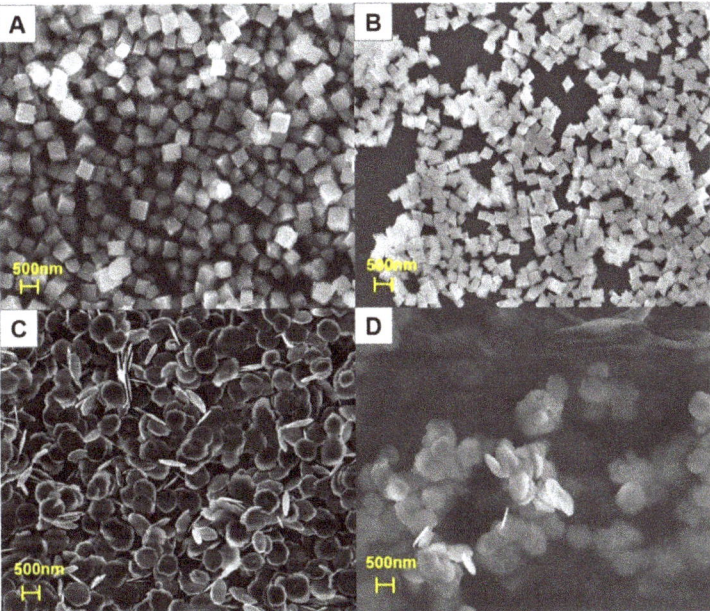

Figure 1. Scanning electron microscopy (SEM) images of α-Fe$_2$O$_3$ NPs with cubic (**A**), thorhombic (**B**) and discal (**C**) morphology, discal α-Fe$_2$O$_3$ NPs/GO nanohybrid (**D**).

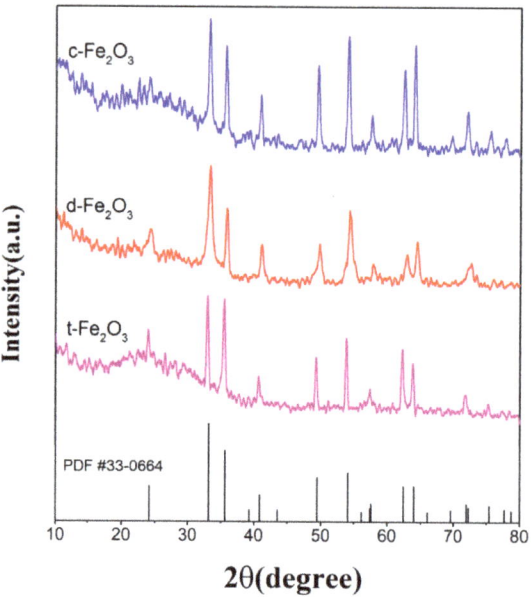

Figure 2. X-ray diffraction (XRD) patterns of α-Fe$_2$O$_3$ NPs with cubic (c-Fe$_2$O$_3$), discal (d-Fe$_2$O$_3$), and thorhombic (t-Fe$_2$O$_3$) morphologies.

For comparison purposes, α-Fe$_2$O$_3$ NPs with various morphologies were utilized for the simultaneous detection of 10 µM DA and UA in 0.1 M PBS through CV and DPV technique, and the results are shown in Figure 3A,B, respectively. On all the electrodes, two anodic peaks for DA and UA are independent and evident (Figure 3A), suggesting high selectivity. However, a very small cathodic peaks relating to UA are observed on all the electrode at reverse scanning, demonstrating the reversibility of UA redox is very poor. Note that a pair of redox peaks can be observed on all the electrodes with I_{pa}/I_{pc} ranging from 1.08 to 1.39, indicating that DA undergoes a quasi-reversible electrode process. The increased anodic peak heights in the CVs indicates enhanced electrocatalytic activity. Among the three morphologies of α-Fe$_2$O$_3$ NPs modified electrodes, d-Fe$_2$O$_3$/GCE exhibits highest electrocatalytic activity toward DA and UA, with well-shaped anodic peaks ($I_{pa(DA)}$ = 0.1764 µA, $I_{pc(UA)}$ = 0.6626 µA) appear at 0.238 and 0.395 V for DA and UA, respectively. Compared to c-Fe$_2$O$_3$/GCE and t-Fe$_2$O$_3$/GCE, the anodic peak current for DA at d-Fe$_2$O$_3$/GCE increases by 65.8% and 30.0% and the anodic peak current for UA increases by 177% and 73.1%, respectively. Liu et al. systematically compared the photocatalytic activity among cubic, discal and thorhombic α-Fe$_2$O$_3$ NPs, d-Fe$_2$O$_3$ NPs were found to have more surface defects and rougher surface which result in highest photocatalytic activity [47]. Generally, the electrochemical active sites will be more accessible on the more defective surface, which can greatly facilitate the electrocatalytic oxidation of target analysts. Moreover, the rougher surface means a larger specific surface area, which is also favorable for electrochemical sensing. These facts explain well the remarkable electrocatalytic activity for d-Fe$_2$O$_3$. For the DPVs in Figure 3B further confirmed discal α-Fe$_2$O$_3$ NPs have the strongest electrocatalytic ability for the oxidation of DA and UA. Therefore, there is solid proof that the electrochemical activity of α-Fe$_2$O$_3$ NPs can be tailored by their morphologies. Considering the excellent sensitivity for the simultaneous determination of DA and UA, high electrocatalytic active discal α-Fe$_2$O$_3$ NPs were chosen as sensing materials in subsequent experiments. To further enhance the sensitivity, discal α-Fe$_2$O$_3$ NPs were coupled with GO (d-Fe$_2$O$_3$/GO) aiming to yield synergistic effect toward the electrooxidation of DA and UA. The SEM image of d-Fe$_2$O$_3$/GO is shown in Figure 1D. Evidently, discal α-Fe$_2$O$_3$ NPs are warped by silky graphene oxide sheets.

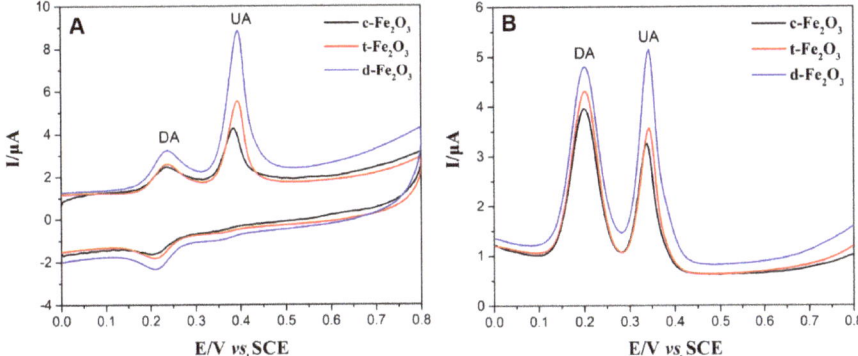

Figure 3. Cyclic voltammetry (CV) (**A**) and differential pulse voltammetry (DPV) (**B**) curves of c-Fe$_2$O$_3$, t-Fe$_2$O$_3$, and d-Fe$_2$O$_3$ NPs modified GCEs in 0.1 M PBS (pH = 7) containing 10 µM DA and UA.

3.2. Electrochemical Performance of d-Fe$_2$O$_3$/GO/GCE

The electrochemical performance of different modified GCEs were investigated by CV using 0.5 mM [Fe(CN)$_6$]$^{3-/4-}$ as probe solution. The corresponding CV curves are plotted in Figure 4. A pair of weak redox peaks (I_{pa} = 12.73 µA, I_{pc} = 12.16 µA) are observed on the bare GCE with peak potential separation (ΔE_p) of 0.156 V, suggesting that the electron transfer rate is very slow. At the GO/GCE, a pair of wide and weak redox peak occurs with the lowest redox peak currents (I_{pa} = 3.942 µA, I_{pc} = 3.768 µA) and maximum ΔE_p (ΔE_p = 0.182 V). It was mainly due to the presence of poor electrical conductivity of GO. At the d-Fe$_2$O$_3$/GCE, a pair of well-shaped and sharp redox peak appears. Both anodic peak current (I_{pa}) and cathodic peak current (I_{pc}) increase greatly (I_{pa} = 3.942 µA, I_{pc} = 3.768 µA) while the ΔE_p decreases to 0.136 V, indicating d-Fe$_2$O$_3$ can effectively promote the redox process due to its high electrocatalytic activity. As expected, d-Fe$_2$O$_3$/GO/GCE displays remarkable electrochemical properties with the highest redox peak currents (I_{pa} = 47.45 µA, I_{pc} = 46.8 µA) and minimum potential separation (ΔE_p = 0.104 V), attributing to the synergistic effect from d-Fe$_2$O$_3$ NPs and GO nanosheets. As well known, electrochemical active area is highly related to the electrochemical sensing properties. The electrochemical active areas of various electrodes were also estimated using the Randles-Sevcik equation [20,21,24]. The electrochemical active areas of bare GCE, GO/GCE, d-Fe$_2$O$_3$/GCE and d-Fe$_2$O$_3$/GO/GCE are estimated as 0.1037, 0.0321, 0.3991 and 1.1181 cm^2, respectively. The electrochemical active area of d-Fe$_2$O$_3$/GO/GCE is approximately 2-fold and 10-fold higher than that of d-Fe$_2$O$_3$/GCE and bare GCE, respectively.

EIS is a valuable tool to acquire interfacial properties, which has extensively used in the various electrochemical sensors [20,21,57–59]. Nyquist plots for bare GCE, d-Fe$_2$O$_3$/GCE, GO/GCE and d-Fe$_2$O$_3$/GO/GCE are presented in Figure 4B. The Nyquist plots reveals electron transfer kinetics and diffusion characteristic. Typically, Nyquist plots include two portions, namely semicircular and linear portions. The semicircular at the higher frequency domain and linear part at the lower frequency domain represents the electron transfer limited and diffusion limited electrode processes, respectively. The semicircle diameter indicates the charge transfer resistance (R_{ct}). Obviously, the largest semicircle diameters (4500 ohm) are obtained at GO/GCE, indicating the poor electrical conductivity of GO retard electron transfer process. Compared to bare GCE (948 ohm), the R_{ct} values for d-Fe$_2$O$_3$/GCE and d-Fe$_2$O$_3$/GO/GCE decrease to 64 ohm and 35 ohm, respectively. The lowest R_{ct} value is obtained at d-Fe$_2$O$_3$/GO/GCE since numerous electrocatalytic active sites occurred, demonstrating that an obvious acceleration of redox kinetics of [Fe(CN)$_6$]$^{3-/4-}$ happened on the surface of d-Fe$_2$O$_3$/GO nanohybrids. The results verify that the d-Fe$_2$O$_3$/GO nanohybrids can effectively lower the R_{ct}.

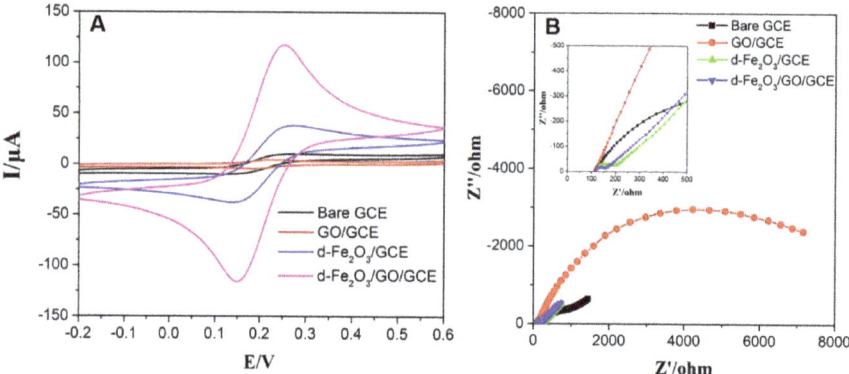

Figure 4. (**A**) CVs curves of bare GCE, GO/GCE, d-Fe$_2$O$_3$/GCE and d-Fe$_2$O$_3$/GO/GCE in the 0.5 mM [Fe(CN)$_6$]$^{3-/4-}$; (**B**) Nyquist plots of different electrodes for the EIS measurements in the presence of 0.5 mM [Fe(CN)$_6$]$^{3-/4-}$ (1:1) in 0.1 M KCl. The inset is the magnification of Nyquist plots at the higher frequency region.

3.3. Electrochemical Behaviors of DA and UA on the d-Fe$_2$O$_3$/GO/GCE

Electrochemical responses of 10 µM DA and UA (1:1) at different electrodes were recorded by DPV in 0.1 M PBS (pH = 5.65). As presented in Figure 5A, the anodic peaks of DA and UA are well separated, indicating simultaneous detection of DA and UA is very feasible. On the bare GCE, the I_{pa} of DA and UA is 0.2724 µA and 0.1555 µA, respectively. After the modification of GO nanosheets, the I_{pa} of DA and UA increased a little ($I_{pa(DA)}$ = 0.2915 µA, $I_{pa(UA)}$ = 0.2356 µA). The large specific area of GO could provide abundant adsorption sites for target analytes, which enhances the electrochemical response signals. But the poor electrical conductivity of GO severely retards the redox process, which impedes electrochemical oxidation of DA and UA. When d-Fe$_2$O$_3$ NPs were modified on the GCE, the I_{pa} of both DA and UA boosts greatly ($I_{pa(DA)}$ = 0.5519 µA, $I_{pa(UA)}$ = 0.4770 µA), suggesting d-Fe$_2$O$_3$ NPs have high electrocatalytic activity toward DA and UA. Among all the electrodes, d-Fe$_2$O$_3$/GO nanohybrids show a remarkable electrocatalytic capacity toward the electrochemical oxidation of DA and UA, with the highest I_{pa} of 1.077 and 1.234 µA for DA and UA, respectively. The synergistic effect from d-Fe$_2$O$_3$ NPs and GO nanosheets contributed to the significant enhancement in the response currents. Specifically, abundant active sites presented in the GO surface greatly facilities the adsorption of target analytes; d-Fe$_2$O$_3$ can efficiently catalyzed the electrooxidation of DA and UA. Besides, the d-Fe$_2$O$_3$ NPs/GO nanohybrids remarkably decrease the electron transfer resistant (confirmed by EIS, Figure 4B), which greatly accelerate the redox kinetics. Moreover, the peak-to-peak separations of bare GCE, GO/GCE, d-Fe$_2$O$_3$/GCE and d-Fe$_2$O$_3$/GO/GCE are 0.132, 0.140, 0.144, and 0.148 V, respectively. The maximum ΔE_p is obtained at d-Fe$_2$O$_3$/GO/GCE, demonstrating good distinguish capacity.

The electrochemical responses of various electrodes were also measured by CV. As shown in Figure 5B, two anodic peaks belonging to DA and UA are observed at all the electrodes. However, a negligible cathodic peak corresponding to the reduction of UA appears at reverse scanning, implying a poor reversible electrode reaction for UA. The ratio of I_{pa}/I_{pc} approaches 1 for all the electrodes, meaning the electrode undergoes a quasi-reversible process. Among all electrodes, d-Fe$_2$O$_3$/GO nanohybrids also have largest response peak currents for DA and UA. The CV result is in good agreement with the results obtained by DPVs. It is worth pointing out that DPV is more sensitive than CV, which is very suitable for multiple analystes detection. Therefore, DPV was used for the quantitative analysis of DA and UA.

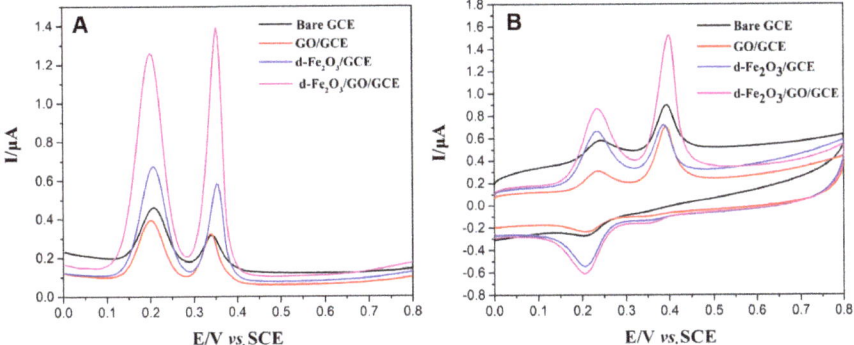

Figure 5. CVs (**A**) and DPVs (**B**) of 10 µM DA and UA (1:1) mixture solution at the bare GCE, GO/GCE, d-Fe$_2$O$_3$/GCE and d-Fe$_2$O$_3$/GO/GCE.

3.4. Optimation of Voltammetric Parameters

3.4.1. Influence of pH

It is well known that pH is a vital parameter that directly influences the electrochemical responses. The DPVs of 10 µM DA and UA mixture solution (1:1) at different pH are depicted in Figure 6A. Notably, the anodic peaks shift to the negative direction as the increasing of pH. The influence of pH on the response peak currents of DA and UA is shown in Figure 6B. In the pH range of 3.10 to 6.90, the anodic peak currents always increase with the increase of pH. The anodic peak currents of UA slowly increase when pH varied from 3.10 to 4.56, then gradually decrease with the further increase of pH. The highest response anodic peak is obtained at pH = 4.56. Trading off the oxidation peak current of DA and UA, pH 5.65 is selected as optimum pH in the following measurements. Moreover, the anodic peak potentials linearly decrease with the increase of pH (Figure 6C). The linear plots of peak potential versus pH can be expressed as $E_{DA} = -0.06079$ pH $+ 0.5762$ ($R^2 = 0.9988$) and $E_{UA} = -0.06130$ pH $+ 0.7184$ ($R^2 = 0.9986$). The slopes of these linear plots (−60.8 mV/pH and −61.3 mV/pH for DA and UA, respectively) are very close to the theoretical value from the Nernst equation (−59 mV/pH), indicating that the equal numbers of protons (H$^+$) and electrons (e$^-$) participate in the electrooxidation process of DA and UA, which is in good agreement with previous works.

3.4.2. Influence of Accumulation Parameters

To enhance the response electrochemical peak currents, accumulation was applied before electrochemical measurements. The influence of accumulation time on the response anodic peak currents of DA and UA was explored (Figure 7A). The response peak currents of DA and UA slowly increase during the first 60 s, then gradually attenuate with further extending of accumulation time. The highest response peaks are obtained at 60 s. Furthermore, the dependence of the anodic peak currents on the accumulation potential was also investigated (Figure 7B). When accumulation potentials vary from −0.4 to −0.3 V, the anodic peak currents of DA and UA enhance gradually. However, the anodic peak currents decline with accumulation potential varying from −0.3 to 0 V. The maximum anodic peaks are obtained at −0.3 V. Therefore, 60 s and −0.3 V were chosen as the optimum accumulation parameters.

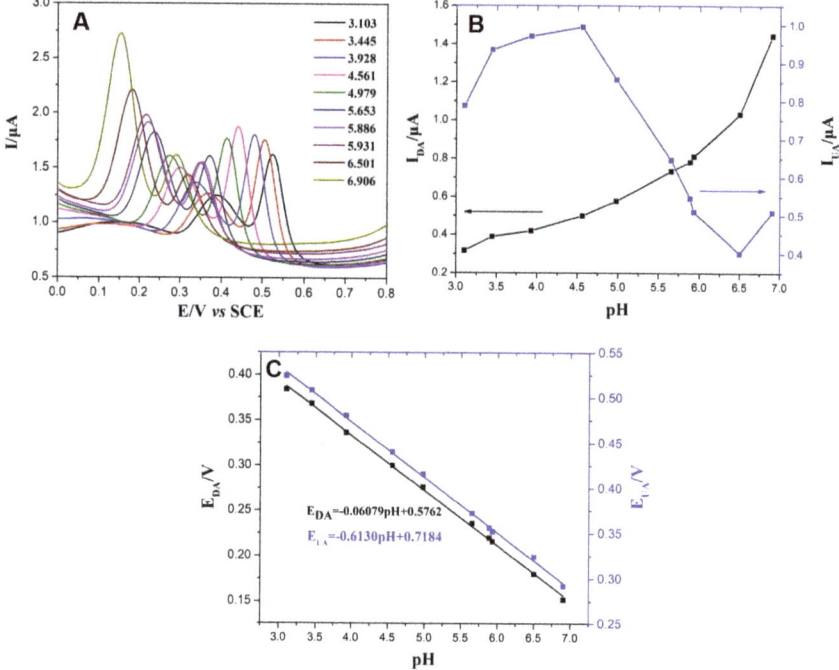

Figure 6. (**A**) DPVs of 10 μM DA and UA (1:1) mixture solution recorded at different pH; (**B**) Effect of pH on the response anodic peak current of DA and UA; (**C**) Linear plots of the anodic peak potential of DA and UA against pH.

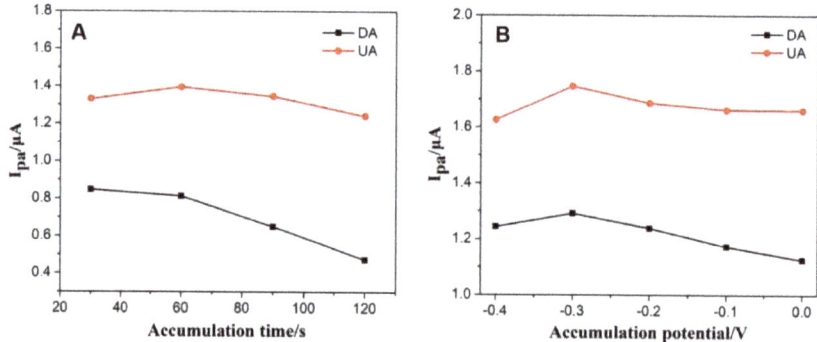

Figure 7. Influence of accumulation time (**A**) and accumulation potential (**B**) on the response anodic peak currents of DA and UA.

3.5. Reaction Mechanism of DA and UA

In order to provide a deeper insight into the electrochemical oxidation mechanism, CVs of 10 μM DA and UA (1:1) mixture solution were recorded at various scanning rates using d-Fe$_2$O$_3$/GO/GCE. CVs at different scanning rates are plotted in Figure 8A. Notably, the redox peaks increase with the increasing of scanning rate. Meanwhile the background currents also increased, probably because high scanning rates can increase the charging current of double layer. Moreover, the anodic peaks for DA and UA shift positivity as the increase of scanning rate while the cathodic peaks for UA shift to negative direction. As shown in Figure 8B, the anodic and cathodic peak currents of DA are highly correlated with the square root of scanning rate ($v^{1/2}$), suggesting the electrooxidation of DA is a diffusion-limited process. The linear regression equations are $I_{pa(DA)} = 13.6968\ v^{1/2} - 0.5756$ and $I_{pc(DA)} = -8.7663\ v^{1/2} + 0.0129$, with the correlation coefficient (R^2) of 0.985. The anodic peak currents of UA are in positive proportion to the square root of scanning rate ($v^{1/2}$), indicating the oxidation of UA limited by the diffusion in bulk solution. The linear regression equation is $I_{pa(UA)} = 11.9383\ v^{1/2} - 0.7370$ with the correlation coefficient (R^2) of 0.999 (Figure 8C). Since the electrooxidation of DA and UA is an equal number of electrons (e$^-$) coupled with protons (H$^+$) process, a two electron and two proton (2H$^+$, 2e$^-$) mechanism was proposed for the oxidation of DA and UA (Scheme 1).

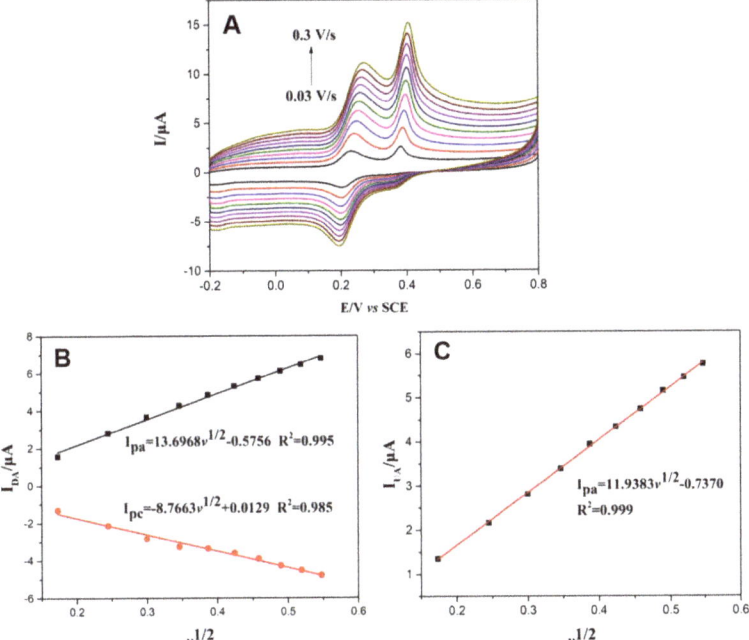

Figure 8. (**A**) CVs of 10 µM DA and UA (1:1) mixture solution recorded at various scanning rates; (**B**) Linear relationship between the redox peak current of DA and the square root of scanning rate ($v^{1/2}$); (**C**) Linear relationship between the anodic peak current of UA and the square root of scanning rate ($v^{1/2}$).

3.6. Individual and Simultaneous Detection of DA and UA

For the individual detection of DA and UA on the d-Fe$_2$O$_3$/GO/GCE, DPVs were performed in the potential range from 0–0.8 V in 0.1 M PBS (pH 5.65). In this case, only the concentrations of the target species were changed, while the concentrations of the other species remained unaltered. As plotted in Figure 9, the response anodic peak currents of DA increases linearly with the DA concentrations increasing from 0.04 to 4 µM. However, the response anodic peak currents of DA are highly correlation to Napierian logarithm of DA concentrations (lnC$_{DA}$) at higher concentration domain (4–100 µM). For individual detection of UA, the response anodic peak currents of UA are in proportion to the all working concentrations from 0.1 to 100 µM (Figure 10). The limit of detection (LOD) for the individual detection of DA and UA are estimated to 4.8 and 12 nM at S/N = 3, respectively. It is noteworthy that addition of the target species does not have obvious interference on the electrochemical responses (i.e., the anodic peak current and potential) of the other species. The results strongly imply that DA and UA can be sensitively and selectively detected on d-Fe$_2$O$_3$/GO/GCE in the DA and UA mixture (specific concentrations (C$_{DA}$/C$_{UA}$) range of 0.01–100).

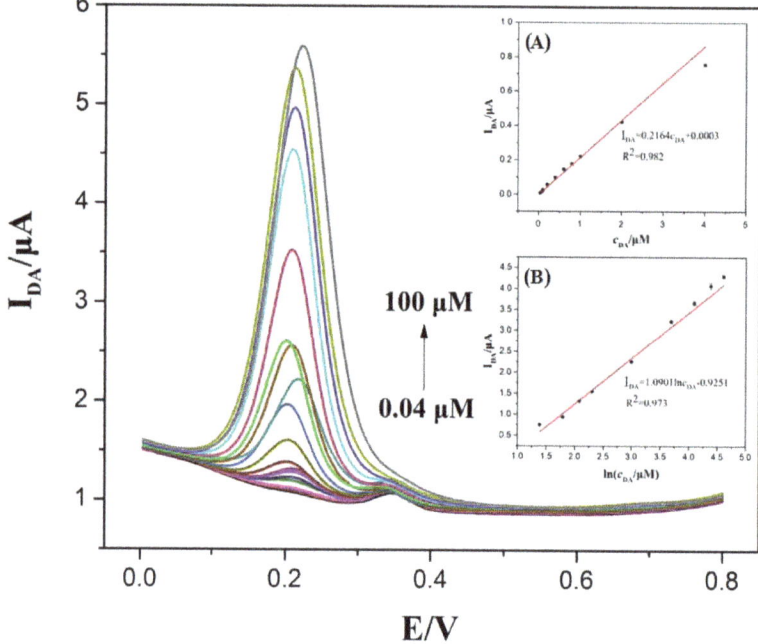

Figure 9. DPVs on the d-Fe$_2$O$_3$/GO/GCE in 0.1 M PBS (pH 5.65) containing 1 µM UA and various concentrations of DA from 0.04 to 100 µM; The inset (**A**) represents the linear plot of the anodic peak currents versus the DA concentrations varying from 0.04 to 4 µM; The inset (**B**) represents the linear plot of the anodic peak currents versus the Napierian logarithm of DA concentrations with DA concentration ranging from 4 to 100 µM.

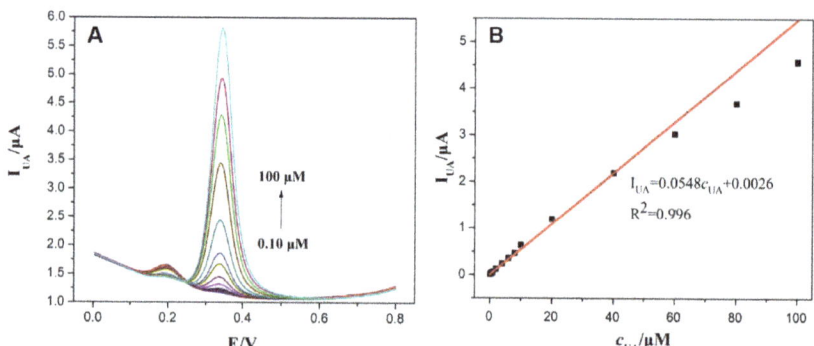

Figure 10. (**A**) DPVs on the d-Fe$_2$O$_3$/GO/GCE in 0.1 M PBS (pH 5.65) containing 1 µM DA and various concentrations of UA from 0.1 to 100 µM; (**B**) The linear plot of the anodic peak currents of UA versus the UA concentrations varying from 0.1 to 100 µM.

Superior electrocatalytic activity of d-Fe$_2$O$_3$/GO also showed simultaneous detection of DA and UA using DPV technique in 0.1 M PBS (pH 5.65) (Figure 11). Two well-separated anodic peaks relating to the electrooxidation of DA and UA occur on DPV curves using d-Fe$_2$O$_3$/GO/GCE. Moreover, DPV responses are resolved into two peaks at 0.22 and 0.36 V, which can be attributed to the oxidations of DA and UA, respectively. These results indicating that simultaneous distinguish from the two substances is feasible in mixture solutions. As expected, all anodic peak currents increase linearly with increasing concentrations of DA and UA. Two linear response regions for both DA and UA are obtained within concentration ranges of 0.02–10 μM and 10–100 μM, respectively (Figure 11B–E). The linear regression equations for DA can be expressed as $I_{DA}(\mu M) = 0.1062 C_{DA}(\mu M) + 0.0122$ ($R^2 = 0.975$) and $I_{DA}(\mu M) = 0.0186 C_{DA}(\mu M) + 0.7835$ ($R^2 = 0.985$). The linear regression equations for UA are $I_{UA}(\mu M) = 0.0862 C_{UA}(\mu M) + 0.0036$ and $I_{UA}(\mu M) = 0.0320 C_{DA}(\mu M) + 0.4804$, with correlation coefficient of 0.995 and 0.990, respectively. The LODs are calculated as 3.2 and 2.5 nM for DA and UA, respectively. All results demonstrate that the proposed d-Fe$_2$O$_3$/GO/GCEs feature wider linear response ranges and lower LOD for the electrochemical oxidation of DA and UA. Hence, the simultaneous detection of DA and UA can be realized on d-Fe$_2$O$_3$/GO/GCE with high sensitivity and good selectivity. The sensing performances are compared to those in previous reports (Table 1). Clearly, the sensing parameters (i.e., linear response ranges and LOD) of the proposed sensor are comparable to, or even better than most previous reported modified electrodes [7,55,60–70]. The high sensitivity of our proposed sensor is closely related to the synergistic electrocatalytic effect from the d-Fe$_2$O$_3$ and GO.

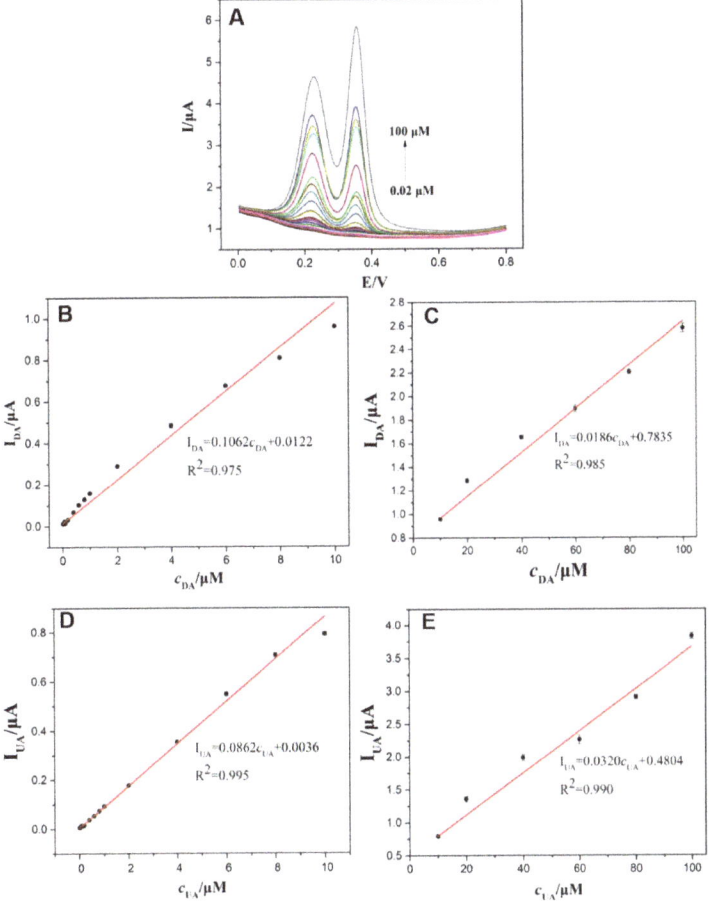

Figure 11. (A) DPVs on the d-Fe$_2$O$_3$/GO/GCE in 0.1 M PBS (pH 5.65) containing different concentrations of DA and UA ranging from 0.02–100 μM; Plots of the anodic peak currents as the function of DA concentrations in the range of 0.02–10 μM (B) and 10–100 μM (C); Plots of the anodic peak currents as the function of UA concentrations in the range of 0.02–10 μM (D) and 10–100 μM (E).

Table 1. Comparison analytical performance between previous reports and the proposed d-Fe$_2$O$_3$/GO/GCE for the simultaneous detection of DA and UA.

Electrodes	Methods	Detection Range (μM)		LOD (μM)		Ref.
		DA	UA	DA	UA	
Au/Cu$_2$O/rGO/GCE	DPV	10–90	100–900	3.9	6.5	[7]
NSG–Fe$_2$O$_3$/GCE	DPV	0.3–210		0.035		[55]
Pd/RGO/GCE	DPV	0.45–71	6–469.5	0.18	1.6	[60]
Pt/RGO/GCE	DPV	10–170	10–130	0.25	0.45	[61]
ZnO/SPCE	DPV	0.1–374	0.1–169	0.004	0.00849	[62]
Fe$_3$O$_4$/rGO/GCE	DPV	0.5–100		0.12		[63]
AuPtNPs/S-NS-GR/GCE	DPV	0.01–400	1–1000	0.006	0.0038	[64]
Au–Pt/GO–ERGO	DPV	0.0682–49,800	0.125–82,800	0.0207	0.0407	[65]
Fe$_2$O$_3$/NrGO/GCE	Amperometry	0.5–340		0.49		[66]
ZnO/PANI/rGO/GCE	DPV	0.1–90	0.5–90	0.017	0.12	[67]
pCu$_2$O NS-rGO/GCE	DPV	0.05–109	1–138	0.015	0.112	[68]
Zn-NiAl LDH/rGO/GCE	DPV	0.0001–1	0.0011–0.95	0.0001	0.0009	[69]
α-Fe$_2$O$_3$@Au-Pd/GCE	SWV	0.1–1000	1–1000	0.0000138	0.97	[70]
d-Fe$_2$O$_3$/GO/GCE	DPV	0.02–10; 10–100	0.02–10; 10–100	0.0032	0.0025	This work

3.7. Practical Application

Prior to the simultaneous determination of DA and UA in actual samples, the selectivity, repeatability, stability and reproducibility were also investigated. In order to assess the anti-interference of the proposed d-Fe$_2$O$_3$/GO/GCE toward the DA and UA, the DPV responses of the DA and UA in the presence of potential interfering species were recorded. Negligible interference with accepted relative error (less than 5.70%) even in the presence of 100-fold ascorbic acid, citric acid, alanine, glutamic acid, and lysine (Figure 12). To assess electrode reproducibility, the variation on anodic peak currents in the 1 μM DA and UA (1:1) mixture solution was measured at room temperature using five d-Fe$_2$O$_3$/GO/GCE, which were fabricated by the same procedure. The relative standard deviations (RSD) for the anodic peak current are 4.56% and 3.03%, respectively, indicating that the electrode fabrication has high reproducibility. To evaluate the repeatability, seven successive measurements for detection of 1 μM DA and UA (1:1) were carried out. The RSD for DA and UA are 5.07% and 4.81%, respectively, suggesting good repeatability.

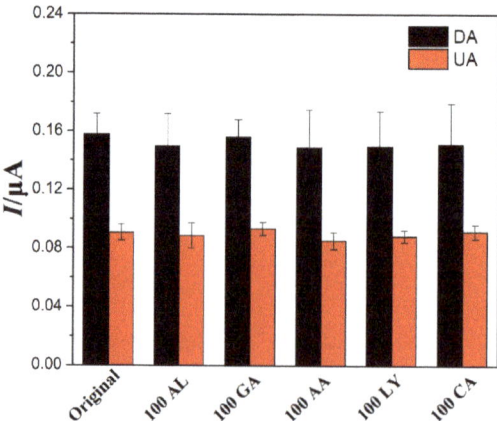

Figure 12. The anodic peak currents of 1 μM DA and UA in the presence of 100-fold alanine (AL), glutamic acid (GA), ascorbic acid (AA), lysine (LY) and citric acid (CA).

To verify applicability, the concentrations of DA and UA in actuate samples were also detected on d-Fe$_2$O$_3$/GO/GCE. Since DA and UA always coexist in most biological fluids, the practicability of d-Fe$_2$O$_3$/GO/GCE in actual sample detection was assessed with human urine and serum samples. The determined results listed in Table 2 were carefully estimated from the standard curves. To further validate the accuracy and precision of the proposed sensor, a series of known concentration solutions of DA and UA were spiked to the actual samples to calculate the recovery. The recoveries of 94.1–106.4% and 95.2–108.2% are obtained for DA and UA, respectively. DPVs for the analysis of human serum samples with 100-fold dilution are presented in Figure S1. Notably, the shape and position of the DPVs are not affected when using real samples. These results confirm that the biological matrixes such as human urine and serum does not affect the simultaneous determination of DA and UA.

Table 2. Detection results of DA and UA in human serum and urine samples using d-Fe$_2$O$_3$/GO/GCE.

Samples [a]		Detected (μM)	Added (μM)	Found (μM)	RSD (%)	Recovery (%)
Serum	DA	ND [b]	20	18.82	3.75	94.1
			40	38.26	2.86	95.7
	UA	20.6	20	41.76	2.75	105.8
			40	63.86	2.21	108.2
Urine	DA	ND [b]	20	21.06	4.98	105.3
			40	42.56	3.67	106.4
	UA	35.24	20	54.27	2.62	95.2
			40	74.52	1.39	98.2

[a] The human urine and serum samples were detected at 100-fold and 10-fold dilution, respectively. [b] Not detected.

4. Conclusions

In summary, the influence of Fe$_2$O$_3$/GO morphologies on electrochemical sensing performances was studied systematically. Cubic, thorhombic and discal Fe$_2$O$_3$ NPs with a uniform size and controllable structure were successfully prepared by a facile meta-ion mediated hydrothermal route. When coupled with GO nanosheets and then worked as sensing films for the simultaneous detection of DA and UA, the α-Fe$_2$O$_3$ NPs with discal morphology displayed the highest electrocatalytic activity. This remarkable electrocatalytic behavior was highly correlated with the discal shapes with more surface defects and a rougher surface. The synergistic effect from d-Fe$_2$O$_3$ NPs and GO nanosheets contributed to a significant enhancement in the response currents. As a consequence, two wide detection ranges (0.02–10 μM and 10–100 μM) were obtained for DA and UA, with very low limit of detection (LOD) of 3.2 and 2.5 nM for DA and UA, respectively. Moreover, the d-Fe$_2$O$_3$/GO nanohybrids showed good selectivity and reproducibility. The proposed d-Fe$_2$O$_3$/GO nanohybrids have become one of the most competitive candidates as sensing materials for the simultaneous determination of DA and UA in various real samples.

Supplementary Materials: The following are available online at http://www.mdpi.com/2079-4991/9/6/835/s1, Figure S1: DPVs for the analysis of human serum samples with 100-fold dilution.

Author Contributions: Conceptualization, Z.C., Y.Y. and G.L.; methodology, Z.C., Y.Y. and X.W.; experiment, Z.C., Y.Y., X.W., and S.Y.; data analysis, Z.C., Y.Y., and Y.X.; writing—original draft preparation, Z.C., Y.Y.; writing—review and editing, G.L. and Q.H.; supervision, G.L. and Q.H.; funding acquisition, G.L., J.L. and Q.H.

Funding: This research was funded by the Undergraduates' Innovation Experiment Program of Hunan Province (No. 2018649), National Natural Science Foundation of China (No. 61703152), Natural Science Foundation of Hunan Province (No. 2019JJ50127, 2018JJ3134), Scientific Research Foundation of Hunan Provincial Education Department (18A273, 18C0522), Project of Science and Technology Plan of Zhuzhou (201707201806), and the Doctoral Program Construction of Hunan University of Technology.

Acknowledgments: We sincerely express our thanks to Zhuzhou People's Hospital for offering human serum samples.

Conflicts of Interest: The authors declare no conflict of interest.

References

1. He, Q.; Liu, J.; Liang, J.; Liu, X.; Li, W.; Liu, Z.; Ding, Z.; Tuo, D. Towards improvements for penetrating the blood–brain barrier—recent progress from a material and pharmaceutical perspective. *Cells* **2018**, *7*, 24. [CrossRef] [PubMed]
2. Dalley, J.W.; Roiser, J.P. Dopamine, serotonin and impulsivity. *Neuroscience* **2012**, *215*, 42–58. [CrossRef] [PubMed]
3. Carlsson, A. Does dopamine play a role in schizophrenia? *Psychol. Med.* **1977**, *7*, 583–597. [CrossRef]
4. Zhang, M.; Liao, C.; Yao, Y.; Liu, Z.; Gong, F.; Yan, F. High-Performance Dopamine Sensors Based on Whole-Graphene Solution-Gated Transistors. *Adv. Funct. Mater.* **2014**, *24*, 978–985. [CrossRef]
5. Wightman, R.M.; May, L.J.; Michael, A.C. Detection of Dopamine Dynamics in the Brain. *Anal. Chem.* **1988**, *60*, 769A–793A. [CrossRef] [PubMed]
6. Jindal, K.; Tomar, M.; Gupta, V. Nitrogen-doped zinc oxide thin films biosensor for determination of uric acid. *Analyst* **2013**, *138*, 4353–4362. [CrossRef]
7. Aparna, T.K.; Sivasubramanian, R.; Dar, M.A. One-pot synthesis of Au-Cu$_2$O/rGO nanocomposite based electrochemical sensor for selective and simultaneous detection of dopamine and uric acid. *J. Alloys Compd.* **2018**, *741*, 1130–1141. [CrossRef]
8. Sharaf El Din, U.A.A.; Salem, M.M.; Abdulazim, D.O. Uric acid in the pathogenesis of metabolic, renal, and cardiovascular diseases: A review. *J. Adv. Res.* **2017**, *8*, 537–548. [CrossRef]
9. Xiang, L.W.; Li, J.; Lin, J.M.; Li, H.F. Determination of gouty arthritis' biomarkers in human urine using reversed-phase high-performance liquid chromatography. *J. Pharm. Anal.* **2014**, *4*, 153–158. [CrossRef]
10. Lin, L.; Qiu, P.; Yang, L.; Cao, X.; Jin, L. Determination of dopamine in rat striatum by microdialysis and high-performance liquid chromatography with electrochemical detection on a functionalized multi-wall carbon nanotube electrode. *Anal. Bioanal. Chem.* **2006**, *384*, 1308–1313. [CrossRef]
11. Zhao, D.; Song, H.; Hao, L.; Liu, X.; Zhang, L.; Lv, Y. Luminescent ZnO quantum dots for sensitive and selective detection of dopamine. *Talanta* **2013**, *107*, 133–139. [CrossRef]
12. Moghadam, M.R.; Dadfarnia, S.; Shabani, A.M.H.; Shahbazikhah, P. Chemometric-assisted kinetic–spectrophotometric method for simultaneous determination of ascorbic acid, uric acid, and dopamine. *Anal. Biochem.* **2011**, *410*, 289–295. [CrossRef]
13. Huang, C.; Chen, X.; Lu, Y.; Yang, H.; Yang, W. Electrogenerated chemiluminescence behavior of peptide nanovesicle and its application in sensing dopamine. *Biosens. Bioelectron.* **2015**, *63*, 478–482. [CrossRef]
14. Kumbhat, S.; Shankaran, D.R.; Kim, S.J.; Gobi, K.V.; Joshi, V.; Miura, N. Surface plasmon resonance biosensor for dopamine using D3 dopamine receptor as a biorecognition molecule. *Biosens. Bioelectron.* **2007**, *23*, 421–427. [CrossRef]
15. Liu, S.; Yan, J.; He, G.; Zhong, D.; Chen, J.; Shi, L.; Zhou, X.; Jiang, H. Layer-by-layer assembled multilayer films of reduced graphene oxide/gold nanoparticles for the electrochemical detection of dopamine. *J. Electroanal. Chem.* **2012**, *672*, 40–44. [CrossRef]
16. Qi, S.; Zhao, B.; Tang, H.; Jiang, X. Determination of ascorbic acid, dopamine, and uric acid by a novel electrochemical sensor based on pristine graphene. *Electrochim. Acta* **2015**, *161*, 395–402. [CrossRef]
17. Lian, Q.; He, Z.; He, Q.; Luo, A.; Yan, K.; Zhang, D.; Lu, X.; Zhou, X. Simultaneous determination of ascorbic acid, dopamine and uric acid based on tryptophan functionalized graphene. *Anal. Chim. Acta* **2014**, *823*, 32–39. [CrossRef]
18. Hou, J.; Xu, C.; Zhao, D.; Zhou, J. Facile fabrication of hierarchical nanoporous AuAg alloy and its highly sensitive detection towards dopamine and uric acid. *Sens. Actuators B Chem.* **2016**, *225*, 241–248. [CrossRef]
19. He, Q.; Liu, J.; Liu, X.; Xia, Y.; Li, G.; Deng, P.; Chen, D. Novel Electrochemical Sensors Based on Cuprous Oxide-Electrochemically Reduced Graphene Oxide Nanocomposites Modified Electrode toward Sensitive Detection of Sunset Yellow. *Molecules* **2018**, *23*, 2130. [CrossRef]
20. He, Q.; Liu, J.; Liu, X.; Li, G.; Chen, D.; Deng, P.; Liang, J. A promising sensing platform toward dopamine using MnO$_2$ nanowires/electro-reduced graphene oxide composites. *Electrochim. Acta* **2019**, *296*, 683–692. [CrossRef]
21. He, Q.; Liu, J.; Liu, X.; Li, G.; Deng, P.; Liang, J. Manganese dioxide Nanorods/electrochemically reduced graphene oxide nanocomposites modified electrodes for cost-effective and ultrasensitive detection of Amaranth. *Colloids Surf. B* **2018**, *172*, 565–572. [CrossRef] [PubMed]

22. He, Q.; Liu, J.; Liu, X.; Li, G.; Deng, P.; Liang, J.; Chen, D. Sensitive and Selective Detection of Tartrazine Based on TiO2-Electrochemically Reduced Graphene Oxide Composite-Modified Electrodes. *Sensors* **2018**, *18*, 1911. [CrossRef]
23. He, Q.; Liu, J.; Liu, X.; Li, G.; Chen, D.; Deng, P.; Liang, J. Fabrication of Amine-Modified Magnetite-Electrochemically Reduced Graphene Oxide Nanocomposite Modified Glassy Carbon Electrode for Sensitive Dopamine Determination. *Nanomaterials* **2018**, *8*, 194. [CrossRef]
24. He, Q.; Liu, J.; Liu, X.; Li, G.; Deng, P.; Liang, J. Preparation of Cu$_2$O-Reduced Graphene Nanocomposite Modified Electrodes towards Ultrasensitive Dopamine Detection. *Sensors* **2018**, *18*, 199. [CrossRef] [PubMed]
25. Chen, L.X.; Zheng, J.-N.; Wang, A.J.; Wu, L.J.; Chen, J.R.; Feng, J.J. Facile synthesis of porous bimetallic alloyed PdAg nanoflowers supported on reduced graphene oxide for simultaneous detection of ascorbic acid, dopamine, and uric acid. *Analyst* **2015**, *140*, 3183–3192. [CrossRef]
26. Zhao, L.; Li, H.; Gao, S.; Li, M.; Xu, S.; Li, C.; Guo, W.; Qu, C.; Yang, B. MgO nanobelt-modified graphene-tantalum wire electrode for the simultaneous determination of ascorbic acid, dopamine and uric acid. *Electrochim. Acta* **2015**, *168*, 191–198. [CrossRef]
27. Zhang, X.; Yan, W.; Zhang, J.; Li, Y.; Tang, W.; Xu, Q. NiCo-embedded in hierarchically structured N-doped carbon nanoplates for the efficient electrochemical determination of ascorbic acid, dopamine, and uric acid. *RSC Adv.* **2015**, *5*, 65532–65539. [CrossRef]
28. Chen, J.; Zhang, J.; Lin, X.; Wan, H.; Zhang, S. Electrocatalytic Oxidation and Determination of Dopamine in the Presence of Ascorbic Acid and Uric Acid at a Poly (4-(2-Pyridylazo)-Resorcinol) Modified Glassy Carbon Electrode. *Electroanalysis* **2007**, *19*, 612–615. [CrossRef]
29. Li, Y.; Lin, X. Simultaneous electroanalysis of dopamine, ascorbic acid and uric acid by poly (vinyl alcohol) covalently modified glassy carbon electrode. *Sens. Actuators B* **2006**, *115*, 134–139. [CrossRef]
30. Sheng, Z.H.; Zheng, X.-Q.; Xu, J.Y.; Bao, W.J.; Wang, F.B.; Xia, X.H. Electrochemical sensor based on nitrogen doped graphene: Simultaneous determination of ascorbic acid, dopamine and uric acid. *Biosens. Bioelectron.* **2012**, *34*, 125–131. [CrossRef]
31. Liu, Y.; Huang, J.; Hou, H.; You, T. Simultaneous determination of dopamine, ascorbic acid and uric acid with electrospun carbon nanofibers modified electrode. *Electrochem. Commun.* **2008**, *10*, 1431–1434. [CrossRef]
32. Cui, R.; Wang, X.; Zhang, G.; Wang, C. Simultaneous determination of dopamine, ascorbic acid, and uric acid using helical carbon nanotubes modified electrode. *Sens. Actuators B Chem.* **2012**, *161*, 1139–1143. [CrossRef]
33. Harraz, F.A.; Ismail, A.A.; Al-Sayari, S.A.; Al-Hajry, A.; Al-Assiri, M.S. Highly sensitive amperometric hydrazine sensor based on novel α-Fe$_2$O$_3$/crosslinked polyaniline nanocomposite modified glassy carbon electrode. *Sens. Actuators B Chem.* **2016**, *234*, 573–582. [CrossRef]
34. Ahmad, R.; Ahn, M.A.; Hahn, Y. A Highly Sensitive Nonenzymatic Sensor Based on Fe$_2$O$_3$ Nanoparticle Coated ZnO Nanorods for Electrochemical Detection of Nitrite. *Adv. Mater. Interfaces* **2017**, *4*, 1700491. [CrossRef]
35. Larsen, G.K.; Farr, W.; Murph, S.E.H. Multifunctional Fe$_2$O$_3$–Au Nanoparticles with Different Shapes: Enhanced Catalysis, Photothermal Effects, and Magnetic Recyclability. *J. Phys. Chem. C* **2016**, *120*, 15162–15172. [CrossRef]
36. Quan, H.; Cheng, B.; Xiao, Y.; Lei, S. One-pot synthesis of α-Fe$_2$O$_3$ nanoplates-reduced graphene oxide composites for supercapacitor application. *Chem. Eng. J.* **2016**, *286*, 165–173. [CrossRef]
37. Zhang, Z.J.; Wang, Y.X.; Chou, S.L.; Li, H.J.; Liu, H.K.; Wang, J.Z. Rapid synthesis of α-Fe$_2$O$_3$/rGO nanocomposites by microwave autoclave as superior anodes for sodium-ion batteries. *J. Power Sources* **2015**, *280*, 107–113. [CrossRef]
38. Wang, H.; Xu, Z.; Yi, H.; Wei, H.; Guo, Z.; Wang, X. One-step preparation of single-crystalline Fe$_2$O$_3$ particles/graphene composite hydrogels as high performance anode materials for supercapacitors. *Nano Energy* **2014**, *7*, 86–96. [CrossRef]
39. Cummings, C.Y.; Bonné, M.J.; Edler, K.J.; Helton, M.; Mckee, A.; Marken, F. Direct reversible voltammetry and electrocatalysis with surface-stabilised Fe$_2$O$_3$ redox states. *Electrochem. Commun.* **2008**, *10*, 1773–1776. [CrossRef]
40. Fu, Y.; Wang, R.; Xu, J.; Chen, J.; Yan, Y.; Narlikar, A.; Zhang, H. Synthesis of large arrays of aligned α-Fe$_2$O$_3$ nanowires. *Chem. Phys. Lett.* **2003**, *379*, 373–379. [CrossRef]
41. Liu, L.; Kou, H.-Z.; Mo, W.; Liu, H.; Wang, Y. Surfactant-assisted synthesis of α-Fe$_2$O$_3$ nanotubes and nanorods with shape-dependent magnetic properties. *J. Phys. Chem. B* **2006**, *110*, 15218–15223. [CrossRef]

42. Mou, X.; Zhang, B.; Li, Y.; Yao, L.; Wei, X.; Su, D.S.; Shen, W.J. Rod-shaped Fe_2O_3 as an efficient catalyst for the selective reduction of nitrogen oxide by ammonia. *Angew. Chem. Int. Ed.* **2012**, *51*, 2989–2993. [CrossRef] [PubMed]
43. Liu, X.; Wang, H.; Su, C.; Zhang, P.; Bai, J. Controlled fabrication and characterization of microspherical $FeCO_3$ and α-Fe_2O_3. *J. Colloid Interface Sci.* **2010**, *351*, 427–432. [CrossRef]
44. Fu, X.; Bei, F.; Wang, X.; Yang, X.; Lu, L. Surface-enhanced Raman scattering of 4-mercaptopyridine on sub-monolayers of α-Fe_2O_3 nanocrystals (sphere, spindle, cube). *J. Raman Spectrosc.* **2009**, *40*, 1290–1295. [CrossRef]
45. Yan, W.; Fan, H.; Zhai, Y.; Yang, C.; Ren, P.; Huang, L. Low temperature solution-based synthesis of porous flower-like α-Fe_2O_3 superstructures and their excellent gas-sensing properties. *Sens. Actuators B* **2011**, *160*, 1372–1379. [CrossRef]
46. Yang, S.; Zhou, B.; Ding, Z.; Zheng, H.; Huang, L.; Pan, J.; Wu, W.; Zhang, H. Tetragonal hematite single crystals as anode materials for high performance lithium ion batteries. *J. Power Sources* **2015**, *286*, 124–129. [CrossRef]
47. Liu, J.; Yang, S.; Wu, W.; Tian, Q.; Cui, S.; Dai, Z.; Ren, F.; Xiao, X.; Jiang, C. 3D Flowerlike α-Fe_2O_3@TiO_2 Core–Shell Nanostructures: General Synthesis and Enhanced Photocatalytic Performance. *ACS Sustain. Chem. Eng.* **2015**, *3*, 2975–2984. [CrossRef]
48. Wu, W.; Yang, S.; Pan, J.; Sun, L.; Zhou, J.; Dai, Z.; Xiao, X.; Zhang, H.; Jiang, C. Metal ion-mediated synthesis and shape-dependent magnetic properties of single-crystalline α-Fe_2O_3 nanoparticles. *CrystEngComm* **2014**, *16*, 5566–5572. [CrossRef]
49. Yin, C.Y.; Minakshi, M.; Ralph, D.E.; Jiang, Z.T.; Xie, Z.; Guo, H.J. Hydrothermal synthesis of cubic α-Fe_2O_3 microparticles using glycine: Surface characterization, reaction mechanism and electrochemical activity. *J. Alloys Compd.* **2011**, *509*, 9821–9825. [CrossRef]
50. Chen, A.; Liang, X.; Zhang, X.; Yang, Z.; Yang, S. Improving Surface Adsorption via Shape Control of Hematite α-Fe_2O_3 Nanoparticle for Sensitive Dopamine Sensors. *ACS Appl. Mater. Interfaces* **2016**, *8*, 33765–33774. [CrossRef]
51. Mitra, S.; Das, S.; Mandal, K.; Chaudhuri, S. Synthesis of a α-Fe_2O_3 nanocrystal in its different morphological attributes: growth mechanism, optical and magnetic properties. *Nanotechnology* **2007**, *18*, 275608. [CrossRef]
52. Jagadeesan, D.; Mansoori, U.; Mandal, P.; Sundaresan, A.; Eswaramoorthy, M. Hollow Spheres to Nanocups: Tuning the Morphology and Magnetic Properties of Single-Crystalline α-Fe_2O_3 Nanostructures. *Angew. Chem. Int. Ed.* **2008**, *47*, 7685–7688. [CrossRef]
53. Liu, X.; Chen, T.; Chu, H.; Niu, L.; Sun, Z.; Pan, L.; Sun, C.Q. Fe_2O_3-reduced graphene oxide composites synthesized via microwave-assisted method for sodium ion batteries. *Electrochim. Acta* **2015**, *166*, 12–16. [CrossRef]
54. Zhang, Y.; Gao, W.; Zuo, L.; Zhang, L.; Huang, Y.; Lu, H.; Fan, W.; Liu, T. In situ growth of Fe_2O_3 nanoparticles on highly porous graphene/polyimide-based carbon aerogel nanocomposites for Effectively selective detection of dopamine. *Adv. Mater. Interfaces* **2016**, *3*, 1600137. [CrossRef]
55. Yasmin, S.; Ahmed, M.S.; Jeon, S. Determination of dopamine by dual doped graphene-Fe_2O_3 in Presence of Ascorbic Acid. *J. Electrochem. Soc.* **2015**, *162*, B363–B369. [CrossRef]
56. Gan, T.; Shi, Z.; Deng, Y.; Sun, J.; Wang, H. Morphology–dependent electrochemical sensing properties of manganese dioxide–graphene oxide hybrid for guaiacol and vanillin. *Electrochim. Acta* **2014**, *147*, 157–166. [CrossRef]
57. Li, G.; Wang, S.; Duan, Y.Y. Towards conductive-gel-free electrodes: Understanding the wet electrode, semi-dry electrode and dry electrode-skin interface impedance using electrochemical impedance spectroscopy fitting. *Sens. Actuators B Chem.* **2018**, *277*, 250–260. [CrossRef]
58. Li, G.; Wang, S.; Duan, Y.Y. Towards gel-free electrodes: A systematic study of electrode-skin impedance. *Sens. Actuators B Chem.* **2017**, *241*, 1244–1255. [CrossRef]
59. Li, G.; Zhang, D.; Wang, S.; Duan, Y.Y. Novel passive ceramic based semi-dry electrodes for recording electroencephalography signals from the hairy scalp. *Sens. Actuators B Chem.* **2016**, *237*, 167–178. [CrossRef]
60. Wang, J.; Yang, B.; Zhong, J.; Yan, B.; Zhang, K.; Zhai, C.; Shiraishi, Y.; Du, Y.; Yang, P. Dopamine and uric acid electrochemical sensor based on a glassy carbon electrode modified with cubic Pd and reduced graphene oxide nanocomposite. *J. Colloid Interface Sci.* **2017**, *497*, 172–180. [CrossRef] [PubMed]

61. Xu, T.Q.; Zhang, Q.L.; Zheng, J.N.; Lv, Z.Y.; Wei, J.; Wang, A.J.; Feng, J.J. Simultaneous determination of dopamine and uric acid in the presence of ascorbic acid using Pt nanoparticles supported on reduced graphene oxide. *Electrochim. Acta* **2014**, *115*, 109–115. [CrossRef]
62. Kogularasu, S.; Akilarasan, M.; Chen, S.-M.; Chen, T.W.; Lou, B.S. Urea-based morphological engineering of ZnO; for the biosensing enhancement towards dopamine and uric acid in food and biological samples. *Mater. Chem. Phys.* **2019**, *227*, 5–11. [CrossRef]
63. Teo, P.S.; Alagarsamy, P.; Huang, N.M.; Lim, H.N.; Yusran, S. Simultaneous electrochemical detection of dopamine and ascorbic acid using an iron oxide/reduced graphene oxide modified glassy carbon electrode. *Sensors* **2014**, *14*, 15227–15243.
64. Zhang, K.; Chen, X.; Li, Z.; Wang, Y.; Sun, S.; Wang, L.N.; Guo, T.; Zhang, D.; Xue, Z.; Zhou, X. Au-Pt bimetallic nanoparticles decorated on sulfonated nitrogen sulfur co-doped graphene for simultaneous determination of dopamine and uric acid. *Talanta* **2018**, *178*, 315–323. [CrossRef] [PubMed]
65. Liu, Y.; She, P.; Gong, J.; Wu, W.; Xu, S.; Li, J.; Zhao, K.; Deng, A. A novel sensor based on electrodeposited Au–Pt bimetallic nano-clusters decorated on graphene oxide (GO)–electrochemically reduced GO for sensitive detection of dopamine and uric acid. *Sens. Actuators B Chem.* **2015**, *221*, 1542–1553. [CrossRef]
66. Yang, Z.; Zheng, X.; Zheng, J. A facile one-step synthesis of Fe 2 O 3 /nitrogen-doped reduced graphene oxide nanocomposite for enhanced electrochemical determination of dopamine. *J. Alloys Compd.* **2017**, *709*, 581–587. [CrossRef]
67. Ghanbari, K.; Moloudi, M. Flower-like ZnO decorated polyaniline/reduced graphene oxide nanocomposites for simultaneous determination of dopamine and uric acid. *Anal. Biochem.* **2016**, *512*, 91–102. [CrossRef] [PubMed]
68. Mei, L.P.; Feng, J.J.; Wu, L.; Chen, J.R.; Shen, L.; Xie, Y.; Wang, A.J. A glassy carbon electrode modified with porous Cu_2O nanospheres on reduced graphene oxide support for simultaneous sensing of uric acid and dopamine with high selectivity over ascorbic acid. *Microchim. Acta* **2016**, *183*, 2039–2046. [CrossRef]
69. Asif, M.; Aziz, A.; Wang, H.; Wang, Z.; Wang, W.; Ajmal, M.; Xiao, F.; Chen, X.; Liu, H. Superlattice stacking by hybridizing layered double hydroxide nanosheets with layers of reduced graphene oxide for electrochemical simultaneous determination of dopamine, uric acid and ascorbic acid. *Microchim. Acta* **2019**, *186*, 61. [CrossRef]
70. Sumathi, C.; Venkateswara Raju, C.; Muthukumaran, P.; Wilson, J.; Ravi, G. Au–Pd bimetallic nanoparticles anchored on α-Fe_2O_3 nonenzymatic hybrid nanoelectrocatalyst for simultaneous electrochemical detection of dopamine and uric acid in the presence of ascorbic acid. *J. Mater. Chem. B* **2016**, *4*, 2561–2569. [CrossRef]

© 2019 by the authors. Licensee MDPI, Basel, Switzerland. This article is an open access article distributed under the terms and conditions of the Creative Commons Attribution (CC BY) license (http://creativecommons.org/licenses/by/4.0/).

Article

Effective Modulation of Optical and Photoelectrical Properties of SnS₂ Hexagonal Nanoflakes via Zn Incorporation

Ganesan Mohan Kumar [1], Pugazhendi Ilanchezhiyan [1,*], Hak Dong Cho [2], Shavkat Yuldashev [1], Hee Chang Jeon [2], Deuk Young Kim [3] and Tae Won Kang [1]

[1] Nano-Information Technology Academy (NITA), Dongguk University-Seoul, Seoul 04623, Korea
[2] Quantum Functional Semiconductor Research Center, Dongguk University-Seoul, Seoul 04623, Korea
[3] Division of Physics and Semiconductor Science, Dongguk University-Seoul, Seoul 04623, Korea
* Correspondence: ilancheziyan@dongguk.edu

Received: 30 May 2019; Accepted: 21 June 2019; Published: 27 June 2019

Abstract: Tin sulfides are promising materials in the fields of photoelectronics and photovoltaics because of their appropriate energy bands. However, doping in SnS₂ can improve the stability and robustness of this material in potential applications. Herein, we report the synthesis of SnS₂ nanoflakes with Zn doping via simple hydrothermal route. The effect of doping Zn was found to display a huge influence in the structural and crystalline order of as synthesized SnS₂. Their optical properties attest Zn doping of SnS₂ results in reduction of the band gap which benefits strong visible-light absorption. Significantly, enhanced photoresponse was observed with respect to pristine SnS₂. Such enhancement could result in improved electronic conductivity and sensitivity due to Zn doping at appropriate concentration. These excellent performances show that $Sn_{1-x}Zn_xS_2$ nanoflakes could offer huge potential for nanoelectronics and optoelectronics device applications.

Keywords: SnS₂ nanoflakes; semiconductor; zinc doping; photoelectronics

1. Introduction

Metal sulfides have received considerable interest due to their unique optoelectronic properties while processed at micro-nano level [1–4]. In particular, two-dimensional (2D) metal sulfides nanostructures such as nanoplates, nanoflakes and nanosheets have received much attention for their potential application in photodetectors, photovoltaic devices and light-emitting diodes [5–20]. 2D form of the material offers high specific surface area, making it advantageous for electrochemical, catalytic and photoelectrical activities. Another advantage in 2D materials is that they are more compatible and can easily be integrated into nano-microscale structures for developing new optoelectronic devices [21–24].

Meanwhile, SnS₂ is considered as one of the promising layered materials with excellent visible light absorption and electrical properties. It possesses band gap (2.1–2.3 eV), n-type characteristics, high sensitivity and high surface activity for applications in Li-ion batteries [25], photovoltaic devices [26] and photodetector [27,28]. Variety of nanostructures such as nanoflakes, nanosheets and nanoplates through physical and chemical techniques including chemical vapor deposition, solvothermal and hydrothermal methods have been reported by several groups [29,30]. Among them, nanoflakes preparation via hydrothermal method have attracted considerable interest due to its low cost and large-scale production at low temperatures. Similarly, many efforts have also been made in controlling morphology and enhancing the photoelectrical, chemical and physical properties for improving the device performance. Moreover, dopants in semiconductor could lead to reduction in particle size, narrowing of band gap and enhance the photoelectrical properties of SnS₂ [31]. Recently, V and

Ti doped SnS$_2$ was reported to be an intermediate band material for application in wider solar absorption [32,33]. Recently doping SnS$_2$ with Fe resulted in room temperature ferromagnetism [34]. Similarly, in our previous work, we reported enhanced optical and electrical properties of SnS$_2$ nanoflakes via Cu doping [35]. More recently, Liu et al. reported enhanced photoresponsivity in Sb doped SnS$_2$ monolayer [36]. Based on the above literatures we test the ability of doping Zn ions in SnS$_2$ to significantly enhance conductivity and sensitivity favorable for its performance in photoelectronics.

The present work reports on hydrothermal synthesis of Zn doped SnS$_2$ nanoflakes at low temperatures. The properties of Sn$_{1-x}$Zn$_x$S$_2$ nanoflakes have been intensively studied through structural, optical and photoelectrical methods. The results show that the Zn doping results in enhanced sensitivity, conductivity and efficiency of charge transfer kinetics. As a proof of concept, Sn$_{1-x}$Zn$_x$S$_2$ nanoflakes were integrated into a patterned indium tin oxide (ITO) substrate (as active material) for photoelectronic device architecture. The results showcased excellent on-off ratio and photoresponse properties than that of pristine counterpart. Our investigations presents Zn doped SnS$_2$ could be a potential candidate for future nano electronic and photoelectronic applications.

2. Experiment

2.1. Synthesis of Sn$_{1-x}$Zn$_x$S$_2$ Nanoflakes

SnS$_2$ and Sn$_{1-x}$Zn$_x$S$_2$ nanoflakes were prepared via low cost hydrothermal route reported previously [35]. In brief, 0.1753 g SnCl$_4$·5H$_2$O (Tin (IV) chloride pentahydrate) and 0.15 g thioacetamide (TAA) were dissolved in 80 mL distilled water, stirred for 1 h to result in homogeneous solution. The prepared solution was transferred to 100 mL Teflon-line autoclave, sealed and heated up to 160 °C for 12 h and finally cooled to room temperature. The prepared SnS$_2$ nanoflakes were then washed with ethanol and deionized water repeatedly and finally dried at 60 °C for 12 h in electric oven. For the synthesis of Sn$_{1-x}$Zn$_x$S$_2$ nanoflakes, 1 and 3 mmol% of Zinc chloride was added to the precursor solution.

2.2. Characterization

The morphological evolution of the sample was examined using field-emission scanning electron microscopy (FESEM, Philips, Model: XL-30, Amsterdam, The Netherland) and field-emission transmission electron microscopy (FE-TEM, JEM-2100F HR, Tokyo, Japan). The phase purity and crystal structure of SnS$_2$ and Sn$_{0.97}$Zn$_{0.03}$S$_2$ nanoflakes was inferred through X-ray diffractometer (SmartLab, Rigaku Corporation, Tokyo, Japan). The Raman measurements were performed in a micro-Raman spectrometer (DawoolAttonics, Model: Micro Raman System, Seongnam, Korea) using an excitation wavelength of 532 nm. The chemical composition of Sn$_{0.97}$Zn$_{0.03}$S$_2$ was obtained using X-ray photoelectron spectroscopy (K-Alpha+, ThermoFisher Scientific, Waltham, MA, USA). In order to avoid charging effect, during the measurement, charge neutralization was performed with an electron flood gun (K-Alpha+, ThermoFisher Scientific, USA). The absorbance spectrum was recorded using a UV/VIS spectrophotometer (K LAB, Model: Optizen POP, Daejeon, Korea). A Keithley 617 semiconductor parameter analyzer (Tektronix, Beaverton, OR, USA; Model: Keithley 617) was employed to study the photo-response of the device under solar simulator (Newport, OR, USA; AM1.5) (SERIC, Model: XIL-01B50KP).

2.3. Device Fabrication

Initially, 2 mg of samples SnS$_2$ and Sn$_{0.97}$Zn$_{0.03}$S$_2$ were added in 10 mL methoxy-ethanol solvent separately and magnetic stirred for 30 min followed by sonication of about 30 min to form colloidal suspension. The resulting suspension was then spin casted on cleaned and patterned ITO/glass substrate at 1000 rpm and dried at 100 °C for 5 min. Several cycles of spin casting process was repeated to obtain a continuous film.

3. Results and Discussions

The morphological features of SnS_2, $Sn_{0.99}Zn_{0.01}S_2$ (Figure S1) and $Sn_{0.97}Zn_{0.03}S_2$ products were examined with the aid of field-emission scanning electron microscope (FESEM) technique. The image seen from Figure 1a–c confirms hexagonal nanoflakes with smooth surface and homogeneous distribution in case of pristine SnS_2. However, on doping with Zinc the morphology appears to be similar with that of pristine nanoflakes with some random aggregates on the surface of SnS_2 (Figure 1d,e). Additionally, transmission electron microscope (TEM) was employed to further investigate the detailed morphological information of SnS_2 and $Sn_{0.97}Zn_{0.03}S_2$ products. Figure 2 shows TEM images of pristine SnS_2 and $Sn_{0.97}Zn_{0.03}S_2$ nanoflakes with different magnifications. From the Figure 2a–c, it is clear that pristine SnS_2 possess typical nanoflakes like structures with hexagonal stacking. Similarly the $Sn_{0.97}Zn_{0.03}S_2$ nanoflakes (Figure 2d–f) also possess indistinguishable hexagonal morphology of pristine SnS_2. The inset of Figure 2c,f displays the selected area electron diffraction (SAED) pattern revealing polycrystalline structure of the obtained samples. Energy dispersive spectroscopy (EDS) analysis was further employed in TEM mode to study the homogeneous distribution of Zn element in $Sn_{0.97}Zn_{0.03}S_2$ nanoflakes. Figure 3a–d displays the TEM image and TEM-EDS mapping of $Sn_{0.97}Zn_{0.03}S_2$ nanoflakes. As seen from Figure 3d, Zn element is distributed evenly throughout the whole structure of $Sn_{0.97}Zn_{0.03}S_2$ nanoflakes.

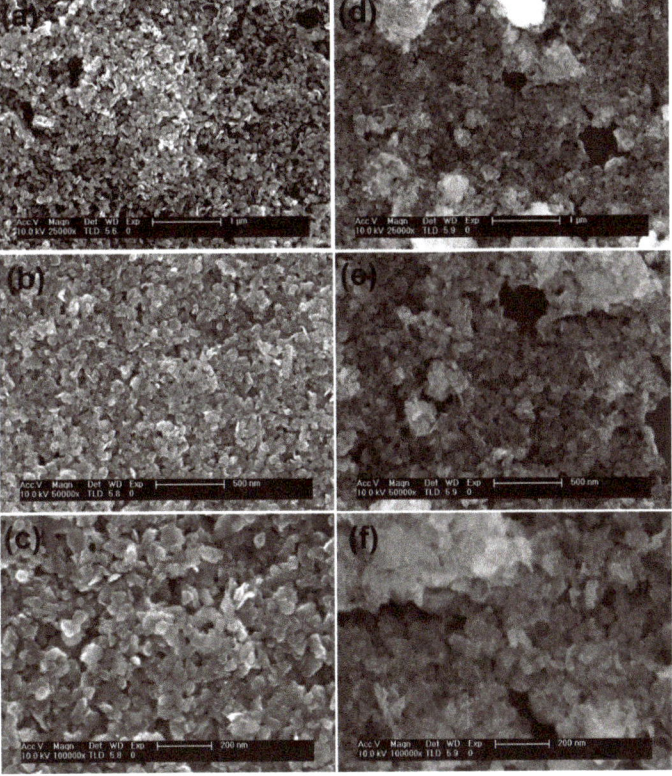

Figure 1. Morphological and structural characterization of SnS_2 and $Sn_{0.97}Zn_{0.03}S_2$ nanoflakes. (**a–c**) low magnification and high magnification scanning electron microscopy (SEM) image of SnS_2; (**d–f**) low magnification and high magnification SEM image of $Sn_{0.97}Zn_{0.03}S_2$ nanoflakes showing their hexagonal structure.

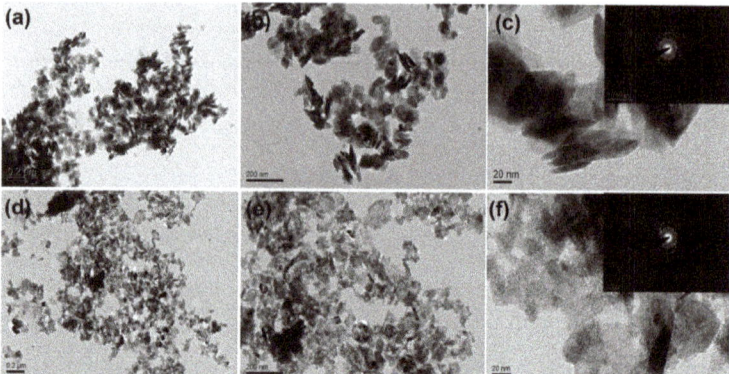

Figure 2. (a–c) Transmission electron microscopy (TEM) images of SnS$_2$ and inset in Figure 2c shows selected area electron diffraction (SAED) pattern of SnS$_2$ nanoflakes; (d–f) TEM images of a typical Sn$_{0.97}$Zn$_{0.03}$S$_2$ nanoflakes with SAED pattern in inset of Figure 2f, revealing polycrystalline structure.

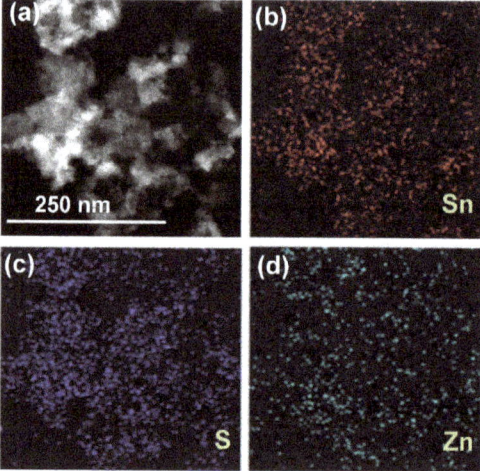

Figure 3. (a) TEM image of Sn$_{0.97}$Zn$_{0.03}$S$_2$ nanoflakes and Energy dispersive spectroscopy (EDS) elemental mapping of Sn (b), S (c) and Zn (d) from selected area for 2D Sn$_{0.97}$Zn$_{0.03}$S$_2$.

The crystallographic pattern of as synthesized SnS$_2$, Sn$_{0.99}$Zn$_{0.01}$S$_2$ and Sn$_{0.97}$Zn$_{0.03}$S$_2$ nanoflakes are investigated by XRD analysis and presented in Figure 4a. Here, the strong diffraction peak observed at 2θ = 14.92° belongs to (001) diffraction, is an indication of the hexagonal structure of SnS$_2$ [37]. However, the diffraction peak (001) tends to shift towards smaller angle on Zn doping. This shifting indicates that Zn ions replace Sn sites in the SnS$_2$ crystal matrix. Furthermore, no peaks related to other compounds namely, ZnS and ZnSnS$_3$ are observed in the XRD pattern. Additionally, Raman measurement was further analyzed to study detailed information about the structural properties of Zn doped SnS$_2$ nanoflakes. Raman spectrum for sample SnS$_2$, Sn$_{0.99}$Zn$_{0.01}$S$_2$ and Sn$_{0.97}$Zn$_{0.03}$S$_2$ nanoflakes are displayed in Figure 4b. Here, in case of pristine SnS$_2$, Sn$_{0.99}$Zn$_{0.01}$S$_2$ and Sn$_{0.97}$Zn$_{0.03}$S$_2$ nanoflakes, a strong signal was observed at 312 cm^{-1}, which is related to A$_{1g}$ phonon vibration mode of SnS$_2$ [38–40].

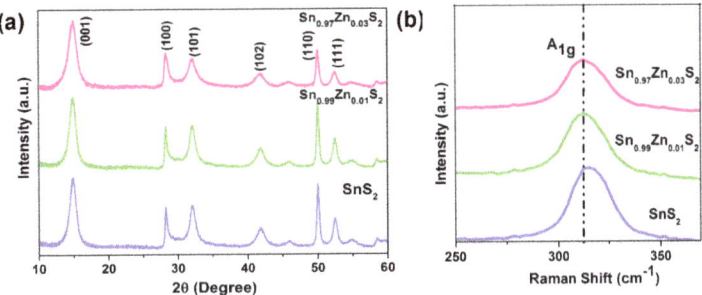

Figure 4. Structure properties of SnS_2, $Sn_{0.99}Zn_{0.01}S_2$ and $Sn_{0.97}Zn_{0.03}S_2$ nanoflakes. (**a**) X-ray diffraction pattern of SnS_2, $Sn_{0.99}Zn_{0.01}S_2$ and $Sn_{0.97}Zn_{0.03}S_2$ nanoflakes; (**b**) Raman spectrum of SnS_2, $Sn_{0.99}Zn_{0.01}S_2$ and $Sn_{0.97}Zn_{0.03}S_2$ nanoflakes at excitation wavelength of 532 nm.

To elucidate the chemical composition of pristine and $Sn_{0.97}Zn_{0.03}S_2$ nanoflakes, XPS measurements have been carried out and shown in Figure 5a. XPS full survey spectrum (Figure 5a) confirms the presence of Zn doping in SnS_2. Figure 5b,c displays the XPS spectra of Sn 3d and S 2p peaks for $Sn_{0.97}Zn_{0.03}S_2$ nanoflakes. As observed in Figure 5b,c, the peaks of Sn 3d at 486.33 and 494.4 eV of Sn 3d is ascribed to $Sn3d_{3/2}$ and $Sn3d_{5/2}$ and peaks at 161.2 and 163.3 eV correspond to S 2p peaks of SnS_2. These results are consistent with those reported for SnS_2 [41,42]. The binding energies of Sn $3d_{5/2}$ peak corresponding to pristine SnS_2 was observed at 486.47 eV. Subsequently doping with Zn on SnS_2, peaks of Sn $3d_{5/2}$ shifts to lower energy position to 486.33 eV. The shifting in the binding energy value of Sn $3d_{5/2}$ peak was about 0.14 eV compared to pristine SnS_2. This shift might be due to Zn ion replace Sn sites in the SnS_2 crystal lattice. Figure 5d shows the XPS spectrum for Zn in SnS_2 nanoflakes. Besides, the Zn $2p_{3/2}$ peak appeared at 1021.3 eV is attributed to Zn^{2+} bonding state [43], confirming Zn^{2+} ions have been incorporated into the SnS_2.

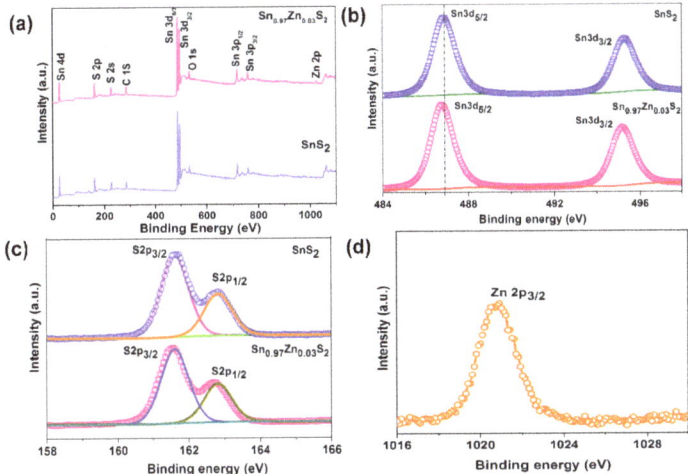

Figure 5. (**a**) Full survey spectra of SnS_2 and $Sn_{0.97}Zn_{0.03}S_2$ sample. (**b**) X-ray photoelectron spectroscopy (XPS) core level Sn 3d spectra of SnS_2 and $Sn_{0.97}Zn_{0.03}S_2$ nanoflakes. (**c**) S 2p core level spectra of SnS_2 and $Sn_{0.97}Zn_{0.03}S_2$ nanoflakes. (**d**) Zn 2p core level spectra of $Sn_{0.97}Zn_{0.03}S_2$.

Figure 6a shows UV–visible absorption spectrum of SnS_2, $Sn_{0.99}Zn_{0.01}S_2$ and $Sn_{0.97}Zn_{0.03}S_2$ in the range of 300–750 nm. SnS_2 displays a strong absorption in visible part of the solar spectrum. However, in contrast the samples $Sn_{0.99}Zn_{0.01}S_2$ and $Sn_{0.97}Zn_{0.03}S_2$ displayed a broad light absorption in 300 to

750 nm, which indicates that doping Zn ion can result in extending of absorption edge of SnS_2. This results suggests that samples $Sn_{0.97}Zn_{0.03}S_2$ possess greater potential than that of pristine sample SnS_2 to drive photo excited charge carriers under the light irradiation. The values estimated was found to be 2.24 eV for sample SnS_2 which is consistent with our previous result (Figure 6b). However, the values was found to be 2.19 and 2.09 eV for sample $Sn_{0.99}Zn_{0.01}S_2$ and $Sn_{0.97}Zn_{0.03}S_2$. It shows band gap becomes narrower than pristine SnS_2 as the Zn content increases [44,45]. This reduction in the band gap might be due to modification in the electronic structures of SnS_2 due to Zn doping, which results in creating energy levels in the band gap. This band gap could result in better absorption in visible region and can increase photo excited charge carriers under illumination.

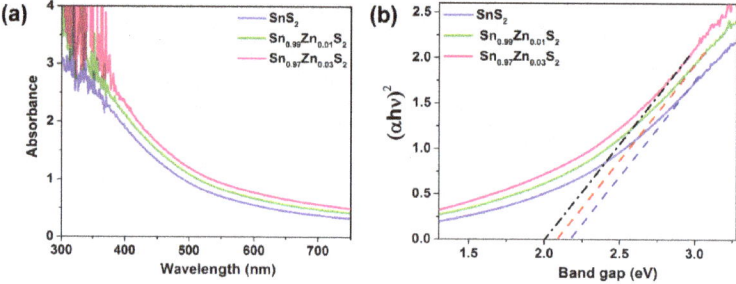

Figure 6. Properties of SnS_2, $Sn_{0.99}Zn_{0.01}S_2$ and $Sn_{0.97}Zn_{0.03}S_2$ nanoflakes. (**a**) UV–vis absorption spectrum of the SnS_2, $Sn_{0.99}Zn_{0.01}S_2$ and $Sn_{0.97}Zn_{0.03}S_2$ nanoflakes. (**b**) Tauc's plot extracted from the absorption spectrum revealing their direct band gap.

Mott–Schottky (M–S) analysis was made to study the electrical properties of pristine SnS_2, $Sn_{0.99}Zn_{0.01}S_2$ and $Sn_{0.97}Zn_{0.03}S_2$ nanoflakes. Generally, Mott-Schottky plot was employed to determine the donor density (N_d) and flat band potential (V_{fb}) of the materials. M–S analysis are generally expressed by [46–48]

$$1/C^2 = (2/e\varepsilon\varepsilon_o N_d)[(V_{fb} - V) - k_B T/e] \quad (1)$$

where e is the electronic charge, ε is the dielectric constant of SnS_2, ε_0 is the relative permittivity, N_d dopant density, V the applied potential, C the specific capacitance, k_B the Boltzmann constant and V_{fb} the flat band potential. The M–S plots of pristine SnS_2, $Sn_{0.99}Zn_{0.01}S_2$ (Figure S2) and $Sn_{0.97}Zn_{0.03}S_2$ nanoflakes are displayed in Figure 7. Here V_{fb} was determined from intercept between the extrapolated linear plot of the curve and was estimated to be ~0.67 V for pristine SnS_2 and 0.64 V for $Sn_{0.97}Zn_{0.03}S_2$ nanoflakes. Additionally the difference in the slope reflects the variation in the carrier density (N_d). The values of carrier density was estimated from the Equation (1) to be about 1.46×10^{19} and 0.47×10^{19} and in case of SnS_2 and $Sn_{0.97}Zn_{0.03}S_2$ nanoflakes.

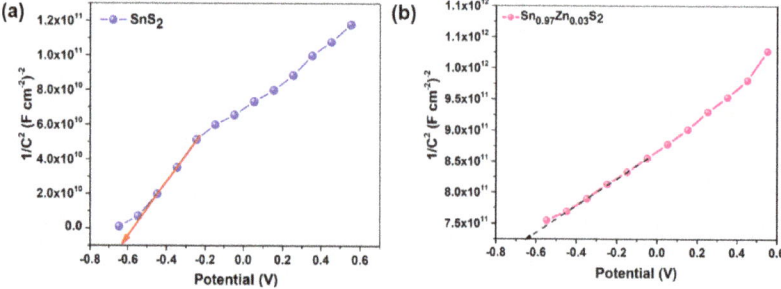

Figure 7. Mott–Schottky plots of (**a**) SnS_2 and (**b**) $Sn_{0.97}Zn_{0.03}S_2$ nanoflakes.

A photoelectronic device was constructed on samples SnS_2 and $Sn_{0.97}Zn_{0.03}S_2$ to study its potential for optoelectronics applications (Figure 8a), (for the details of fabrication process refer Expt. sections). I-V curves of pristine SnS_2 nanoflakes at various illumination intensities and dark condition is displayed in Figure 8b. Inset shows I-V curves of the pristine SnS_2 nanoflakes under dark and illumination. Here, the I-V curve shows a roughly symmetric behavior indicating Schottky-like junction established at ITO and SnS_2 contacts. The dark current was noted to be 0.29 µA at a bias of 3 V. In contrast, the enhancement of current was measured and the value reaches to 0.98 µA under illumination, demonstrating excellent photosensitivity of the SnS_2 samples. I-V curves of $Sn_{0.97}Zn_{0.03}S_2$ nanoflakes device under illumination and dark is displayed in Figure 8c. Here, the value of dark current was found to increase than that of pristine SnS_2, which suggests reduction in resistance of SnS_2 after Zn doping. However, a notable enhancement in photocurrent under illumination was noted compared to that of dark current at same bias voltage in $Sn_{0.97}Zn_{0.03}S_2$ nanoflakes device, indicating their excellent sensitivity. Moreover, photo to dark current (I_{light}/I_{dark}) ratio for $Sn_{0.97}Zn_{0.03}S_2$ device (~10.1) tends to increase compared to pristine SnS_2 (~3.37). The high sensitivity and enhancement in photocurrent of $Sn_{0.97}Zn_{0.03}S_2$ nanoflakes reveal the effective separation of photoexcited carriers in samples, which are actually promoted after Zn-doping. Figure 8d shows I-V curves of the $Sn_{0.97}Zn_{0.03}S_2$ device measured at room temperature under different light intensities. The photocurrent increases with increasing light intensities revealing strong and clear photon-induced currents phenomena, indicating excellent photoresponse ability of the device. Under illumination, photoexcited charge carriers are mainly generated in $Sn_{0.97}Zn_{0.03}S_2$. Then the charge carriers are quickly separated and driven towards the nearby electrodes due to built-in electric field created at the interface, resulting in photocurrent generation.

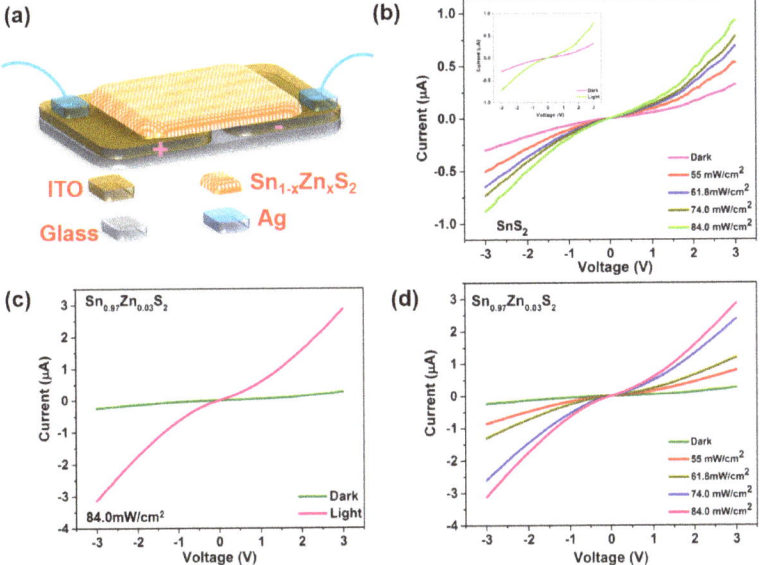

Figure 8. (a) Schematic representation of the photoelectronic device. (b) I-V characteristics of SnS_2 device under different illumination intensities (Inset shows the I-V characteristics under dark and illumination intensity 84.0 mW/cm^2). (c) I-V characteristics of $Sn_{0.97}Zn_{0.03}S_2$ device under illumination conditions. (d) I-V characteristics of $Sn_{0.97}Zn_{0.03}S_2$ device under different light intensities (55, 61.8, 74.0, 84.0 mW/cm^2).

Figure 9a shows light intensity-dependent photocurrent values of pristine SnS_2 and $Sn_{0.97}Zn_{0.03}S_2$ device. The observed photocurrent value to illumination intensities suggest that the charge carrier photo-generation efficiency is proportional to the number of photons absorbed by the pristine SnS_2 and

$Sn_{0.97}Zn_{0.03}S_2$ nanoflakes. Reliable response speed and stability to illumination conditions are crucial for the photoelectronic device. To address this concern, time related photoresponse of pristine SnS_2 and $Sn_{0.97}Zn_{0.03}S_2$ device was measured with turning light on/off condition for a period of 10 seconds for multiple cycles. Figure 9b,c shows time related photoresponse of the pristine and $Sn_{0.97}Zn_{0.03}S_2$ device under several switch on and switch off conditions. Here, the photocurrent of pristine SnS_2 was found to be 0.8 µA. Interestingly the photocurrent is improved by two fold in case of $Sn_{0.97}Zn_{0.03}S_2$ nanoflakes (1.75 µA) compared to pristine SnS_2 (Figure 9c). The photoresponse enhancement could be related to Zn ions which acts as an effective dopant and enhance charge separation taking place at the interface. The rise/decay time was measured to be 0.2 and 0.2 s. The reason for the relative longer response speed in our case is probably related to the formation of interface states between the $Sn_{0.97}Zn_{0.03}S_2$ nanoflakes and ITO substrate, which can block the photo-generated carriers, resulting in long life time of the photo-generated carriers. Meanwhile, the device shows no fluctuation under illumination for several repetitive cycles, inferring the excellent stability of the $Sn_{0.97}Zn_{0.03}S_2$ device. The time related response of the $Sn_{0.97}Zn_{0.03}S_2$ device under varied light intensities are displayed in Figure 9d. Here, the photocurrent value varies with different light intensities demonstrating excellent reproducibility of $Sn_{0.97}Zn_{0.03}S_2$ based device. Such high and stable photoresponse behavior may come from the fact that Zn ions act as an effective dopant and result in increased light absorption, which enhances photogenerated charge carriers and leads to an enhanced photocurrent of the device. Thus, photoelectrical studies on $Sn_{0.97}Zn_{0.03}S_2$ nanoflakes illustrates that Zn doping in SnS_2 results in significant enhancement of their optoelectronic properties, which leads to improved conductivity and sensitivity.

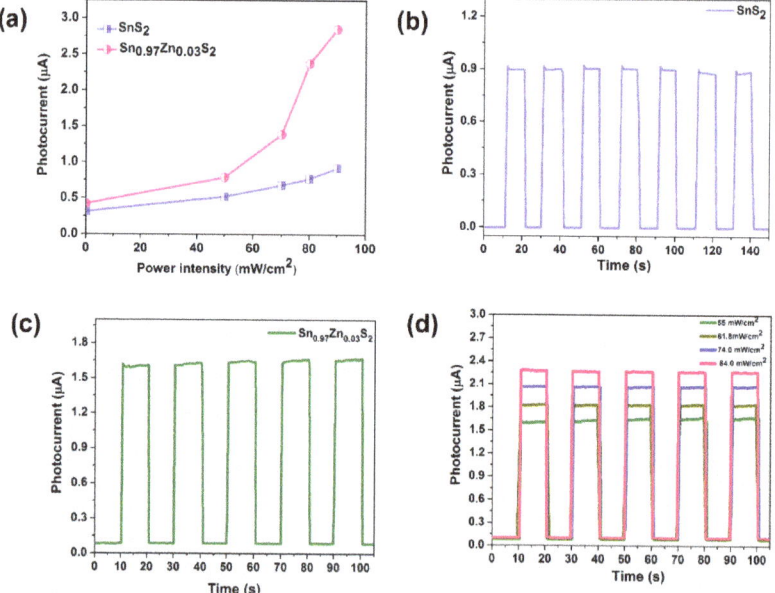

Figure 9. (**a**) Light intensity-dependent photocurrent values of pristine SnS_2 and $Sn_{0.97}Zn_{0.03}S_2$ device. Time-dependent photocurrent response of (**b**) SnS_2 device and (**c**) $Sn_{0.97}Zn_{0.03}S_2$. (**d**) Time-dependent photocurrent response of $Sn_{0.97}Zn_{0.03}S_2$ device under different illumination intensities.

The mechanism involved in the enhanced photoresponse of $Sn_{0.97}Zn_{0.03}S_2$/ITO structure was explained through energy band diagram in Figure 10. Since the work function between ITO and $Sn_{0.97}Zn_{0.03}S_2$ is different, a Schottky-type behavior is established at $Sn_{0.97}Zn_{0.03}S_2$/ITO interface

(Figure 10). Due to this behavior, an electric field was established at the $Sn_{0.97}Zn_{0.03}S_2$/ITO interface. This electric field then accelerates the separation of the photoexcited charge carriers without the application of any applied bias. When illuminated, photoexcited charge carriers produced in $Sn_{0.97}Zn_{0.03}S_2$ are then separated at the $Sn_{0.97}Zn_{0.03}S_2$/ITO interface. This charge carriers separation which was induced due to the electric field results in band bending at the $Sn_{0.97}Zn_{0.03}S_2$/ITO interface. As a result, the photoexcited charge carriers are swept towards ITO electrodes, involving in enhancement of photocurrent (Figure 10b).

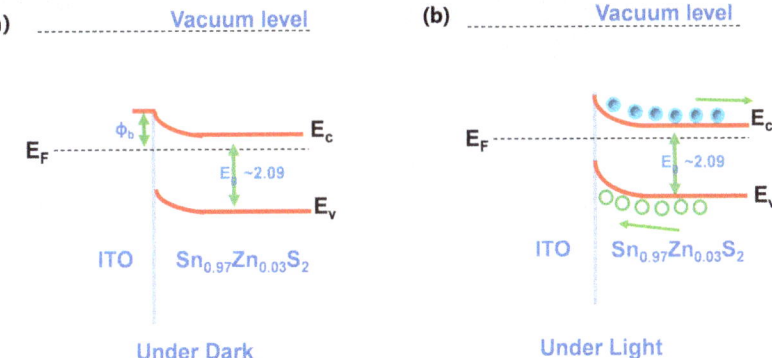

Figure 10. Energy diagram of the $Sn_{0.97}Zn_{0.03}S_2$/ITO Schottky junction under (a) dark and (b) illumination conditions.

4. Conclusions

In summary, $Sn_{0.97}Zn_{0.03}S_2$ nanoflakes were prepared via low temperature hydrothermal synthesis. The modulation of the structural and photoelectrical properties in SnS_2 via doping Zinc have been discussed in detail. A shift in XPS peak of Sn $3d_{5/2}$ and S $2p_{3/2}$ has been observed in $Sn_{0.97}Zn_{0.03}S_2$ nanoflakes due to Zn ion replaced Sn sites in the SnS_2 crystal lattice. Optical properties studies show that $Sn_{0.97}Zn_{0.03}S_2$ nanoflakes possess higher visible-light absorption than that of pristine SnS_2. Photoelectrical properties based on $Sn_{0.97}Zn_{0.03}S_2$ nanoflakes reveal that Zn doping leads to significant improvement in conductivity and sensitivity to illuminations compared to pristine SnS_2. Such an excellent performance of $Sn_{0.97}Zn_{0.03}S_2$ nanoflakes may endow it as a potential candidate for emerging 2D materials in optoelectronic applications.

Supplementary Materials: The following are available online at http://www.mdpi.com/2079-4991/9/7/924/s1, Figure S1: SEM image of $Sn_{0.99}Zn_{0.01}S_2$ nanoflakes, Figure S2: Mott–Schottky plot of $Sn_{0.99}Zn_{0.01}S_2$ nanoflakes.

Author Contributions: Conceptualization: G.M.K. and P.I.; data curation, G.M.K., H.D.C., H.C.J. and P.I.; supervision, T.W.K.; validation, G.M.K. and S.Y.; visualization, T.W.K. and D.Y.K.; writing—original draft, P.I.; all authors read and approved the final manuscript.

Acknowledgments: This work was supported by the Basic Science Research Program through the National Research Foundation of Korea (NRF) grant funded by the Ministry of Education (no. 2018R1D1A1B07051461, no. 2018R1D1A1B07051474, no. 2018R1D1A1B07051406, no. 2018R1D1A1B07050237, no. 2018R1D1A1B07051095, no. 2017R1D1A1B03032759 and no. 2016R1A6A1A1A01012877).

Conflicts of Interest: The authors declare no conflict of interest.

References

1. Wang, Q.H.; Kalantar-Zadeh, K.; Kis, A.; Coleman, J.N.; Strano, M.S. Electronics and optoelectronics of two-dimensional transition metal dichalcogenides. *Nat. Nanotechnol.* **2012**, *7*, 699–712. [CrossRef] [PubMed]
2. Najmaei, S.; Liu, Z.; Zhou, W.; Zou, X.; Shi, G.; Lei, S.; Yakobson, B.I.; Idrobo, J.; Ajayan, P.M.; Lou, J. Vapour phase growth and grain boundary structure of molybdenum disulphide atomic layers. *Nat. Mater.* **2013**, *12*, 754–759. [CrossRef] [PubMed]
3. Chhowalla, M.; Shin, H.S.; Eda, G.; Li, L.-J.; Loh, K.P.; Zhang, H. The Chemistry of two-dimensional layered transition metal dichalcogenides nanoflakes. *Nat. Chem.* **2013**, *5*, 263–275. [CrossRef] [PubMed]
4. Van der Zande, A.M.; Huang, P.Y.; Chenet, D.A.; Berkelbach, T.C.; You, Y.; Lee, G.-H.; Heinz, T.F.; Reichman, D.R.; Muller, D.A.; Hone, J.C. Grains and grain boundaries in highly crystalline monolayer molybdenum disulphide. *Nat. Mater.* **2013**, *15*, 554–561. [CrossRef] [PubMed]
5. Cheng, L.; Huang, W.; Gong, Q.; Liu, C.; Liu, Z.; Li, Y.; Dai, H. Ultrathin WS_2 nanoflakes as a high-performance electrocatalyst for the hydrogen evolution reaction. *Angew. Chem. Int. Ed.* **2014**, *53*, 7860–7863. [CrossRef] [PubMed]
6. Peimyoo, N.; Yang, W.; Shang, J.; Shen, X.; Wang, Y.; Yu, T. Chemically driven tunable light emission of charged and neutral excitons in monolayer WS_2. *ACS Nano* **2014**, *8*, 11320–11329. [CrossRef] [PubMed]
7. Late, D.J.; Liu, B.; Ramakrishna Matte, H.S.S.; Dravid, V.P.; Rao, C.N.R. Hysteresis in single-layer MoS_2 field effect transistors. *ACS Nano* **2012**, *6*, 5635–5641. [CrossRef] [PubMed]
8. Wang, H.; Yuan, H.; Hong, S.S.; Li, Y.; Cui, Y. Physical and chemical tuning of two-dimensional transition metal dichalcogenides. *Chem. Soc. Rev.* **2015**, *44*, 2664–2680. [CrossRef] [PubMed]
9. Sun, Y.; Sun, Z.; Gao, S.; Cheng, H.; Liu, Q.; Piao, J.; Yao, T.; Wu, C.; Hu, S.; Wei, S.; et al. Fabrication of flexible and freestanding zinc chalcogenide single layers. *Nat. Commun.* **2012**, *3*, 1057–1063. [CrossRef]
10. Lei, F.; Sun, Y.; Liu, K.; Gao, S.; Liang, L.; Pan, B.; Xie, Y. Oxygen vacancies confined in ultrathin indium oxide porous sheets for promoted visible-light water splitting. *J. Am. Chem. Soc.* **2014**, *136*, 6826–6829. [CrossRef]
11. Yang, J.; Son, J.S.; Yu, J.H.; Joo, J.; Hyeon, T. Advances in the colloidal synthesis of two-dimensional semiconductor nanoribbons. *Chem. Mater.* **2013**, *25*, 1190–1198. [CrossRef]
12. Huang, J.-K.; Pu, J.; Hsu, C.-L.; Chiu, M.-H.; Juang, Z.-Y.; Chang, Y.-H.; Chang, W.-H.; Iwasa, Y.; Takenobu, T.; Li, L.-J. Large area synthesis of highly crystalline WSe_2 monolayers and device applications. *ACS Nano* **2014**, *8*, 923–930. [CrossRef] [PubMed]
13. Zeng, Z.; Yin, Z.; Huang, X.; Li, H.; He, Q.; Lu, G.; Boey, F.; Zhang, H. Single-layer semiconducting nanoflakes: High-yield preparation and device fabrication. *Angew. Chem. Int. Ed.* **2011**, *50*, 11093–11097. [CrossRef]
14. Xia, J.; Zhu, D.; Wang, L.; Huang, B.; Huang, X.; Meng, X.-M. Large-scale growth of two-dimensional SnS_2 crystals driven by screw dislocations and application to photodetectors. *Adv. Funct. Mater.* **2015**, *25*, 4255–4261. [CrossRef]
15. Fu, X.; Ilanchezhiyan, P.; Mohan Kumar, G.; Cho, H.D.; Zhang, L.; Sattar Chan, A.; Lee, D.J.; Panin, G.N.; Kang, T.W. Tunable UV-visible absorption of SnS_2 layered quantum dots produced by liquid phase exfoliation. *Nanoscale* **2017**, *9*, 1820–1826. [CrossRef] [PubMed]
16. Tao, Y.; Wu, X.; Wang, W.; Wang, J. Flexible photodetector from ultraviolet to near infrared based on a SnS_2 nanosheet microsphere film. *J. Mater. Chem. C* **2015**, *3*, 1347–1353. [CrossRef]
17. Huang, Y.; Deng, H.-X.; Xu, K.; Wang, Z.-X.; Wang, Q.-S.; Wang, F.-M.; Wang, F.; Zhan, X.-Y.; Li, S.-S.; Luo, J.-W.; et al. Highly sensitive and fast phototransistor based on large size CVD-grown SnS_2 nanoflakes. *Nanoscale* **2015**, *7*, 14093–14099. [CrossRef]
18. Mohan Kumar, G.; Fu, X.; Ilanchezhiyan, P.; Yuldashev, S.U.; Lee, D.J.; Cho, H.D.; Kang, T.W. Highly sensitive flexible photodetectors based on self-assembled tin monosulfide nanoflakes with graphene electrodes. *ACS Appl. Mater. Interfaces* **2017**, *9*, 32142–32150. [CrossRef]
19. Mohan Kumar, G.; Xiao, F.; Ilanchezhiyan, P.; Yuldashev, S.U.; Kang, T.W. Enhanced photoelectrical performance of chemically processed SnS_2 nanoplates. *RSC Adv.* **2016**, *6*, 99631–99637. [CrossRef]
20. De, D.; Manongdo, J.; See, S.; Zhang, V.; Guloy, A.; Peng, H.B. High on/off ratio field effect transistors based on exfoliated crystalline SnS_2 nano-membrane. *Nanotechnology* **2013**, *24*, 025202. [CrossRef]
21. Ye, X.; Chen, J.; Engel, M.; Millan, J.A.; Li, W.; Qi, L.; Xing, G.; Collins, J.E.; Kagan, C.R.; Li, J.; et al. Competition of shape and interaction patchiness for self-assembling nanoplates. *Nat. Chem.* **2013**, *5*, 466–473. [CrossRef] [PubMed]

22. Ye, X.; Collins, J.E.; Kang, Y.; Chen, J.; Chen, D.T.; Yodh, A.G.; Murray, C.B. Morphologically controlled synthesis of colloidal upconversion nanophosphors and their shape-directed self-assembly. *Proc. Natl. Acad. Sci. USA* **2010**, *107*, 22430–22435. [CrossRef] [PubMed]
23. Novoselov, K.S.; Fal'ko, V.I.; Colombo, L.; Gellert, P.R.; Schwab, M.G.; Kim, K. A roadmap for graphene. *Nature* **2012**, *490*, 192–200. [CrossRef] [PubMed]
24. Zhang, K.; Zhang, T.N.; Cheng, G.H.; Li, T.X.; Wang, S.X.; Wei, W.; Zhou, X.H.; Yu, W.W.; Sun, Y.; Wang, P.; et al. Interlayer transition and infrared photodetection in atomically thin type-II $MoTe_2/MoS_2$ van der Waals heterostructures. *ACS Nano* **2016**, *10*, 3852–3858. [CrossRef] [PubMed]
25. Zhang, Y.; Zhu, P.; Huang, L.; Xie, J.; Zhang, S.; Cao, G.; Zhao, X. Few-layered SnS_2 on few-layered reduced graphene oxide as Na-Ion battery anode with ultralong cycle life and superior rate capability. *Adv. Funct. Mater.* **2015**, *25*, 481–489. [CrossRef]
26. Tan, F.R.; Qu, S.C.; Wu, J.; Liu, K.; Zhou, S.Y.; Wang, Z.G. Preparation of SnS_2 colloidal quantum dots and their application in organic/inorganic hybrid solar cells. *Nanoscale Res. Lett.* **2011**, *6*, 1–8. [CrossRef] [PubMed]
27. Su, G.; Hadjiev, V.G.; Loya, P.E.; Zhang, J.; Lei, S.; Maharjan, S.; Dong, P.; Ajayan, P.M.; Lou, J.; Peng, H. Chemical vapor deposition of thin crystals of layered semiconductor SnS_2 for fast photodetection application. *Nano Lett.* **2015**, *15*, 506–513. [CrossRef]
28. Zhou, X.; Zhang, Q.; Gan, L.; Li, H.; Zhai, T. Large-size growth of ultrathin SnS_2 nanoflakes and high performance for phototransistors. *Adv. Funct. Mater.* **2016**, *26*, 4405–4413. [CrossRef]
29. Wei, R.; Hu, J.; Zhou, T.; Zhou, X.; Liu, J.; Li, J. Ultrathin SnS_2 nanoflakes with exposed {001} facets and enhanced photocatalytic properties. *Acta Mater.* **2014**, *66*, 163–171. [CrossRef]
30. Wang, J.; Liu, J.; Xu, H.; Ji, S.; Wang, J.; Zhou, Y.; Hodgson, P.; Li, Y. Gram-scale and template-free synthesis of ultralong tin disulfide nanobelts and their lithium ion storage performances. *J. Mater. Chem. A* **2013**, *1*, 1117–1122. [CrossRef]
31. Cui, X.; Xu, W.; Xie, Z.; Dorman, J.A.; Gutierrez-Wing, M.T.; Wang, Y. Effect of dopant concentration on visible light driven photocatalytic activity of $Sn_{1-x}Ag_xS_2$. *Dalton Trans.* **2016**, *45*, 16290–16297. [CrossRef] [PubMed]
32. Wahnón, P.; Conesa, J.C.; Palacios, P.; Lucena, R.; Aguilera, I.; Seminovski, Y.; Fresno, F. V-doped SnS_2: A new intermediate band material for a better use of the solar spectrum. *Phys. Chem. Chem. Phys.* **2011**, *13*, 20401–20407. [CrossRef] [PubMed]
33. Hu, K.; Wang, D.; Zhao, W.; Gu, Y.; Bu, K.; Pan, J.; Qin, P.; Zhang, X.; Huang, F. Intermediate band material of titanium-doped tin disulfide for wide spectrum solar absorption. *Inorg. Chem.* **2018**, *57*, 3956–3962. [CrossRef] [PubMed]
34. Li, B.; Xing, T.; Zhong, M.; Huang, L.; Lei, N.; Zhang, J.; Li, J.; Wei, Z. A two-dimensional Fe-doped SnS_2 magnetic semiconductor. *Nat. Commun.* **2017**, *8*, 1958. [CrossRef] [PubMed]
35. Mohan Kumar, G.; Fu, X.; Ilanchezhiyan, P.; Yuldashev, S.U.; Madhan Kumar, A.; Cho, H.D.; Kang, T.W. High performance photodiodes based on chemically processed Cu doped SnS_2 nanoflakes. *Appl. Surf. Sci.* **2018**, *455*, 446–454. [CrossRef]
36. Liu, J.; Liu, X.; Chen, Z.; Miao, L.; Liu, X.; Li, B.; Tang, L.; Chen, K.; Liu, Y.; Li, J.; et al. Tunable Schottky barrier width and enormously enhanced photoresponsivity in Sb doped SnS_2 monolayer. *Nano Res.* **2018**, *12*, 463–468. [CrossRef]
37. Yu, J.; Xu, C.-Y.; Ma, F.-X.; Hu, S.-P.; Zhang, Y.-W.; Zhen, L. Monodisperse SnS_2 nanoflakes for high-performance photocatalytic hydrogen generation. *ACS Appl. Mater. Interfaces* **2014**, *6*, 22370–22377. [CrossRef]
38. Qu, B.; Ma, C.; Ji, G.; Xu, C.; Xu, J.; Meng, Y.S.; Wang, T.; Lee, J.Y. Layered SnS_2-reduced graphene oxide composite—a high-capacity, high-rate, and long-cycle life sodium-ion battery anode material. *Adv. Mater.* **2014**, *26*, 3854–3859. [CrossRef]
39. Du, Y.; Yin, Z.; Rui, X.; Zeng, Z.; Wu, X.J.; Liu, J.; Zhu, Y.; Zhu, J.; Huang, X.; Yan, Q.; et al. A facile, relative green, and inexpensive synthetic approach toward large-scale production of SnS_2 nanoplates for high-performance lithium-ion batteries. *Nanoscale* **2013**, *5*, 1456–1459. [CrossRef]
40. Chen, Q.; Lu, F.; Xia, Y.; Wang, H.; Kuang, X. Interlayer expansion of few layered Mo-doped SnS_2 nanoflakes grown on carbon cloth with excellent lithium storage performance for lithium ion batteries. *J. Mater. Chem. A* **2017**, *5*, 4075–4083. [CrossRef]

41. Liu, X.; Zhao, H.L.; Kulka, A.; Trenczek-Zajac, A.; Xie, J.Y.; Chen, N.; Swierczek, K. Characterization of the physicochemical properties of novel SnS$_2$ with cubic structure and diamond-like Sn sublattice. *Acta Mater.* **2015**, *82*, 212–223. [CrossRef]
42. Ilanchezhiyan, P.; Kumar, G.M.; Kang, T.W. Electrochemical studies of spherically clustered MoS$_2$ nanostructures for electrode applications. *J. Alloys Compd.* **2015**, *634*, 104–108. [CrossRef]
43. Liu, X.; Bai, H. Hydrothermal synthesis of visible light active zinc-doped tin disulfide photocatalyst for the reduction of aqueous Cr(VI). *Powder Technol.* **2013**, *237*, 610–615. [CrossRef]
44. An, X.; Yu, J.C.; Tang, J. Biomolecule-assisted fabrication of Copper doped SnS$_2$ nanosheet–reduced graphene oxide junctions with enhanced visible-light photocatalytic activity. *J. Mater. Chem. A* **2014**, *2*, 1000–1005. [CrossRef]
45. Yassin, O.A.; Abdelaziz, A.A.; Jaber, A.Y. Structural and optical characterization of V-and W-doped SnS$_2$ thin films prepared by spray pyrolysis. *Mater. Sci. Semicond. Process.* **2015**, *38*, 81–86. [CrossRef]
46. Patel, M.; Chavda, A.; Mukhopadhyay, I.; Kim, J.; Ray, A. Nanostructured SnS with inherent anisotropic optical properties for high photoactivity. *Nanoscale* **2016**, *8*, 2293–2303. [CrossRef] [PubMed]
47. Wang, L.; Xia, L.; Wu, Y.; Tian, Y. Zr-Doped β-In$_2$S$_3$ ultrathin nanoflakes as photoanodes: Enhanced visible-light-driven photoelectrochemical water splitting. *ACS Sustain. Chem. Eng.* **2016**, *4*, 2606–2614. [CrossRef]
48. Mohan Kumar, G.; Ilanchezhiyan, P.; Madhan Kumar, A.; Yuldashev, S.U.; Kang, T.W. Electrical property studies on chemically processed polypyrolle/aluminum doped ZnO based hybrid heterostructures. *Chem. Phys. Lett.* **2016**, *649*, 130–134. [CrossRef]

© 2019 by the authors. Licensee MDPI, Basel, Switzerland. This article is an open access article distributed under the terms and conditions of the Creative Commons Attribution (CC BY) license (http://creativecommons.org/licenses/by/4.0/).

Article

Nanohybrid Assemblies of Porphyrin and Au$_{10}$ Cluster Nanoparticles

Mariachiara Trapani [1], Maria Angela Castriciano [1,*], Andrea Romeo [1,2], Giovanna De Luca [2], Nelson Machado [3], Barry D. Howes [3], Giulietta Smulevich [3] and Luigi Monsù Scolaro [1,2,*]

1. CNR-ISMN, Istituto per lo Studio dei Materiali Nanostrutturati c/o Dipartimento di Scienze Chimiche, Biologiche, Farmaceutiche ed Ambientali, University of Messina V. le F. Stagno D'Alcontres, 3198166 Messina, Italy
2. Dipartimento di Scienze Chimiche, Biologiche, Farmaceutiche ed Ambientali and C.I.R.C.M.S.B., University of Messina V. le F. Stagno D'Alcontres, 3198166 Messina, Italy
3. Dipartimento di Chimica "Ugo Schiff", Università di Firenze, Via della Lastruccia 3-13, 50019 Sesto Fiorentino (Fi), Italy
* Correspondence: maria.castriciano@cnr.it (M.A.C.); lmonsu@unime.it (L.M.S.)

Received: 26 June 2019; Accepted: 16 July 2019; Published: 18 July 2019

Abstract: The interaction between gold sub-nanometer clusters composed of ten atoms (Au$_{10}$) and tetrakis(4-sulfonatophenyl)porphyrin (TPPS) was investigated through various spectroscopic techniques. Under mild acidic conditions, the formation, in aqueous solutions, of nanohybrid assemblies of porphyrin J-aggregates and Au$_{10}$ cluster nanoparticles was observed. This supramolecular system tends to spontaneously cover glass substrates with a co-deposit of gold nanoclusters and porphyrin nanoaggregates, which exhibit circular dichroism (CD) spectra reflecting the enantiomorphism of histidine used as capping and reducing agent. The morphology of nanohybrid assemblies onto a glass surface was revealed by atomic force microscopy (AFM), and showed the concomitant presence of gold nanoparticles with an average size of 130 nm and porphyrin J-aggregates with lengths spanning from 100 to 1000 nm. Surface-enhanced Raman scattering (SERS) was observed for the nanohybrid assemblies.

Keywords: gold clusters; plating; porphyrin; chirality; SERS

1. Introduction

Gold metal nanoparticles with diameters smaller than ~2 nm have received considerable interest in nanoscience because they can be controlled with atomic precision, and, due to discrete energy levels and molecular-like HOMO–LUMO transitions, exhibit optical features fundamentally different from those of larger nanoparticles [1]. Commonly, the term nanoparticles (NPs) is used to refer to entities with diameters greater than 2 nm and an indefinite structure, whereas defined stoichiometric species with a diameter smaller than 2 nm are described as nanoclusters (NCs) [2]. The knowledge of the exact number of atoms present in gold nanoclusters is a very important aspect, as these clusters constitute the link between atomic and AuNP behavior. In this respect, AuNCs show discrete, size-dependent absorption and fluorescence emission from the UV to the near-IR spectral regions, together with significant quantum yield values [1,3,4]. Nonlinear optical (NLO) properties of gold NCs, such as third-order optical nonlinearity, two photon absorption (TPA), two-photon excited emission, and hyperpolarizabilities have been investigated [5–9]. Furthermore, AuNCs exhibit high photostability and biocompatibility, characteristics advantageous in biomedical applications, which can be improved by refining the synthesis, processing, and surface coating of the NCs [10–12]. Surface ligands, such as dendrimers, polymers, proteins, and oligomers play a key role in the stability of AuNCs in solution, and also affect their structure and optical properties [1,3]. In this framework, the use of biomolecules as capping ligands improves biocompatibility, and is considered an efficient way to transfer chirality to nanoclusters [13–15]. Indeed, three possible mechanisms for chirality transfer have been reported: (i) chiral growth, due to the formation of a chiral metal core influenced by the presence of chiral ligands, (ii) chiral polarization, due to electronic interaction between an achiral metal core and chiral ligands, and (iii) chiral footprint, due to the arrangement of the ligands on an achiral metal core [16,17]. In this respect, chiral emitting gold nanoclusters, in which the two enantiomers of histidine (His) have been used as both reducing agent and protecting ligands, have been easily synthesized by a one-step approach [18,19]. The presence of suitable ligands bearing –NH$_2$, –COOH, or polymerizable substituents open the way to further functionalization of NCs for applications in sensing, bioimaging, and energy transfer [20–22]. A central focus throughout materials research is control of the organization of the clusters on nanostructured surfaces, as well as understanding of the growth mechanisms at an atomic scale in order to obtain well defined and uniform architectures [23]. In fact, the quality of the metallic layers has a strong impact on the mechanical, electrical, and optical properties of the films. Several procedures to obtain gold nanostructure thin films have been reported, spanning from sputtering [24], lithographic methods [25], chemical vapor deposition (CVD) [26], electroless deposition (ELD) [27,28], and self-assembly approaches, which involve a variety of substrates [29,30]. In this framework, it has been reported that self-assembly of monodisperse gold colloid particles into monolayers on polymer-coated substrates produces highly reproducible macroscopic surfaces active for surface-enhanced Raman scattering (SERS) [31]. In fact, many spectroscopic studies on the interaction of AuNPs with porphyrins and their aggregates in solution and on surfaces have been reported [32–36]. In this particular area, our interest has focused on the J-aggregates of tetrakis(4-sulfonatophenyl)porphyrin (TPPS) [37–42]. These aggregates are formed under acidic conditions and the partially protonated porphyrins are arranged in a lateral stacking geometry, which leads to the occurrence of very peculiar optical properties [43,44]. Due to their manifold applications, various reports have dealt with the immobilization of J-aggregated porphyrins on substrates, highlighting the importance of the deposition step [45]. In particular, microscopic analysis has revealed the presence of differently shaped aggregates, such as rod, and nanotubular structures, depending on the experimental conditions [35,43,46]. Moreover, a detailed study has revealed that porphyrin concentration strongly affects the amount of J-aggregate adsorbed onto glass surfaces, showing a higher number of adsorbed entities at lower porphyrin concentrations [47]. In the past, J-aggregated porphyrin has been exploited for the design of inorganic/organic nanocomposites in combination with gold nanoparticles and spermine [48] or with gold nanorods [36,49]. Herein, we report on the ability of sub-nanometer sized gold clusters (Au$_{10}$) capped with L- or D-histidine to induce the formation of chiral TPPS J-aggregates under rather mild acidic conditions. We describe a simple procedure to co-deposit gold nanoparticles and chiral porphyrin J-aggregates onto glass substrates by simple acidification of an aqueous solution of such Au$_{10}$ clusters. Furthermore, these films have been shown to be active substrates for SERS, as demonstrated by the observation of intensified Raman signals for the co-deposited porphyrin J-aggregates. To the best of our knowledge, this is the first time that gold clusters composed of ten atoms (Au$_{10}$) have been used in combination with porphyrins to build a nanohybrid assembly (Au$_{10}$@Jagg) able to self-organize on a solid that can be potentially used as a SERS-active substrate for the detection of chemically- and biologically-relevant species.

2. Materials and Methods

Chemicals. Hydrogen tetrachloroaurate(III) hydrate (99.9%) was supplied by Strem Chemicals (Bischheim, France). D- and L-histidine (98%) were obtained from Sigma-Aldrich (Milan, Italy). The 5,10,15,20-tetrakis(4 sulfonatophenyl)porphyrin (TPPS) was purchased from Aldrich Chemicals (Milan, Italy), and its solutions of known concentration were prepared using the extinction coefficient at the Soret maximum ($\varepsilon = 5.33 \times 10^5$ M^{-1} cm^{-1} at $\lambda = 414$ nm). Hydrogen peroxide (30%, Sigma Aldrich, Milan, Italy), NaBH$_4$ (98%, Aldrich, Milan, Italy), potassium nitrate, and sulfuric (98%), hydrochloric (37%), and nitric (69%) acids (Fluka, Milan, Italy) were used. All the reagents were used without further purification and the solutions were prepared in dust free Milli-Q water (Merck, Darmstadt, Germany).

Gold cluster synthesis. Synthesis of gold clusters (Au$_{10}$) was carried out according to a literature procedure [18]. Briefly, a solution of L- or D-histidine (0.1 M, 6 mL) was added to the solution of the metal precursor (HAuCl$_4$, 10 mM, 2 mL) under continuous stirring at 25 °C for 2 h. After this time, the solution turned pale yellow and was used in this form.

Plating procedure. Small pieces of glass cover slides were thoroughly washed in an acid piranha solution (H$_2$SO$_4$:H$_2$O$_2$ 4:1 v/v). After being rinsed in pure water, the slides were vertically immersed in 3 mL of solution containing gold clusters (1.5 mL) and HCl (pH 2.0). The metallic structures were left to grow on the substrates for 24 h at room temperature. Care was taken to remove any excess histidine possibly adsorbed onto the surface by dipping the slides in acidic water (pH 2.0). Finally, the substrates were dried under a gentle nitrogen flow.

Au$_{10}$@Jagg assemblies. Glass slides, carefully cleaned according to the procedure described above, were vertically immersed in 3 mL of solution containing gold clusters (1.5 mL), porphyrin (up to a concentration of 5 µM), and HCl (pH 2.0). After an aging time of four days at room temperature, the slides were dried under a gentle nitrogen flow and the excess histidine was removed by quick immersion in water at pH 2.

Spectroscopic and morphological characterization. UV–vis spectra were collected on a diode-array spectrophotometer Agilent model 8452, subtracting the spectrum of a clean glass slide. Circular dichroism (CD) spectra were recorded with a Jasco model J-720 spectropolarimeter. Atomic force microscopy (AFM) measurements were performed using a NanoSurf easyScan2 microscope operating in non-contact mode at room temperature, with a resolution of 512 × 512 pixels and a moderate scan rate (1–2 lines/s). Commercial Si-N-type tips (AppNano mod. ACLA) with resonance frequencies of 145–230 kHz were used. Fluorescence emission and resonance light scattering (RLS) experiments were performed on a Jasco mod. FP-750 spectrofluorimeter. A synchronous scan protocol with a right angle geometry was adopted for collecting RLS spectra [50], which were not corrected for the absorption of the samples. Raman and SERS spectra were obtained at room temperature using a Renishaw RM2000 single-grating spectrograph apparatus, equipped with an Ar$^+$ laser at 514.5 nm and a near-infrared diode laser at 785 nm. Measurements were made in backscattered geometry using a 50× microscope objective filtered by a notch holographic filter, dispersed by a single grating (1200 lines mm^{-1}), into a charge-coupled device (CCD) detector cooled to −70 °C by the Peltier effect. Spatial resolution was 2 mm and the spectral resolution was 3 cm^{-1}. Laser power at the sample was in the range 17–185 µW (514.1 nm exc.) and 200 µW–2 mW (785 nm exc.). No sample degradation was observed under these conditions. To improve the signal-to-noise ratio, a number of spectra were accumulated and summed only if no spectral differences were noted. Raman and SERS spectra were calibrated with indene and CCl$_4$ as standards. In order to measure Raman intensities, nitrate and sulfate were used as internal standards. Therefore, to this end, porphyrin aggregation was fostered by means of the addition of KNO$_3$ (150 mM) and HNO$_3$ (32 mM) to the porphyrin solution (5 µM). Due to the overlap of the 993 cm^{-1} band of sulfate and a porphyrin mode, the 1053 cm^{-1} band of NO$_3^-$ was used to normalize the spectra.

3. Results

Au$_{10}$ clusters capped with D- or L-histidine were obtained according to a previously reported procedure [18]. The imidazole group of histidine plays a key role in AuNCs formation, as it serves as both a reducing agent and a protecting ligand. The clusters showed an absorption edge at around 450 nm that rose very steeply below 320 nm (full line spectrum, inset of Figure 1), together with the typical fluorescence emission centered at 500 nm, as previously reported [19]. No significant spectroscopic differences were observed for the Au$_{10}$ sample stabilized by the two different histidine enantiomers. The chiroptical properties of histidine AuNCs were investigated by circular dichroism spectroscopy, confirming that the chirality of the histidine ligands was transferred to the metal core [19]. When TPPS (5 µM) was added to the AuNCs solution, the electronic spectrum for both histidine enantiomers showed the presence of a porphyrin Soret band centered at 418 nm and four bands located in the visible region at 517, 554, 587, and 642 nm (red spectrum, Figure 1). The positions of both the B and visible bands were bathochromically shifted compared to those of the free base TPPS in aqueous solution in the presence of histidine [51].

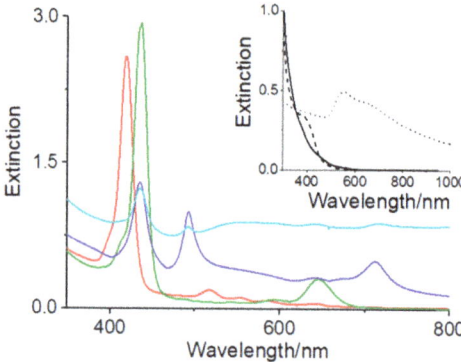

Figure 1. UV–vis spectra of tetrakis(4-sulfonatophenyl)porphyrin (TPPS)–Au$_{10}$ in aqueous solution (red line): instantaneously upon HCl addition (pH 2) (green line), after 1 h (blue line), after 24 h (cyano line). The inset shows the crude Au$_{10}$ NCs for the same experimental conditions in aqueous solution (black full line) upon HCl addition (pH 2) instantaneously (black dashed line) and after 24 h (black dotted line). Experimental conditions: [Au$_{10}$] = 125 μM, [TPPS] = 5 μM.

Furthermore, the CD spectra of the two enantiomeric forms showed a signal in the porphyrin absorption region, indicating the occurrence of an asymmetrical perturbation resulting from the chiral AuNCs. The CD spectra (inset of Figure 2) performed for L- and D-His-AuNCs enantiomers showed signals in the porphyrin region that were correlated with the enantiomorphism of the stereocenter present on the AuNCs protecting ligand, with a negative and positive bisignate Cotton effect for L- and D-histidine, respectively. This suggests that chiral information was successfully expressed on the porphyrin chromophore at the supramolecular level. The low induced CD intensity was in line with the monomeric nature of the chromophore. Recently, it has been reported that no interaction between TPPS and histidine has been detected in aqueous solution, mainly due to electrostatic repulsions between the negatively charged porphyrins and the amino acids [51].

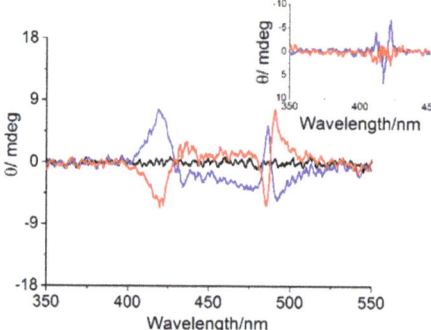

Figure 2. Circular dichroism (CD) spectra of Au$_{10}$ in aqueous solution (black line), Au$_{10}$–TPPS Jagg nanohybrid assemblies for L-(blue line) and D-histidine (red line) before (inset) and 1 h after HCl addition. ([Au$_{10}$] = 125 μM, [TPPS] = 5 μM, pH 2).

This was confirmed by the absence of any CD signal induction in the porphyrin region. Nevertheless, as has been observed for other systems, the absorption band shifted and the induced

CD signal could be ascribable to electrostatic interactions or hydrogen bonds between the negative sulfonated groups of the macrocycle and the protonated amino groups of histidine, and/or to the localization of the porphyrin in a hydrophobic microenvironment due to the histidine surrounding the metal nanostructures [52–55]. Moreover, the formation of positively charged histidine oligomers with Cl$^-$ as a counter anion has been reported [18]. Therefore, it also cannot be excluded that electrostatic interactions between the positively charged histidine oligomers and the negatively charged sulphonate groups present in the periphery of the macrocycle were the origin of the spectral variations. It should be mentioned that, upon addition of the positively charged tetra N-methylpyridinium porphyrin (H$_2$TMPyP^{4+}) to the Au$_{10}$ solution, the Soret band red-shifted to 427 nm with respect to the free base (422 nm) (ESI, red spectrum, Figure S1). This shift can be explained by an interaction of the carboxylic groups of histidine with the positively charged N-methylpyridinium groups at the meso position of the macrocycle. Since histidine Au$_{10}$ clusters can be successfully prepared over a wide pH range (pH 2–12) [18], we decided to foster porphyrin aggregation by lowering the pH. When HCl (pH 2) was added to the TPPS Au$_{10}$ cluster solution, the UV–vis spectra showed an almost instantaneous formation of diacid TPPS with a Soret band at 435 nm (green spectrum, Figure 1), which slowly interconverted into J-aggregates characterized by a narrow peak located at 492 nm. Furthermore, the electronic spectrum at the end of the aggregation process showed a very broad band at around 550 nm not ascribable to porphyrin features (blue spectrum, Figure 1). For comparison, a control experiment was performed in the same conditions in the absence of porphyrin. In this case, acidification of the Au$_{10}$ cluster aqueous solution instantaneously induced the formation of a new band at 400 nm (dashed line spectrum, inset Figure 1) that broadened and red-shifted within 1 h to around 550 nm (ESI, red spectrum, Figure S2), and eventually it led to a general increase of the baseline within 24 h (dotted line, inset Figure 1). This effect is ascribable to the growth of AuNcs to form AuNPs, which show the typical plasmon resonance band [56]. The process was accompanied by a color variation of the solution from pale yellow to pink. It is reasonable to expect that upon acidification of the Au$_{10}$ cluster solution, larger metallic structures were produced by aggregation of the clusters, which resulted from hydrogen bond formation between the carboxylic and protonated amino groups of adjacent units of the histidine residues covering the gold surface, as already reported for AuNPs stabilized by histidine [57]. In fact, in the case of heterocycle amines such as pyrrole or tryptophan, the occurrence of macroscopic segregation of the polymer due to the amine oxidation process and the metal phase has been reported as a function of the heterocycle amines/gold stoichiometric ratio [58]. After removing the solution from the cuvette, a purple deposit was observed on the quartz surface (ESI, Figure S3). The corresponding electronic absorption spectrum, obtained from washing the cuvette with water, exhibited a broad band that extended from ca. 550 nm to higher wavelengths very similar to that observed 24 h after HCl addition (inset Figure 1). This effect is ascribable to the adhesion of the metal nanoparticles to the quartz surface. A similar experiment was carried out at the same pH value, reducing the amount of gold clusters. In this case, the UV–vis spectra evolved from the initial 400 nm band to a broad feature extending from ca. 550 nm to the near-IR, but neither the 530 nm band due to AuNPs nor flocculated material were detected (ESI, Figure S4). This suggests the direct growth of small gold clusters into larger entities with nucleation on the surface, without the formation of larger metal colloidal suspensions in solution. This observation may result from slower kinetics related to a lower amount of Au$_{10}$ clusters, thus improving the homogeneity of the deposited material onto the quartz surface. At a higher Au$_{10}$ load, the formation of solid material could be ascribed to a faster growth process causing the formation of much larger metal structures that become unstable in solution and eventually precipitate. It is noteworthy that no metallization of the cuvette wall occured when Au nanoparticles synthesized by standard reduction with NaBH$_4$ in the presence of histidine as capping reagent were used under the same experimental conditions (ESI, Figure S5) [59]. In the presence of TPPS, a further broadening of the plasmonic band was observed over a period of 24 h, together with a drastic decrease of band intensities. This was due to the formation of a dark precipitate, as confirmed by the electronic absorption spectrum recorded the day after, in which an increased baseline was evident (cyano spectrum, Figure 1).

The resonance light scattering (RLS) spectrum (black spectrum, Figure 3) of the Au_{10} clusters in aqueous solution in the absence of porphyrin displayed a Rayleigh scattering profile. After addition of TPPS, a well at 418 nm appeared due to absorption resulting from the presence of porphyrin in its monomeric form (green spectrum, Figure 3).

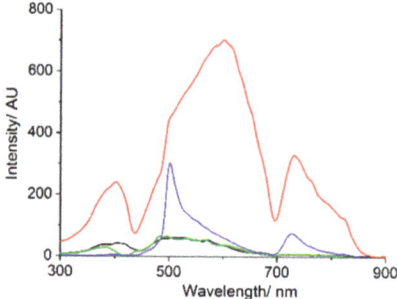

Figure 3. Resonance light scattering (RLS) spectra of Au_{10} in aqueous solution (black line), TPPS@Au_{10} in aqueous solution (green line), upon addition of HCl (pH 2) after 1 h (blue line) and after 24 h (red line). ([Au_{10}] = 125 µM, [TPPS] = 5 µM).

Upon addition of HCl, the RLS spectrum of the sample showed a peak at 500 nm due to the presence in solution of large ordered J-aggregates, stabilized by a network of electrostatic and solvophobic interactions among porphyrins (blue spectrum, Figure 3). After 24 h, the Rayleigh scattering due to the presence of larger gold structures in solution formed by aggregation phenomena occurring in acidic medium, modulated by the absorbance of the sample, was the only detectable component present (red spectrum, Figure 3).

As the Au_{10} clusters can be stabilized by either L- or D-histidine, in order to confirm the involvement of the chiral capping agent in the aggregated samples, we performed circular dichroism experiments. As previously described, chirality is transferred from the histidine capping agent to the monomeric porphyrin. Upon acidification and aggregation, chirality induced by the two histidine enantiomers was observed, analogously to data already reported for similar systems in the absence of gold nanoparticles [51]. The CD spectra for solutions of both cluster types showed the presence of a weak bisignate Cotton effect, centered in the J-aggregate absorption region (490 nm). D- and L-histidine led to almost mirror image spectra, characterized by a negative and positive Cotton effect (Figure 2). Moreover, a strong light scattering contribution due to the presence of porphyrin aggregates and gold nanostructures in solution broadened the bands.

As already observed for crude AuNCs, after removing the TPPS–AuNP solution from the cuvette, a purple deposit was observed on the quartz surface. The corresponding electronic absorption spectra obtained for both enantiomers showed, in addition to the spectroscopic features of the AuNPs, the typical J-aggregate band at 491 nm (Figure 4).

The RLS signals confirmed the presence of nanostructured material attached to the quartz substrate (Figure 4, inset). The preferential adsorption of the porphyrin J-aggregates on the cuvette surface was further proven by the observation of a residual amount of diacid monomeric TPPS in solution. After transferring the solution into a new cuvette, only the spectrum corresponding to this species could be detected, with the Soret band at 435 nm and with no CD features detectable (ESI, Figure S6). Interestingly, the CD spectra of these films in the J-aggregate spectroscopic region showed an induced bisignate Cotton effect related to the configuration of the amino acid used in the synthesis of the metallic clusters (Figure 5). These dichroic signals cannot be ascribed to linear dichroism, as they do not depend on the cuvette orientation with respect to incident light (ESI, Figure S7). No CD signal was detected for the corresponding AuNP deposit grown from D- and L-histidine (Figure 5, inset).

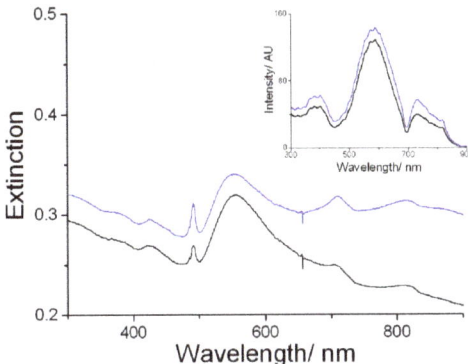

Figure 4. UV–vis spectra of the TPPS@AuNPs deposit left on the cuvette surface, obtained after washing with water, for D- (**black line**) and L- (**blue line**) histidine samples. The inset shows the corresponding RLS spectra.

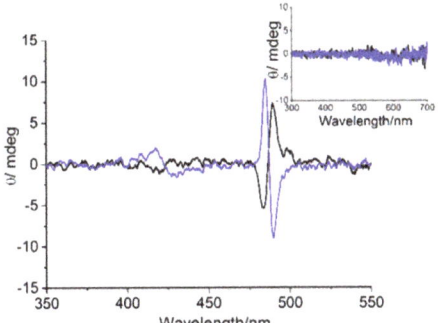

Figure 5. CD spectra of TPPS@AuNP co-deposit for D-(**black line**) and L-(**blue line**) histidine samples. The inset shows the corresponding AuNP deposit grown from D-(**black line**) and L-(**blue line**) histidine.

Raman and SERS measurements. The co-deposition of J-aggregated porphyrins and gold nanoparticles prompted us to investigate the possibility of detecting surface-enhanced Raman scattering (SERS) effects. In order to quantify the SERS effect, nitrate was added to the solution and the intensity of the Raman band at 1053 cm^{-1} was taken as an internal standard. No significant spectral changes were introduced due to the presence of the internal standard. Figure 6 shows representative Raman spectra of the J-aggregates obtained with and without Au$_{10}$ clusters for excitation at 514.5 and 785 nm. The spectra of the samples with (Figure 6a,c) and without (Figure 6b,d) Au$_{10}$ clusters were obtained at the same point on the surface of each sample for the two excitation wavelengths. The presence of Au$_{10}$ clusters did not alter the frequency or the relative intensity of the J-aggregate bands, as the spectra were very similar to those previously reported in solution [60]. This result confirms that co-deposition of the J-aggregated porphyrins and the gold nanoparticles generally does not alter their structural integrity, as was previously found for the deposition of the solution phase aggregates onto a gold substrate [46]. However, an intensification of the overall porphyrin spectrum was evident in the samples containing Au compared to those without Au. The effect was considerably more evident in the spectra obtained with 785 nm excitation (Figure 6c,d) than those obtained with 514.5 nm (Figure 6a,b). In fact, in the latter case, the resonance Raman intensification of the porphyrin bands due to the vicinity of the excitation wavelength with the 490 nm aggregate band precluded an accurate evaluation of the SERS effect. In order to quantify the intensification of the porphyrin spectrum, we evaluated the intensity ratio (R) of the band of the internal standard and isolated bands of the aggregate. The spectra were collected on NO_3^- crystals, which displayed both NO_3^- and the J-aggregate Raman signals. Two isolated bands of the porphyrin were used, at 317 and 1232 cm^{-1}, corresponding to the out-of-plane vibration involving motion of the pyrrolic hydrogens and to the totally symmetric C_m-phenyl stretch, respectively [32]. Similar results were also obtained using the 993 cm^{-1} band of SO_4^{2-}, however, in this case, the band of the internal standard overlapped with the band at 986 cm^{-1} of the J-aggregate (data not shown). In the presence of Au_{10} clusters, the J-aggregate spectrum was intensified compared to the 1053 cm^{-1} band of NO_3^-.

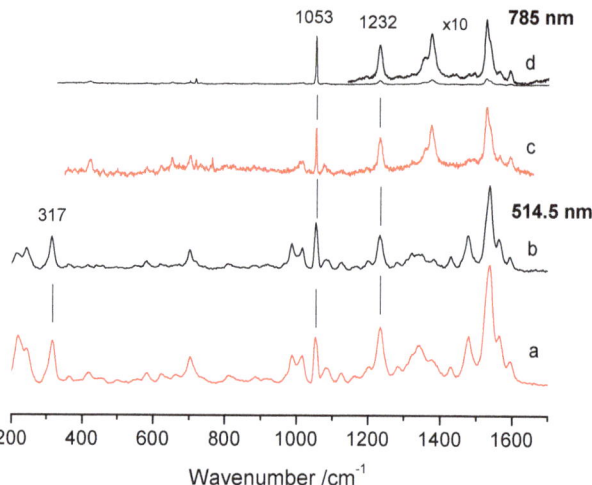

Figure 6. Representative Raman spectra of porphyrin J-aggregates in the presence (**a**,**c**) and absence (**b**,**d**) of Au_{10} clusters with excitation at 514.5 and 785 nm. The spectra of each sample (**a**,**c**) and (**b**,**d**) were obtained at the same point on the sample surface for the two excitation wavelengths. The inset of spectrum (**d**) has been multiplied 10-fold. Experimental conditions: (514.5 nm): (**a**) average of twelve spectra with 10 s integration time; (**b**) average of eleven spectra with 10 s integration time; (785 nm): (**c**) average of seventeen spectra with 30 s integration time; (**d**) average of nine spectra with 30 s integration time. All the spectra have been normalized with respect to the band of nitrate at 1053 cm^{-1}.

The SERS intensification is evident from the marked reduction of the R values observed in the presence of Au_{10} (Table 1). The scatter of the R_{785} values for the sample without Au was likely due to variation in crystal size of the internal standard; the bigger/thicker the crystal, the more intense the salt band.

AFM investigations. To obtain insight into the structures of the J-aggregates and of the gold entities grown on the glass surfaces, the morphology of the deposited materials was examined by AFM. The samples consisted mainly of small AuNPs, which were aggregates formed by several sub-nanometer AuNCs. Irregular agglomerates, due to clustering of these small objects, can be also detected with average height ca. 60 nm and diameter 130 nm (ESI, Figure S8). These were accompanied by flat and regular objects, probably single AuNCs, of heights in the tens of nm and mean widths of a few hundred nm. In addition, elongated TPPS J-aggregates formed on the glass surface, displaying lengths that ranged between 100 and 1000 nm, 3–4 nm height, and ca. 80 nm width (Figure 7 and ESI, Figure S9). These results are consistent with bundles of collapsed J-aggregates due to the solvent evaporation from the inner compartment of the nanotube [61].

Table 1. Comparison of the intensity ratio (R) of the 1053 cm^{-1} NO$_3^-$ band and the 317 and 1232 cm^{-1} bands of the J-aggregates obtained with 785 and 514.5 nm excitation.

Point	$R_{1053/1232}$	$R_{1053/1232}$	$R_{1053/317}$
	785 nm	514.5 nm	
Without Au			
9	5.9	0.7	0.9
8	8.9	1.2	1.3
7	34.5	2.4	2.7
6	13.3	1.4	1.4
5	24.3	2.2	2.7
4	16.2	1.4	1.5
With Au			
12	1.0	0.9	1.0
11	1.3	0.9	1.0
10	1.5	0.5	0.7
3	1.3	0.8	1.0
2	2.0	0.5	0.6
1	3.5	2.8	3.1

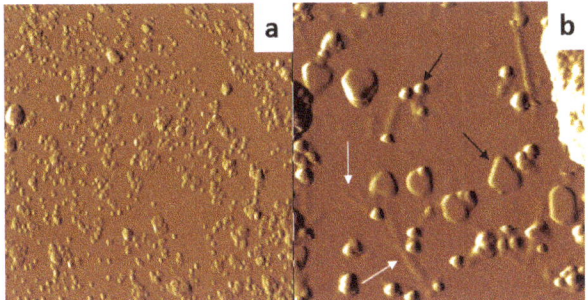

Figure 7. AFM topography images (x gradient) of the deposits formed on a glass substrate immersed in a Au$_{10}$ and TPPS solution at pH 2.0, showing Au nanostructures (dark arrows) with J-aggregates (white arrows). Experimental conditions: (**a**) 9 × 9 µm, Z range = 478 nm; (**b**) 2 × 2 µm, Z range = 145 nm.

4. Conclusions

Nanoassemblies of gold clusters and J-aggregated porphyrins were synthesized and characterized by spectroscopic techniques. The aggregated porphyrin showed a chirality related to the amino acid used in the synthesis of the metal nanostructures. A simple procedure for the deposition of gold nanoparticles on glass substrates was described, which starts with the preparation of easily synthesized and highly reactive Au$_{10}$ clusters. In fact, we found that acidification of aqueous solutions of Au$_{10}$ sub-nanometer clusters led to the formation of plate-like and regular nano-objects on glass surfaces. Our procedure offers the advantage of not requiring the use of expensive and sophisticated equipment and hazardous reagents. The growth of the metallic nanostructures with the co-deposition of TPPS J-aggregates on the substrates was exploited as a test system for the potential use of these nanoparticles in SERS applications. In fact, a significant increase of the Raman signals for the porphyrin J-aggregates deposited on gold plated surfaces was observed compared with J-aggregates deposited on bare glass. Furthermore, these J-aggregates maintained their peculiar CD signals on the glass surface, showing a chirality that is related to the configuration of the amino acid (D- and L-histidine) used in the metal cluster synthesis. This latter observation suggests that the Au$_{10}$ clusters, analogously to other templating systems [62,63], have a role as chiral seeds in the growth of J-aggregates.

Supplementary Materials: The following are available online at http://www.mdpi.com/2079-4991/9/7/1026/s1, Figure S1: UV-vis spectra of Au_{10} solution after addition of the positively charged H_2TMPyP^{4+}. Figure S2: Spectral changes of Au_{10} clusters upon addition of HCl (pH 2) within 1 h. Figure S3: Photographs of gold plated cuvettes on dark and white backgrounds. Figure S4: UV-vis spectra of Au_{10} solutions upon acidification. Figure S5: AuNPs prepared from histidine and $NaBH_4$ and of a glass dipped in the corresponding solutions at acidic pH. Figure S6: UV-vis and CD spectra of residual TPPS solution after formation of co-deposit. Figure S7: Circular dichroism (CD) spectra of co-deposit J-aggregates and AuNPs. Figure S8: Histograms of the height and diameter distributions, height profiles of Au nanostructures and TPPS J-aggregates deposited on glass surfaces. Figure S9. AFM image and relative profile.

Author Contributions: Conceptualization, L.M.S. and G.S.; investigation, M.T, N.M, B.D.H and G.D.L; data curation, M.T, A.R. and M.A.C.; writing—original draft preparation, M.A.C., M.T. and B.D.H.; writing—review and editing, L.M.S., A.R., G.S.; visualization, A.R.; Authorship must be limited to those who have contributed substantially to the work reported.

Funding: This research received no external funding.

Conflicts of Interest: The authors declare no conflict of interest.

References

1. Jin, R.; Zeng, C.; Zhou, M.; Chen, Y. Atomically Precise Colloidal Metal Nanoclusters and Nanoparticles: Fundamentals and Opportunities. *Chem. Rev.* **2016**, *116*, 10346–10413. [CrossRef] [PubMed]
2. Schmid, G. The relevance of shape and size of Au55 clusters. *Chem. Soc. Rev.* **2008**, *37*, 1909–1930. [CrossRef] [PubMed]
3. Chakraborty, I.; Pradeep, T. Atomically Precise Clusters of Noble Metals: Emerging Link between Atoms and Nanoparticles. *Chem. Rev.* **2017**, *117*, 8208–8271. [CrossRef] [PubMed]
4. Zheng, J.; Zhang, C.; Dickson, R.M. Highly Fluorescent, Water-Soluble, Size-Tunable Gold Quantum Dots. *Phys. Rev. Lett.* **2004**, *93*, 077402. [CrossRef] [PubMed]
5. Van Steerteghem, N.; Van Cleuvenbergen, S.; Deckers, S.; Kumara, C.; Dass, A.; Häkkinen, H.; Clays, K.; Verbiest, T.; Knoppe, S. Symmetry breaking in ligand-protected gold clusters probed by nonlinear optics. *Nanoscale* **2016**, *8*, 12123–12127. [CrossRef]
6. Ramakrishna, G.; Varnavski, O.; Kim, J.; Lee, D.; Goodson, T. Quantum-Sized Gold Clusters as Efficient Two-Photon Absorbers. *J. Am. Chem. Soc.* **2008**, *130*, 5032–5033. [CrossRef] [PubMed]
7. Philip, R.; Chantharasupawong, P.; Qian, H.; Jin, R.; Thomas, J. Evolution of Nonlinear Optical Properties: From Gold Atomic Clusters to Plasmonic Nanocrystals. *Nano Lett.* **2012**, *12*, 4661–4667. [CrossRef] [PubMed]
8. Bertorelle, F.; Russier-Antoine, I.; Calin, N.; Comby-Zerbino, C.; Bensalah-Ledoux, A.; Guy, S.; Dugourd, P.; Brevet, P.-F.; Sanader, Ž.; Krstić, M.; et al. Au10(SG)10: A Chiral Gold Catenane Nanocluster with Zero Confined Electrons. Optical Properties and First-Principles Theoretical Analysis. *J. Phys. Chem. Lett.* **2017**, *8*, 1979–1985. [CrossRef]
9. Russier-Antoine, I.; Bertorelle, F.; Vojkovic, M.; Rayane, D.; Salmon, E.; Jonin, C.; Dugourd, P.; Antoine, R.; Brevet, P.-F. Non-linear optical properties of gold quantum clusters. The smaller the better. *Nanoscale* **2014**, *6*, 13572–13578. [CrossRef]
10. Polavarapu, L.; Manna, M.; Xu, Q.-H. Biocompatible glutathione capped gold clusters as one- and two-photon excitation fluorescence contrast agents for live cells imaging. *Nanoscale* **2011**, *3*, 429–434. [CrossRef]
11. Derfus, A.M.; Chan, W.C.W.; Bhatia, S.N. Probing the Cytotoxicity of Semiconductor Quantum Dots. *Nano Lett.* **2004**, *4*, 11–18. [CrossRef] [PubMed]
12. Zhang, X.-D.; Luo, Z.; Chen, J.; Shen, X.; Song, S.; Sun, Y.; Fan, S.; Fan, F.; Leong, D.T.; Xie, J. Ultrasmall Au10−12(SG)10−12 Nanomolecules for High Tumor Specificity and Cancer Radiotherapy. *Adv. Mater.* **2014**, *26*, 4565–4568. [CrossRef] [PubMed]
13. Noguez, C.; Garzón, I.L. Optically active metal nanoparticles. *Chem. Soc. Rev.* **2009**, *38*, 757–771. [CrossRef] [PubMed]
14. Nieto-Ortega, B.; Burgi, T. Vibrational Properties of Thiolate-Protected Gold Nanoclusters. *Acc. Chem. Res.* **2018**, *51*, 2811–2819. [CrossRef] [PubMed]
15. Knoppe, S.; Buergi, T. Chirality in Thiolate-Protected Gold Clusters. *Acc. Chem. Res.* **2014**, *47*, 1318–1326. [CrossRef] [PubMed]
16. Fan, Z.; Govorov, A.O. Plasmonic Circular Dichroism of Chiral Metal Nanoparticle Assemblies. *Nano Lett.* **2010**, *10*, 2580–2587. [CrossRef] [PubMed]

17. Govorov, A.O.; Fan, Z.; Hernandez, P.; Slocik, J.M.; Naik, R.R. Theory of Circular Dichroism of Nanomaterials Comprising Chiral Molecules and Nanocrystals: Plasmon Enhancement, Dipole Interactions, and Dielectric Effects. *Nano Lett.* **2010**, *10*, 1374–1382. [CrossRef] [PubMed]
18. Yang, X.; Shi, M.; Zhou, R.; Chen, X.; Chen, H. Blending of HAuCl4 and histidine in aqueous solution: A simple approach to the Au10 cluster. *Nanoscale* **2011**, *3*, 2596–2601. [CrossRef] [PubMed]
19. Guo, Y.; Zhao, X.; Long, T.; Lin, M.; Liu, Z.; Huang, C. Histidine-mediated synthesis of chiral fluorescence gold nanoclusters: Insight into the origin of nanoscale chirality. *RSC Adv.* **2015**, *5*, 61449–61454. [CrossRef]
20. Wu, Z.; Suhan, J.; Jin, R. One-pot synthesis of atomically monodisperse, thiol-functionalized Au25 nanoclusters. *J. Mater. Chem.* **2009**, *19*, 622–626. [CrossRef]
21. Bain, D.; Maity, S.; Patra, A. Opportunities and challenges in energy and electron transfer of nanocluster based hybrid materials and their sensing applications. *Phys. Chem. Chem. Phys.* **2019**, *21*, 5863–5881. [CrossRef] [PubMed]
22. Mondal, N.; Paul, S.; Samanta, A. Photoinduced 2-way electron transfer in composites of metal nanoclusters and semiconductor quantum dots. *Nanoscale* **2016**, *8*, 14250–14256. [CrossRef] [PubMed]
23. Jang, G.G.; Hawkridge, M.E.; Roper, D.K. Silver disposition and dynamics during electroless metal thin film synthesis. *J. Mater. Chem.* **2012**, *22*, 21942–21953. [CrossRef]
24. Siegel, J.; Lyutakov, O.; Rybka, V.; Kolská, Z.; Švorčík, V. Properties of gold nanostructures sputtered on glass. *Nanoscale Res. Lett.* **2011**, *6*, 96. [CrossRef] [PubMed]
25. Claudia Manuela, M.; Flavio Carlo Filippo, M.; Ralph, S. Ordered arrays of faceted gold nanoparticles obtained by dewetting and nanosphere lithography. *Nanotechnology* **2008**, *19*, 485306.
26. Grodzicki, A.; Łakomska, I.; Piszczek, P.; Szymańska, I.; Szłyk, E. Copper(I), silver(I) and gold(I) carboxylate complexes as precursors in chemical vapour deposition of thin metallic films. *Coord. Chem. Rev.* **2005**, *249*, 2232–2258. [CrossRef]
27. Livshits, P.; Inberg, A.; Shacham-Diamand, Y.; Malka, D.; Fleger, Y.; Zalevsky, Z. Precipitation of gold nanoparticles on insulating surfaces for metallic ultra-thin film electroless deposition assistance. *Appl. Surf. Sci.* **2012**, *258*, 7503–7506. [CrossRef]
28. De Leo, M.; Pereira, F.C.; Moretto, L.M.; Scopece, P.; Polizzi, S.; Ugo, P. Towards a Better Understanding of Gold Electroless Deposition in Track-Etched Templates. *Chem. Mater.* **2007**, *19*, 5955–5964. [CrossRef]
29. Thorkelsson, K.; Bai, P.; Xu, T. Self-assembly and applications of anisotropic nanomaterials: A review. *Nano Today* **2015**, *10*, 48–66. [CrossRef]
30. Zhang, S.-Y.; Regulacio, M.D.; Han, M.-Y. Self-assembly of colloidal one-dimensional nanocrystals. *Chem. Soc. Rev.* **2014**, *43*, 2301–2323. [CrossRef]
31. Freeman, R.G.; Grabar, K.C.; Allison, K.J.; Bright, R.M.; Davis, J.A.; Guthrie, A.P.; Hommer, M.B.; Jackson, M.A.; Smith, P.C.; Walter, D.G.; et al. Self-Assembled Metal Colloid Monolayers: An Approach to SERS Substrates. *Science* **1995**, *267*, 1629–1632. [CrossRef] [PubMed]
32. Rich, C.C.; McHale, J.L. Resonance Raman Spectra of Individual Excitonically Coupled Chromophore Aggregates. *J. Phys. Chem. C* **2013**, *117*, 10856–10865. [CrossRef]
33. Hajduková-Šmídová, N.; Procházka, M.; Osada, M. SE(R)RS excitation profile of free-base 5,10,15,20-tetrakis(1-methyl-4-pyridyl) porphyrin on immobilized gold nanoparticles. *Vib. Spectrosc.* **2012**, *62*, 115–120. [CrossRef]
34. Leishman, C.W.; McHale, J.L. Illuminating Excitonic Structure in Ion-Dependent Porphyrin Aggregates with Solution Phase and Single-Particle Resonance Raman Spectroscopy. *J. Phys. Chem. C* **2016**, *120*, 12783–12795. [CrossRef]
35. Friesen, B.A.; Rich, C.C.; Mazur, U.; McHale, J.L. Resonance Raman Spectroscopy of Helical Porphyrin Nanotubes. *J. Phys. Chem. C* **2010**, *114*, 16357–16366. [CrossRef]
36. Trapani, M.; De Luca, G.; Romeo, A.; Castriciano, M.A.; Scolaro, L.M. Spectroscopic investigation on porphyrins nano-assemblies onto gold nanorods. *Spectrochim. Acta Mol. Biomol. Spectrosc.* **2017**, *173*, 343–349. [CrossRef] [PubMed]
37. Romeo, A.; Castriciano, M.A.; Zagami, R.; Pollicino, G.; Monsu Scolaro, L.; Pasternack, R.F. Effect of zinc cations on the kinetics for supramolecular assembling and the chirality of porphyrin J-aggregates. *Chem. Sci.* **2016**, *8*, 961–967. [CrossRef]

38. Occhiuto, I.G.; Zagami, R.; Trapani, M.; Bolzonello, L.; Romeo, A.; Castriciano, M.A.; Collini, E.; Monsù Scolaro, L. The role of counter-anions in the kinetics and chirality of porphyrin J-aggregates. *Chem. Commun.* **2016**, *52*, 11520–11523. [CrossRef]
39. Micali, N.; Villari, V.; Castriciano, M.A.; Romeo, A.; Scolaro, L.M. From fractal to nanorod porphyrin J-aggregates. Concentration-induced tuning of the aggregate size. *J. Phys. Chem. B* **2006**, *110*, 8289–8295. [CrossRef]
40. Castriciano, M.A.; Romeo, A.; Villari, V.; Micali, N.; Scolaro, L.M. Structural rearrangements in 5,10,15,20-tetrakis(4-sulfonatophenyl)porphyrin J-aggregates under strongly acidic conditions. *J. Phys. Chem. B* **2003**, *107*, 8765–8771. [CrossRef]
41. Zagami, R.; Romeo, A.; Castriciano, M.A.; Monsù Scolaro, L. Inverse kinetic and equilibrium isotopic effect on self-assembly and supramolecular chirality of porphyrin J-aggregates. *Chem. Eur.J.* **2017**, *23*, 70–74. [CrossRef] [PubMed]
42. Zagami, R.; Castriciano, M.A.; Romeo, A.; Trapani, M.; Pedicini, R.; Monsù Scolaro, L. Tuning supramolecular chirality in nano and mesoscopic porphyrin J-aggregates. *Dyes Pigments* **2017**, *142*, 255–261. [CrossRef]
43. Schwab, A.D.; Smith, D.E.; Rich, C.S.; Young, E.R.; Smith, W.F.; de Paula, J.C. Porphyrin nanorods. *J. Phys. Chem. B* **2003**, *107*, 11339–11345. [CrossRef]
44. Schwab, A.D.; Smith, D.E.; Bond-Watts, B.; Johnston, D.E.; Hone, J.; Johnson, A.T.; de Paula, J.C.; Smith, W.F. Photoconductivity of Self-Assembled Porphyrin Nanorods. *Nano Lett.* **2004**, *4*, 1261–1265. [CrossRef]
45. Castriciano, M.A.; Gentili, D.; Romeo, A.; Cavallini, M.; Scolaro, L.M. Spatial control of chirality in supramolecular aggregates. *Sci. Rep.* **2017**, *7*. [CrossRef] [PubMed]
46. Friesen, B.A.; Nishida, K.R.A.; McHale, J.L.; Mazur, U. New Nanoscale Insights into the Internal Structure of Tetrakis(4-sulfonatophenyl) Porphyrin Nanorods. *J. Phys. Chem. C* **2009**, *113*, 1709–1718. [CrossRef]
47. Arai, Y.; Segawa, H. Significantly Enhanced Adsorption of Bulk Self-Assembling Porphyrins at Solid/Liquid Interfaces through the Self-Assembly Process. *J. Phys. Chem. B* **2012**, *116*, 13575–13581. [CrossRef]
48. Villari, V.; Mazzaglia, A.; Trapani, M.; Castriciano, M.A.; De Luca, G.; Romeo, A.; Scolaro, L.M.; Micali, N. Optical enhancement and structural properties of a hybrid organic-inorganic ternary nanocomposite. *J. Phys. Chem. C* **2011**, *115*, 5435–5439. [CrossRef]
49. Zhang, L.; Chen, H.; Wang, J.; Li, Y.F.; Wang, J.; Sang, Y.; Xiao, S.J.; Zhan, L.; Huang, C.Z. Tetrakis(4-sulfonatophenyl)porphyrin-Directed Assembly of Gold Nanocrystals: Tailoring the Plasmon Coupling Through Controllable Gap Distances. *Small* **2010**, *6*, 2001–2009. [CrossRef]
50. Pasternack, R.F.; Collings, P.J. Resonance Light-Scattering—A New Technique for Studying Chromophore Aggregation. *Science* **1995**, *269*, 935–939. [CrossRef]
51. Randazzo, R.; Gaeta, M.; Gangemi, C.M.A.; Fragalà, M.E.; Purrello, R.; D'Urso, A. Chiral Recognition of L- and D- Amino Acid by Porphyrin Supramolecular Aggregates. *Molecules* **2018**, *24*, 84. [CrossRef] [PubMed]
52. Castriciano, M.A.; Donato, M.G.; Villari, V.; Micali, N.; Romeo, A.; Scolaro, L.M. Surfactant-like Behavior of Short-Chain Alcohols in Porphyrin Aggregation. *J. Phys. Chem. B* **2009**, *113*, 11173–11178. [CrossRef] [PubMed]
53. Zagami, R.; Trapani, M.; Castriciano, M.A.; Romeo, A.; Mineo, P.G.; Scolaro, L.M. Synthesis, characterization and aggregation behavior of room temperature ionic liquid based on porphyrin-trihexyl(tetradecyl)phosphonium adduct. *J. Mol. Liq.* **2017**, *229*, 51–57. [CrossRef]
54. Castriciano, M.A.; Leone, N.; Cardiano, P.; Manickam, S.; Scolaro, L.M.; Lo Schiavo, S. A new supramolecular polyhedral oligomeric silsesquioxanes (POSS)-porphyrin nanohybrid: Synthesis and spectroscopic characterization. *J. Mater. Chem. C* **2013**, *1*, 4746–4753. [CrossRef]
55. Castriciano, M.A.; Romeo, A.; Villari, V.; Angelini, N.; Micali, N.; Scolaro, L.M. Aggregation behavior of tetrakis(4-sulfonatophenyl)porphyrin in AOT/water/decane microemulsions. *J. Phys. Chem. B* **2005**, *109*, 12086–12092. [CrossRef] [PubMed]
56. Daniel, M.-C.; Astruc, D. Gold Nanoparticles: Assembly, Supramolecular Chemistry, Quantum-Size-Related Properties, and Applications toward Biology, Catalysis, and Nanotechnology. *Chem. Rev.* **2004**, *104*, 293–346. [CrossRef] [PubMed]
57. Liu, Z.; Zu, Y.; Fu, Y.; Meng, R.; Guo, S.; Xing, Z.; Tan, S. Hydrothermal synthesis of histidine-functionalized single-crystalline gold nanoparticles and their pH-dependent UV absorption characteristic. *Coll. Surf. B Biointerf.* **2010**, *76*, 311–316. [CrossRef] [PubMed]

58. Selvan, T.; Spatz, J.P.; Klok, H.-A.; Möller, M. Gold–Polypyrrole Core–Shell Particles in Diblock Copolymer Micelles. *Adv. Mater.* **1998**, *10*, 132–134. [CrossRef]
59. Lakshminarayana, P.; Qing-Hua, X. A single-step synthesis of gold nanochains using an amino acid as a capping agent and characterization of their optical properties. *Nanotechnology* **2008**, *19*, 075601.
60. Akins, D.L.; Zhu, H.R.; Guo, C. Absorption and Raman Scattering by Aggregated meso-Tetrakis(p-sulfonatophenyl)porphine. *J. Phys. Chem.* **1994**, *98*, 3612–3618. [CrossRef]
61. Rotomskis, R.; Augulis, R.; Snitka, V.; Valiokas, R.; Liedberg, B. Hierarchical structure of TPPS4 J-aggregates on substrate revealed by atomic force microscopy. *J. Phys. Chem. B* **2004**, *108*, 2833–2838. [CrossRef]
62. Đorđević, L.; Arcudi, F.; D'Urso, A.; Cacioppo, M.; Micali, N.; Bürgi, T.; Purrello, R.; Prato, M. Design principles of chiral carbon nanodots help convey chirality from molecular to nanoscale level. *Nat. Commun.* **2018**, *9*, 3442. [CrossRef] [PubMed]
63. Gaeta, M.; Raciti, D.; Randazzo, R.; Gangemi, C.M.A.; Raudino, A.; D'Urso, A.; Fragalà, M.E.; Purrello, R. Chirality Enhancement of Porphyrin Supramolecular Assembly Driven by a Template Preorganization Effect. *Angew. Chem. Int. Ed.* **2018**, *57*, 10656–10660. [CrossRef] [PubMed]

© 2019 by the authors. Licensee MDPI, Basel, Switzerland. This article is an open access article distributed under the terms and conditions of the Creative Commons Attribution (CC BY) license (http://creativecommons.org/licenses/by/4.0/).

Article

SnSe$_2$ Quantum Dots: Facile Fabrication and Application in Highly Responsive UV-Detectors

Xiangyang Li [†], Ling Li [*,†], Huancheng Zhao, Shuangchen Ruan, Wenfei Zhang, Peiguang Yan, Zhenhua Sun, Huawei Liang and Keyu Tao

Shenzhen Key Laboratory of Laser Engineering, College of Physics and Optoelectronic Engineering, Shenzhen University, Shenzhen 518060, China; 2170285209@email.szu.edu.cn (X.L.); 1800281011@email.szu.edu.cn (H.Z.); scruan@szu.edu.cn (S.R.); zhangwf@szu.edu.cn (W.Z.); yanpg@szu.edu.cn (P.Y.); szh@szu.edu.cn (Z.S.); hwliang@szu.edu.cn (H.L.); taokeyu@szu.edu.cn (K.T.)
* Correspondence: liling@szu.edu.cn; Tel.: +86-755-8653-2505
† These authors contributed equally to this work.

Received: 19 August 2019; Accepted: 10 September 2019; Published: 15 September 2019

Abstract: Synthesizing quantum dots (QDs) using simple methods and utilizing them in optoelectronic devices are active areas of research. In this paper, we fabricated SnSe$_2$ QDs via sonication and a laser ablation process. Deionized water was used as a solvent, and there were no organic chemicals introduced in the process. It was a facile and environmentally-friendly method. We demonstrated an ultraviolet (UV)-detector based on monolayer graphene and SnSe$_2$ QDs. The photoresponsivity of the detector was up to 7.5×10^6 mAW^{-1}, and the photoresponse time was ~0.31 s. The n–n heterostructures between monolayer graphene and SnSe$_2$ QDs improved the light absorption and the transportation of photocarriers, which could greatly increase the photoresponsivity of the device.

Keywords: SnSe$_2$ quantum dots; graphene; phototransistor; UV-detector

---

1. Introduction

Graphene-based electronic and optoelectronic devices have attracted extensive attention [1–3]. Mueller et al. demonstrated a vertical incidence metal–graphene–metal photodetector with an external responsivity of 6.1 mAW^{-1} at 1.55 µm [4]. The photoresponsivity was limited by the low absorption of the graphene. Quantum dots (QDs) can break this limitation. They can act as light absorption spots. The photo-induced carriers in them can transfer into the graphene film, and the charges in the graphene film transport to the electrodes quickly. Thus, the responsivity of a graphene-based device is improved [5–7]. Cheng et al. showed a phototransistor based on graphene and graphene QDs with a photoresponsivity of up to 4×10^{10} mAW^{-1}, but the response time was 10 s [8]. Sun et al. constructed an infrared photodetector based on graphene and PbS QDs with a responsivity of up to 10^{10} mAW^{-1} and a response time of 0.26 s [9]. Sun et al. demonstrated a UV phototransistor based on graphene and ZnSe/ZnS core/shell QDs. Its responsivity was up to 10^6 mAW^{-1} and the response time was 0.52 s [10].

In order to fabricate the QD solution with uniform distribution, the wet chemical method was commonly used. Some organic solvents, such as toluene or pyridine, were used in the process [9,10]. The chemical groups can cap the surface of the QDs and modify their charge transfer property, thus influencing the photo responsivity of the device. Synthesizing QDs using facile and green methods and utilizing them in optoelectronic devices are active areas of research.

Two-dimensional transition-metal dichalcogenides (TMDCs) have been applied in fluorescent imaging [11], biological sensing [12], and photocatalytic [13] due to their unique optoelectronic properties. Tin diselenide (SnSe$_2$) is a semiconductor in the TMDCs family. SnSe$_2$ QDs can be used in fast and highly responsive phototransistors since they have a tunable bandgap and high quantum efficiency. In this paper, SnSe$_2$ QDs were fabricated via sonication and a laser ablation process.

The deionized water was used as a solvent, and there were no organic chemicals introduced in the process. It was a facile and environmentally-friendly method. The phototransistor based on monolayer graphene and SnSe$_2$ quantum dots was demonstrated. The photoresponse time was ~0.31 s, and the photoresponsivity was up to 7.5×10^6 mAW^{-1}. The n–n heterostructures between monolayer graphene and SnSe$_2$ quantum dots enhance the light absorption and the generation of photocarriers. The photocarriers can transfer quickly from SnSe$_2$ QDs to graphene, thus improving the photoresponsivity of the device.

2. Experiment

SnSe$_2$ QDs were fabricated by sonication and the laser ablation process, as shown in Figure 1. The SnSe$_2$ bulk was bought from Six Carbon Technology. We put the SnSe$_2$ bulk in an agate mortar and manually ground it for 15 min to get SnSe$_2$ powders. Then, we dispersed 20 mg of powder in 30 mL of deionized water. The mixture was sonicated with a sonic tip for 2 h at the output power of 650 W in an ice-bath. The power was on for 4 s and off for 2 s. After sonication, the solution was a mixture of SnSe$_2$ small particles and flakes. The solution was transferred into a quartz cuvette and irradiated under a 1064 nm pulsed Nd:YAG laser for 10 min (6 ns, 10 Hz). The laser output power was 2.2 W. When the solution was irradiated by the laser pulses, the small particles and flakes absorbed the incident photon energies and formed extreme non-equilibrium conditions (high pressure and temperature) in a short time (~ns). After sustainable irradiation, the particles and nanosheets broke into tiny pieces. Then, the solution was centrifuged for 30 min at a speed of 6000 rpm. After that, the supernatant containing SnSe$_2$ QDs was collected.

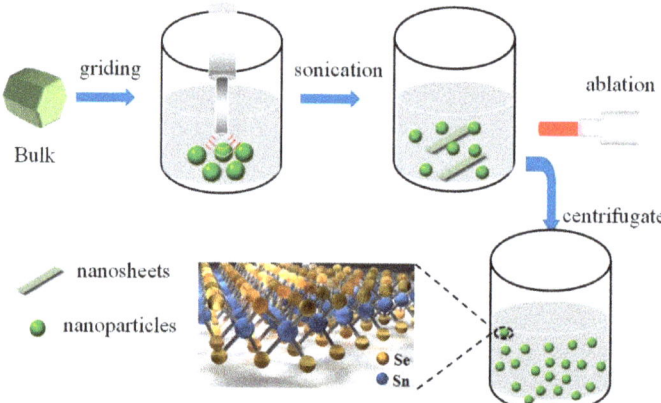

Figure 1. Schematic show of the SnSe$_2$ structure and the quantum dot (QD) fabrication process.

The morphology of the SnSe$_2$ QDs was studied using a high-resolution transmission electron microscope (TEM, FEI Tecnai G2 F30). The structure of the SnSe$_2$ QDs was characterized by X-ray diffraction spectroscopy (XRD, Bruker D8 Advance) and the Raman spectra (Horiba Labram HR Evolution). The absorption spectra were measured by a UV-vis spectrometer (Shimadzu UV-1700).

The chemical vapor deposition (CVD)-grown monolayer graphene was wet-transferred onto a p$^+$Si/SiO$_2$ substrate [14,15]. The thickness of SiO$_2$ was 285 nm. The highly doped p-type silicon served as the back-gate electrode. Then, the Cr/Au (10 nm/90 nm) source and drain electrodes were deposited on top of the graphene film by the thermal evaporation method. The channel length and width were 0.2 mm and 2 mm, respectively. The optoelectronic properties were studied using a probe station equipped with a semiconductor parameter analyzer (Keithley 4200). The illumination LED light wavelength was 405 nm.

3. Results and Discussion

Figure 2a shows the transmission electron microscope (TEM) image of SnSe$_2$ QDs as-fabricated. It shows an e- a size distribution in the range of 5–11 nm, and the average size is 9.8 nm, as indicated in Figure 2b. The average size of the QDs comes from the statistical analysis of the sizes of 200 QDs measured from TEM images. A high-resolution TEM image of a single SnSe$_2$ QD is shown in the inset of Figure 2a. The lattice spacing is about 0.33 nm, which corresponds to the (1010) planes of a hexagonal-phase SnSe$_2$ [16]. The result shows that the SnSe$_2$ QDs are crystalline.

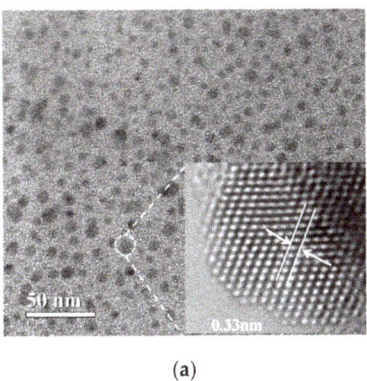
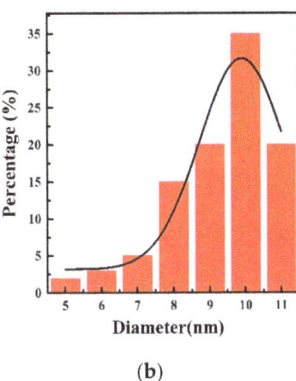

(a) (b)

Figure 2. (a) TEM image of SnSe$_2$ QDs with a centrifugal speed of 6000 rpm. The inset shows the detailed crystal structure of a single QD; (b) the size distribution of the SnSe$_2$ QDs.

Figure 3a shows the XRD patterns of the SnSe$_2$ bulk and QDs. The SnSe$_2$ bulk has an obvious diffraction peak at $2\theta = 14.4°$ which corresponds to the (001) faces. In addition, some lower peaks located at $2\theta = 29.1°$, $2\theta = 31.2°$, $2\theta = 44.3°$, and $2\theta = 60.4°$ are assigned to the (002), (101), (003), and (004) faces. In SnSe$_2$ QDs, these diffraction peaks almost disappear except for a tiny peak at $2\theta = 29.1°$. After sonication and laser ablation, the SnSe$_2$ bulk was cracked into nanoparticles, and there was no constructive interference from the aligned crystal planes [13,17]. The tiny peak at $2\theta = 29.1°$ corresponds to the (002) face, which may come from the partial restacking of QDs in the process of drying. Figure 3b shows the Raman spectra of the SnSe$_2$ bulk and QDs. The incident laser wavelength is 514 nm and the spot size is around 2 μm. For the bulk SnSe$_2$, two Raman active vibration modes are observed at 110.3 cm^{-1} and 183.6 cm^{-1}, which correspond to the in-plane E_g and out-of-plane A_{1g} modes [18]. For the SnSe$_2$ QDs, the peak of the E_g mode is very weak, but the peak of the A_{1g} mode is observable and has a small blue-shift of ~1 cm^{-1}, which may be due to the surface effect and decrease of SnSe$_2$ thickness [19].

Figure 3c shows the absorption spectra for SnSe$_2$ QDs and SnSe$_2$ nanosheets solutions in the range of 250–1000 nm. The absorption band of the SnSe$_2$ nanosheets solution is broad, covering regions from the ultraviolet to near-infrared. It is similar to the absorption band reported for the SnSe$_2$ powders [20]. For the SnSe$_2$ QDs solution, only strong absorption from 250 nm to 420 nm is observed. The bulk SnSe$_2$ has an indirect bandgap of 1.0 eV [20]. When the particle size is reduced, the emergence of the quantum confinement effects leads to the discretization of energy levels. As a result, the SnSe$_2$ QDs show a larger band gap [21].

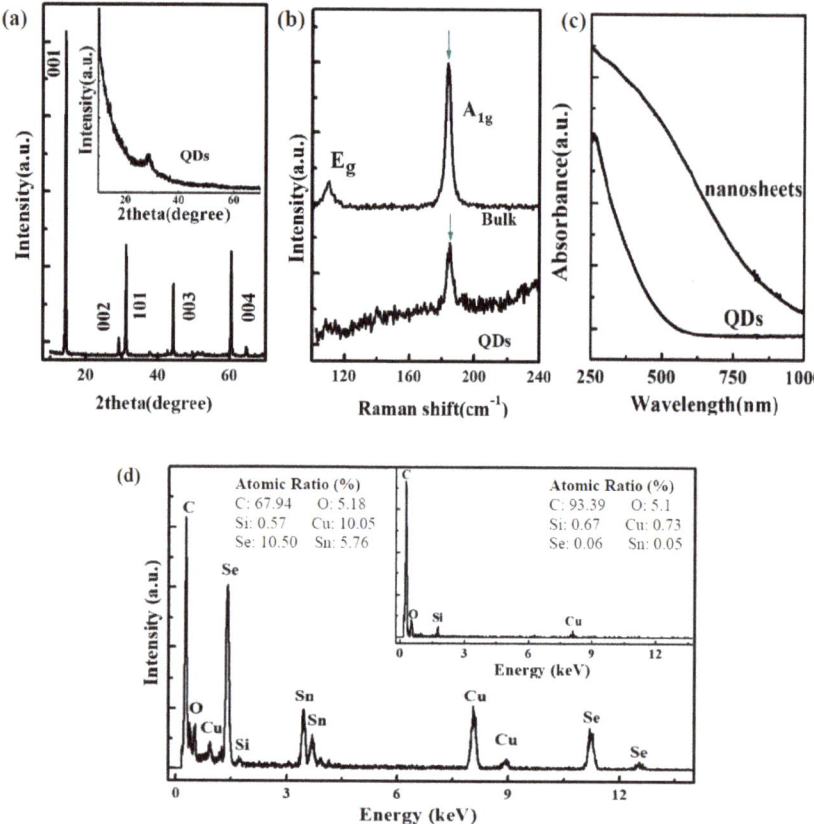

Figure 3. Spectroscopic characterizations. (a) XRD pattern of the SnSe$_2$ bulk and QDs; (b) Raman spectra of the SnSe$_2$ bulk and QDs; (c) absorption spectra of the SnSe$_2$ QDs and nanosheet solutions; (d) TEM energy dispersive spectra (TEM-EDS) of the SnSe$_2$ QDs. The inset shows the EDS of the TEM substrate without QDs.

Figure 3d shows the TEM energy dispersive spectra (TEM–EDS) of the SnSe$_2$ QDs. Tin and selenium can be clearly observed, and their atomic ratios are 5.76% (Sn) and 10.50% (Se), respectively. For comparison, the EDS of the TEM substrate (carbon film-coated copper grid without QDs) is shown in the inset of Figure 3d. The Cu, C, and Si signals come from the TEM grid and sample holder. The O peak arises from the oxygen adsorbed on the surface of the grid. The results show that the QDs are composed of tin and selenium.

Figure 4a schematically shows the photodetector decorated with SnSe$_2$ QDs on a p$^+$Si/SiO$_2$ substrate. The Raman spectra of the pure graphene on a p$^+$Si/SiO$_2$ substrate is shown in Figure 4b. Two Raman peaks at 1582 cm^{-1} (G line) and 2698 cm^{-1} (2D line) are observed. The ratio of the integrated intensities of the G line and 2D line is ~0.25. The peak at 1350 cm^{-1} (D line) in the spectra is very weak, indicating that the graphene is a monolayer with good quality. The I–V curves for the monolayer graphene phototransistor in the dark and with illumination under zero gate voltage ($V_G = 0$ V) are shown in Figure 4c. The illumination density is 350 μW/cm^2. As shown in the figure, there is no change between the current in the dark and under illumination, indicating that the photoresponse of pure graphene is negligible.

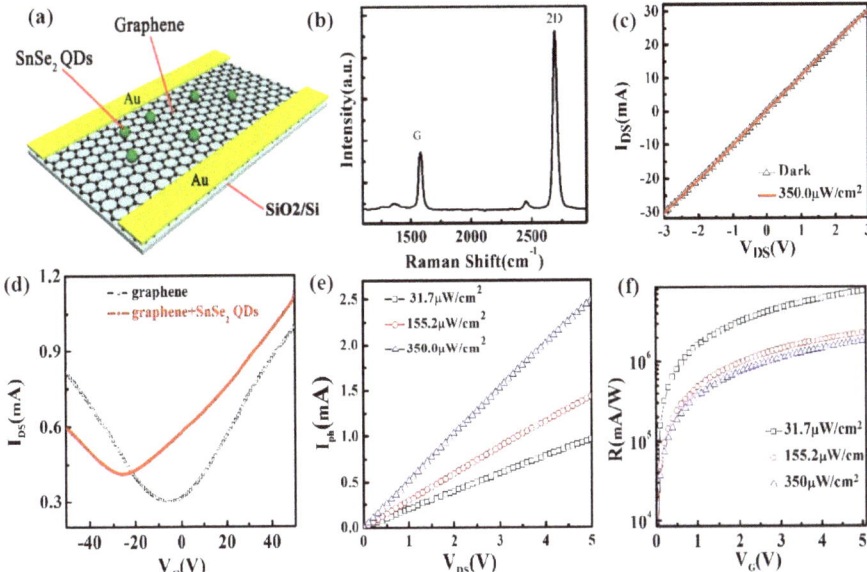

Figure 4. (a) Schematic diagram of a graphene photodetector decorated with SnSe$_2$ QDs; (b) Raman spectra of the pure graphene on a p$^+$Si/SiO$_2$ substrate; (c) the I–V curves for the single-layer graphene phototransistor in the dark and with illumination under zero-gate voltage (V$_G$ = 0 V); (d) transfer characteristics (I$_{DS}$-V$_G$, V$_{DS}$ = 0.5 V) of the phototransistor with and without SnSe$_2$ QDs on the graphene film; (e) photocurrent and (f) responsivity of a SnSe$_2$ QD-decorated graphene photodetector as functions of drain voltages at different illumination densities. The illumination wavelength is 405 nm.

Figure 4d shows the transfer curves (I$_{DS}$-V$_G$, V$_{DS}$ = 0.5 V) of the device with and without SnSe$_2$ QDs in which the light is absent. The transfer curve of the device without SnSe$_2$ QDs exhibits a typical V-shape. The field-effect mobilities are ~230 cm^{2}V^{-1}s^{-1} for electrons and ~220 cm^{2}V^{-1}s^{-1} for holes. The negative, neutral charge point (about −5 V) of single-layer graphene is observed in Figure 4d, indicating an electron dominated conduction in the graphene. The same behavior was also observed by Sun et al. [10]. Graphene is very sensitive to the surroundings. The defects in the SiO$_2$ substrate, residues from processing and handling, charged impurities, and substrate surface roughness can cause the shift of the neutral charge point [22]. The SnSe$_2$ QDs solution was dropped on the top of graphene film and heated at 40 °C for 30 min in a glove box filled with N$_2$ gas. The transfer curve of the photodetector with SnSe$_2$ QDs becomes asymmetric, and the Dirac point converts to a negative gate voltage (about −22 V). The shift indicates that the SnSe$_2$ QDs are n-type semiconductors, which are the same type as the bulk SnSe$_2$ [20]. The electron and hole mobilities decrease to ~160 cm^{2}V^{-1}s^{-1} and ~130 cm^{2}V^{-1}s^{-1}, respectively.

In order to study the optoelectronic properties of the device, we measured the photocurrents at different illumination densities with zero gate voltage (V$_G$ = 0 V). Figure 4e shows the relationship between the photocurrent (I$_{Ph}$ = I$_{Light}$ − I$_{Dark}$) and the applied drain voltages. I$_{Light}$ is the drain current under illumination, and I$_{Dark}$ is the drain current without illumination. The photocurrent increases while increasing the illumination density. Figure 4f represents the responsivity (R = I$_{ph}$/(WLE$_e$)) of the photodetector as functions of drain voltages at different illumination densities. The responsivity decreases while increasing the illumination density, which is consistent with the reported UV-detectors [23]. The maximum responsivity of the device is about 7.5 × 10^6 mAW^{-1} (V$_{DS}$ = 5 V) at an incident power density of 31.7 μW/cm^2, which is higher than that reported in graphene-based UV phototransistors [24–26].

Figure 5a shows the transfer curves of the photodetector at different illumination densities. The Dirac point of the device shifts to a lower negative gate voltage while increasing illumination density. The shift of the transfer curves (ΔV_G) is plotted as a function of illumination densities in Figure 5b. The shift of the transfer curve (ΔV_G) changes linearly with the light illumination density (Ee), indicating that the photo-induced carrier density in SnSe$_2$ QDs increases with increasing illumination density. This illumination density-dependent shift does not appear in the pure graphene phototransistor. The existence of SnSe$_2$ QDs leads to this photoresponse behavior. As shown in Figure 5a, the electron mobility in the SnSe$_2$ QD-decorated device is higher than that of holes at different illumination densities. The photo-induced electron-hole pairs are separated at the interface between SnSe$_2$ QDs and monolayer graphene. The SnSe$_2$ QDs/graphene heterojunction facilitates the injection of photo-generated electrons from SnSe$_2$ QDs into the graphene, leading a local n-doping in the graphene channel. Since the transfer rate of holes is lower that of electrons, net positive charges remain in the SnSe$_2$ QDs. Then, a lower negative gate voltage is required to obtain the charge neutral point (Dirac point) in the detector. A similar process was reported in a p-doped graphene/PbS QDs phototransistor by Sun et al. [9].

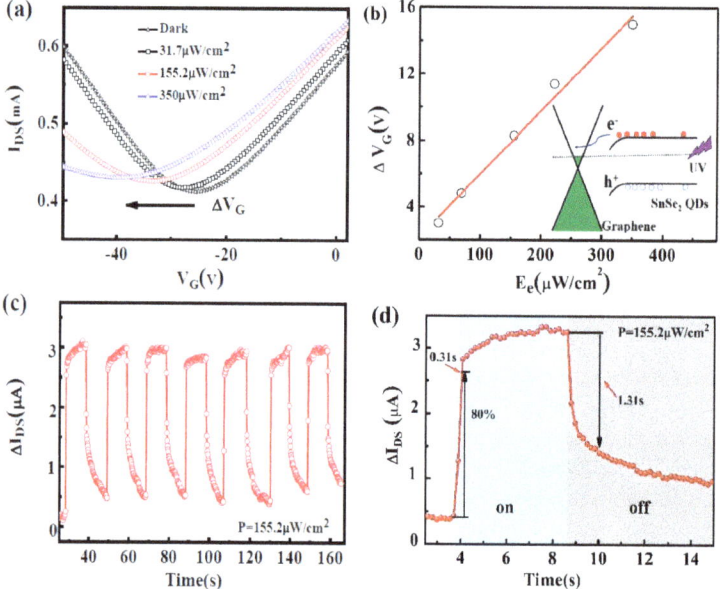

Figure 5. (a) Transfer characteristics of a graphene photodetector decorated with SnSe$_2$ quantum dots at different illumination densities (wavelength: 405 nm, V_{DS} = 0.5 V); (b) horizontal shift of transfer curves as functions of illumination densities. The inset shows the charge transfer between SnSe$_2$ QDs and graphene; (c) current response to on/off light illumination for several cycles; (d) photocurrent response time of the device. (V_{DS} = 0.05 V, illumination density: 155.2 µW/cm^2).

Figure 5c shows the current response to on/off light illumination and Figure 5d shows the photocurrent response time of the device (V_G = 0 V, V_{DS} = 0.05 V, illumination density: 155.2 µW/cm^2). The photocurrent increases with time when the illumination is on and decreases with time when the illumination is off. As shown in Figure 5d, the photocurrent increases to 80% with a response time of 0.31 s, which is faster than that reported in graphene devices [9,10,24,26,27]. The response time includes charge generation time, charge transfer time in heterojunctions, and charge collection time. In our experiment, the measured graphene charge mobility is smaller than the value for perfect graphene (up to 200,000 cm^{2}V^{-1}s^{-1}), which may be due to the defects induced in the graphene film while transferring to the substrate, and the response time can be improved by optimizing the graphene transfer process. When the light is turned out, the photocurrent decreases to 20% with a time of 1.31 s.

The photocurrent of the detector is influenced by the SnSe$_2$ QDs density. We have measured the AFM pictures and photocurrents for detectors with different SnSe$_2$ QDs densities. As shown in Figure 6, the photocurrent increases with an increase of the SnSe$_2$ QDs density under the same irradiation density (illumination density: 350 µW/cm^2). When the SnSe$_2$ QDs thickness is larger than 40 nm, the photocurrent tends to decrease, which may be due to the decrease of the charge transfer between the QDs layers.

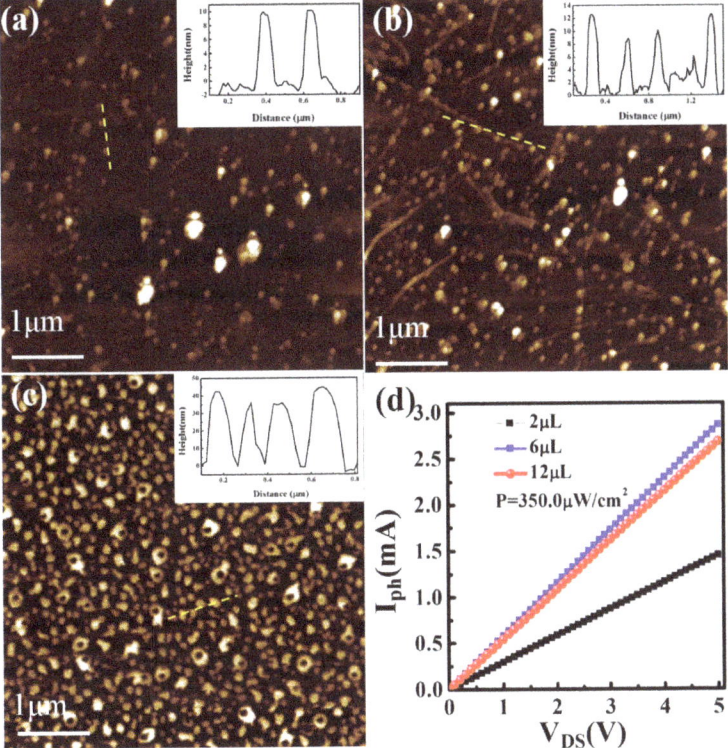

Figure 6. The AFM images of SnSe$_2$ QDs with different densities (a) 2 µL, (b) 6 µL, and (c) 12 µL. The insets show their height profiles. (d) The photocurrents with different SnSe$_2$ QDs densities at the irradiation density of 350.0 µW/cm^2.

4. Conclusions

In summary, uniformly distributed SnSe$_2$ quantum dots were synthesized at room temperature using a facile and environment-friendly method. The UV-detector based on monolayer graphene and SnSe$_2$ quantum dots was demonstrated. The device showed fast photoresponse time of ~0.31 s, and its photoresponsivity was up to 7.5×10^6 mAW^{-1}. The n–n heterostructures between monolayer graphene and SnSe$_2$ QDs improved the light absorption and the transportation of photocarriers, which have promising applications in optoelectronic devices.

Author Contributions: Data curation, X.L. and H.Z.; formal analysis, L.L., H.L., and K.T.; resources, S.R., W.Z., P.Y., and Z.S.; writing—review and editing, L.L.

Funding: This work was supported by Science and Technology Projects of Shenzhen (JCYJ20180305125000525, JCYJ20170302151033006, and JCYJ20160328144942069).

Acknowledgments: The authors thank Min Zhang from Peking University Shenzhen Graduate School for the help in the device fabrication and the Electron Microscope Center of Shenzhen University for the help in the TEM-EDS measurements.

Conflicts of Interest: The authors declare no conflict of interest.

References

1. Bonaccorso, F.; Sun, Z.; Hasan, T.; Ferrari, A.C. Graphene Photonics and Optoelectronics. *Nat. Photonics* **2010**, *4*, 611–622. [CrossRef]
2. Xia, F.; Thomas, M.; Lin, Y.; Valdes, Y.; Phaedon, A. Ultrafast Graphene Phototransistor. *Nat. Nanotechnol.* **2009**, *4*, 839–843. [CrossRef] [PubMed]
3. Li, X.; Li, T.; Chen, Z.; Fang, H.; Li, X.; Wang, X.; Xu, J.; Zhu, H. Graphene and Related Two-Dimensional Materials: Structure-Property Relationships for Electronics and Optoelectronics. *Appl. Phys. Rev.* **2017**, *4*, 021306. [CrossRef]
4. Mueller, T.; Xia, F.; Phaedon, A. Graphene Phototransistors for High-Speed Optical Communications. *Nat. Photonics* **2010**, *4*, 297–301. [CrossRef]
5. Li, X.M.; Zhu, H. The Graphene–Semiconductor Schottky Junction. *Phys. Today* **2016**, *69*, 46–51. [CrossRef]
6. Liu, C.; Chang, Y.; Theodore, B.; Zhong, Z. Graphene Phototransistors with Ultra-Broadband and High Responsivity at Room Temperature. *Nat. Nanotechnol.* **2014**, *9*, 273–278. [CrossRef] [PubMed]
7. Fitzmorris, B.; Pu, Y.; Jason, K.; Lin, Y.; Hu, Y.; Li, U.; Zhang, J. Optical Properties and Exciton Dynamics of Alloyed Core/Shell/Shell $Cd_{1-x}Zn_xSe/ZnSe/ZnS$ Quantum Dots. *ACS Appl. Mater. Interfaces* **2013**, *5*, 2893–2900. [CrossRef]
8. Cheng, S.; Weng, T.; Lu, M.; Tan, W.; Chen, J.; Chen, Y. All Carbon-Based Phototransistors: An Eminent Integration of Graphite Quantum Dots and Two Dimensional Graphene. *Sci. Rep.* **2013**, *3*, 2694. [CrossRef]
9. Sun, Z.; Liu, Z.; Li, J.; Tai, G.; Lan, S.; Yan, F. Infrared Phototransistors Based on CVD-Grown Graphene and PbS Quantum Dots with Ultrahigh Responsivity. *Adv. Mater.* **2012**, *24*, 5878–5883. [CrossRef]
10. Sun, Y.; Dan, X.; Sun, M.; Teng, C.; Liu, Q.; Chen, R.; Lan, X.; Ren, T. Hybrid Graphene/Cadmium-Free ZnSe/ZnS Quantum Dots Phototransistors for UV Detection. *Sci. Rep.* **2018**, *8*, 1–8. [CrossRef]
11. Long, H.; Li, T.; Chun, P.; Chun, Y.; Kin, H.; Yang, C.; Tsang, Y. The WS_2 Quantum Dot: Preparation, Characterization and Its Optical Limiting Effect in Polymethylmethacrylate. *Nanotechnology* **2016**, *27*, 414005. [CrossRef] [PubMed]
12. Han, Y.; Zhang, C.; Gu, W.; Ding, C.; Li, X.; Xian, Y. Facile Synthesis of Water-Soluble WS_2 Quantum Dots for Turn-on Fluorescent Measurement of Lipoic Acid. *J. Phys. Chem. C* **2016**, *120*, 12170–12177.
13. Xu, S.; Li, D.; Wu, P. One-Pot, Facile, and Versatile Synthesis of Monolayer MoS_2/WS_2 Quantum Dots as Bioimaging Probes and Efficient Electrocatalysts for Hydrogen Evolution Reaction. *Adv. Funct. Mater.* **2015**, *7*, 1127–1136. [CrossRef]
14. He, R.; Peng, L.; Liu, Z.; Zhu, H.; Zhao, X.; Chan, H.; Yan, F. Solution-Gated Graphene Field Effect Transistors Integrated in Microfluidic Systems and Used for Flow Velocity Detection. *Nano Lett.* **2012**, *12*, 1404–1409. [CrossRef] [PubMed]
15. Chiang, C.W.; Haider, G.; Tan, W.C.; Chen, Y.F. Highly Stretchable and Sensitive Photodetectors Based on Hybrid Graphene and Graphene Quantum Dots. *ACS Appl. Mater. Interfaces.* **2016**, *8*, 466–471. [CrossRef] [PubMed]
16. Zhou, X.; Lin, G.; Tian, W.; Zhang, Q.; Jin, S.; Li, H.; Yoshio, B.; Dmitri, G.; Zhai, T. Ultrathin $SnSe_2$ Flakes Grown by Chemical Vapor Deposition for High-Performance Phototransistors. *Adv. Mater.* **2015**, *27*, 8035–8041. [CrossRef]
17. Zankat, C.; Pratik, P.; Solanki, G.K.; Patel, K.V.; Pathak, M.; Som, N.; Prafulla, J. Investigation of Morphological and Structural Properties of V Incorporated $SnSe_2$ Single Crystals. *Mater. Sci. Semicond. Process.* **2018**, *8*, 137–142. [CrossRef]
18. Zhou, W.; Yu, Z.; Song, H.; Fang, R.; Wu, Z.; Li, L.; Ni, Z.; Wei, R.; Lin, W.; Ruan, S. Lattice Dynamics in Monolayer and Few-Layer $SnSe_2$. *Physcal Rev. B* **2017**, *96*, 035401. [CrossRef]
19. Qiao, X.; Wu, J.; Zhou, L.; Qiao, J.; Shi, W.; Chen, T.; Zhang, X.; Zhang, J.; Ji, W.; Tan, P.H. Polytypism and Unexpected Strong Interlayer Coupling of Two-Dimensional Layered ReS_2. *Adv. Funct. Mater.* **2015**, *8*, 8324–8332. [CrossRef]

20. Borges, Z.V.; Poffo, C.M.; de Lima, J.C.; de souza, S.M.; Trichês, D.M.; Nogueira, T.P.O.; Manzato, L.; de Biasi, R.S. Study of Structural, Optical and Thermal Properties of Nanostructured SnSe$_2$ Prepared by Mechanical Alloying. *Mater. Chem. Phys.* **2016**, *169*, 47–54. [CrossRef]
21. Solati, E.; Laya, D.; Davoud, D. Effect of Laser Pulse Energy and Wavelength on the Structure, Morphology and Optical Properties of ZnO Nanoparticles. *Opt. Laser Technol.* **2014**, *58*, 26–32. [CrossRef]
22. Kathalingam, A.; Senthilkumar, V.; Rhee, J.-K. Hysteresis I-V nature of mechanically exfoliated graphene FET. *J. Mater. Sci. Mater. Electron.* **2014**, *25*, 1303–1308. [CrossRef]
23. Cheng, W.; Tang, L.; Xiang, J.; Ji, R.; Zhao, J. An Extreme High-Performance Ultraviolet Photovoltaic Detector Based on Zno Nanorods/Phenanthrene Heterojunction. *RSC Adv.* **2016**, *6*, 12076–12080. [CrossRef]
24. Babichev, A.V.; Zhang, H.; Pierre, L.; Julien, F.H.; Egorov, Y.; Lin, Y.; Tu, L.; Maria, T. Gan Nanowire Ultraviolet Phototransistor with a Graphene Transparent Contact. *Appl. Phys. Lett.* **2013**, *103*, 201103. [CrossRef]
25. Boruah, B.; Anwesha, M.; Abha, M. Sandwiched Assembly of Zno Nanowires between Graphene Layers for a Self-Powered and Fast Responsive Ultraviolet Phototransistor. *Nanotechnology* **2016**, *27*, 095205. [CrossRef] [PubMed]
26. Lu, Y.; Wu, Z.; Xu, W.; Lin, S. ZnO Quantum Dot-Doped Graphene/H-BN/Gan-Heterostructure Ultraviolet Phototransistor with Extremely High Responsivity. *Nanotechnology* **2016**, *26*, 48LT03. [CrossRef] [PubMed]
27. Tang, J.; Kyle, K.; Sjoerd, H.; Kwang, J.; Liu, H.; Larissa, L.; Melissa, F.; Wang, X.; Ratan, D.; Dongkyu, C.; et al. Colloidal-Quantum-Dot Photovoltaics Using Atomic-Ligand Passivation. *Nat. Mater.* **2011**, *10*, 765–771. [CrossRef] [PubMed]

© 2019 by the authors. Licensee MDPI, Basel, Switzerland. This article is an open access article distributed under the terms and conditions of the Creative Commons Attribution (CC BY) license (http://creativecommons.org/licenses/by/4.0/).

Article

Two-Step Exfoliation of WS$_2$ for NO$_2$, H$_2$ and Humidity Sensing Applications

Valentina Paolucci [1,*], Seyed Mahmoud Emamjomeh [1], Michele Nardone [2], Luca Ottaviano [2,3] and Carlo Cantalini [1]

1. Department of Industrial and Information Engineering and Economics, Via G. Gronchi 18, University of L'Aquila, I-67100 L'Aquila, Italy; seyedmahmoud.emamjomeh@graduate.univaq.it (S.M.E.); carlo.cantalini@univaq.it (C.C.)
2. Department of Physical and Chemical Sciences, Via Vetoio 10, University of L'Aquila, I-67100 L'Aquila, Italy; michele.nardone@univaq.it (M.N.); luca.ottaviano@aquila.infn.it (L.O.)
3. CNR-SPIN Uos L'Aquila, Via Vetoio 10, I-67100 L'Aquila, Italy
* Correspondence: valentina.paolucci2@graduate.univaq.it

Received: 30 August 2019; Accepted: 17 September 2019; Published: 24 September 2019

Abstract: WS$_2$ exfoliated by a combined ball milling and sonication technique to produce few-layer WS$_2$ is characterized and assembled as chemo-resistive NO$_2$, H$_2$ and humidity sensors. Microstructural analyses reveal flakes with average dimensions of 110 nm, "aspect ratio" of lateral dimension to the thickness of 27. Due to spontaneous oxidation of exfoliated WS$_2$ to amorphous WO$_3$, films have been pre-annealed at 180 °C to stabilize WO$_3$ content at ≈58%, as determined by X-ray Photoelectron Spectroscopy (XPS), Raman and grazing incidence X-ray Diffraction (XRD) techniques. Microstructural analysis repeated after one-year conditioning highlighted that amorphous WO$_3$ concentration is stable, attesting the validity of the pre-annealing procedure. WS$_2$ films were NO$_2$, H$_2$ and humidity tested at 150 °C operating Temperature (OT), exhibiting experimental detection limits of 200 ppb and 5 ppm to NO$_2$ and H$_2$ in dry air, respectively. Long-term stability of the electrical response recorded over one year of sustained conditions at 150 °C OT and different gases demonstrated good reproducibility of the electrical signal. The role played by WO$_3$ and WS$_2$ upon gas response has been addressed and a likely reaction gas-mechanism presented. Controlling the microstructure and surface oxidation of exfoliated Transition Metal Dichalcogenides (TMDs) represents a stepping-stone to assess the reproducibility and long-term response of TMDs monolayers in gas sensing applications.

Keywords: 2D-materials; WS$_2$; exfoliation; gas sensors; NO$_2$; H$_2$; cross sensitivity

1. Introduction

In recent years, layered materials such as two-dimensional (2D) transition metal dichalcogenides (TMDs) have attracted a high level of interest due to their features, which make them appealing for potential applications in gas sensing [1,2], photo-electro-catalytic hydrogen evolution [3,4] optical and electronic devices [5,6] and energy storage [7,8].

Few-layer 2D TMDs can be produced through different methodologies like mechanical and liquid phase exfoliation of bulk crystals, classified as top-down routes, or via direct bottom-up routes like chemical vapor deposition [9,10]. High-yield liquid exfoliation methods comprising ion intercalation [11,12] and ultrasonic cleavage [13,14] have also been widely employed to exfoliate bulk-layered materials. Besides liquid phase exfoliation, low energy ball milling as a newly explored high-yield mechanical exfoliation method has been utilized for scalable production of mono and few-layer graphene [15,16] and TMDs nano-sheets [17,18]. More recently, enhanced mixed methods comprising assisted grinding and sonication have been shown to produce higher concentrations of TMDs nano-sheets and a reduced amount of defects [19].

Regarding gas sensing applications, 2D single layer MoS$_2$ Field Effect Transistor [20,21] Pd-doped WS$_2$ films [22] have been shown to be alternative substitutes for traditional metal oxides sensors. Moreover, considering that the possibility to find practical applications of those materials is generally dependent on the reproducibility of the preparation with respect to both microstructure (i.e., number of layers, lateral size, surface area, etc.) and chemical composition (i.e., defects concentration and surface oxidation), the need to find a practical, high-reproducible, and easy way to exfoliate TMDs is always under investigation. We already presented the fabrication of chemo-resistive thin films gas sensor, by drop casting suspensions of few flakes graphene oxide [23], phosphorene [24] and more recently TMDs utilizing both liquid-exfoliated MoS$_2$ [25] and commercially exfoliated suspensions of WS$_2$ [26]. The aim of this work is to apply a low energy ball milling and sonication method to achieve a reproducible and high-yield exfoliation methodology and to test the gas sensing performances of the obtained material. Starting from commercial WS$_2$ powders, we have firstly performed the exfoliation process based on the grinding and sonication method and investigated the morphology of few flakes WS$_2$. Secondly, we have determined that with exposing the material to mild air annealing at 180 °C, a controlled partial oxidation of WS$_2$ flakes to amorphous WO$_3$ is achievable. Lastly, we investigated the sensing responses to NO$_2$, H$_2$ and humidity of drop casted exfoliated WS$_2$ chemo-resistive thin films, discussing the likely gas-response mechanism.

2. Materials and Methods

WS$_2$ exfoliation: According to the flow sheet shown in Supplementary Figure S1, 2 g of WS$_2$ commercial powder (Sigma–Aldrich 243639-50G, St. Louis, MO, USA) were dispersed in 4 mL Acetonitrile (ACN—VWR 83639.320, Radnor, PA, USA) with 30 g Zirconium Oxide balls (D = 3 mm), and ball milled in a planetary milling machine (Fritsch—Planetary Micro Mill Pulverisette 7, Idar-Oberstein, Germany) at 400 rpm, for 2 h in ambient air.

To evaporate ACN residuals after milling, the collected slurry was left overnight at 23 ± 2 °C temperature and 40% ± 3% Relative Humidity (RH) (ATP DT-625 High Accuracy Thermo-hygrometer, Ashby-de-la-Zouch, UK). After ACN evaporation, 0.05 g of the dried powder was dispersed in 100 mL of pure ethanol (99.94% VWR 20821.330, Radnor, PA, USA) and probe sonicated (Sonics VC 505 ultrasonic processor, Newtown, CT, USA) at 250 W for 90 min in a thermostat bath to prevent temperature rise (T 25 °C). Finally, the solution was centrifuged at 2500 rpm for 40 min in a refrigerated (20 ± 2 °C) micro-centrifuge (Eppendorf 5417R, Hamburg, Germany) and the supernatant collected.

Microstructural and chemical characterization: Air tapping mode Atomic Force Microscopy (AFM) was performed with a Veeco Digital D5000 system. Using silicon tips with spring constant of 3 N·m^{-1} and resonance frequencies between 51 and 94 kHz. Samples for AFM investigations were prepared via spinning (at 2000 rpm for 30 s) 10 μL of centrifuged WS$_2$/Ethanol solution on a Si$_3$N$_4$ substrate. The substrates have been previously cleaned in a piranha base solution (3:1:3 mixture of ammonium hydroxide NH$_4$OH with hydrogen peroxide and milli-Q water) to enhance their wettability.

Exfoliated flakes were investigated using High Resolution Transmission Electron Microscopy HRTEM—JEOL 2100 Field Emission electron microscope (Tokyo, Japan) operated at 200 kV. Samples prepared by drop casting the WS$_2$/Eth solution on Si$_3$N$_4$ substrate were investigated by X-Ray Photoelectron Spectroscopy (XPS) using a PHI 1257 spectrometer (Perkin Elmer, Norwalk, CT, USA) equipped with a monochromatic Al Kα source (hν = 1486.6 eV) with a pass energy of 11.75 eV (93.9 eV survey), corresponding to an overall experimental resolution of 0.25 eV. Raman spectra were acquired using a Micro Raman Spectrometer (μRS) (LABRAM spectrometer, λ = 633 nm, 1 μm spatial resolution, and ≈2 cm^{-1} spectral resolution, Horiba-Jobin Yvon, Kyoto, Japan) equipped with a confocal optical microscope (100 × MPLAN objective with 0.9 numerical aperture and 0.15 mm work distance). 10 μL of the WS$_2$/Eth solution was deposited on a 270 nm SiO$_2$ substrate.

Gas sensing measurements: Electrical properties were determined by a volt-amperometric technique (AGILENT 34970A), as reported in Supplementary Figure S2, utilizing WS$_2$ thin films prepared by multiple drop casting and air annealing at 180 °C for 1 h the centrifuged WS$_2$/Eth suspension on Si$_3$N$_4$

substrates provided with 30 µm-spaced Pt interdigitated electrodes on the front side and a Pt resistor acting as a heater on the back side. Different gas concentrations in the range 1 ppm–250 ppm H_2 and 40 ppb–5 ppm NO_2 were obtained by mixing certified H_2, and NO_2 mixtures with dry air carrier, by means of an MKS147 multi gas mass controller. Different relative humidity (RH) air streams in the 10–80% RH range were obtained by mixing dry with saturated water-vapor air. The following definitions apply to discuss the gas response properties: base line resistance (BLR): the resistance in dry air at equilibrium before gas exposure, relative response (RR): the ratio (R_A/R_G) where R_A represents the resistance in air and R_G the one in gas at equilibrium for a given gas concentration, and sensor sensitivity (S): is the slope of the calibration curve in the sensitivity plot.

3. Results and Discussion

3.1. Microstructural Properties of Exfoliated WS_2

The microstructure of exfoliated WS_2 obtained by the combined ball milling and sonication process is characterized. In our case, acetonitrile (ACN) as the milling solvent, with surface tension of 29.5 mJ m^{-2} [27], has been selected as a trade-off between surface tension and moderate boiling point, enabling complete removal of the solvent at room temperature after grinding.

Regarding the influence of the grinding time, particles' size distribution of the starting WS_2 powder (blue plot), determined by Dynamic Light Scattering (DLS) technique (see experimental section) and shown in Figure 1a, downshifts towards smaller average sizes after 72 h grinding (red plot) and slightly further after 90 min sonication (green plot). The particle size distribution of the WS_2 starting powder displays an average particle size of $\approx$ 8 µm (blue plot) whereas the 72 h ball milled shows a bimodal distribution (red plot), with larger aggregates centered at $\approx$ 20 µm and smaller ones at $\approx$0.7 µm. After 90 min sonication, the bimodal distribution of the grinded powder disappears (green plot) and the average particle dimension places at $\approx$ 0.6 µm.

It may be concluded that grinding has an effective influence to reduce the particle size, while sonication, beside its effectiveness to suppress the bimodal distribution (presumably by separating agglomerated WS_2 particles), shows only minor effects to further decrease the particle size of the grinded powder, confirming the dominant role of the milling step. Figure 1b shows the AFM image of the 72 h ball milled and 90 min sonicated WS_2 sample. The inset of this image depicts a rough thickness profile along the selected line, with an average height from the substrate of 2 nm. Low and high magnification TEM images illustrated in Figure 1c,d, show that long term ball milling for 72 h results in a fragmented structure, which was eventually revealed to be amorphous by fast Fourier electron diffraction measurements. These features can be explained considering the two main forces induced by ball milling. The primary force is the shear force provided by rolling of balls on the surface of layers, which causes the removal and the exfoliation of surface layers. The secondary force is the vertical impact from the balls which combined with longer grinding times can fragment the larger exfoliated sheets into smaller ones, eventually collapsing of the crystal structure [15,28].

With the aim to minimize the fragmentation effect, we have reduced the ball milling duration time from 72 to 2 h, maintaining the sonication time at 90 min. Decreasing the milling time to 2 h, the flake's fragmentation sharply decreases, enforcing the formation of well-defined terraced structures comprising stacked WS_2 flakes, as shown in Figure 1e,f.

Figure 2 shows the main microstructural features of the 2 h ball milled and 90 min sonicated WS_2 powders. The AFM image shown in Figure 2a depicts the formation of a well-shaped 2D-flake with a large flat surface of 1 µm length. The corresponding thickness profile drawn in Figure 2b highlights a clear formation of a stacked structure comprising a 3 nm thick basal plane, 6 nm thick secondary plane and a third one at the top. Considering that the slight step on top of the profile, 0.6 nm high, corresponds to 1 layer thickness WS_2 [29], it is shown that the first step is made of 5 layers and the second one of 10 layers, respectively. The 2D character of the actual stacked structure shown in Figure 2b, as defined by the "aspect ratio" (i.e., the ratio of lateral dimension to the thickness), is high, with an associated value of 250, attesting to the successful optimization of the grinding time for exfoliation. The low-resolution TEM image depicted in Figure 2c also illustrates a well-shaped exfoliated flake with edge angles of 120°, confirming the preservation of the typical crystalline WS$_2$ hexagonal geometry.

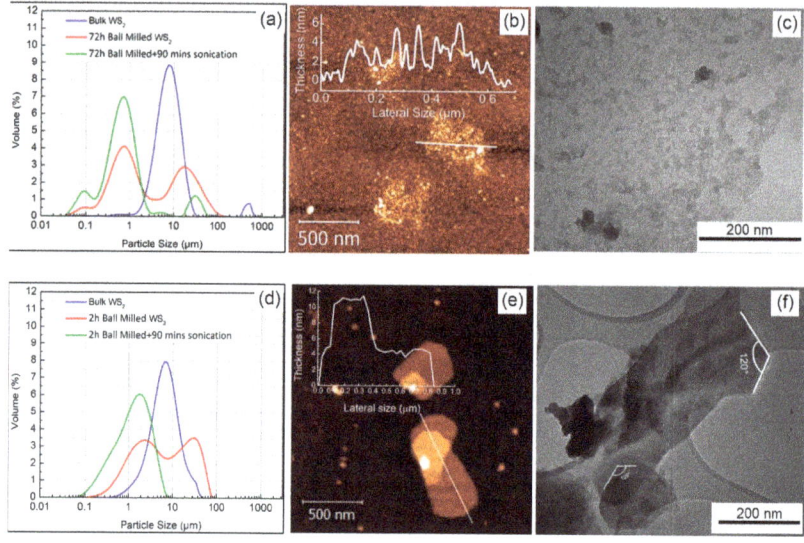

Figure 1. Comparison of the effect of long-time ball milling on WS$_2$ exfoliation: (**a**) Comparison of the particle size distribution of the starting WS$_2$ commercial powder (blue), 72 h ball milled (red) and 72 h ball milled and 90 min sonicated (green), (**b**) AFM picture of 72 h ball milled and 90 min sonicated and associated thickness profile along the white line, (**c**) TEM picture of the 72 h ball milled and 90 min sonicated WS$_2$, (**d**) Comparison of the particle size distribution in case of 2 h ball milling, (**e**) AFM picture of 2 h ball milled and 90 min sonicated and associated thickness profile, (**f**) TEM picture of the flakes obtained by 2 h ball milling and 90 min sonication of WS$_2$.

Figure 3 shows HRTEM images of both edges and surfaces of the flakes. Figure 3b (the magnification of Figure 3a), exhibits two layers with associated interlayer distances of ≈ 0.63 nm which is in good agreement with the AFM thickness measurements illustrated in Figure 2b. This interlayer displacement could also be observed at the flake's edges depicted in Figure 3d, where almost 11 layers can be clearly counted on the 7 nm thick edge. The atoms arrangement displayed in Figure 3b exhibits the hexagonal atomic structure, with lattice spacing of 0.27 nm and 0.25 nm, that are characteristics of (100) and (101) crystal planes of 2H-WS$_2$ flakes, respectively [30,31]. Moreover, the Fast Fourier Transforms (FFTs) shown as the inset of Figure 3d, further confirms the hexagonal crystalline structure of the flake.

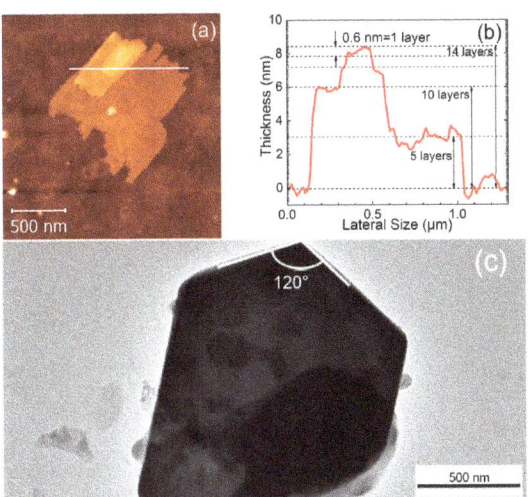

Figure 2. (a) AFM picture of the 2 h ball milled and 90 min sonicated WS$_2$, (b) thickness profile of the stacked flake along the white line of Figure (a), (c) low-resolution TEM picture of an exfoliated flake.

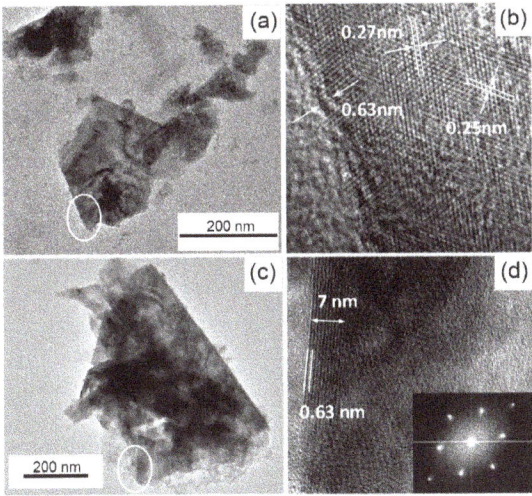

Figure 3. (a,c) TEM of 2 h ball milled and 90 min sonicated WS$_2$, (b) HRTEM corresponding to the circled area shown in figure (a) with highlighted the interlayer distance (0.63 nm) and lattice spacing (0.27 nm and 0.25 nm), corresponding to (100) and (101) planes of WS$_2$ respectively, (d) HRTEM of the edge of the flake corresponding to the circled area shown in figure (c) with highlighted the 7 nm thick edge corresponding to 11 layers. The inset shows the Selected Area Electron Diffraction (SAED) of the flake.

To give a statistical insight of the reproducibility of the preparation, four different suspensions were prepared after 2 h milling and 90 min sonication and the corresponding centrifuged suspensions collected, and spin coated on Si$_3$N$_4$ substrates (see experimental section). Figure 4a–d shows the AFM images of each prepared sample covering an area of 10 × 10 µm^2 and corresponding to a total population of ≈ 220 flakes. Overall, flakes' thickness follows a log-normal distribution as shown in Figure 4e, indicating that almost 30% of the flakes are ≤3.0 nm thick (i.e., ≈ 5 layers) and that about

75% are ≤6 nm (i.e., ≈ 10 layers). Moreover, as displayed in Figure 4f, average flake lateral dimensions are approximately ≈ 110 nm, yielding a surface coverage of ≈ 6%, as shown in Figure 4g. The overall calculated "aspect ratio" is 27.5, which is comparable to the ones previously reported for MoS$_2$ and WS$_2$, given the same preparation methodology [27,32]. The reduced standard deviations shown in Figure 4f,g attest to the high reproducibility of the exfoliation process.

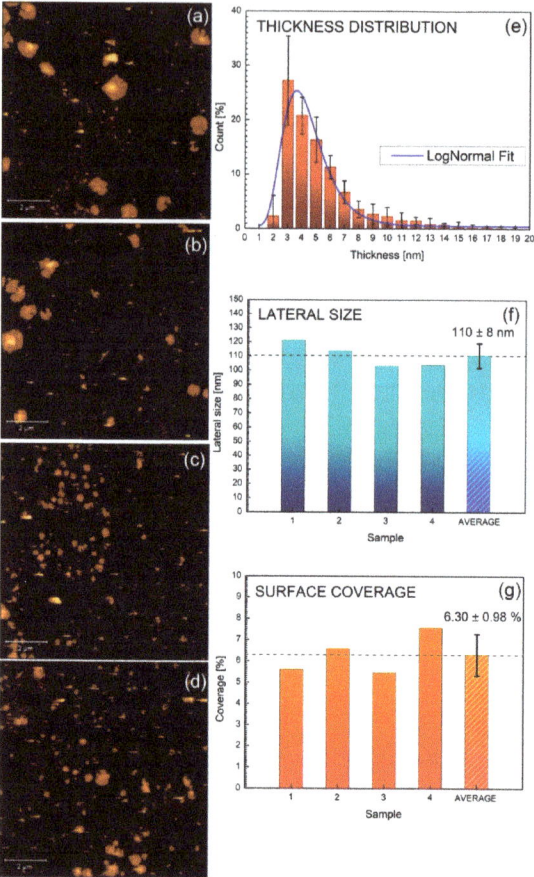

Figure 4. (**a**–**d**) AFM images of WS$_2$ exfoliated corresponding to four different samples prepared under the same conditions (i.e., 2 h ball milling and 90 min sonication). Statistical analysis corresponding to thickness distribution (**e**), Lateral dimensions (**f**) and surface area coverage (**g**).

3.2. Chemical Composition of the Exfoliated WS$_2$

Chemical issues related to both the evolution of point defects and oxidation phenomena of TMDs highlight important challenges associated with the practical utilization of TMDs monolayers in electronic and optoelectronic devices. Sulphur vacancy is one of the most typical point defects in 2D MoS$_2$ and WS$_2$ monolayers [33], eventually leading to active sites for gas adsorption. On the other hand, spontaneous oxidation of metal sulphides into their metal oxide counterparts [34], may result in poor reproducibility of the gas-sensing response over the long-term, as discussed in our previous publication [26].

Figure 5, panel (a), reports the W 4f core level XPS spectra of the pristine commercial WS$_2$ powder (PWD). The two doublets corresponding to 4f$_{7/2}$ peaks are assigned, according to the literature, to WS$_2$

and WO$_3$, respectively [35,36]. It turns out that the pristine powder is already oxidized, with a WO$_3$ content of ≈ 18%. This phenomenon, is not surprising considering that spontaneous oxidation of MoS$_2$ at room temperature after 6–12 months has been already reported in the literature, demonstrating the occurrence of ageing phenomena of MoS$_2$ [25] and WS$_2$ monolayers in ambient air [26]. Figure 5b shows that after exfoliation, the WO$_3$ content decreases to ≈ 16%. This could be explained considering that by grinding and sonicating WS$_2$ powders, newly not-yet-oxidized surfaces are formed, resulting in a smaller content of WO$_3$, as evidenced by XPS measurements. Furthermore, as discussed in the next paragraph, given the optimum operating temperature for gas sensing at 150 °C, exfoliated WS$_2$ suspensions, were therefore drop-casted and pre-annealed at 180 °C for 1 h in air to stabilize the oxidation levels. As shown in Figure 5c, after annealing, the WO$_3$ content increased to ≈ 58%.

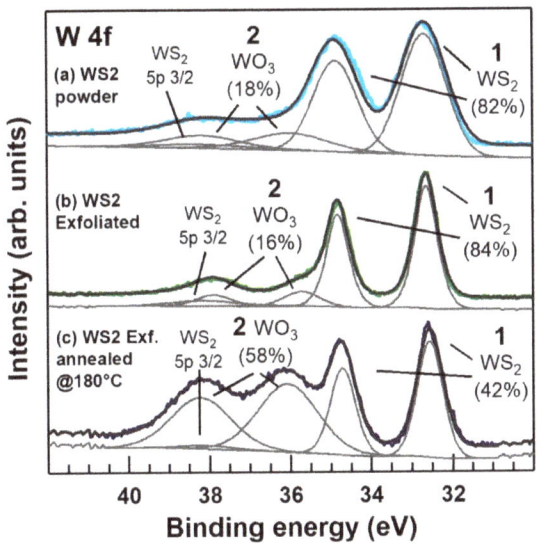

Figure 5. X-Ray Photoemission Spectroscopy (XPS) spectra of W 4f core level acquired respectively on (a) pristine WS$_2$ commercial powder (WS$_2$ PWD), (b) Exfoliated WS$_2$ by ball milling and sonication at 25 °C, (c) WS$_2$ exfoliated and post-annealed at 180 °C. All the components and their relative atomic percentages are labelled in the figure.

Regarding sulphur vacancies formation, considering 1% the detection limit of the XPS measurement, we found no clear evidence of defects of sulfur vacancies-related components (typically at binding energies of ≈ 36.1 in the W 4f core level XPS spectra in Figure 5). The analysis of the S 2p core level XPS spectra, reported in Supplementary Figure S3, is in line with the analysis reported for the W 4f.

Figure 6 displays the Raman spectra of bulk powder, exfoliated flakes and 180 °C annealed film. Peaks located at 350 and 419 cm^{-1} refer to crystalline WS$_2$. The peak at 520 cm^{-1} corresponds to the substrate (i.e., crystalline SiO$_2$). Raman spectra reveal that neither crystalline nor amorphous WO$_3$ are formed. According to the literature [37], the displacement of crystalline WO$_3$ is excluded considering that no peaks corresponding to the dashed lines at 719 and 807 cm^{-1} are observed. No signals of amorphous WO$_3$ were shown, as attested by the absence of a broad peak between 600 and 900 cm^{-1}, attributed to the W-O stretching vibration of amorphous WO$_3$. The lack from the Raman spectra of any WO$_3$ signal, as opposed to XPS, can be explained considering that Raman spectroscopy penetrates more deeply inside the material, suggesting that the overall amount of WO$_3$ throughout the whole flakes is negligible.

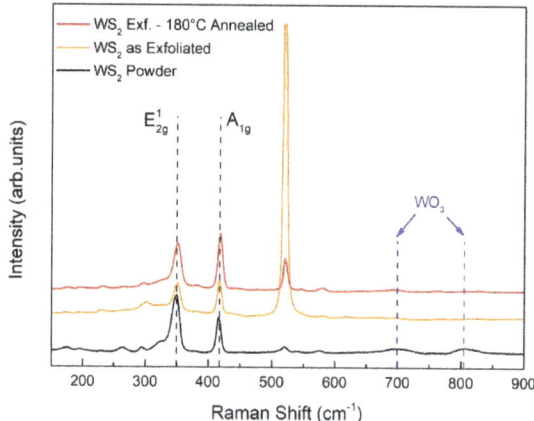

Figure 6. Raman spectra of WS_2 bulk powder, WS_2 as-exfoliated and WS_2 flakes post-annealed at 180 °C.

In order to have a better understanding where the as-formed WO_3 is located, firstly we have to consider that both XPS and Raman techniques give information on the chemical bonding of the elements, secondly that XPS information comes from photoelectrons escaping maximum up to 10 nm below the material surface, lastly that Raman spectroscopy, compared to XPS, is a "bulk" technique (given the negligible attenuation of visible light at the length scale of microns). It turns out that the WS_2/WO_3 percentage content measured by XPS and shown in Figure 4 (i.e., ≈ 58%), represents the average chemical compositions of a portion of the material confined within at last 10 nm from the material surface. This region, for simplicity, can be referred as a "surface layer", which represents the reacting surface to interfering gases. Regarding the crystallographic nature of the "surface layer", grazing incidence XRD diffraction carried on exfoliated and 200 °C annealed WS_2 flakes (shown in Supplementary Figure S4) revealed the occurrence of peaks belonging only to WS_2, thus excluding the formation of any crystalline WO_3 in the 180 °C annealed film. It may be concluded that the "surface layer" is a composite structure comprising amorphous WO_3 and pristine crystalline WS_2. These experimental results are in line with that previously discussed in the literature [38], that the oxidation of bulk TMDs provides two parallel steps. Oxygen atoms rapidly exchange with surface sulphur forming an amorphous oxide layer, whilst WS_2 interlayer channels provide a path for inward-oxygen and backward-sulphur diffusion, resulting in the formation of amorphous WO_3, which propagates over time inside the TMD flakes.

3.3. Gas Sensing Response

It has been reported that TMDs gas sensors operating at room temperature have shown remarkable limitations, largely related to irreversible desorption of the gas molecules, displaying incomplete recovery of the baseline at 25 °C [39,40]. The selection of the best operating temperature (OT), which it corresponds the complete base line recovery, within reasonable response times, and acceptable gas relative responses (RR), is limited in TMDs by the intensifying of the oxidation processes with increasing the OT, as previously discussed. Baseline recovery and response times depend on the adsorption-desorption kinetic of gases with the sensor surface, which eventually improves with increasing the OT. RR is mostly related to microstructure (i.e., surface area, grain size), concentration of surface defects (i.e., oxygen or sulphur vacancies) and structure of the reacting surface (crystalline or amorphous). To find the best OT exfoliated WS_2 drop casted on Si_3N_4, substrates were previously air annealed at 180 °C to stabilize the WO_3 content up to ≈ 58%, as shown in Figure 5c. Afterwards, different OT in the 50–150 °C temperature range where tested, resulting in 150 °C as the best OT, as shown in Supplementary Figure S5.

The Scanning Electron Microscopy (SEM) picture of the sensing device shown in Figure 7 highlights a homogeneous distribution of annealed flakes, enabling current percolation paths between adjacent flakes, covering an area of 1.4 × 0.6 mm² over 30 µm spaced Pt interdigitated electrodes.

Figure 7. SEM image of sensor obtained by drop casting exfoliated WS_2 and annealing at 180 °C on Si_3N_4 substrate provided with Pt finger-type electrodes (30 microns apart).

Figure 8a shows the normalized dynamic resistance changes, at 150 °C OT of few-layers WS_2 thin films to NO_2 and H_2 in the 100 ppb–5 ppm and 1 ppm–250 ppm gas concentration ranges, respectively. WS_2 films respond as n-type semiconductors with decreasing/increasing resistance upon exposure to H_2/NO_2, respectively. Degassing with dry air at 150 °C OT, the baseline resistance (BLR), as indicated by the black dotted line in the figure, is almost recovered. WS_2 flakes are more sensitive to NO_2 than H_2 gas, with associated low detection limits (LDL) of 200 ppb and 5 ppm respectively, confirming what has been previously reported in the literature.

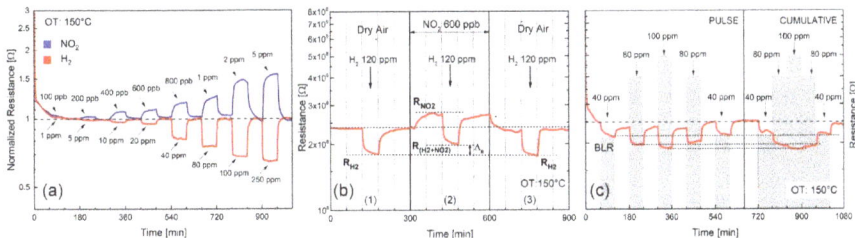

Figure 8. Electrical responses of the exfoliated WS_2 post-annealed at 180 °C, at 150 °C operating temperature in dry air. (a) Comparison of the normalized dynamic response to NO_2 (100 ppb–5 ppm) and H_2 (1–250 ppm), (b) NO_2 cross-sensitivity to H_2: first panel, the response to 120 ppm H_2 in dry air, second panel, response to 120 ppm H_2 with 600 ppb NO_2, third panel, response to 120 ppm H_2 (as to first panel) for comparison, (c) Reproducibility and baseline recovery by exposing the film to both pulse and cumulative H_2 concentrations in the range 40–100 ppm. H_2 concentrations are highlighted in the figure by grey shadowed rectangular plots.

Cross-sensitivity, which represent the ability of WS_2 to detect H_2 in the presence of NO_2 interfering gas, has been shown in Figure 8b. Panel (1) of Figure 8b shows the WS_2 response to H_2 alone in dry air carrier at 150 °C OT. The "cross-sensitivity" produced by interfering NO_2 to the measure of H_2 is displayed in panel (2) of Figure 8b. By exposing to 600 ppb NO_2, sensor resistance initially increases, yielding at equilibrium the resistance value R_{NO2}. As soon as 120 ppm H_2 is introduced, the resistance decreases, yielding the equilibrium value shown as $R_{(H2+NO2)}$. The cross-sensitivity effect is displayed in the picture as (ΔR), to indicate the gap between the electrical resistance to 120 ppm H_2 alone (i.e., R_{H2}) and that in the presence of 120 ppm H_2 and 600 ppb NO_2 (i.e., $R_{(H2+NO2)}$). These results imply that the response to 120 ppm H_2 is affected by the presence of a small amount (i.e., 600 ppb), of NO_2 interfering gas, confirming the stronger affinity of WS_2 to detect NO_2 as respect to H_2.

Reproducibility of the electrical response to pulse (on/off) and cumulative modes H_2 gas adsorption is shown in Figure 8c, demonstrating acceptable response characteristics to H_2. Under pulsed conditions, the baseline resistance (BLR) fully regains its initial value after completion of each desorption cycle in dry air. Under cumulative stepwise adsorption/desorption mode, the H_2 gas resistance increases/decreases steadily, matching almost the same H_2 resistance values obtained under pulsed conditions (black lines at saturation correspond to 40, 80 and 100 ppm H_2).

Selectivity tests to both oxidizing and reducing gases carried out at 150 °C operating temperature with respect to 5 ppm NO_2, exhibit satisfactory WS_2 selectivity to both 5 ppm H_2 and NH_3 gases and to 250 ppm ethanol and acetone, as shown in Supplementary Figure S6. These data were demonstrated to be in line with previous research on WS_2 nanoflakes synthesized by electrospinning [41].

Long-term stability properties of the electrical response of both baseline and saturation resistances to 800 ppb NO_2 over a period of 12 months (corresponding to approximately 5 months of cumulative operations at 150 °C operating temperature) were also recorded. Figure 9a shows baseline resistances (lower curve) and saturation resistances corresponding to 800 ppb NO_2 (upper curve), randomly collected over a period of 52 weeks. Average resistances with associated standard deviations are calculated over a set of 5 consecutive measurements. Relative responses (RR) taken over the investigated period are also highlighted in the figure. No remarkable fluctuations of both baseline and resistances at saturation were detectable, attesting good long-term stability of the electrical properties of the WS_2 films. To validate the electrical responses shown in Figure 8a, we also investigated the oxidation state of the sensor surface, measuring the WO_3 content before and after 52 weeks conditioning. Figure 9b compares the XPS W 4f signal of the as-exfoliated WS_2 180 °C annealed film before (lower curve) and after 52 weeks long-term conditioning to different gases and 150 °C OT (upper curve). Beside an increase of the signal noise after long-term conditionings, no substantial increase of the WO_3 content was detected. These observations imply that exfoliated WS_2 films, previously stabilized at 180 °C, can satisfactorily respond to different gases under sustained conditions at 150 °C operating temperature.

Finally, the influence of humidity on NO_2 and H_2 gas response at 150 °C OT is also reported. Figure 10 shows the dynamic response of the films exposed to air with increasing amounts of humidity, in the 10–80% relative humidity (RH) range. The inset of Figure 10a displays the related sensitivity plot. Considering that at 150 °C operating temperature physisorbed water is reasonably evaporated, it is unlikely that the decrease of the resistance is induced by a protonic surface charge-transfer mechanism, as reported for WS_2 humidity sensors operating at room temperature [42]. More reasonably, water vapor at 150 °C OT behaves like a reducing gas inducing a steady resistance decrease in WS_2 n-type semiconductor, which does not saturate with increasing the RH, as shown in the inset of Figure 10a.

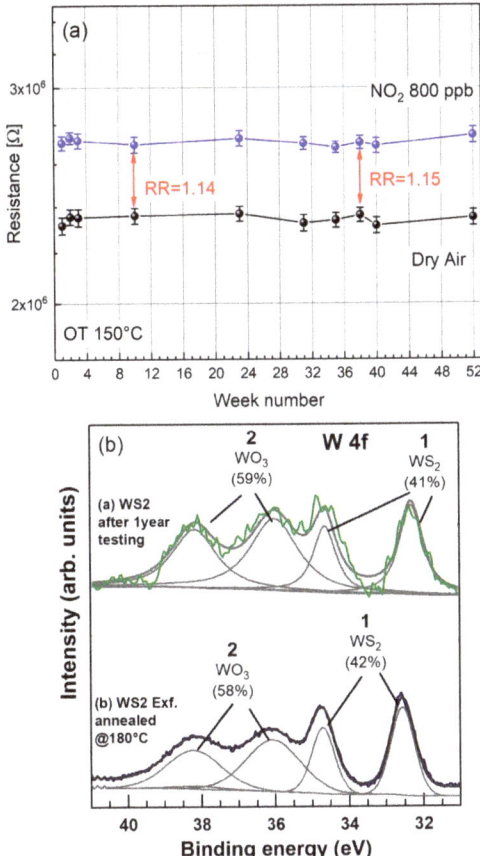

Figure 9. WS$_2$ exfoliated and post-annealed at 180 °C. (**a**) Long-term stability properties of the electrical resistances of the baseline (lower curve) and 800 ppb NO$_2$ over a period of 12 months (equivalent to approximately 5 months of continuous operation at 150 °C operating temperature). Average resistance values with associated standard deviations are calculated over a set of 5 consecutive measurements. (**b**) Comparison of the XPS signals of the as-exfoliated WS$_2$ annealed at 180 °C (lower curve) and the same sample after one-year conditioning to various gases and 150 °C operating temperature.

In order to evaluate the influence of humidity to NO$_2$ and H$_2$ gases response, cross-sensitivity tests have been performed. Figure 10b, c shows the dynamic responses to different NO$_2$ and H$_2$ concentrations, measured in humid air at 40% RH. As soon as water vapor is introduced, a downshift of the baseline from BLR$_{DRY}$ to BLR$_{H2O}$ is shown. NO$_2$ and H$_2$ dynamic responses in dry and 40% RH are almost similar, as shown in Figures 8a and 10b,c. Comparison of the NO$_2$ and H$_2$ gases sensitivities in dry and 40% RH air are shown in Figure 10d. Notably, no significant differences are displayed as respect to slopes of the calibrating curves (i.e., sensitivity) and relative responses values (i.e., Ra/Rg), attesting that both NO$_2$ and H$_2$ measurements are not affected by the presence of moisture.

These results, indeed, demonstrate the possibility to produce efficient and reproducible gas sensors, able to detect NO$_2$ and H$_2$ with no significant cross-sensitivity effects induced by humid air in the 10–80% RH range and 150 °C operating temperature.

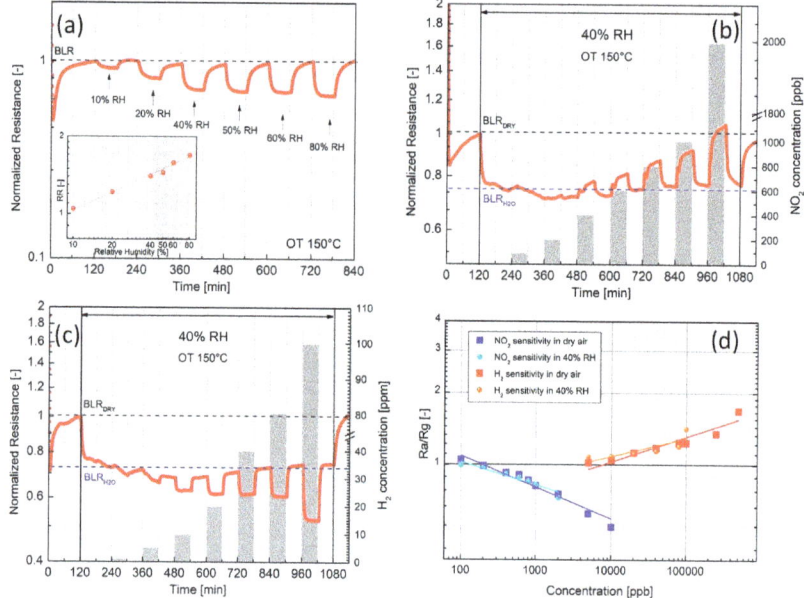

Figure 10. Electrical responses of the exfoliated WS_2 post-annealed at 180 °C (150 °C operating temperature) to different Relative Humidity (RH) conditions. (**a**) Normalized dynamic response to humidity (10–80% RH). The inset depicts the corresponding sensitivity plot. (**b**) Dynamic response to increasing NO_2 concentrations in air with 40% RH, (**c**) Dynamic response to increasing H_2 concentrations in in air with 40% RH, (**d**) Comparison of the sensitivity plots to NO_2 and H_2 in dry air and 40% RH, respectively.

3.4. Gas Response Mechanism

As previously discussed, by annealing the exfoliated WS_2 at 180 °C a "surface layer" containing ≈58% of amorphous WO_3, penetrating at last 10 nm from the surface, is formed. It turns out that both structure and chemical composition of the "Surface layer", comprising crystalline WS_2 and amorphous WO_3, strongly influence the gas response mechanism.

HRTEM investigations of the pre-annealed 180 °C sample shown in Figure 11, display the occurrence of a complex surface patchwork comprising amorphous WO_3 regions (located inside the green square of Figure 11b), possibly rearranging as not-connected, amorphous, isolated-clusters which are eventually embedded in crystalline WS_2 phase (located inside the red square of Figure 11b).

To investigate the contribution of single WO_3 on to the electrical properties of WS_2/WO_3 pre-annealed composite, we have prepared a fully-oxidized WS_2, containing ≈ 99% amorphous WO_3 and tested to NO_2. This sample has been prepared by the same exfoliation method by grinding and drying WS_2 powders at 25 °C, but setting the sonicating temperature at 60 °C, instead of 25 °C for 90 min, and finally pre-annealing in dry air at 180 °C.

Figure 12a,b shows the XPS and grazing incidence XRD patterns of the fully oxidized film, attesting that the chemical composition of the "surface layer" is ≈ 99% WO_3 (Figure 12a) and that the as-formed WO_3 is amorphous, as highlighted by the absence of WO_3 peaks inside the inset of the XRD pattern of Figure 12b. It turns out that the fully oxidized region possibly covers the whole surface of the flakes, not extending to the core, which maintain the crystalline structure of pristine exfoliated WS_2 (as attested by the presence of WS_2 peaks in the XRD pattern of Figure 12b). The electrical response of the fully oxidized amorphous WO_3 to 800 ppb NO_2 and different operating temperatures in dry air is shown in Figure 12c. According to Figure 12c, amorphous WO$_3$ is not responsive to NO$_2$ gas in dry air, in the operating temperature range 75–150 °C.

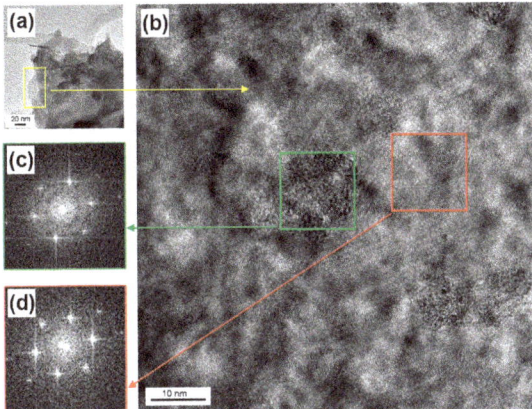

Figure 11. HRTEM images of the WS$_2$ film pre-annealed at 180 °C. (**b**) Magnification of the yellow area of Figure 11 (**a**) displaying the presence of ordered structures (i.e., inside the red square) referred to crystalline WS$_2$ and the presence of disordered ones (i.e., inside the green square) attributed to amorphous WO$_3$. Related Selected Area Electron Diffraction (SAED) patterns are shown in (**c**) and (**d**), highlighting the occurrence of sharper reflections (**d**) associated to crystalline WS$_2$.

Literature reports discussing the gas sensing properties of amorphous WO$_3$ are very scarce. Some authors found no gas response to NO$_x$ of amorphous WO$_3$ deposited by sputtering [43] whereas others demonstrated negligible NO$_2$ response using photochemically-produced amorphous WO$_3$ [44]. In most cases, the NO$_2$ gas response of amorphous WO$_3$ is smaller with respect to crystalline WO$_3$, frequently associated with baseline drift phenomena, with few exceptions, mostly related to the preparation conditions. In our case, we demonstrated that the interaction of NO$_2$ with amorphous 99% WO$_3$ has no effects altogether.

Having shown in Figure 8a the substantial NO$_2$ and H$_2$ gas response of the WO$_3$/WS$_2$ composite, we conclude that it is crystalline WS$_2$ which primarily respond to NO$_2$ gas. The predominant gas sensing role played by crystalline WS$_2$ with respect to amorphous WO$_3$, is also supported by the decrease of the electrical resistance with increasing relative Humidity (RH), as shown in Figure 10a. Considering that WO$_3$ interacts with humidity by increasing the resistance, due to WO$_3$ lattice oxidation induced by humidity, as reported in the literature [45], the resistance decrease displayed in Figure 10a rules out any significant contribution of WO$_3$ to the overall humidity response. Moreover, the hypothesis that WS$_2$ is likely to be the responding material is also supported by our previous research, which demonstrated that humidity decreases the sensor resistance in MoS$_2$-based exfoliated [25].

Discussing the contribution of crystalline WS$_2$ to the overall electrical resistance, it was recently demonstrated by first-principles calculations on single MoS$_2$ sulphur-defective layer that O$_2$ irreversibly chemisorbs on sulphur vacancies [46] and that the "heal" of these defects by substitutional O atoms is thermodynamically favorable [47]. Furthermore, in case of direct NO$_2$ molecules interaction with sulphur vacancies, a dissociative adsorption of NO$_2$, leading to O atoms passivating the vacancies, and NO molecules physisorbed on the MoS$_2$ surface, was also proposed [48,49]. Given these premises, we hypothesize that both O$_2$ and NO$_2$ suppress sulphur vacancies, supporting a gas response mechanism based only on physisorption of NO$_2$ and H$_2$ molecules on passivated (i.e., defect-free) WS$_2$ surface. This hypothesis is sustained by theoretical studies on the adsorption of NO$_2$, H$_2$, O$_2$, H$_2$O, NH$_3$ and CO gases on defect-free single layer MoS$_2$ and WS$_2$ [50,51]. According to this physisorption model, the size and sign of the resistance changes, when exposing few-flakes of WS$_2$ to oxidizing (NO$_2$) and reducing (H_2) gases, depend on the number of exchanged carriers (i.e., electrons) and their direction. NO_2 being more electronegative than H_2 induces a large electron withdrawal, whereas H_2 results in weak electron injection, explaining the increase/decrease of electrical resistance in *n*-type WS_2, as well as the smaller detection limit measured for NO_2 (200 ppb) as compared to the one found for H_2 (i.e., 5 ppm).

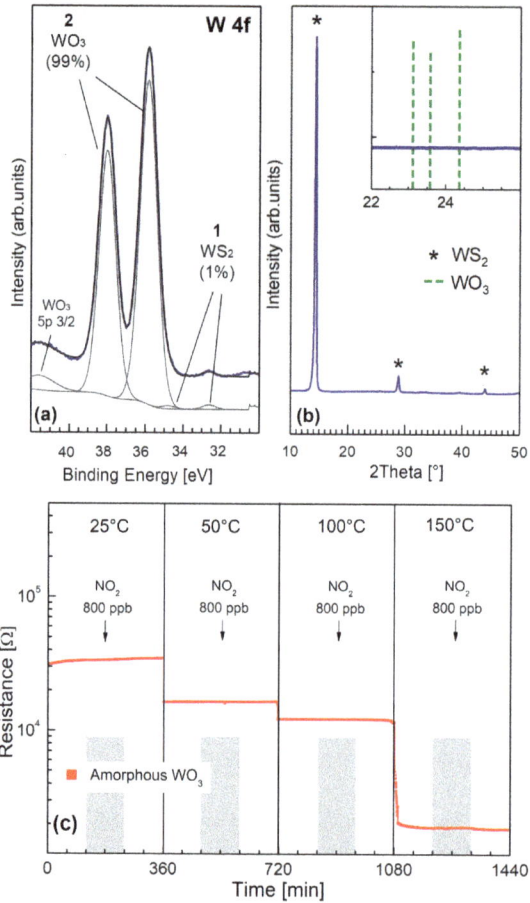

Figure 12. Chemical composition, crystalline structure and microstructural features of a fully oxidized WO_3 thin film. (**a**) W 4f core level XPS spectra, (**b**) XRD grazing incidence spectra. Top right inset shows the close up of the 2θ region characteristic of crystalline WO_3 (corresponding peaks of crystalline WO_3, according to ICDS 98-001-7003, are highlighted by dashed green lines), (**c**) electrical response of the fully oxidized WO_3 amorphous film to NO_2 and different OTs.

Lastly, a question to be resolved is why both NO_2 and H_2 sensitivities are not affected by the presence of moisture, as shown in Figure 10d. This behavior may suggest that water vapor adsorbs on to the WS_2 surface with a different and non-competitive mechanism with respect to NO_2 and H_2 gases. Clearly this interaction is a complex issue, and is yet to be clarified based on specific theoretical and experimental studies.

4. Conclusions

In conclusion, we have demonstrated an effective, reproducible and high-yield exfoliation process, obtained by enhanced low energy ball milling and sonication. Specifically, the two-step exfoliation followed by drop casting the centrifuged suspension leads to the deposition of thin films of well-packed and interconnected WS_2 flakes with controlled and reproducible microstructure over large areas, thus representing a fast, simple and scalable method, compatible with standard microelectronic fabrication techniques. We found that a spontaneous oxidation of WS_2 leading to the formation of amorphous WO_3 on the surface of the exfoliated WS_2 takes place, addressing the crucial drawback of surface oxidation of TMDs. We also found that by pre-annealing the WS_2 films at 180 °C, a reproducible surface oxidation of WS_2 to amorphous WO_3 takes place, which stabilize from further oxidation the WS_2 layers. Reproducible gas sensing responses to NO_2 and H_2 and humidity at 150 °C operating temperature were achieved with detection limits of 200 ppb and 5 ppm to NO_2 and H_2, respectively. The cross-sensitivity test highlighted a weak interference played by NO_2 to the H_2 gas response. Water vapor at 40% RH also resulted in having no interference to the measure of NO_2 and H_2 gases, attesting promising characteristics of WS_2 exfoliated films for gas sensing applications.

Supplementary Materials: The following are available online at http://www.mdpi.com/2079-4991/9/10/1363/s1, Figure S1: Schematic illustration of the exfoliation process, Figure S2: Schematic illustration of the gas sensing equipment, Figure S3: XPS spectra of S 2p core level, Figure S4: Grazing incidence XRD spectra of the as-exfoliated WS_2; as-exfoliated WS_2—200 °C annealed for 1 h, Figure S5: The electrical response of WS_2 post-annealed at 180 °C at different operating temperatures and 800 ppb NO_2 in dry air, Figure S6: Selectivity response of WS_2 post-annealed at 180 °C at 150 °C operating temperature.

Author Contributions: Conceptualization, L.O., C.C., V.P. and S.M.E.; methodology, L.O. and C.C.; validation, L.O. and C.C.; formal analysis, V.P.; investigation, V.P. and S.M.E.; resources, C.C., L.O. and M.N.; data curation, V.P. and S.M.E.; Supervision, L.O. and C.C.; writing—original draft preparation, V.P., S.M.E. and C.C.; writing—review and editing, C.C., L.O. and V.P.

Funding: This research was funded by REGIONE ABRUZZO Dipartimento Sviluppo Economico, Politiche del Lavoro, Istruzione, Ricerca e Università Servizio Ricerca e Innovazione Industriale for financial support through progetto POR FESR Abruzzo 2018–2020 Azione 1.1.1 e 1.1.4—"Studio di soluzioni innovative di prodotto e di processo basate sull'utilizzo industriale dei materiali avanzati" CUP n. C17H18000100007.

Conflicts of Interest: The authors declare no conflict of interest. The funders had no role in the design of the study, in the collection, analyses, or interpretation of data, in the writing of the manuscript, or in the decision to publish the results.

References

1. Yang, W.; Gan, L.; Li, H.; Zhai, T. Two-dimensional layered nanomaterials for gas-sensing applications. *Inorg. Chem. Front.* **2016**, *3*, 433–451. [CrossRef]
2. Late, D.J.; Huang, Y.K.; Liu, B.; Acharya, J.; Shirodkar, S.N.; Luo, J.; Yan, A.; Charles, D.; Waghmare, U.V.; Dravid, V.P.; et al. Sensing behavior of atomically thin-layered MoS_2 transistors. *ACS Nano* **2013**, *7*, 4879–4891. [CrossRef] [PubMed]
3. Voiry, D.; Yamaguchi, H.; Li, J.; Silva, R.; Alves, D.C.B.; Fujita, T.; Chen, M.; Asefa, T.; Shenoy, V.B.; Eda, G.; et al. Enhanced catalytic activity in strained chemically exfoliated WS_2 nanosheets for hydrogen evolution. *Nat. Mater.* **2013**, *12*, 850–855. [CrossRef] [PubMed]
4. Voiry, D.; Yang, J.; Chhowalla, M. Recent Strategies for Improving the Catalytic Activity of 2D TMD Nanosheets Toward the Hydrogen Evolution Reaction. *Adv. Mater.* **2016**, *28*, 6197–6206. [CrossRef] [PubMed]
5. Mak, K.F.; Shan, J. Photonics and optoelectronics of 2D semiconductor transition metal dichalcogenides. *Nat. Photonics* **2016**, *10*, 216–226. [CrossRef]
6. Shim, J.; Park, H.-Y.; Kang, D.-H.; Kim, J.-O.; Jo, S.-H.; Park, Y.; Park, J.-H. Electronic and Optoelectronic Devices based on Two-Dimensional Materials: From Fabrication to Application. *Adv. Electron. Mater.* **2017**, *3*, 1600364. [CrossRef]
7. Pumera, M.; Sofer, Z.; Ambrosi, A. Layered transition metal dichalcogenides for electrochemical energy generation and storage. *J. Mater. Chem. A* **2014**, *2*, 8981–8987. [CrossRef]
8. Wang, H.; Feng, H.; Li, J. Graphene and graphene-like layered transition metal dichalcogenides in energy conversion and storage. *Small* **2014**, *10*, 2165–2181. [CrossRef]

9. Gupta, A.; Sakthivel, T.; Seal, S. Recent development in 2D materials beyond graphene. *Prog. Mater. Sci.* **2015**, *73*, 44–126. [CrossRef]
10. Bhimanapati, G.R.; Lin, Z.; Meunier, V.; Jung, Y.; Cha, J.; Das, S.; Xiao, D.; Son, Y.; Strano, M.S.; Cooper, V.R.; et al. Recent Advances in Two-Dimensional Materials beyond Graphene. *ACS Nano* **2015**, *9*, 11509–11539. [CrossRef]
11. Yang, D.; Frindt, R.F. Li-intercalation and exfoliation of WS_2. *J. Phys. Chem. Solids* **1996**, *57*, 1113–1116. [CrossRef]
12. Ambrosi, A.; Sofer, Z.; Pumera, M. Lithium intercalation compound dramatically influences the electrochemical properties of exfoliated MoS_2. *Small* **2015**, *11*, 605–612. [CrossRef] [PubMed]
13. Nicolosi, V.; Chhowalla, M.; Kanatzidis, M.G.; Strano, M.S.; Coleman, J.N. Liquid Exfoliation of Layered Materials. *Science* **2013**, *340*, 1226419. [CrossRef]
14. Niu, L.; Coleman, J.N.; Zhang, H.; Shin, H.; Chhowalla, M.; Zheng, Z. Production of Two-Dimensional Nanomaterials via Liquid-Based Direct Exfoliation. *Small* **2016**, *12*, 272–293. [CrossRef] [PubMed]
15. Yi, M.; Shen, Z. A review on mechanical exfoliation for the scalable production of graphene. *J. Mater. Chem. A* **2015**, *3*, 11700–11715. [CrossRef]
16. Kumar, G.R.; Jayasankar, K.; Das, S.K.; Dash, T.; Dash, A.; Jena, B.K.; Mishra, B.K. Shear-force-dominated dual-drive planetary ball milling for the scalable production of graphene and its electrocatalytic application with Pd nanostructures. *RSC Adv.* **2016**, *6*, 20067–20073. [CrossRef]
17. Abdelkader, A.M.; Kinloch, I.A. Mechanochemical Exfoliation of 2D Crystals in Deep Eutectic Solvents. *ACS Sustain. Chem. Eng.* **2016**, *4*, 4465–4472. [CrossRef]
18. Krishnamoorthy, K.; Pazhamalai, P.; Veerasubramani, G.K.; Kim, S.J. Mechanically delaminated few layered MoS_2 nanosheets based high performance wire type solid-state symmetric supercapacitors. *J. Power Sources* **2016**, *321*, 112–119. [CrossRef]
19. Yao, Y.; Tolentino, L.; Yang, Z.; Song, X.; Zhang, W.; Chen, Y.; Wong, C.P. High-concentration aqueous dispersions of MoS_2. *Adv. Funct. Mater.* **2013**, *23*, 3577–3583. [CrossRef]
20. Perkins, F.K.; Friedman, A.L.; Cobas, E.; Campbell, P.M.; Jernigan, G.G.; Jonker, B.T. Chemical Vapor Sensing with Monolayer MoS_2. *Nano Lett.* **2013**, *13*, 668–673. [CrossRef]
21. Liu, B.; Chen, L.; Liu, G.; Abbas, A.N.; Fathi, M.; Zhou, C. High-Performance Chemical Sensing Using Schottky-Contacted Chemical Vapor Deposition Grown Monolayer MoS_2 Transistors. *ACS Nano* **2014**, *8*, 5304–5314. [CrossRef] [PubMed]
22. Kuru, C.; Choi, D.; Kargar, A.; Liu, C.H.; Yavuz, S.; Choi, C.; Jin, S.; Bandaru, P.R. High-performance flexible hydrogen sensor made of WS_2 nanosheet–Pd nanoparticle composite film. *Nanotechnology* **2016**, *27*, 195501. [CrossRef] [PubMed]
23. Prezioso, S.; Perrozzi, F.; Giancaterini, L.; Cantalini, C.; Treossi, E.; Palermo, V.; Nardone, M.; Santucci, S.; Ottaviano, L. Graphene Oxide as a Practical Solution to High Sensitivity Gas Sensing. *J. Phys. Chem. C* **2013**, *117*, 10683–10690. [CrossRef]
24. Donarelli, M.; Ottaviano, L.; Giancaterini, L.; Fioravanti, G.; Perrozzi, F.; Cantalini, C. Exfoliated black phosphorus gas sensing properties at room temperature. *2D Mater.* **2016**, *3*, 025002. [CrossRef]
25. Donarelli, M.; Prezioso, S.; Perrozzi, F.; Bisti, F.; Nardone, M.; Giancaterini, L.; Cantalini, C.; Ottaviano, L. Response to NO_2 and other gases of resistive chemically exfoliated MoS_2-based gas sensors. *Sens. Actuators B Chem.* **2015**, *207*, 602–613. [CrossRef]
26. Perrozzi, F.; Emamjomeh, S.M.M.; Paolucci, V.; Taglieri, G.; Ottaviano, L.; Cantalini, C. Thermal stability of WS_2 flakes and gas sensing properties of WS_2/WO_3 composite to H_2, NH_3 and NO_2. *Sens. Actuators B Chem.* **2017**, *243*, 812–822. [CrossRef]
27. Nguyen, E.P.; Carey, B.J.; Daeneke, T.; Ou, J.Z.; Latham, K.; Zhuiykov, S.; Kalantar-zadeh, K. Investigation of two-solvent grinding-assisted liquid phase exfoliation of layered MoS_2. *Chem. Mater.* **2015**, *27*, 53–59. [CrossRef]
28. Xing, T.; Mateti, S.; Li, L.H.; Ma, F.; Du, A.; Gogotsi, Y.; Chen, Y. Gas Protection of Two-Dimensional Nanomaterials from High-Energy Impacts. *Sci. Rep.* **2016**, *6*, 35532. [CrossRef]
29. Gutiérrez, H.R.; Perea-López, N.; Elías, A.L.; Berkdemir, A.; Wang, B.; Lv, R.; López-Urías, F.; Crespi, V.H.; Terrones, H.; Terrones, M.; et al. Extraordinary room-temperature photoluminescence in triangular WS_2 monolayers. *Nano Lett.* **2013**, *13*, 3447–3454. [CrossRef]
30. Ghorai, A.; Midya, A.; Maiti, R.; Ray, S.K. Exfoliation of WS_2 in the semiconducting phase using a group of lithium halides: A new method of Li intercalation. *Dalt. Trans.* **2016**, *45*, 14979–14987. [CrossRef]

31. Zhou, P.; Xu, Q.; Li, H.; Wang, Y.; Yan, B.; Zhou, Y.; Chen, J.; Zhang, J.; Wang, K. Fabrication of Two-Dimensional Lateral Heterostructures of WS$_2$/WO$_3$·H$_2$O Through Selective Oxidation of Monolayer WS$_2$. *Angew. Chem. Int. Ed.* **2015**, *54*, 15226–15230. [CrossRef] [PubMed]
32. Carey, B.J.; Daeneke, T.; Nguyen, E.P.; Wang, Y.; Zhen Ou, J.; Zhuiykov, S.; Kalantar-zadeh, K. Two solvent grinding sonication method for the synthesis of two-dimensional tungsten disulphide flakes. *Chem. Commun.* **2015**, *51*, 3770–3773. [CrossRef] [PubMed]
33. Hong, J.; Hu, Z.; Probert, M.; Li, K.; Lv, D.; Yang, X.; Gu, L.; Mao, N.; Feng, Q.; Xie, L.; et al. Exploring atomic defects in molybdenum disulphide monolayers. *Nat. Commun.* **2015**, *6*, 6293. [CrossRef] [PubMed]
34. Gao, J.; Li, B.; Tan, J.; Chow, P.; Lu, T.M.; Koratkar, N. Aging of Transition Metal Dichalcogenide Monolayers. *ACS Nano* **2016**, *10*, 2628–2635. [CrossRef] [PubMed]
35. Di Paola, A.; Palmisano, L.; Venezia, A.M.; Augugliaro, V. Coupled Semiconductor Systems for Photocatalysis. Preparation and Characterization of Polycrystalline Mixed WO$_3$/WS$_2$ Powders. *J. Phys. Chem. B* **1999**, *103*, 8236–8244. [CrossRef]
36. Wong, K.C.; Lu, X.; Cotter, J.; Eadie, D.T.; Wong, P.C.; Mitchell, K.A.R. Surface and friction characterization of MoS$_2$ and WS$_2$ third body thin films under simulated wheel/rail rolling-sliding contact. *Wear* **2008**, *264*, 526–534. [CrossRef]
37. Shigesato, Y.; Murayama, A.; Kamimori, T.; Matsuhiro, K. Characterization of evaporated amorphous WO$_3$ films by Raman and FTIR spectroscopies. *Appl. Surf. Sci.* **1988**, *34*, 804–811. [CrossRef]
38. Margolin, A.; Rosentsveig, R.; Albu-Yaron, A.; Popovitz-Biro, R.; Tenne, R. Study of the growth mechanism of WS$_2$ nanotubes produced by a fluidized bed reactor. *J. Mater. Chem.* **2004**, *14*, 617–624. [CrossRef]
39. Long, H.; Harley-Trochimczyk, A.; Pham, T.; Tang, Z.; Shi, T.; Zettl, A.; Carraro, C.; Worsley, M.A.; Maboudian, R. High Surface Area MoS$_2$/Graphene Hybrid Aerogel for Ultrasensitive NO$_2$ Detection. *Adv. Funct. Mater.* **2016**, *26*, 5158–5165. [CrossRef]
40. Long, H.; Chan, L.; Harley-Trochimczyk, A.; Luna, L.E.; Tang, Z.; Shi, T.; Zettl, A.; Carraro, C.; Worsley, M.A.; Maboudian, R. 3D MoS$_2$ Aerogel for Ultrasensitive NO$_2$ Detection and Its Tunable Sensing Behavior. *Adv. Mater. Interfaces* **2017**, *4*, 1700217. [CrossRef]
41. Wang, K.; Feng, W.L.; Qin, X.; Deng, D.S.; Feng, X.; Zhang, C. Tungsten sulfide nanoflakes: Synthesis by electrospinning and their gas sensing properties. *Zeitschrift für Naturforschung A J. Phys. Sci.* **2017**, *72*, 375–381. [CrossRef]
42. Jha, R.K.; Guha, P.K. Liquid exfoliated pristine WS$_2$ nanosheets for ultrasensitive and highly stable chemiresistive humidity sensors. *Nanotechnology* **2016**, *27*, 475503. [CrossRef] [PubMed]
43. Kim, T.S.; Kim, Y.B.; Yoo, K.S.; Sung, G.S.; Jung, H.J. Sensing characteristics of dc reactive sputtered WO$_3$ thin films as an NOx gas sensor. *Sens. Actuators B Chem.* **2000**, *62*, 102–108. [CrossRef]
44. Chu, C.W.; Deen, M.J.; Hill, R.H. Sensors for detecting sub-ppm NO$_2$ using photochemically produced amorphous tungsten oxide. *J. Electrochem. Soc.* **1998**, *145*, 4219–4225. [CrossRef]
45. Staerz, A.; Berthold, C.; Russ, T.; Wicker, S.; Weimar, U.; Barsan, N. The oxidizing effect of humidity on WO$_3$ based sensors. *Sens. Actuators B Chem.* **2016**, *237*, 54–58. [CrossRef]
46. Ma, D.; Wang, Q.; Li, T.; He, C.; Ma, B.; Tang, Y.; Lu, Z.; Yang, Z. Repairing sulfur vacancies in the MoS$_2$ monolayer by using CO, NO and NO$_2$ molecules. *J. Mater. Chem. C* **2016**, *4*, 7093–7101. [CrossRef]
47. Kc, S.; Longo, R.C.; Wallace, R.M.; Cho, K. Surface oxidation energetics and kinetics on MoS$_2$ monolayer. *J. Appl. Phys.* **2015**, *117*, 135301. [CrossRef]
48. Barsan, N.; Weimar, U. Conduction model of metal oxide gas sensors. *J. Electroceramics* **2001**, *7*, 143–167. [CrossRef]
49. Li, H.; Huang, M.; Cao, G. Markedly different adsorption behaviors of gas molecules on defective monolayer MoS$_2$: A first-principles study. *Phys. Chem. Chem. Phys.* **2016**, *18*, 15110–15117. [CrossRef]
50. Yue, Q.; Shao, Z.; Chang, S.; Li, J. Adsorption of gas molecules on monolayer MoS$_2$ and effect of applied electric field. *Nanoscale Res. Lett.* **2013**, *8*, 425. [CrossRef]
51. Zhou, C.; Yang, W.; Zhu, H. Mechanism of charge transfer and its impacts on Fermi-level pinning for gas molecules adsorbed on monolayer WS$_2$. *J. Chem. Phys.* **2015**, *142*, 1–8. [CrossRef] [PubMed]

© 2019 by the authors. Licensee MDPI, Basel, Switzerland. This article is an open access article distributed under the terms and conditions of the Creative Commons Attribution (CC BY) license (http://creativecommons.org/licenses/by/4.0/).

Article

Enhancing the Relative Sensitivity of V^{5+}, V^{4+} and V^{3+} Based Luminescent Thermometer by the Optimization of the Stoichiometry of $Y_3Al_{5-x}Ga_xO_{12}$ Nanocrystals

Karolina Kniec *, Karolina Ledwa and Lukasz Marciniak *

Institute of Low Temperature and Structure Research, Polish Academy of Sciences, Okólna 2, 50-422 Wroclaw, Poland; k.ledwa@intibs.pl
* Correspondence: k.kniec@intibs.pl (K.K.); l.marciniak@intibs.pl (L.M.)

Received: 17 August 2019; Accepted: 24 September 2019; Published: 25 September 2019

Abstract: In this work the influence of the Ga^{3+} concentration on the luminescent properties and the abilities of the $Y_3Al_{5-x}Ga_xO_{12}$: V nanocrystals to noncontact temperature sensing were investigated. It was shown that the increase of the Ga^{3+} amount enables enhancement of V^{4+} emission intensity in respect to the V^{3+} and V^{5+} and thus modify the color of emission. The introduction of Ga^{3+} ions provides the appearance of the crystallographic sites, suitable for V^{4+} occupation. Consequently, the increase of V^{4+} amount facilitates $V^{5+} \rightarrow V^{4+}$ interionic energy transfer throughout the shortening of the distance between interacting ions. The opposite thermal dependence of V^{4+} and V^{5+} emission intensities enables to create the bandshape luminescent thermometr of the highest relative sensitivity of V-based luminescent thermometers reported up to date (S_{max}, 2.64%/°C, for $Y_3Al_2Ga_3O_{12}$ at 0 °C). An approach of tuning the performance of $Y_3Al_{5-x}Ga_xO_{12}$: V nanocrystals to luminescent temperature sensing, including the spectral response, maximal relative sensitivity and usable temperature range, by the Ga^{3+} doping was presented and discussed.

Keywords: vanadium; gallium; garnets; inorganic nanocrystals; luminescence; luminescent nanothermometry

---

1. Introduction

Inorganic nanocrystals, due to their high mechanical, thermal and chemical stability, have garnered an immense interest from the point of view of their potential implementation in biomedical application, i.e., optical and magnetic resonance imaging, drug delivery, light-induced hyperthermia generation etc. [1–4]. Their optical properties may be in a facile way modified by the introduction of the appropriate optically active ions like lanthanide (Ln^{3+}) and/or transition metals (TM) ions [5–13] to the host material. Besides unique chemical and physical features, they reveal size- and shape-dependent spectroscopic properties, which are not observed for organic-based nanomaterials [1]. Due to the fact that the optical properties of such nanoparticles are strongly affected by the temperature, their luminescence may be employed to non-contact temperature sensing (luminescent thermometry, LT). In LT, temperature readout relies on the analysis of thermally-affected spectroscopic parameters like emission intensity, luminescence lifetime, peak position, band shape and polarization anisotropy [14–16]. One of the most important advantages of LT in respect to other temperature measurement techniques is the fact that it provides a real-time temperature readout with unprecedented spatial and thermal resolution [15,17,18]. Additionally, temperature readout is provided in an electrically passive mode what enables to achieve the information about, i.e., the condition of living organisms where even small temperature fluctuations are usually accompanied by serious health diseases and improper cellular biochemical processes [16,19–22]. The use of the nanosized LTs enables the improvement of the spatial resolution of temperature readout. However, in order to obtain high thermal resolution of temperature measurement, different approaches, which enable to increase the relative sensitivity of LT to temperature changes, were proposed up to date. As was recently demonstrated, the utilization of transition metal ions luminescence with lanthanide co-dopant as a luminescent reference enables the enhancement of temperature sensing sensitivity, luminescence brightness and the broadening of usable temperature range in which LT operates [23–25]. For this purpose, optical properties of different TM were investigated, such as $V^{3+}/V^{4+}/V^{5+}$ [23,26], Co^{2+} [27], Ti^{3+}/Ti^{4+} [28], Cr^{3+} [24,25], Mn^{3+}/Mn^{4+} [29] and Ni^{2+} [30]. Another advantage of using TM is the susceptibility of their optical properties to the modification of the crystal field strength via host stoichiometry due to the fact that d electrons, located on the valence shell, are exposed to the local environment and crystal field changes. This phenomenon was investigated in detailed in case of temperature sensing performance of Cr^{3+} ions where the structure of host materials were varying from $Gd_3Al_5O_{12}$ (GAG) to $Gd_3Ga_5O_{12}$ (GGG), and from $Y_3Al_5O_{12}$ (YAG) to $Y_3Ga_5O_{12}$ (YGG) via changing the Al^{3+} to Ga^{3+} ratio [24,25]. As was recently shown for Cr^{3+} ions, such modification enables not only enhancement of the sensitivity of LT but also tuning of the spectral position of emission band [25]. These kinds of studies have not yet been conducted for V-based luminescent thermometers.

Therefore, in this work, we present for the first time a strategy that enables the improvement of temperature-sensing properties of V-based luminescent nanothermometers via modification of the host material composition. This approach bases on the gradual substitution of Al^{3+} ions by Ga^{3+} ions into YAG nanocrystals. The introduction of gallium ions, which possess larger ionic radii in respect to Al^{3+} ones leads to the lowering of crystal field (CF) strength. This arises from the elongation of the metal-oxygen (M-O) distance along with the enhancement of the contribution of Ga^{3+} ions. The modification of the crystal field strength should strongly influence the temperature-dependent luminescent properties of V ions of different oxidation state (V^{5+}, V^{4+}, V^{3+}). Moreover, the introduction of the gallium ions facilitates the stabilization of V^{4+} oxidation state that possesses favorable performance for luminescent thermometry. However, these expectations have not yet been experimentally verified. Therefore, the aim of this work is to study the influence of the Ga^{3+} ions concentration of the temperature dependent luminescent properties of vanadium ions in $Y_3Al_{5-x}Ga_xO_{12}$:V nanocrystals, with the special emphasize put on their application in luminescent thermometry.

2. Materials and Methods

2.1. Synthesis of V-doped $Y_3Al_{5-x}Ga_xO_{12}$

The $Y_3Al_{5-x}Ga_xO_{12}$ nanocrystals doped with 0.1% concentration of V ions were synthesized via a modified Pechini method, where the Ga^{3+} amount was set to x = 1, 2, 3, 4 and 5. The amount of V ions was set to 0.1% due to the fact that this V concentration provides the most significant temperature sensing properties of YAG:V, Ln^{3+} luminescent nanothermometers [23]. The first step was the creation of yttrium nitrate from yttrium oxide (Y_2O_3, 99.995% purity from Stanford Materials Corporation, Lake Forest, CA, USA) using the recrystallization process, including the dissolution in distillated water and ultrapure nitric acid (65%). All nitrates, namely appropriate amounts of $Ga(NO_3)_3 \cdot 9H_2O$ (Puratronic 99.999% purity from Alfa Aesar, Kandel, GERMANY), $Al(NO_3)_3 \cdot 9H_2O$ (Puratronic 99.999% purity from Alfa Aesar, Kandel, GERMANY) and $Y(NO_3)_3$ were dissolved in water and mixed together. After that, NH_4VO_3 (99% purity from Alfa Aesar, Kandel, GERMANY) were added to the solution. To enable the dissolution of ammonium metavanadate and the complexation of each metal, calculated quantity of citric acid (CA, $C_6H_8O_7$ with 99.5+% purity from Alfa Aesar, Kandel, GERMANY), used in six-fold excess in respect to the total amount of metal ions, was mixed with all reagents and heated up to 90 °C for 1 h. Next, PEG-200 (poly(ethylene glycol), from Alfa Aesar, Kandel, GERMANY) was added dropwise to the CA-metal complex and stirred for 2 h at 90 °C (CA: PEG-200 was 1:1) to conduct the polyestrification reaction. Then, the resin was obtained by heating at 90 °C for 1 week. In turn, the nanopowders were received via annealing of resin at 1100 °C for 16 h in air atmosphere.

2.2. Characterization

Powder X-ray diffraction (XRD) studies were carried out on PANalytical X'Pert Pro diffractometer equipped with Anton Paar TCU 1000 N Temperature Control Unit using Ni-filtered Cu $K\alpha$ radiation ($V = 40$ kV, $I = 30$ mA).

Transmission electron microscope images were taken using transmission electron microscopy (TEM) Philips CM-20 SuperTwin with 160 kV of accelerating voltage and 0.25 nm of optical resolution.

The hydrodynamic size of the nanoparticles was determined by dynamic light scattering (DLS), conducted in Malvern ZetaSizer at room temperature in polystyrene cuvette, using distilled water as a dispersant.

The emission spectra were measured using the 266 nm excitation line from a laser diode (LD) and a Silver-Nova Super Range TEC Spectrometer form Stellarnet (1 nm spectral resolution) as a detector. The temperature of the sample was controlled using a THMS600 heating stage from Linkam (0.1 1C temperature stability and 0.1 1C set point resolution).

Luminescence decay profiles were recorded using FLS980 Fluorescence Spectrometer from Edinburgh Instruments with µFlash lamp as an excitation source and R928P side window photomultiplier tube from Hamamatsu as a detector.

3. Results and Discussion

The yttrium aluminum/gallium garnets crystallize in a cubic structure of Ia3d space group. The general formula of garnets is expressed as follows: $A_3B_2C_3O_{12}$, where three different metallic sites are represented by dodecahedral site (A), octahedral site (B) and tetrahedral site (C), which in our case are occupied by eight-fold coordinated Y^{3+} ions, six-fold coordinated Al^{3+}/Ga^{3+} ions and four-fold coordinated Al^{3+}/Ga^{3+} ions, respectively. The optically active ions introduced to the structure may occupy different crystallographic sites, which results from the similarities in the coordination number, ionic radii and ionic charge between the host and dopant metal. Therefore, lanthanides (Ln^{3+}) prefer to replace A site, while (TM) mainly substitute B and C sites. Additionally, depending on the size of TM ion, they occupy larger (B) (ionic radii 0.67 Å for Al^{3+} and 0.76 Å for Ga^{3+}) or smaller (C) (0.53 Å for Al^{3+} and 0.61 Å for Ga^{3+}) metallic sites. An XRD analysis was used to verify the phase purity of synthesized materials. It is evident that the obtained diffraction peaks of V-doped $Y_3Al_{5-x}Ga_xO_{12}$ nanocrystals correspond to the reference patterns of cubic structures of adequate host materials (Figure 1a). Observed peaks broadening can be assigned to the small size of the nanoparticles. The a cell parameter increases linearly as the Ga^{3+}-dopant concentration increased, which results from the enlargement of the crystallographic cell associated with the difference in the ionic radii of Al^{3+} and Ga^{3+} ions ($rAl^{3+} < rGa^{3+}$) (Figure 1b). However, it was found that Ga^{3+} ions preferentially occupy four-fold coordinated sites of Al^{3+} rather than the octahedral counterpart. This phenomenon can be explained based on the stronger covalency of Ga^{3+}-O^{2-} bonds with respect to the Al^{3+}-O^{2-} ones and the lowering of repulsive forces between cations, providing stabilization of the crystal structure [31,32]. On the other hand, the slight shift of the XRD peaks with respect to the reference pattern arises from the implementation of V ions into $Y_3Al_{5-x}Ga_xO_{12}$ lattice. It was found that $Y_3Al_{5-x}Ga_xO_{12}$ matrix is a suitable host material for three different V oxidation states, namely V^{3+} and V^{5+} [23,26,33]. The replacement of Ga^{3+} and Al^{3+} ions by V ions is possible due to their comparable ionic radii, which in the case of four-fold coordinated V^{5+} and V^{3+} ions are 0.54 Å, 0.64 Å, respectively, and for six-fold coordinated V^{5+}, V^{4+} and V^{3+} ions are 0.68 Å, 0.72 Å and 0.78 Å, respectively. As can be seen from the TEM images, synthesized powders consist of well-crystallized and highly agglomerated nanocrystals (Figure 1c,e,g,i,k). The hydrodynamic sizes of the aggregates of $Y_3Al_{5-x}Ga_xO_{12}$ nanocrystals examined using DLS analysis were found to be around 300 nm (Figure 1d,f,h,j,l).

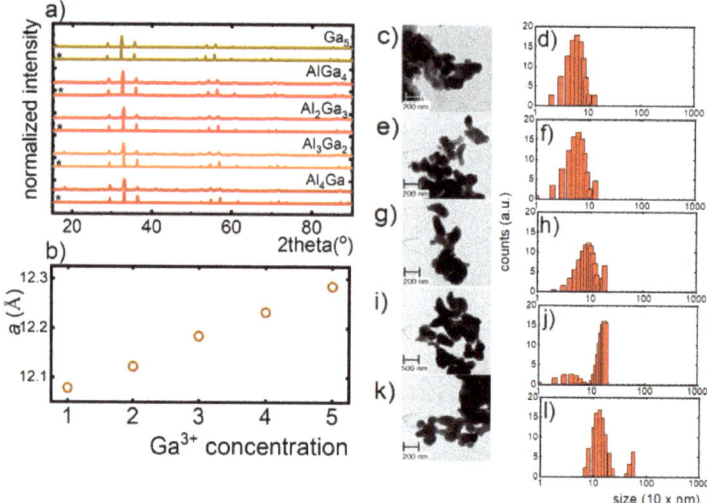

Figure 1. (a) XRD patterns of $Y_3Al_{5-x}Ga_xO_{12}$ nanocrystals, doped with 0.1% V; (b) influence of the Ga^{3+} concentration on the *a* cell parameter; (c,e,g,i,k): the morphology of $Y_3Al_4GaO_{12}$, $Y_3Al_3Ga_2O_{12}$, $Y_3Al_2Ga_3O_{12}$, $Y_3AlGa_4O_{12}$, $Y_3Ga_5O_{12}$, respectively; (d,f,h,j,l): the distribution of the hydrodynamic size of aggregates.

Luminescent properties of V- doped $Y_3Al_{5-x}Ga_xO_{12}$ nanocrystals were investigated upon 266 nm of excitation in the −150 °C to 300 °C (123.15 K to 573.15 K) temperature range (Figure 2a). The emission spectrum obtained at −150 °C consists of three transition bands, for materials with Ga^{3+} concentration from 1 to 4, and of two emission bands for YGG, being related to the presence of different V oxidations states - V^{5+}, V^{4+} and V^{3+}. In the course of our previous investigation, it was found that due to the difference in the ionic radii and the charge, V^{5+} ions preferentially occupy surface sites of Al^{3+}, while V^{3+} and V^{4+} are mainly located in the core part of the nanoparticles [26,33]. The first broad emission band at 520 nm is attributed to the charge transfer transition of V^{5+}($V^{4+} \to O^{2-}$). The second band at 640 nm originates from $^2E \to {}^2T_2$ radiative transition of V^{4+} ions, while the band at 820 nm is associated with $^1E_2 \to {}^3T_{1g}$ transition of V^{3+} ions. As can be seen, the addition of Ga^{3+} ions significantly affects the luminescent properties of $Y_3Al_{5-x}Ga_xO_{12}$:V nanocrystals (Figure 2b). The presented results stay in agreement with the observations obtained for the vanadium doped yttrium aluminum oxide and lanthanum gallium oxide nanoparticles [23,26]. The representative emission spectra measured at −150 °C indicate that the increase of Ga^{3+} concentration caused the enhancement of the V^{4+} emission intensity in respect to the V^{5+} and V^{3+} ones. This effect results from the large ionic radii of V^{4+}, which significantly exceeds Al^{3+} ones. Therefore, V^{4+} cannot efficiently replace Al^{3+} in the structure. However, when the concentration of Ga^{3+} ions gradually increases, the number of the crystallographic sites that can be occupied by V^{4+} rises up, leading to the enhancement of $^2E \to {}^2T_2$ emission intensity. Moreover, the Ga-doping induces the reduction of the distance between V^{4+} and V^{5+} ions facilitating the energy transfer between them, which contributed to the V^{4+} luminescent intensity increase. It is worth noticing that the emission of trivalent V dominates in the spectrum up to x = 4, while in the case of YGG V^{4+}, the emission band prevails. To quantify these changes the histogram presenting the contribution of the emission intensities (calculated as an integral emission intensity in appropriate spectral range) of particular oxidation state of vanadium ions to the overall emission intensity as a function of Ga^{3+} concentration is presented in Figure 2c. The observed enhancement of V^{4+} emission intensity with respect to the V^{5+} with an increase of Ga^{3+} concentration causes tuning of the emission color toward red emission (Figure 2d). However, for YGG:V, orange emission was found. As has been already proven, the V^{5+} ions are located mainly in the surface part of the nanocrystals [23]. Since the morphology and the size of the nanoparticle is independent on the Ga^{3+} concentration, the number of V^{5+} can be assumed to be constant. The confirmation of this hypothesis is the fact that its lifetime ($<\tau_{V5+}>$ = 6.4 ms) is independent on the host stoichiometry (Figure S1). On the other hand, the average lifetime of V^{3+} and V^{4+} shortens consequently from 7.6 ms to 7.0 ms and 1.2 ms to 0.5 ms, respectively, with Ga^{3+} concentration (x changed from 1 to 5).

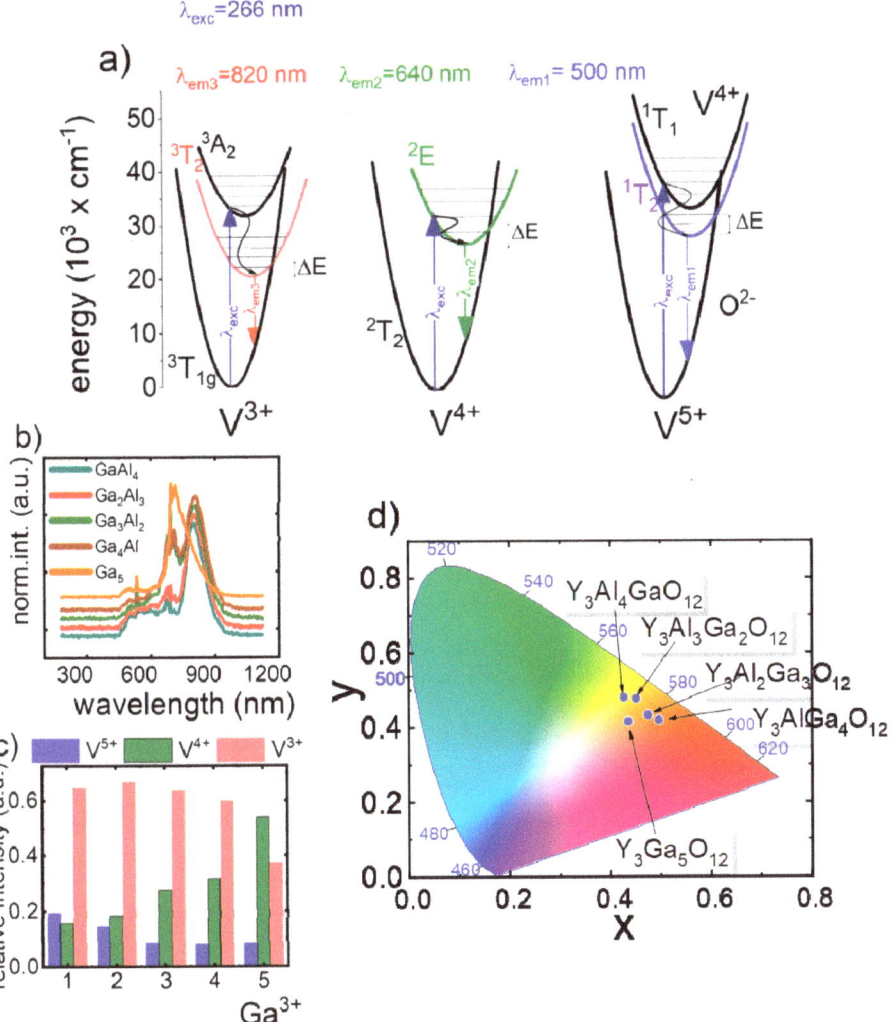

Figure 2. (a) The energy diagram of V ions at different oxidation states; (b) the influence of Ga-doping on the V emission spectrum (at −150 °C under 266 nm) in $Y_3Al_{5-x}Ga_xO_{12}$ nanomaterials at 0 °C; (c) the contribution of emission intensity of particular oxidation state of V ions into the overall emission spectrum of V-doped $Y_3Al_{5-x}Ga_xO_{12}$ nanocrystals; (d) the Commission internationale de l'éclairage CIE 1931 chromatic coordinates calculated for V:$Y_3Al_{5-x}Ga_xO_{12}$ nanocrystals at 0 °C.

In order to evaluate how the spectral changes of $Y_3Al_{5-x}Ga_xO_{12}$ nanocrystal, induced by the stoichiometry modification, affect the performance of analyzed nanoparticles for noncontact temperature sensing, their luminescence spectra were analyzed in a wide range of temperature (from $-150\,°C$ to $300\,°C$) (Figure 3a, Figure S2). In the course of these studies, it was found that emission intensity of each V ion is quenched by temperature; however, their luminescence thermal quenching rates differ (Figure 3b–d). In the case of V^{5+}, emission intensity is gradually quenched by almost two orders of magnitude with temperature. However, correlation between Ga^{3+} introduction and temperature of thermal quenching was not observed. This effect is understandable, since, as has been shown before, V^{5+} occupy mainly surface part of the nanoparticles. In turn, the emission intensity of V^{4+} initially decreases with temperature and above some critical temperature, it significantly increases as the temperature grows, which results from the efficient $V^{5+} \rightarrow V^{4+}$ energy transfer. It was found that the threshold temperature above which rise up of intensity was observed lowers with Ga^{3+} concentration (from around $10\,°C$ for $Y_3Al_4GaO_{12}$ to $-100\,°C$ for $Y_3AlGa_4O_{12}$ and YGG). Additionally the magnitude of the intensity increase growths with Ga^{3+} content. This phenomenon can be explained by the increase of the $V^{5+} \rightarrow V^{4+}$ energy transfer probability. Higher numbers of Ga^{3+} sites in the structures promote the stabilization of the V^{4+} ions, which, as a consequence, shortens the average distance between V^{5+} and V^{4+} facilitating interionic interactions. Due to the fact that energy of V^{5+} excited state is higher than that of V^{4+}, the energy transfer between them occurs with the assistance of the phonon. According to the Miyakava-Dexter theory, the probability of this process is strongly dependent on temperature, which is in agreement with our data [34]. It needs to be noted that although V^{5+} ions serve as a sensitizers for V^{4+}, there is no correlation between Ga^{3+} concentration and the V^{5+} luminescence thermal quenching. This comes from the fact that in the case of V^{5+} intensity the luminescence thermal quenching process plays dominant role over $V^{5+} \rightarrow V^{4+}$ energy transfer. The correlation between Ga^{3+} concentration and the luminescent thermal quenching rate is also evident in the case of V^{3+} ions. The higher the amount of Ga^{3+}, the lower the thermal quenching rate of the $^1E_2 \rightarrow {}^3T_{1g}$ emission band. Above $100\,°C$, the V^{4+} emission intensity becomes so efficient that its intensity dominates over the V^{3+} ones and thus hinders its emission intensity analysis. In the case of YGG, the V^{3+} emission is impossible to detect.

Since the emission intensity of V ions in $Y_3Al_{5-x}Ga_xO_{12}$ nanocrystals is strongly affected by the temperature changes, a quantitative analysis, which verify their performance for non-contact temperature sensing, was performed. For this purpose, the relative sensitivities (S) of three different intensity-based luminescent thermometers were calculated according to the following Equation (1):

$$S = \frac{1}{\Omega} \frac{\Delta \Omega}{\Delta T} \cdot 100\%, \qquad (1)$$

where Ω corresponds to the temperature dependent spectroscopic parameter, which in this case is represented by emission of adequate V ions (S_1 for V^{5+}, S_2 for V^{4+} and S_3 for V^{3+}), and $\Delta\Omega$ and ΔT indicate to the change of Ω and temperature, respectively.

The maximal values of relative sensitivity (S_1) of V^{5+}-based luminescent thermometer, which exceed $2\%/°C$, were found at temperatures below $-100\,°C$ and with increase of temperature S_1 gradually decreases reaching $1.34\%/°C$, $1.12\%/°C$, $1.13\%/°C$, $1.30\%/°C$ and $0.76\%/°C$ for $Y_3Al_4GaO_{12}$, $Y_3Al_3Ga_2O_{12}$, $Y_3Al_2Ga_3O_{12}$, $Y_3AlGa_4O_{12}$ and $Y_3Ga_5O_{12}$, respectively, in the biological temperature range ($0\,°C$–$50\,°C$). The highest value of the S_1 was found at $-150\,°C$ for $Y_3Ga_5O_{12}$, which is in agreement with our expectation that short distance between V^{5+} and V^{4+} facilitates the interionic energy transfer between them. The presented correlations confirm that relative sensitivity of temperature sensors based on V^{5+} emission intensity can be modulated by varying the Ga^{3+}-concentration (Figure 3e). In case of $Y_3Al_{5-x}Ga_xO_{12}$:V^{4+} temperature sensors, the highest value of sensitivity reveal the YGG nanocrystals ($S_{2max} = 1.34\%/°C$ at $-15\,°C$), and its value gradually decreases with the lowering of Ga^{3+} concentration. Moreover, the temperature at which maximal S_2 was found decreases with Ga^{3+} concentration from $75\,°C$ for $Y_3Al_4GaO_{12}$ to $-15\,°C$ for YGG. This phenomenon is also observed in the case of biological temperature range, where reducing the Ga^{3+} concentration the S value decreases from 1.32%/°C at 0 °C to 0.2%/°C at 30 °C for $Y_3Ga_5O_{12}$ to $Y_3Al_4GaO_{12}$ (Figure 3f). It should be mentioned here that usable temperature range for this luminescent thermometer (temperature range in which Ω reveals monotonic change) is limited, and the most narrow one was found for YGG (from −100 °C to 120 °C). The negative values of S_2 come from the fact of the intensity trend reversal. Hence, the balance between relative sensitivity and the usable temperature range can be optimized by the appropriate host material composition. Therefore, depending on the type of application of such luminescent thermometer, including required relative sensitivity and operating temperatures range, different stoichiometry of host material can be proposed. Since the V^{3+} emission intensity monotonically decreases in the temperature range below 200 °C the relative sensitivity S_3 reveals positive values with the single maxima at temperature which is dependent on the Ga^{3+} concentration (Figure 3g). The increase of Ga^{3+} amount causes the reduction of both value of the S_3 and the temperature of S max from 1.08%/°C at 152 °C for $Y_3Al_4GaO_{12}$ to 0.45%/°C at 51 °C for $Y_3AlGa_4O_{12}$.

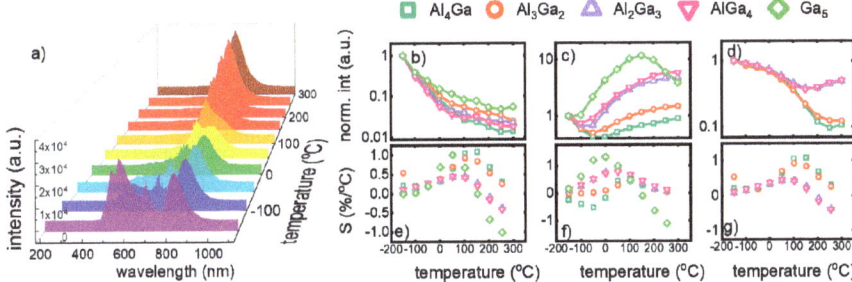

Figure 3. (a) Thermal evolution of emission spectrum of $Y_3AlGa_4O_{12}$:V nanocrystals; (b–d) the influence of local temperature on the emission intensity of V^{5+}, V^{4+} and V^{3+}, respectively; (e–g) corresponding relative sensitivities.

Although the performance of the intensity-based luminescent thermometer, which take advantage from V^{5+}, V^{4+} and V^{3+} emission, are very promising, the reliability of accurate temperature readout is limited due to the fact that emission intensity of a single band may be affected by the number of experimental and physical parameters. Therefore, most of the studies concern the bandshape luminescent thermometer, for which relative emission intensity of two bands is used for temperature sensing. Taking advantage of the fact that emission intensities of V^{5+} and V^{4+} ions reveal opposite temperature dependence, their luminescence intensity ratio (LIR) can be used as a sensitive thermometric parameter:

$$LIR = \frac{V^{5+}(V^{4+} \rightarrow O^{2-})}{V^{4+}(^2E \rightarrow {}_2T_2)}, \qquad (2)$$

Analysis of the thermal evolution of LIR reveals that for each stoichiometry of the host material the decrease of LIR's value by over three orders of magnitude can be found for −150–300 °C temperature range (Figure 4a). Observed thermal changes of LIR significantly exceed those noticed for single ion emission. The relative sensitivities of LIR-based luminescent thermometers (S_4) were defined as follows:

$$S_4 = \frac{1}{LIR} \cdot \frac{\Delta LIR}{\Delta T} \cdot 100\%, \qquad (3)$$

Thereby, the relative sensitivities calculated for LIR-based luminescent thermometers reached values that exceed 2%/°C (Figure 4b). Thermal evolution of S_4 attains single maxima at temperature T_{Smax}. As was shown before, both the S_{4max} and T_{Smax} can be successfully modified by the incorporation of the Ga^{3+} ions. The increase of the Ga^{3+} concentration causes the lowering of the T_{Smax} from 20 °C for $Y_3Al_4GaO_{12}$ to −100 °C for $Y_3Ga_5O_{12}$, while the maximal relative sensitivity increases from 1.47%/°C for $Y_3Al_4GaO_{12}$ to 2.48%/°C for $Y_3Ga_5O_{12}$ (Figure 4c,d). However, the maximal value of S_4 = 2.64%/°C was found for $Y_3Al_2Ga_3O_{12}$. It needs to be mentioned here that, to the best of our knowledge, described nanocrystals reveal the highest values of relative sensitivity for vanadium-based luminescent thermometers up to date. Moreover, it was found that the higher the Ga^{3+} content ($Y_3Al_4GaO_{12}$–$Y_3AlGa_4O_{12}$), the more significant the change of CIE 1931 chromatic coordinates is (Figure 4e,f).

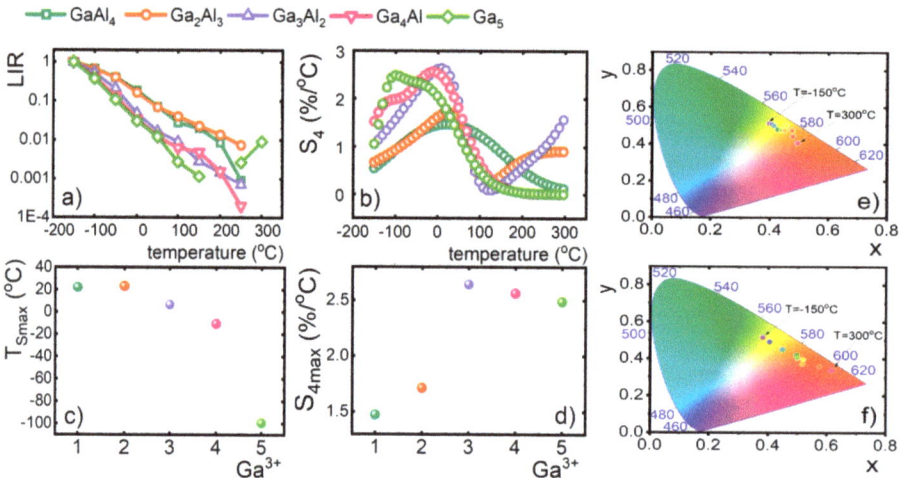

Figure 4. (a) Thermal evolution of luminescence intensity ratio (LIR); (b) their relative sensitivities for $Y_3Al_{5-x}Ga_xO_{12}$ nanocrystals; (c) the temperature at which the maximal value of S_4 was observed; (d) S_{4max} as a function of Ga^{3+} concentration; (e,f) the CIE 1931 chromatic coordinates calculated for $Y_3Al_4GaO_{12}$:V and $Y_3AlGa_4O_{12}$:V nanocrystals, respectively.

4. Conclusions

In this work, the impact of the host material composition on the temperature-dependent luminescent properties of vanadium-doped nanocrystalline garnets was investigated. It was demonstrated that the incorporation of Ga^{3+} ions into the $Y_3Al_{5-x}Ga_xO_{12}$:V structure enables modification of the emission color of the phosphor by the stabilization of the vanadium ions on the V^{4+} oxidation state. Taking advantage from the fact that V^{4+} ions, due to their similar ionic radii, mainly occupy the octahedral site of Ga^{3+} ions, the enlargement of their amount leads to the increase of their emission intensity. Moreover, a growing number of V^{4+} ions cause a shortening of the average V^{5+}–V^{4+} distance facilitating interionic energy transfer between them. Conducted studies regarding the influence of temperature on the emission intensities of the vanadium ions at different oxidation states reveal that the most susceptible to thermal quenching is the V^{5+} emission intensity. On the other hand, due to the $V^{5+} \to V^{4+}$ energy transfer, the V^{4+} emission intensity increases with temperature. The higher the amount of Ga^{3+} ions in the host, the more evident the enhancement of V^{4+} emission intensity and the lower the threshold temperature above which this enhancement occurs. Taking advantage form the fact of opposite temperature dependence of V^{5+} and V^{4+} emission intensities, their ratio was used to create the bandshape luminescent thermometer, to the best of our knowledge, the highest relative sensitivity of V-based luminescent thermometers up to date S_{max}, 2.64%/°C, 2.56%/°C and 2.49%/°C for $Y_3Al_2Ga_3O_{12}$ (at 0 °C), $Y_3AlGa_4O_{12}$ (at −20 °C) and $Y_3Ga_5O_{12}$ (at −100 °C), respectively. With an increase of the Ga^{3+} concentration, the value of the relative sensitivity, as well as the temperature at which S_{max} was observed, can be modified. Additionally, it was found that the higher the contamination of Ga^{3+} ions, the more evident the change of the chromatic coordinates of emitted light with temperature changes in a −150 °C–300 °C temperature range. As was proven in this manuscript, the introduction of the Ga^{3+} ions in the garnet host enables modification of the performance of nanocrystalline luminescent thermometer like: its usable temperature range, maximal value of the relative sensitivity, as well as the temperature at which maximal sensitivity can be obtained. The dominant effect, which is responsible for described modification of the luminescent properties of V doped luminescent thermometers, is the increase of the $V^{5+} \rightarrow V^{4+}$ energy transfer probability associated with the growing number of the crystallographic sites that can be occupied by the V^{4+} ions. This shortens the average distance between the interaction ions, facilitating energy transfer process.

Supplementary Materials: The following are available online at http://www.mdpi.com/2079-4991/9/10/1375/s1, Figure S1: (a), (b), (c) The luminescence decay profile of V^{5+}, V^{4+} and V^{3+} ions for different Ga^{3+}, respectively; Figure S2: Emission spectra of V-doped nanocrystals recorded in the range of −150 °C–300 °C.

Author Contributions: Formal analysis, K.K. and L.M.; Investigation, K.K. and L.M.; Methodology, K.K. and K.L.; Writing—original draft, K.K. and L.M.; Writing—review and editing, L.M.

Funding: The "High sensitive thermal imaging for biomedical and microelectronic application" project is carried out within the First Team programme of the Foundation for Polish Science, co-financed by the European Union under the European Regional Development Fund.

Conflicts of Interest: The authors declare no conflict of interest.

References

1. Berry, C.C. *Applications of Inorganic Nanoparticles for Biotechnology*, 1st ed.; Elsevier: Amsterdam, The Netherlands, 2012; Volume 4, ISBN 9780124157699.
2. Ali, A.; Zafar, H.; Zia, M.; Phull, A.R.; Ali, J.S. Synthesis, characterization, applications, and challenges of iron oxide nanoparticles. *Nanotechnol. Sci. Appl.* **2016**, *9*, 49–67. [CrossRef] [PubMed]
3. Press, D. Nanoparticles in relation to peptide and protein aggregation. *Int. J. Nanomed.* **2014**, *9*, 899–912.
4. Holzinger, M.; Le Goff, A.; Cosnier, S. Nanomaterials for biosensing applications: A review. *Front. Chem.* **2014**, *2*, 1–10. [CrossRef] [PubMed]
5. Wang, F.; Liu, X. 1.18 Rare-Earth Doped Upconversion Nanophosphors. In *Comprehensive Nanoscience and Technology*; Elsevier: Amsterdam, The Netherlands, 2011; pp. 607–635.
6. Smet, P.F.; Moreels, I.; Hens, Z.; Poelman, D. Luminescence in sulfides: A rich history and a bright future. *Materials* **2010**, *3*, 2834–2883. [CrossRef]
7. Cornejo, C.R. Luminescence in Rare Earth Ion-Doped Oxide Compounds. In *Luminescence—An Outlook on the Phenomena and their Applications*; IntechOpen: London, UK, 2016; pp. 33–63.
8. Pott, G.T.; McNicol, B.D. The phosphorescence of Fe^{3+} ions in oxide host lattices. Zero-phonon transitions in $Fe^{3+}/LiAl_5O_8$. *Chem. Phys. Lett.* **1971**, *12*, 62–64. [CrossRef]
9. Lakshminarasimhan, N.; Varadaraju, U.V. Luminescent host lattices, $LaInO_3$ and $LaGaO_3$—A reinvestigation of luminescence of d^{10} metal ions. *Mater. Res. Bull.* **2006**, *41*, 724–731. [CrossRef]
10. Yamamoto, H.; Okamoto, S.; Kobayashi, H. Luminescence of rare-earth ions in perovskite-type oxides: From basic research to applications. *J. Lumin.* **2002**, *100*, 325–332. [CrossRef]
11. Denisov, A.L.; Ostroumov, V.G.; Saidov, Z.S.; Smirnov, V.A.; Shcherbakov, I.A. Spectral and luminescence properties of Cr^{3+} and Nd^{3+} ions in gallium garnet crystals. *J. Opt. Soc. Am. B* **1986**, *3*, 95–101. [CrossRef]
12. Brik, M.G.; Papan, J.; Jovanović, D.J.; Dramićanin, M.D. Luminescence of Cr^{3+} ions in $ZnAl_2O_4$ and $MgAl_2O_4$ spinels: Correlation between experimental spectroscopic studies and crystal field calculations. *J. Lumin.* **2016**, *177*, 145–151. [CrossRef]
13. Tratsiak, Y.; Trusova, E.; Bokshits, Y.; Korjik, M.; Vaitkevičius, A.; Tamulaitis, G. Garnet-type crystallites, their isomorphism and luminescence properties in glass ceramics. *CrystEngComm* **2019**, *21*, 687–693. [CrossRef]
14. Del Rosal, B.; Ruiz, D.; Chaves-Coira, I.; Juárez, B.H.; Monge, L.; Hong, G.; Fernández, N.; Jaque, D. In Vivo Contactless Brain Nanothermometry. *Adv. Funct. Mater.* **2018**, *28*, 1–7. [CrossRef]
15. Jaque, D.; Vetrone, F. Luminescence nanothermometry. *Nanoscale* **2012**, *4*, 4301–4326. [CrossRef] [PubMed]
16. Lozano-Gorrín, A.D.; Rodríguez-Mendoza, U.R.; Venkatramu, V.; Monteseguro, V.; Hernández-Rodríguez, M.A.; Martín, I.R.; Lavín, V. Lanthanide-doped $Y_3Ga_5O_{12}$ garnets for nanoheating and nanothermometry in the first biological window. *Opt. Mater.* **2018**, *84*, 46–51. [CrossRef]
17. Del Rosal, B.; Ximendes, E.; Rocha, U.; Jaque, D. In Vivo Luminescence Nanothermometry: From Materials to Applications. *Adv. Opt. Mater.* **2017**, *5*, 1600508. [CrossRef]

18. Vetrone, F.; Naccache, R.; Zamarrón, A.; De La Fuente, A.J.; Sanz-Rodríguez, F.; Maestro, L.M.; Rodriguez, E.M.; Jaque, D.; Sole, J.G.; Capobianco, J.A. Temperature sensing using fluorescent nanothermometers. *ACS Nano* **2010**, *4*, 3254–3258. [CrossRef] [PubMed]
19. Marciniak, L.; Bednarkiewicz, A. The influence of dopant concentration on temperature dependent emission spectra in $LiLa_{1-x-y}Eu_xTb_yP_4O_{12}$ nanocrystals: Toward rational design of highly-sensitive luminescent nanothermometers. *Phys. Chem. Chem. Phys.* **2016**, *18*, 15584–15592. [CrossRef]
20. Jaque, D.; Martínez Maestro, L.; del Rosal, B.; Haro-Gonzalez, P.; Benayas, A.; Plaza, J.L.; Martín Rodríguez, E.; García Solé, J. Nanoparticles for photothermal therapies. *Nanoscale* **2014**, *6*, 9494–9530. [CrossRef]
21. Jaque, D.; Jacinto, C. Luminescent nanoprobes for thermal bio-sensing: Towards controlled photo-thermal therapies. *J. Lumin.* **2016**, *169*, 394–399. [CrossRef]
22. Jaque, D.; Del Rosal, B.; Rodríguez, E.M.; Maestro, L.M.; Haro-González, P.; Solé, J.G. Fluorescent nanothermometers for intracellular thermal sensing. *Nanomedicine* **2014**, *9*, 1047–1062. [CrossRef]
23. Kniec, K.; Marciniak, L. The influence of grain size and vanadium concentration on the spectroscopic properties of $YAG:V^{3+},V^{5+}$ and $YAG:V,Ln^{3+}$ (Ln^{3+} = Eu^{3+}, Dy^{3+}, Nd^{3+}) nanocrystalline luminescent thermometers. *Sens. Actuators B Chem.* **2018**, *264*, 382–390. [CrossRef]
24. Elzbieciak, K.; Bednarkiewicz, A.; Marciniak, L. Temperature sensitivity modulation through crystal field engineering in Ga^{3+} co-doped $Gd_3Al_{5-x}Ga_xO_{12}:Cr^{3+}$, Nd^{3+} nanothermometers. *Sens. Actuators B Chem.* **2018**, *269*, 96–102. [CrossRef]
25. Elzbieciak, K.; Marciniak, L. The Impact of Cr^{3+} Doping on Temperature Sensitivity Modulation in Cr^{3+} Doped and Cr^{3+}, Nd^{3+} Co-doped $Y_3Al_5O_{12}$, $Y_3Al_2Ga_3O_{12}$, and $Y_3Ga_5O_{12}$ Nanothermometers. *Front. Chem.* **2018**, *6*, 424-1–424-8. [CrossRef] [PubMed]
26. Kniec, K.; Marciniak, L. Spectroscopic properties of $LaGaO_3:V,Nd^{3+}$ nanocrystals as a potential luminescent thermometer. *Phys. Chem. Chem. Phys.* **2018**, *20*, 21598–21606. [CrossRef] [PubMed]
27. Kobylinska, A.; Kniec, K.; Maciejewska, K.; Marciniak, L. The influence of dopant concentration and grain size on the ability for temperature sensing using nanocrystalline $MgAl_2O_4:Co^{2+},Nd^{3+}$ luminescent thermometers. *New J. Chem.* **2019**, *43*, 6080–6086. [CrossRef]
28. Drabik, J.; Cichy, B.; Marciniak, L. New Type of Nanocrystalline Luminescent Thermometers Based on Ti^{3+}/Ti^{4+} and Ti^{4+}/Ln^{3+} (Ln^{3+} = Nd^{3+}, Eu^{3+}, Dy^{3+}) Luminescence Intensity Ratio. *J. Phys. Chem. C* **2018**, *122*, 14928–14936. [CrossRef]
29. Trejgis, K.; Marciniak, L. The influence of manganese concentration on the sensitivity of bandshape and lifetime luminescent thermometers based on $Y_3Al_5O_{12}:Mn^{3+},Mn^{4+},Nd^{3+}$ nanocrystals. *Phys. Chem. Chem. Phys.* **2018**, *20*, 9574–9581. [CrossRef] [PubMed]
30. Matuszewska, C.; Elzbieciak-Piecka, K.; Marciniak, L. Transition Metal Ion-Based Nanocrystalline Luminescent Thermometry in $SrTiO_3:Ni^{2+},Er^{3+}$ Nanocrystals Operating in the Second Optical Window of Biological Tissues. *J. Phys. Chem. C* **2019**, *123*, 18646–18653. [CrossRef]
31. Nakatsuka, A.; Yoshiasa, A.; Yamanaka, T. Cation distribution and crystal chemistry of $Y_3Al_{5-x}Ga_xO_{12}$ ($0 <= x <= 5$) garnet solid solutions. *Acta Cryst.* **1999**, *12*, 266–272. [CrossRef] [PubMed]
32. Yousif, A.; Kumar, V.; Ahmed, H.A.A.S.; Som, S.; Noto, L.L.; Ntwaeaborwa, O.M.; Swart, H.C. Effect of Ga^{3+} Doping on the Photoluminescence Properties of $Y_3Al_{5-x}Ga_xO_{12}:Bi^{3+}$ Phosphor. *ECS J. Solid State SC.* **2014**, *3*, 222–227. [CrossRef]
33. Kniec, K.; Marciniak, L. Different Strategies of Stabilization of Vanadium Oxidation States in $LaGaO_3$ Nanocrystals. *Front. Chem.* **2019**, *7*, 1–8. [CrossRef] [PubMed]
34. Miyakawa, T.; Dexter, D.L. Phonon sidebands, multiphonon relaxation of excited states, and phonon-assisted energy transfer between ions in solids. *Phys. Rev. B* **1970**, *1*, 2961–2969. [CrossRef]

© 2019 by the authors. Licensee MDPI, Basel, Switzerland. This article is an open access article distributed under the terms and conditions of the Creative Commons Attribution (CC BY) license (http://creativecommons.org/licenses/by/4.0/).

Article

Polyfluorene-Based Multicolor Fluorescent Nanoparticles Activated by Temperature for Bioimaging and Drug Delivery

Marta Rubio-Camacho, Yolanda Alacid, Ricardo Mallavia, María José Martínez-Tomé * and C. Reyes Mateo *

Instituto de Investigación Desarrollo e Innovación en Biotecnología Sanitaria de Elche (IDiBE), Universidad Miguel Hernández de Elche (UMH), 03202 Elche, Alicante, Spain; marta.rubioc@umh.es (M.R.-C.); yoli2395@gmail.com (Y.A.); r.mallavia@umh.es (R.M.)
* Correspondence: mj.martinez@umh.es (M.J.M.-T.); rmateo@umh.es (C.R.M.);
  Tel.: +34-966-652-475 (M.J.M.-T.); +34-966-658-469 (C.R.M.)

Received: 27 September 2019; Accepted: 15 October 2019; Published: 18 October 2019

Abstract: Multifunctional nanoparticles have been attracting growing attention in recent years because of their capability to integrate materials with different features in one entity, which leads them to be considered as the next generation of nanomedicine. In this work, we have taken advantage of the interesting properties of conjugated polyelectrolytes to develop multicolor fluorescent nanoparticles with integrating imaging and therapeutic functionalities. With this end, thermosensitive liposomes were coated with three recently synthesized polyfluorenes: copoly-((9,9-bis(6′-*N,N,N*-trimethylammonium)hexyl)-2,7-(fluorene)-alt-1,4-(phenylene)) bromide (HTMA-PFP), copoly-((9,9-bis(6′-*N,N,N*-trimethylammonium)hexyl)-2,7-(fluorene)-alt-4,7-(2-(phenyl)benzo(d) (1,2,3) triazole)) bromide (HTMA-PFBT) and copoly-((9,9-bis(6′-*N,N,N*-trimethylammonium)hexyl)-2,7-(fluorene)-alt-1,4-(naphtho(2,3c)-1,2,5-thiadiazole)) bromide (HTMA-PFNT), in order to obtain blue, green and red fluorescent drug carriers, respectively. The stability, size and morphology of the nanoparticles, as well as their thermotropic behavior and photophysical properties, have been characterized by Dynamic Light Scattering (DLS), Zeta Potential, transmission electron microscope (TEM) analysis and fluorescence spectroscopy. In addition, the suitability of the nanostructures to carry and release their contents when triggered by hyperthermia has been explored by using carboxyfluorescein as a hydrophilic drug model. Finally, preliminary experiments with mammalian cells demonstrate the capability of the nanoparticles to mark and visualize cells with different colors, evidencing their potential use for imaging and therapeutic applications.

Keywords: multifunctional fluorescent nanoparticles; conjugated polyelectrolytes (CPEs); thermosensitive liposomes (TSLs); bioimaging; drug carrier; release experiments

---

1. Introduction

Nanomedicine is an emergent area which results from the application of nanotechnology to medicine. Research in this field has experienced rapid growth during the last decade, extending its applications in bioimaging, disease treatment and diagnosis. A number of nanoformulations for diagnostics and therapeutics have been approved for use in humans, and even more are currently under investigation [1–3]. An important trend in this field is the development of multifunctional nanoplatforms integrating different properties, such as imaging and therapeutic functionalities, in one entity [4,5]. In this regard, a large variety of biocompatible materials including lipids, proteins, carbon and quantum dots, synthetic polymers, fluorophores, dendrimers or metallic nanoparticles have been coupled and organized in nanostructures forming vesicles, micelles, nanorods, dendrimers and more, to be used as nanomedical vehicles in theragnostic applications [6–8].

The main component of these nanostructures is the carrier, which is responsible for transporting the drug and releasing it. Since their discovery in the mid-1960's, liposomes, which are lipid vesicles formed by one or more concentric lipid bilayers surrounding an aqueous core, have been considered to be the most successful nanocarriers for drug delivery, especially in anticancer chemotherapy [9,10]. This success is mainly due to the numerous advantages they offer. On one hand, liposomes are highly versatile, allowing the incorporation of hydrophobic drugs in the lipid bilayer or hydrophilic drugs in the aqueous core, as well as surface modifications in order to control their interactions with biological targets. On the other hand, the lipid membrane can behave as a barrier contributing to protect the encapsulated drug from degradation, thus prolonging its half-life in the bloodstream. In addition, due to the enhanced permeability and retention (EPR) phenomenon displayed by tumor tissues in comparison with normal tissues, liposomes can selectively accumulate in the tumor, enhancing efficacy and minimizing adverse side effects [11–13].

The release of drugs from the liposomal formulation is usually very low; nevertheless, the fact that the physical properties of the lipid membrane respond to a wide range of internal and external stimuli (temperature, pH, light, pressure, ions, magnetic field, etc.) can be used to selectively release the drug in a controlled way [14]. One well-established approach for triggering the encapsulated drug release, benefits from the differences in the permeability of the lipid membrane between the gel and fluid phase. In the gel phase, lipids are closely packed, with the acyl chains extended, and there is little lateral diffusion and low permeability. On the contrary, in the fluid phase, the acyl chains are more kinked, packing is lost, and the permeability is relatively high. The gel–fluid phase transition occurs cooperatively at the transition temperature (T_m), so it is possible to increase the permeability of the lipid bilayer by increasing the temperature beyond T_m, allowing the release of entrapped drugs [15,16]. The permeability is additionally increased just at this temperature, as a consequence of the coexistence of membrane areas in both phases [17].

Liposomes which release their contents at specific temperatures are called thermosensitive liposomes (TSLs) [18,19]. Among the different TSLs reported in the literature, the most clinically preferable are those having their T_m between 39–43 °C (mild hyperthermia), because these temperatures improve drug uptake, increase tumor perfusion and rend cancer cells temporarily sensitive to other treatments [20,21]. Typical TSLs have been mostly prepared from 1,2-dipalmitoyl-sn-glycero-3-phospho-rac-(1-glycerol) sodium salt (DPPG) and 1,2-dipalmitoyl-sn-glycero-3-phosphocholine (DPPC) derivatives, alone or combined with other lipids or polymers, due to its drug encapsulation capacity, excellent biodegradability and because their T_m occurs at 41–42 °C [22,23]. In the hyperthermia treatment, the tumor is locally heated to 40–43 °C for a defined period of time. The drug circulating in the bloodstream is safely entrapped in the TSLs because the liposomal membrane is in the gel phase, but once it reaches the heated tumor, the phase transition takes place and the drug is released, generally by passive transfer across the membrane according to a concentration gradient [24].

Incorporation of reporter groups—such as fluorescent components—in drug delivery carrier systems is of great interest in the fabrication of multifunctional nanoplatforms since they can act as probes for bioimaging and labelling (diagnosis) while monitoring the pathway concerning the drug, providing information from its final location [25,26]. During the last decades, these fluorescent components have been usually small organic fluorophores, fluorescent proteins and inorganic quantum dots (QDs) [27]. In contrast to organic dyes and fluorescent proteins, QDs do not suffer from photobleaching, self-quenching or chemical degradation and show unique properties such as broad absorption bands, high molar extinction coefficients, narrow emission peaks and large Stoke shifts. In addition, QDs made up of the same material have distinct emission wavelengths, depending on their size, shifting to the red as the size of QDs increases. These properties allow to simultaneously excite mixed QDs populations with different emission wavelengths at a single wavelength, facilitating multicolor fluorescence imaging [28]. Nonetheless, the potential cytotoxicity risk associated with the chemical composition of QDs (cadmium, selenium, tellurium, etc.) remains the major limitation for using these fluorescent nanoparticles in biological approaches [29–31]. To overcome this concern, it is necessary to develop safe and more efficient fluorescent carriers which, in addition to encapsulating the drug, incorporate new fluorescent materials with improved characteristics, such as nontoxicity, stability, high sensitivity, etc.

In comparison with common organic dyes, fluorescent proteins and QDs, conjugated polyelectrolytes (CPEs) have unique physiochemical properties. These materials are polymers with highly electron-delocalized backbones, containing ionic side groups which facilitate their water solubilization. CPEs have the optoelectronic properties from their neutral counterpart conjugated polymers (CPs), showing high-fluorescent quantum yields, broad absorption and emission spectra, large Stokes shift, good photostability, and more efficient intramolecular/intermolecular energy transfer than common organic dyes and fluorescent proteins, while having the advantage of being more biocompatible than QDs [32]. Contrary to CPs, CPEs display the typical physicochemical behavior of polyelectrolytes in aqueous solvents, allowing the coupling with different biological systems via electrostatic interactions [33]. In addition, CPEs have easily tunable side chains for bio-conjugation with several recognition elements and show high versatility in their synthesis, enabling fine-tuning of their absorption and emission bands through backbone modification [34]. Given these properties, CPEs have been successfully applied for detection of a wide range of biological and chemical molecules, but also as novel fluorescent probes for bioimaging, and exhibit enormous potential for therapeutic applications and/or diagnostics [35–38].

Fluorene-based conjugated polyelectrolytes, being fluorescent, nontoxic and photostable, provide excellent thermal and chemical stability. Furthermore, fluorene-based CPEs have good synthetic accessibility at the C9 position of the fluorene ring. Polyfluorenes usually emit in the blue spectral region, but copolymerization with other aromatic units allows for shifting the emission spectrum to longer wavelengths [39–41]. On this matter, we have synthesized three cationic polyfluorenes which emit in the blue, green and red regions of the visible spectrum: copoly-((9,9-bis(6'-N,N,N-trimethylammonium)hexyl)-2,7-(fluorene)-alt-1,4-(phenylene)) bromide (HTMA-PFP), which incorporate a phenyl group on fluorene backbone, copoly-((9,9-bis(6'-N,N,N-trimethylammonium)hexyl)-2,7-(fluorene)-alt- 4,7-(2-(phenyl) benzo(d) (1,2,3) triazole)) bromide (HTMA-PFBT), which incorporates the chromophore 2-phenylbenzotriazole on the backbone and copoly-((9,9-bis(6'- N,N,N-trimethylammonium)hexyl)-2,7-(fluorene)-alt-1,4-(naphtho(2,3c)- 1,2,5-thiadiazole)) bromide (HTMA-PFNT), which incorporates a naphtha(2,3c) (1,2,5)thiadiazole group on fluorene backbone [42–44] (Scheme 1, upper part). The three CPEs have been extensively characterized, showing interesting properties as fluorescent membrane markers for bioimaging studies [42–46]. In addition, they have common absorption bands around 330–350 nm (Scheme 1, bottom part), therefore, they could be excited simultaneously upon UV excitation, allowing multicolor fluorescence imaging. Finally, because of their cationic charge, the synthesized CPEs have high affinity to anionic species, forming complexes with certain proteins and lipid vesicles [33,43].

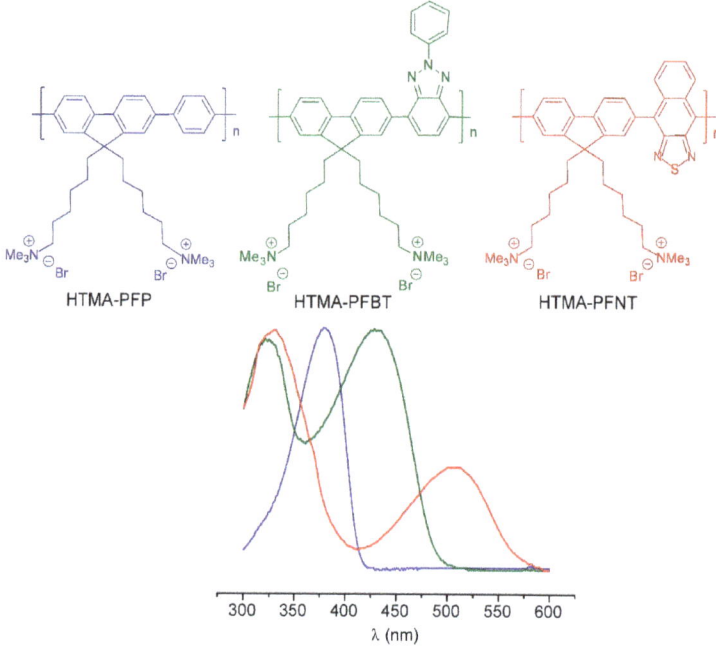

Scheme 1. Chemical structures (**upper part**) and normalized absorption spectra (**bottom part**) of conjugated polyelectrolytes (CPEs) copoly-((9,9-bis(6′-N,N,N-trimethylammonium)hexyl)-2,7-(fluorene)-alt-1,4-(phenylene)) bromide (HTMA-PFP) (in blue), copoly-((9,9-bis(6′-N,N,N-trimethylammonium)hexyl)-2,7-(fluorene)-alt-4,7-(2-(phenyl)benzo(d)(1,2,3) triazole)) bromide (HTMA-PFBT) (in green) and copoly-((9,9-bis(6′-N,N,N-trimethylammonium)hexyl)-2,7-(fluorene)-alt-1,4-(naphtho(2,3c)-1,2,5-thiadiazole)) bromide (HTMA-PFNT) (in red).

In this work, we have taken advantage of the interesting properties of the fluorescent CPEs to develop a multifunctional nanoplatform able to integrate imaging and therapeutic functionalities in one entity. With this end, TSLs composed of the anionic lipid DPPG have been prepared and coated with HTMA-PFP, HTMA-PFBT and HTMA-PFNT, in order to obtain blue, green and red fluorescent drug carriers, respectively. The stability, size and morphology of the nanoparticles have been characterized, as well as their photophysical properties and thermotropic behavior. In addition, the suitability of the nanoparticles as carrier systems to release a drug in response to external mild hyperthermia has been explored using carboxyfluorescein (CF) as a model hydrophilic drug. Finally, preliminary experiments have been carried out to evaluate the capacity of the nanostructures to mark and visualize mammalian cells in different colors.

2. Materials and Methods

2.1. Materials

The synthetic phospholipid 1,2-dipalmitoyl-sn-glycero-3-phospho-rac-(1-glycerol) sodium salt (DPPG) was purchased from Sigma-Aldrich (St. Louis, MO, USA). The polyfluorenes HTMA-PFP (M_n (g·mol^{-1}) = 4170; M_w (g·mol^{-1}) = 8340), HTMA-PFBT (M_n (g·mol^{-1}) = 4584; M_w (g·mol^{-1}) = 8531) and HTMA-PFNT (M_n (g·mol^{-1}) = 4507; M_w (g·mol^{-1}) = 8990) were synthesized and subsequently characterized in our laboratory [43,44,47]. Stock solutions of the polyfluorenes were dissolved in dimethyl sulfoxide (DMSO) with a final concentration of 3.65 × 10^{-4} M for HTMA-PFP and HTMA-PFNT, and 6.24 × 10^{-4} M for HTMA-PFBT (in repeat units), and stored at −20 °C before use. The dye 5(6)-carboxyfluorescein (CF) was purchased from Sigma-Aldrich (St. Louis, MO, USA), as well as the fluorescent membrane probe 1,6-diphenyl-1,3,5-hexatriene (DPH), and the quencher 9,10-anthraquinone-2,6-disulfonic acid (AQS). Stock solutions of the three compounds were prepared in dimethyl sulfoxide (DMSO) (1.25 M), dimethylformamide (DFM) (1 mM) and water (5 mM), respectively. Phosphate buffer solution (50 mM, 0.1 M NaCl, pH 7.4) was prepared in Milli-Q water. The rest of the chemicals were of spectroscopic or analytical reagent grade.

2.2. Methods

2.2.1. Preparation of Thermosensitive Liposomes (TSLs)

Lipids solutions of 2 mg of DPPG were left to dry at room temperature. Straight away, the dried phospholipid was resuspended in sodium phosphate buffer to a final concentration of 0.5 mM in order to obtain multilamellar vesicles (MLVs). MLVs were then heated above the phospholipid phase transition (~41 °C) and vortexed several times. TSLs were obtained from MLVs by pressure extrusion through 0.1 µm polycarbonate filters above T_m (~41 °C) in order to obtain homogenous large unilamellar vesicles (LUVs).

2.2.2. Preparation of Multicolor Fluorescent Nanoparticles

Aliquots of CPEs solubilized in DMSO (3.65×10^{-4} M for HTMA-PFP and HTMA-PFNT, and 6.24×10^{-4} M for HTMA-PFBT, in repeat units) were externally added to the TSLs suspension and incubated for at least 30 min at room temperature, as is depicted in Scheme 2. The proportion of DMSO in final samples was lower than 1% (v/v) in all the cases, and the final CPEs concentration was 3 µM in terms of repeat units.

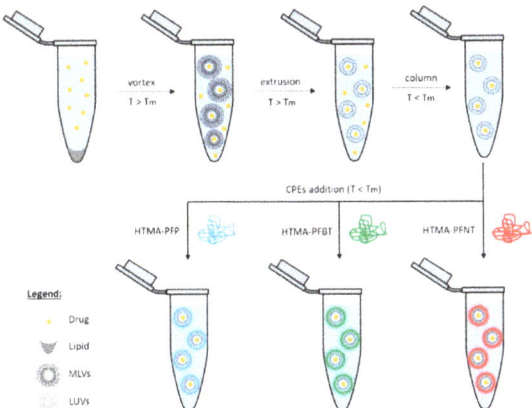

Scheme 2. Preparation of multicolor fluorescent nanoparticles.

2.2.3. Drug Encapsulation and Release Assays

CF was used as a hydrophilic model drug to investigate the encapsulation and controlled release properties of the TSLs and multicolor fluorescent nanoparticles. Encapsulation of the compound was carried out as is depicted in Scheme 2. Briefly, TSLs composed of DPPG were prepared with CF encapsulated in the aqueous core at a concentration of 40 mM in sodium phosphate buffer. The non-encapsulated CF was taken off by using a gel filtration column loaded with Sephadex G-75 and eluted with sodium phosphate buffer. Aliquots of CPEs were externally added to the TSLs loaded with CF, as was previously described. Self-quenching is expected when the CF is entrapped inside the vesicles. The CF release was assayed by treating the CF-loaded vesicles with increasing temperatures up to 60 °C. Samples were excited at 492 nm in order to minimize the absorbance of CPEs, and the emission was collected between 500–550 nm. The amount of CF released during and after thermal treatment was determined by inducing the total breakdown of the vesicles with Triton X-100 at 10%.

2.2.4. Particle Size and Zeta Potential

The size and Zeta Potential of the TSLs and multicolor fluorescent nanoparticles was explored by Dynamic Light Scattering (DLS) technique, with a Malvern Zetasizer Nano-ZS instrument (Worcestershire, UK) equipped with a monochromatic coherent 4 mW Helium Neon laser (λ = 633 nm) light source, where size measurements were performed at angles of 173°. Size was measured in disposable cuvettes, while Zeta Potential measurements were performed in specific Zeta Potential cells. All measurements were carried out in triplicate at room temperature.

2.2.5. Fluorescence Experiments

Fluorescence measurements were carried out in a PTI-QuantaMaster Spectrofluorometer (Birmingham, AL, USA) equipped with a Peltier cell holder. 1 cm path length quartz cuvettes were used to place the samples, which were subsequently excited at 380 nm (HTMA-PFP), 510 nm (HTMA-PFNT), 425 nm (HTMA-PFBT) or 492 nm (CF). Background intensities were checked and removed from the samples when it was necessary.

2.2.6. Anisotropy Experiments

Changes in the anisotropy as a function of temperature were used to explore the phase transition cooperativity and T_m of TSLs. Steady state anisotropy, $<r>$, is defined as:

$$<r> = \frac{(I_{VV} - GI_{VH})}{(I_{VV} + 2GI_{VH})} \qquad (1)$$

where, I_{VV} and I_{VH} correspond to the fluorescence intensities collected with the excitation polarizer oriented in a vertical position, and the emission polarizer oriented in a vertical and horizontal position, respectively. These measurements were obtained using Glan–Thompson polarizers incorporated in the spectrofluorometer. The liposome samples in the presence of DPH were excited at 360 nm and the emission was collected at 430 nm. The G factor ($G = I_{HV}/I_{HH}$) corrects the transmissivity bias introduced by the equipment.

2.2.7. Partition Coefficient Experiments

The partition coefficient, K_p, of the polyfluorenes between the gel-phase DPPG membranes of the TSLs and the aqueous medium was assayed by quantifying fluorescence intensity changes of CPEs in the presence of increasing TSLs concentrations. K_p is defined as:

$$K_p = \frac{n_L/V_L}{n_w/V_w} \qquad (2)$$

where, n_i and V_i correspond with the moles and the volume of phase i, respectively. The phase i could be either lipidic ($i = L$) or aqueous ($i = W$). The determination of K_p was carried out according to Reference [47]:

$$\Delta I = \frac{\Delta I_{max}[L]}{1/(K_p\gamma) + [L]} \qquad (3)$$

where, ΔI ($\Delta I = I - I_0$) corresponds with the difference between either fluorescence intensities or emission spectrum areas of the polyfluorenes measured in the presence (I) and absence (I_0) of TSLs, $\Delta I_{max} = I_\infty - I_0$ represents the highest value of this difference once the lowest value is reached (I_∞) upon increasing the TSLs concentration (L), and γ corresponds with the phospholipid molar volume (0.763 M^{-1}) [48].

2.2.8. Fluorescence Quenching Experiments

Fluorescence emissions of the CPEs in sodium phosphate buffer and integrated in TSLs were studied in the presence and absence of different AQS concentrations. This compound is an electron acceptor which works as a quencher of cationic CPEs, creating static quenching complexes through electrostatic interactions [49,50]. Stern–Volmer analysis was applied to the obtained fluorescence quenching values according to Equation (4):

$$\frac{I_0}{I} = 1 + K_{SV}(Q) \qquad (4)$$

where, I and I_0 correspond with the steady-state fluorescence intensities in the presence and absence of AQS respectively, and (Q) represents the AQS concentration. The meaning of K_{SV} relies on the nature of the quenching process: it could represent the rate of dynamic quenching or the association constant for complex formation (which is the case of AQS and CPEs) [51].

2.2.9. Morphological Observation

The morphological observation of the TSLs and multicolor fluorescent nanoparticles was performed by using a transmission electron microscope (TEM) (JEM-1400 Plus, JEOL, Tokyo, Japan), working at 120 kV. A drop of the samples was placed on to 300-mesh copper grips coated with carbon. In order to visualize the vesicles, a drop of lead citrate was also added. Samples were left dry before being placed under the microscope. A Gatan ORIUS camera was employed to record the images.

2.2.10. Cell Imaging Experiments

The human embryonic kidney cell line HEK293 was kindly donated by Dr. Alberto Falcó Gracia (Instituto de Investigación, Desarrollo e Innovación en Biotecnología Sanitaria de Elche, IDiBE, Elche, Spain). Cells were cultured in Dulbecco's Modified Eagle Medium (DMEM) with 10% (v/v) fetal calfserum (FCS), 2 mM L-glutamine, 100 U/mL penicillin and 100 µg/mL streptomycin and maintained in an incubator with controlled humidity (5% CO_2).

For fluorescence microscopy experiments, 96-well plates were used. The final concentration was 10^5 HEK293 cells per mL. Microscopy images were taken in the presence of multicolor fluorescent nanoparticles up to a final CPEs concentration of 0.365 µM, 0.73 µM and 0.18 µM of HTMA-PFP, HTMA-PFNT and HTMA-PFBT, respectively.

Fluorescence microscopy images were captured by using an inverted microscope (Leica DMI 3000B, Leica, Wetzlar, Germany) equipped with a compact light source (Leica EL6000) and a digital camera (Leica DFC3000G). The recording was carried out by using a 63× objective (0.7 magnification) and the filters: DAPI (Ex BP 350/50, Em BP 460/50), FITR (Ex BP 480/40, Em BP 527/30) or DsRed (Ex BP 555/25, Em BP 620/60). Data acquisition was performed manually with Leica Application Suite AF6000 Module Systems, and the image processing was carried out by using the software ImageJ.

3. Results

3.1. Characterization of Thermosensitive Liposomes (TSLs)

TSLs composed of DPPG were prepared as is described in the Methods Section (Section 2.2.1). Their size and stability were characterized by DLS and Zeta Potential, as shown in Table 1. The lipid was selected because its high ability to associate with cationic polielectrolytes as well as for its transition temperature in the mild hypertermia range. The experiments were done at 24 °C, where DPPG is in the gel phase. The vesicles exhibited hydrodynamic diameters around 130 nm and a high negative Zeta Potential, which should guarantee adequate suspension stability, since it was previously reported that Zeta Potential values higher than |20| mV are sufficient to prevent vesicle coalescence [52]. The polydispersity index value was 0.11, evidencing a narrow distribution of particle size. The morphology of the freshly prepared TSLs was observed under transmission electron microscopy. Results showed well-dispersed spherical-shaped vesicles, with a particle size compatible with the one estimated by DLS (Figure 1).

Table 1. Hydrodynamic diameter (d) and Zeta Potential (ZP) of TLSs and blue, green and red fluorescent nanoparticles.

	TSLs	TSLs + HTMA-PFP	TSLs + HTMA-PFBT	TSLs + HTMA-PFNT
d (nm)	127.3 ± 1.3	133.4 ± 0.9	143.5 ± 0.5	143.1 ± 0.1
ZP (mV)	−35.2 ± 1.7	−32.0 ± 0.7	−28.3 ± 0.7	−30.9 ± 0.9

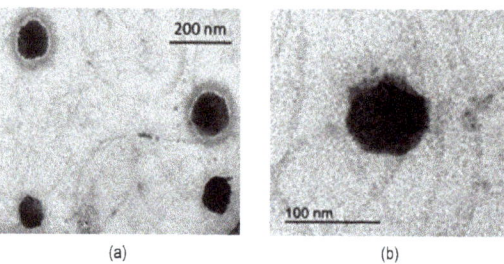

(a) (b)

Figure 1. Transmission electron microscopy (TEM) images (**a**,**b**) of thermosensitive liposomes (TSLs) at different magnifications.

The thermal behavior of the TSLs was explored using light scattering measurements. We analyzed the light scattered by the TSLs suspension as a function of temperature, taking into account that gel phase bilayers scatter more light than fluid membranes, a feature which has been attributed to the higher refractive index of gel membranes, as compared with fluid ones [53]. This experiment was directly made in the spectrofluorometer, by selecting the same wavelength for both excitation and emission monochromators with the smallest slit. The scattered light (430 nm) was collected at an angle of 90° of the incident light. As is shown in Figure 2a, a sharp drop of the scattered light was observed around 41–42 °C, which coincides with the transition temperature reported for DPPG. These results were confirmed by the fluorescence anisotropy measurements using the fluorescent probe DPH incorporated into the TSLs bilayer. This is a commonly used tool for thermotropic characterization of liposomes [54]. The plot of the steady-state fluorescence anisotropy, <r>, of DPH versus temperature is shown in the Supplementary Material (Figure S1), and was a perfect sinusoidal, displaying a sharp transition of anisotropy values around 42 °C.

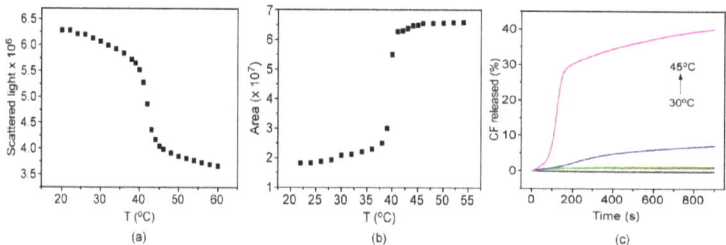

Figure 2. Effect of temperature on (**a**) the scattered light by TSLs and on (**b**) area of the emission spectrum of carboxyfluorescein (CF) encapsulated in TSLs in sodium phosphate buffer. (**c**) CF released in % as function of time (0–900 s) at different temperatures: 30 °C (black), 35 °C (red), 37 °C (green), 40 °C (blue) and 45 °C (magenta).3.2. Encapsulation and Release Assays.

3.2. Encapsulation and Release Assays

The dye carboxyfluorescein (CF) was entrapped in the aqueous cavity of TSLs, as was described in the Methods Section (Section 2.2.3). This marker was chosen for the release assays because of its interesting photophysical properties. When CF is highly concentrated, most of its fluorescence is quenched due to dimerization to a nonfluorescent compound as well as the Förster Resonance Energy Transfer (FRET) process between monomers and dimers [55]. Therefore, it is possible to monitor its release from the TSLs measuring the increase in fluorescence as a function of temperature (Figure 2b) or time (Figure 2c). Figure 2b shows an abrupt increase in the fluorescence of CF above 40 °C, which evidences the release of the dye from the liposome close to the transition temperature. Samples were also exposed to temperatures ranging from 30 to 45 °C, and the time release profile was recorded at each temperature (Figure 2c). Every sample was preheated at 30 °C before being placed in the thermostatized fluorimeter holder. Percent release was calculated from the change in fluorescence intensity, as detailed in the Methods Section (Section 2.2.3), after addition of Triton X-100 10%. Results show that below 37 °C, the dye remains entrapped into the liposome, at least within the experimental time period. At 40 °C, it starts to be slowly removed from the TSLs, probably due to the coexistence of gel and fluid domains at temperatures just below T_m, as has been reported in previous works for DPPG [56]. Finally, a rapid release of CF takes place at 45 °C, especially in the first three minutes, evidencing the ability of the DPPG-TSLs to carry and release hydrophilic compounds triggered by hyperthermia. The fact that only a 40% of CF is released at this temperature after 15 min of incubation could be attributed to dye interaction with the lipid membrane [56].

3.3. Preparation and Characterization of Fluorescent Nanoparticles

Once the physical properties and thermal behavior of the TSLs were characterized, the next step was to obtain the fluorescent nanoparticles. They were prepared by incorporation of the three polyelectrolytes: blue (HTMA-PFP), green (HTMA-PFBT) and red (HTMA-PFNT) in the bilayer of TSLs, at 25 °C (gel phase), as is shown in Scheme 2. This temperature was selected to prevent the release of the encapsulated drug from the TSLs before heating. In previous works, we demonstrated the ability of these CPEs to interact and insert into fluid phase lipid bilayers, by determining their affinity and membrane location [42–44]. But, taking into account the higher lipid packing of the gel phase, we cannot assume that the amount of polyelectrolyte bound to the fluid-phase membrane was the same as that bound to the gel-phase membrane. Therefore, the first experiments were focused to estimate the affinity of the CPEs for the TSLs, by determining their partition coefficient, K_p, defined in Equation (2). With this end, three series of samples containing increasing concentrations of DPPG-TSLs with final lipid concentrations ranging from 0 to 1 mM were prepared in buffer, and a constant concentration (3 µM) of HTMA-PFP, HTMA-PFBT and HTMA-PFNT was added to each series of samples respectively, which were incubated for 30 min at room temperature.

The emission spectra for the three polyelectrolytes, recorded at the different lipid concentrations, are shown in Figure 3a–c. A low fluorescence emission signal was detected for the CPEs in buffer, as a consequence of the formation of metastable aggregates [42–44]. The increase of the lipid concentration induced an enhancement in fluorescence intensity and a blue-shift of the emission spectra (insets in Figure 3), which suggests the breaking of aggregates as a consequence of the interaction of the three polyelectrolytes with TSLs-bilayer. Plotting the area of each spectrum versus lipid concentration (Figure S2) and using Equation (3), it was possible to determine the K_p values for the three CPEs. These values are summarized in Table 2 and indicate that, although the bilayer is in the gel-phase, the three polymers have high affinity for the TSLs. However, the K_p values are one order of magnitude lower than those obtained for anionic membranes in the fluid-phase [42–44]. This result suggests that the mode and nature of the interaction between CPEs and lipid membranes is different in the two phases. Probably, in the gel-phase, the interaction is mainly electrostatic between the quaternary amine groups of CPEs and the negative charge of the lipid head groups. In contrast, in the fluid-phase, the decrease of the lipid packing allows the insertion of the polymer chains and the hydrophobic forces contribute to a better solubilization. The fact that the fluorescence intensity of the CPEs incorporated in the gel-phase was lower than that registered in the fluid-phase at the same conditions (data not shown), supports this hypothesis.

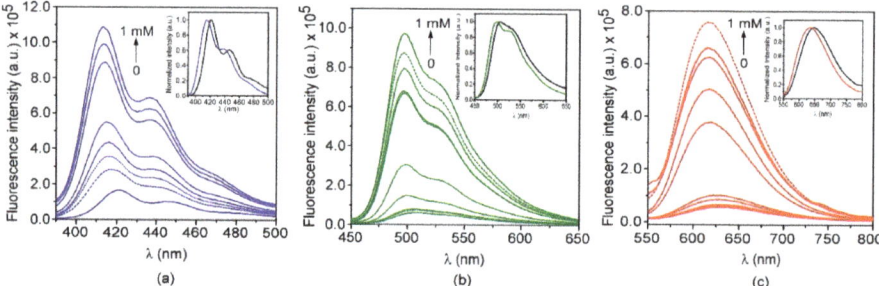

Figure 3. Fluorescence emission spectra of (**a**) HTMA-PFP (3 µM), (**b**) HTMA-PFBT (3 µM) and (**c**) HTMA-PFNT (3 µM) in buffer with increasing concentrations of TSLs. Insets: Normalized fluorescence emission spectra of CPEs in sodium phosphate buffer (black) and incorporated in TSLs (color).

Table 2. Partition coefficient, K_p, values, and Stern–Volmer constants, K_{SV}, for HTMA-PFP, HTMA-PFBT and HTMA-PFNT in TSLs by using 9,10-anthraquinone-2,6-disulfonic acid (AQS) as a quencher.

	K_p	K_{SV} (M^{-1})
HTMA-PFP + TSLs	$3.04 \pm 0.66 \times 10^4$	$1.11 \pm 0.02 \times 10^3$
HTMA-PFP + Buffer		$1.60 + 0.01 \times 10^7$
HTMA-PFBT + TSLs	$5.03 \pm 1.16 \times 10^4$	$1.32 \pm 0.02 \times 10^3$
HTMA-PFBT + Buffer		$6.08 \pm 0.34 \times 10^5$
HTMA-PFNT + TSLs	$2.59 \pm 0.41 \times 10^4$	$1.00 \pm 0.02 \times 10^3$
HTMA-PFNT + Buffer		$4.19 \pm 0.04 \times 10^5$

The K_p values were used to optimize the concentration of each component for the fabrication of the fluorescent nanoparticles. The lipid concentration was fixed to 0.5 mM to limit the turbidity of the samples, which becomes an obstacle for fluorescence measurements. The concentration of the CPEs was 3 µM in order to obtain a good fluorescent signal and to ensure that more than 90% of the polyelectrolyte was bound to the TSLs. The stability of the nanoparticles was assessed by monitoring the fluorescence intensity of the sample in the maximum of the emission spectrum as a function of time, after addition of the CPEs. The signal stabilized in the first seconds for HTMA-PFP and HTMA-PFNT and in ~5 min for HTMA-PFBT and remained stable during the experimental time (Figure S3). In addition, the possibility of simultaneously exciting the fluorescence emission of blue, green and red nanoparticles suspended in the same sample was also studied. The mixture was excited at 335 nm, where the three polyelectrolytes absorb (see Scheme 1). The recorded spectrum, shown in Figure S4, clearly displays the three bands corresponding to the characteristic spectra of HTMA-PFP, HTMA-PFBT and HTMA-PFNT.

The localization of the CPEs in the lipid bilayer of TSLs was investigated by quenching experiments using the anionic acceptor AQS as a fluorescence quencher. This molecule has been observed to be an excellent quencher for cationic conjugated polyelectrolytes, and it is soluble in water but not in lipid bilayer, so the fluorescence only will be deactivated if the CPEs are in the buffer or in the membrane surface [43,46]. When increasing concentrations of AQS were added to three different samples, containing the three polyelectrolytes in buffer, a strong decrease in their fluorescence signal was observed. In contrast, when the same experiment was performed in samples containing the fluorescent nanoparticles, the quenching effect was much less efficient (Figure 4). The Stern–Volmer plots (Equation (4)) were linear in all the studied ranges, with K_{sv} values similar in the three multicolor fluorescent nanoparticles, ranging from 1.00×10^3 M^{-1} for red fluorescent nanoparticles to 1.32×10^3 M^{-1} for green fluorescent nanoparticles. These quenching values are lower than in buffer (Table 2), confirming that the polyelectrolytes are bound to the lipid bilayer but close to the surface, because they are relatively accessible to the quencher. Finally, we compared the K_{sv} values obtained in the fluorescent nanoparticles with those achieved in anionic (PG) lipid vesicles in fluid-phase, which were ~0, especially for HTMA-PFP and HTMA-PFNT [46]. These differences in K_{sv} support the hypothesis previously proposed that the lipid packing affects the mode of interaction between the polyelectrolytes and the lipid membrane, as well as their final location in the bilayer.

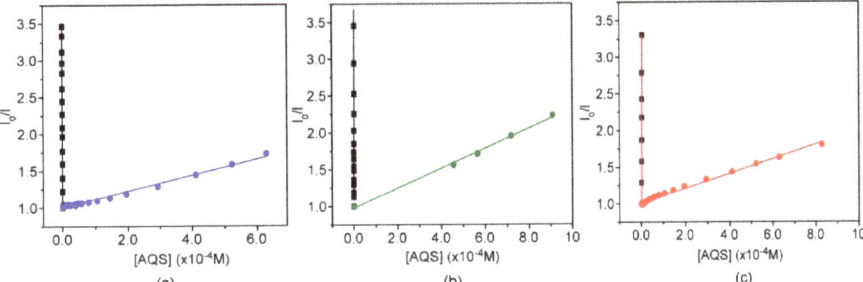

Figure 4. Stern–Volmer plots for quenching of (**a**) HTMA-PFP (3 µM), (**b**) HTMA-PFBT (3 µM) and (**c**) HTMA-PFNT (3 µM) by AQS in sodium phosphate buffer (squares) and in TSLs (circles).

Once optimized and analyzed, the distribution of components of the fluorescent nanoparticles we characterized by their size and colloidal stability, as well as their morphology (Table 1 and Figure 5). DLS results show that the incorporation of the polyelectrolytes slightly increases the size of the TSLs, which is compatible with their location close to the membrane surface. The decrease in the Zeta Potential was minimum and the nanoparticles exhibited good colloidal stability, preserving their spherical shape.

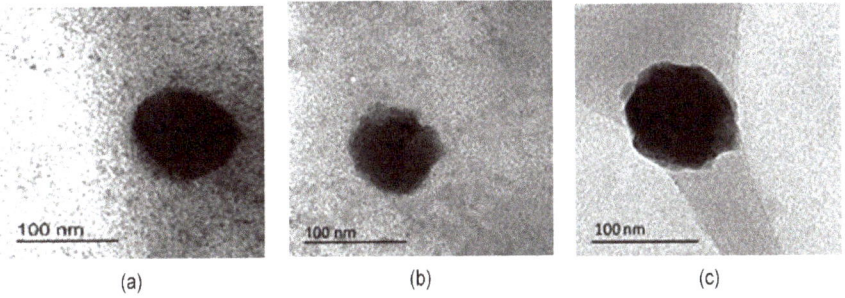

Figure 5. Transmission electron microscopy (TEM) images of blue (**a**), green (**b**) and red (**c**) fluorescent nanoparticles.

The evaluation of the thermosensitive properties of the fluorescent nanoparticles were carried out through two kinds of studies. First, we explored if the incorporation of CPEs modified the thermotropic behavior of the TSLs, previously characterized in Figure 2, and second, we analyzed if the fluorescence of the polyelectrolytes was sensitive to the structural modifications which take place in the vesicle bilayers at the lipid phase transition. For the first purpose, light scattering measurements were performed on the nanoparticles' suspension, as a function of temperature. Figure 6 shows the thermograms recorded for the fluorescent nanoparticles, blue, green and red, as well as for the TSLs in the absence of CPEs. The shape of the obtained curves is similar in the four samples: the light scattered by the nanoparticles slightly decreases as temperature rises, with a sharp drop occurring at T_m, whose value can be obtained from the first derivative plot and is coincident with that obtained in the absence of CPEs (inset in Figure 6). These results indicate that the integration of the polyelectrolytes does not affect the T_m and cooperativity of the lipid transition and thus, the fluorescent nanoparticles display the same thermosensitive properties as the DPPG-TSLs.

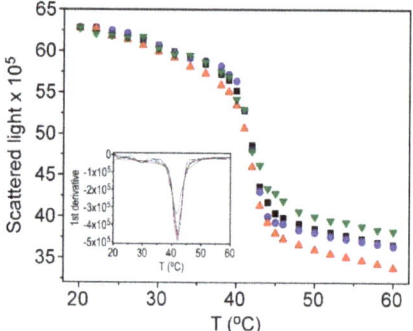

Figure 6. Effect of temperature on the light scattered by TSLs (squares) and by blue (circles), green (down-triangles) and red (triangles) fluorescent nanoparticles. Inset: First derivative of the thermograms.

As for the second purpose, the emission spectra of the three fluorescent nanoparticles were recorded as a function of temperature and the area under each spectrum was plotted between 20 and 60 °C (Figure 7a–c). In these plots, it is possible to distinguish three regions with a different behavior. Between 20 and 32 °C, the changes in fluorescence intensity are not very noticeable. However, above 30 °C, a rise in the fluorescence intensity is observed, reaching a maximum value around 42–45 °C. One possible explanation for this behavior could be that the CPEs are sensitive to the lipid pretransition, which occurs in DPPG at temperatures above 30 °C. The lipid pretransition is a transition of low enthalpy occurring below T_m, in which a flat bilayer in gel-phase becomes a periodically undulated membrane, called the ripple-phase [56]. In this ripple-phase, the hydrocarbon chains remain mainly in their rigid, extended, all-trans conformation, like in the gel-phase. However, some works have proposed the possible existence of fluid regions coupled with the geometry of the ripples [56].

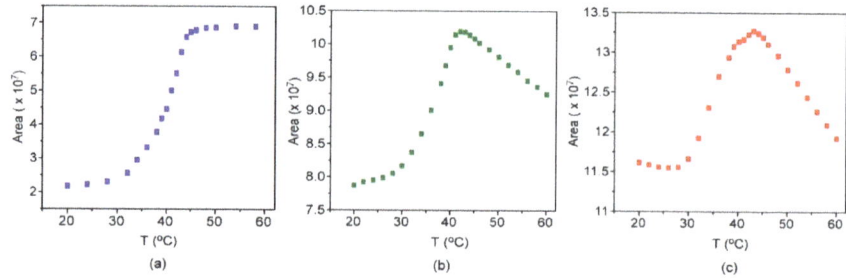

Figure 7. Effect of temperature on the emission spectrum area of (a) blue, (b) green and (c) red fluorescent nanoparticles.

Probably, when the CPEs are added to the DPPG-TSLs at 24 °C, they are mainly adsorbed to the surface instead of being incorporated in the lipid bilayer, as is demonstrated from the quenching experiments. The beginning of pretransition, above 30 °C, allows the polyelectrolytes to go into the lipid membrane, which leads to an increase of fluorescence emission intensity that reaches its plateau close to T_m. Once the fluid phase is reached, the polyelectrolye remains embedded in the bilayer and the fluorescence signal stabilizes or tends to decrease because the probability of nonradiative transitions increases with increasing temperature.

3.4. Nanoparticles as Drug Carriers and Bioimaging Probes

The above experiments indicate that the fluorescent nanoparticles preserve the thermosensitive properties of the TSLs, but they do not inform if the ability to encapsulate and release hydrophilic compounds triggered by hyperthermia is maintained. In fact, the possible internalization of the CPEs in the lipid membrane above 32 °C (as previously suggested), could cause a disruption of the lipid packaging, causing the compounds to be released from the nanoparticle before reaching T_m. To check this possibility, CF was firstly entrapped in the aqueous cavity of TSLs and the CPEs were then added to the suspension, as is described in Scheme 2. To monitor the release of CF from the nanoparticles, we recorded the fluorescence intensity of the dye as a function of temperature (Figure 8). The profile of the curves in Figure 8 was very similar for blue, green and red nanoparticles as well as for TSLs in the absence of CPEs. Above 40 °C, an abrupt increase in the fluorescence intensity of CF was detected, which suggests that most of the dye is released from the nanoparticles when the phase transition is reached. However, to confirm this conclusion, it is necessary to perform release kinetics of the dye at different temperatures. Nanoparticles were then exposed to temperatures ranging from 30 to 45 °C and the time release profile was recorded at each temperature (Figure 9), as was previously performed for the TSLs suspension. Results indicate that the release kinetics from the nanoparticles presents some differences with respect to those obtained in the absence of CPEs (Figure 2c). The most significant difference is that in the case of nanoparticles, an important fraction of CF is released at 40 °C, especially from the green nanoparticles. In addition, ~5% of the dye is released at 37 °C, after different incubation periods, depending on the fluorescent nanoparticles. For the blue one, it was necessary to have more than 900 s of incubation, while for the green and red nanoparticles, the release percentage of 5% was reached at 200 and 400 s, respectively. Therefore, although Figure 8 suggests that the presence of CPEs is not affecting the release properties of the TSLs, the internalization of the polyelectrolytes, which takes place above the pretransition temperature, seems to slightly affect the permeability of the membrane, allowing a small fraction of the dye to be slowly released at temperatures below T_m.

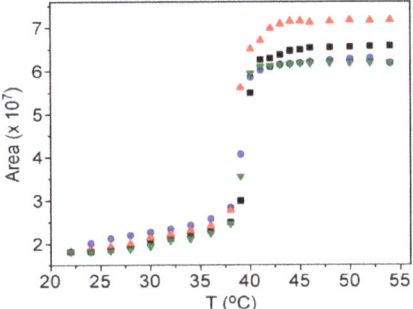

Figure 8. Area of the emission spectrum of CF encapsulated in TSLs (squares) and blue (circles), green (down-triangles) and red (triangles) fluorescent nanoparticles as function of temperature (20–55 °C).

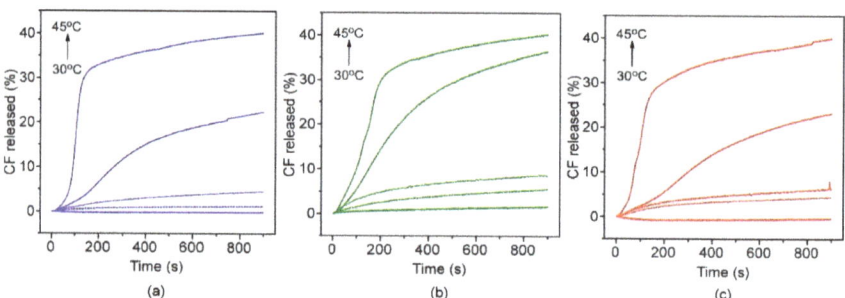

Figure 9. CF released in % as function of time (0–900 s) at different temperatures (30, 35, 37, 40 and 45 °C) in blue (**a**), green (**b**) and red (**c**) fluorescent nanoparticles multicolor fluorescent nanoparticles.

Finally, we have performed preliminary experiments to test the capability of the fluorescent nanoparticles to be employed as bioimaging probes. With this end, phase contrast and fluorescence microscopy images of HEK293 cells were taken before and 30 min after the addition of blue, green and red nanoparticles. Figure 10 shows the HEK293 cells in the presence of the polyfluorene-based fluorescent nanoparticles, observed by phase contrast and fluorescent microscopy. Phase contrast and fluorescence images correspond to the same field for each type of fluorescent nanoparticle. These images clearly show that nanoparticles are able to interact with cells, allowing for their visualization in three different colors under fluorescence microscopy. This result extends the applications of these new fluorescent nanoparticles, which could be used as probes for bioimaging while transporting and monitoring the pathway of a drug, controlling its release.

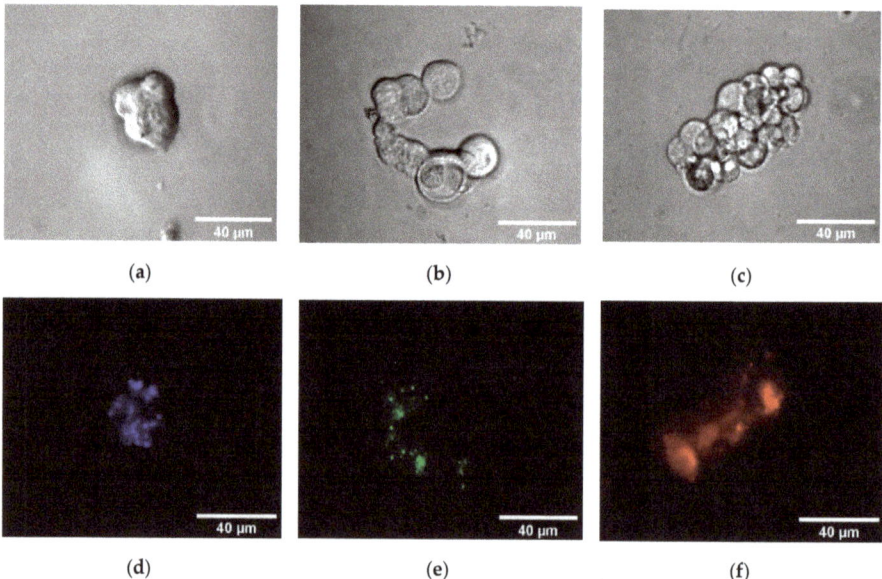

Figure 10. Microscopy images of HEK293 cells in the presence of (**a**, **d**) blue, (**b**, **e**) green and (**c**, **f**) red fluorescent nanoparticles, observed under (**a**–**c**) phase contrast and (**d**–**f**) visible-light using the Leica DAPI filter (Ex BP 350/50, Em BP 460/50), DsRed filter (Ex BP 555/25, Em BP 620/60) and FITR filter (Ex BP 480/40, Em BP 527/30).

4. Conclusions

Blue, green and red fluorescent nanoparticles composed of thermosensitive liposomes (TSLs) of DPPG and the conjugated polyelectrolytes HTMA-PFP, HTMA-PFBT and HTMA-PFNT respectively, have been prepared and characterized in order to obtain fluorescent drug carriers. In addition, the ability of the nanoparticles for bioimaging applications, transport and control drug delivery has been explored, evidencing their potential use as multifunctional nanoplatforms with imaging and therapeutic functionalities.

The nanoparticles exhibited stable fluorescence signals, good colloidal stability, spherical morphology and hydrodynamic diameters slightly higher to those of DPPG-TSLs, suggesting a membrane surface location of polyelectrolytes, which was confirmed by quenching experiments. In addition, their thermosensitive properties (cooperativity and transition temperature close to 42 °C) were similar to those of the DPPG-TSLs, supporting the potential use of these nanoparticles to carry and release drugs triggered by mild hyperthermia. The use of the dye carboxyfluorescein (CF) as a model hydrophilic drug allowed confirmation of this assumption. The dye was entrapped in the aqueous cavity of the fluorescent nanoparticles and was mostly released when nanoparticles were incubated above 40 °C. However, a small fraction of dye was slowly released at temperatures near 37 °C. This behavior has been attributed to a deeper penetration of the polyelectrolytes in the lipid bilayer, occurring above the pre-transition temperature, which could slightly modify the membrane permeability, allowing the slow release of the dye.

Finally, preliminary experiments with mammalian cells showed the capability of the nanoparticles to mark and visualize cells in blue, green and red colors, extending their applications as bioimaging probes. In this respect, it would be possible to encapsulate a different hydrophilic drug in each type of nanoparticle. This result could be of great interest in two-photon excitation microscopy and for dynamic imaging of living cells, due to the possibility of simultaneously exciting the fluorescence emission of the three nanoparticles at a single wavelength. We plan to further expand the multifunctionality of the nanoparticles in the future by linking different tumor-targeting molecules, such as peptides, folic acid, antibodies or other small molecules to the fluorescent lipid vesicles. By this procedure, the nanoparticles can be more effectively targeted in order to perform differential cell marking and active delivery to different tumor sites, which could be visualized in real time.

Supplementary Materials: The following are available online at http://www.mdpi.com/2079-4991/9/10/1485/s1, Figure S1. Anisotropy values, <r>, of DPH in DPPG-TSLs as function of temperature (20–70 °C) in sodium phosphate buffer, Figure S2. Changes in fluorescence intensity (ΔI) of (a) HTMA-PFP (3 µM), (b) HTMA-PFBT (3 µM) and (c) HTMA-PFNT (3 µM) at increasing concentrations of DPPG, Figure S3. Stability kinetics of (a) blue, (b) green and (c) red fluorescent nanoparticles (squares) compared with the stability of the corresponding polyelectrolytes in sodium phosphate buffer (circles), measured at 25 °C by monitoring their fluorescence intensity (blue: λexc = 380 nm, λem = 412 nm; green: λexc = 425 nm, λem = 500 nm; red: λexc = 510 nm, λem = 622 nm), Figure S4. Fluorescence emission spectrum of a sample containing simultaneously blue, green and red nanoparticles, upon excitation at 335 nm.

Author Contributions: Conceptualization, C.R.M. and M.J.M.-T.; methodology, C.R.M. and M.J.M.-T.; validation, M.R.-C. and Y.A.; formal analysis, C.R.M., M.J.M.-T., M.R.-C. and Y.A.; investigation, M.R.-C. and Y.A.; resources, R.M.; data curation, M.J.M.-T., M.R.-C. and Y.A.; writing—original draft, C.R.M. and M.J.M.-T. with the collaboration of M.R.-C.; writing—review and editing, C.R.M. and M.J.M.-T. with the collaboration of M.R.-C.; visualization, M.R.-C., Y.A. and M.J.M.-T.; supervision, C.R.M. and M.J.M.-T.; project administration, C.R.M. and R.M.; funding acquisition, C.R.M. and R.M.

Funding: This research was funded by the Ministerio de Economía, Industria y Competitividad, Gobierno de España (MAT-2017-86805-R), Conselleria d'Educació, Investigació, Cultura i Esport (ACIF/2018/226) and the European Regional Development Fund (IDIFEDER2018/20).

Acknowledgments: The authors gratefully acknowledge Alberto Falcó (IDiBE, Elche, Spain) for the kindly donation of the human embryonic kidney cell line HEK293, and Elisa Perez (IDiBE, Elche, Spain) for her kind help and technical assistance.

Conflicts of Interest: The authors declare no conflict of interest.

References

1. Bao, G.; Mitragotri, S.; Tong, S. Multifunctional Nanoparticles for Drug Delivery and Molecular Imaging. *Annu. Rev. Biomed. Eng.* **2013**, *15*, 253–282. [CrossRef] [PubMed]
2. Farjadian, F.; Ghasemi, A.; Gohari, O.; Roointan, A.; Karimi, M.; Hamblin, M.R. Nanopharmaceuticals and nanomedicines currently on the market: Challenges and opportunities. *Nanomedicine* **2019**, *14*, 93–126. [CrossRef] [PubMed]
3. Bejarano, J.; Navarro-Marquez, M.; Morales-Zavala, F.; Morales, J.O.; Garcia-Carvajal, I.; Araya-Fuentes, E.; Flores, Y.; Verdejo, H.E.; Castro, P.F.; Lavandero, S.; et al. Nanoparticles for diagnosis and therapy of atherosclerosis and myocardial infarction: Evolution toward prospective theranostic approaches. *Theranostics* **2018**, *8*, 4710–4732. [CrossRef] [PubMed]
4. Anderson, S.D.; Gwenin, V.V. Magnetic Functionalized Nanoparticles for Biomedical, Drug Delivery and Imaging Applications. *Nanoscale Res. Lett.* **2019**, *14*, 188. [CrossRef] [PubMed]
5. Cole, J.T.; Holland, N.B. Multifunctional nanoparticles for use in theranostic applications. *Drug Deliv. Transl. Res.* **2015**, *5*, 295–309. [CrossRef]
6. Edelman, R.; Assaraf, Y.G.; Slavkin, A.; Dolev, T.; Shahar, T.; Livney, Y.D. Developing Body-Components-Based Theranostic Nanoparticles for Targeting Ovarian Cancer. *Pharmaceutics* **2019**, *11*, 216. [CrossRef]
7. Lim, E.K.; Kim, T.; Paik, S.; Haam, S.; Huh, Y.M.; Lee, K. Nanomaterials for theranostics: Recent advances and future challenges. *Chem. Rev.* **2015**, *115*, 327–392. [CrossRef]
8. Martínez-Carmona, M.; Gun'ko, Y.; Vallet-Regí, M. ZnO Nanostructures for Drug Delivery and Theranostic Applications. *Nanomaterials* **2018**, *8*, 268. [CrossRef]
9. Xing, H.; Hwang, K.; Lu, Y. Recent developments of liposomes as nanocarriers for theranostic applications. *Theranostics* **2016**, *6*, 1336–1352. [CrossRef]
10. Allen, T.M.; Cullis, P.R. Liposomal drug delivery systems: From concept to clinical applications. *Adv. Drug Deliv. Rev.* **2013**, *65*, 36–48. [CrossRef]
11. Pandey, H.; Rani, R.; Agarwal, V. Liposome and their applications in cancer therapy. *Braz. Arch. Biol. Technol.* **2016**, *59*, e16150477. [CrossRef]
12. Díaz, M.R.; Vivas-Mejia, P.E. Nanoparticles as drug delivery systems in cancer medicine: Emphasis on RNAi-containing nanoliposomes. *Pharmaceuticals* **2013**, *6*, 1361–1380. [CrossRef] [PubMed]
13. Fernandes, L.F.; Bruch, G.E.; Massensini, A.R.; Frézard, F. Recent advances in the therapeutic and diagnostic use of liposomes and carbon nanomaterials in ischemic stroke. *Front. Neurosci.* **2018**, *12*, 453. [CrossRef] [PubMed]
14. Lee, Y.; Thompson, D.H. Stimuli-responsive liposomes for drug delivery. *Wiley Interdiscip. Rev. Nanomed. Nanobiotechnol.* **2017**, *9*, e1450. [CrossRef]
15. Yatvin, M.B.; Weinstein, J.N.; Dennis, W.H.; Blumenthal, R. Design of liposomes for enhanced local release of drugs by hyperthermia. *Science* **1978**, *202*, 1290–1293. [CrossRef]
16. Petrov, R.R.; Chen, W.H.; Regen, S.L. Thermally gated liposomes: A closer look. *Bioconjug. Chem.* **2009**, *20*, 1037–1043. [CrossRef]
17. Hossann, M.; Wiggenhorn, M.; Schwerdt, A.; Wachholz, K.; Teichert, N.; Eibl, H.; Issels, R.D.; Lindner, L.H. In vitro stability and content release properties of phosphatidylglyceroglycerol containing thermosensitive liposomes. *Biochim. Biophys. Acta Biomembr.* **2007**, *1768*, 2491–2499. [CrossRef]
18. Kneidl, B.; Peller, M.; Winter, G.; Lindner, L.H.; Hossann, M. Thermosensitive liposomal drug delivery systems: State of the art review. *Int. J. Nanomed.* **2014**, *9*, 4387–4398.
19. Nardecchia, S.; Sánchez-Moreno, P.; de Vicente, J.; Marchal, J.A.; Boulaiz, H. Clinical Trials of Thermosensitive Nanomaterials: An Overview. *Nanomaterials* **2019**, *9*, 191. [CrossRef]
20. Tagami, T.; Ernsting, M.J.; Li, S.D. Optimization of a novel and improved thermosensitive liposome formulated with DPPC and a Brij surfactant using a robust in vitro system. *J. Control. Release* **2011**, *154*, 290–297. [CrossRef]
21. Willerding, L.; Limmer, S.; Hossann, M.; Zengerle, A.; Wachholz, K.; ten Hagen, T.L.M.; Koning, G.A.; Sroka, R.; Lindner, L.H.; Peller, M. Method of hyperthermia and tumor size influence effectiveness of doxorubicin release from thermosensitive liposomes in experimental tumors. *J. Control. Release* **2016**, *222*, 47–55. [CrossRef] [PubMed]

22. Bi, H.; Xue, J.; Jiang, H.; Gao, S.; Yang, D.; Fang, Y.; Shi, K. Current developments in drug delivery with thermosensitive liposomes. *Asian J. Pharm. Sci.* **2019**, *14*, 365–379. [CrossRef]
23. Hossann, M.; Wang, T.; Wiggenhorn, M.; Schmidt, R.; Zengerle, A.; Winter, G.; Eibl, H.; Peller, M.; Reiser, M.; Issels, R.D.; et al. Size of thermosensitive liposomes influences content release. *J. Control. Release* **2010**, *147*, 436–443. [CrossRef] [PubMed]
24. Hossann, M.; Syunyaeva, Z.; Schmidt, R.; Zengerle, A.; Eibl, H.; Issels, R.D.; Lindner, L.H. Proteins and cholesterol lipid vesicles are mediators of drug release from thermosensitive liposomes. *J. Control. Release* **2012**, *162*, 400–406. [CrossRef] [PubMed]
25. Bruun, K.; Hille, C. Study on intracellular delivery of liposome encapsulated quantum dots using advanced fluorescence microscopy. *Sci. Rep.* **2019**, *9*, 10504. [CrossRef] [PubMed]
26. Weng, K.C.; Noble, C.O.; Papahadjopoulos-Sternberg, B.; Chen, F.F.; Drummond, D.C.; Kirpotin, D.B.; Wang, D.; Horn, Y.K.; Hann, B.; Park, J.W. Targeted tumor cell internalization and imaging of multifunctional quantum dot-conjugated immunoliposomes in vitro and in vivo. *Nano Lett.* **2008**, *8*, 2851–2857. [CrossRef]
27. Resch-Genger, U.; Grabolle, M.; Cavaliere-Jaricot, S.; Nitschke, R.; Nann, T. Quantum dots versus organic dyes as fluorescent labels. *Nat. Methods* **2008**, *5*, 763–775. [CrossRef]
28. Pisanic, T.R.; Zhang, Y.; Wang, T.H. Quantum dots in diagnostics and detection: Principles and paradigms. *Analyst* **2014**, *139*, 2968–2981. [CrossRef]
29. Pelley, J.L.; Daar, A.S.; Saner, M.A. State of academic knowledge on toxicity and biological fate of quantum dots. *Toxicol. Sci.* **2009**, *112*, 276–296. [CrossRef]
30. Zhang, R.; Ding, Z. Recent Advances in Graphene Quantum Dots as Bioimaging Probes. *J. Anal. Test.* **2018**, *2*, 45–60. [CrossRef]
31. Zhang, Y.; Wei, C.; Lv, F.; Liu, T. Real-time imaging tracking of a dual-fluorescent drug delivery system based on doxorubicin-loaded globin- polyethylenimine nanoparticles for visible tumor therapy. *Colloids Surf. B Biointerfaces* **2018**, *170*, 163–171. [CrossRef] [PubMed]
32. Pu, K.Y.; Liu, B. Fluorescent conjugated polyelectrolytes for bioimaging. *Adv. Funct. Mater.* **2011**, *21*, 3408–3423. [CrossRef]
33. Martiínez-Tomé, M.J.; Esquembre, R.; Mallavia, R.; Mateo, C.R. Formation of Complexes between the Conjugated Polyelectrolyte Poly{[9,9-bis(6′,N,N-trimethylammonium)hexyl]fluorene-phenylene} Bromide (HTMA-PFP) and Human Serum Albumin. *Biomacromolecules* **2010**, *11*, 1494–1501. [CrossRef] [PubMed]
34. Hou, M.; Lu, X.; Zhang, Z.; Xia, Q.; Yan, C.; Yu, Z.; Xu, Y.; Liu, R. Conjugated Polymer Containing Organic Radical for Optical/MR Dual-Modality Bioimaging. *ACS Appl. Mater. Interfaces* **2017**, *9*, 44316–44323. [CrossRef] [PubMed]
35. Qian, C.G.; Chen, Y.L.; Feng, P.J.; Xiao, X.Z.; Dong, M.; Yu, J.C.; Hu, Q.Y.; Shen, Q.D.; Gu, Z. Conjugated polymer nanomaterials for theranostics. *Acta Pharmacol. Sin.* **2017**, *38*, 764–781. [CrossRef]
36. Fu, N.; Wang, Y.; Liu, D.; Zhang, C.; Su, S.; Bao, B.; Zhao, B.; Wang, L. A conjugated polyelectrolyte with pendant high dense short-alkyl-chain-bridged cationic ions: Analyte-induced light-up and label-free fluorescent sensing of tumor markers. *Polymers (Basel)* **2017**, *9*, 227.
37. Gaylord, B.S.; Heeger, A.J.; Bazan, G.C. DNA detection using water-soluble conjugated polymers and peptide nucleic acid probes. *Proc. Natl. Acad. Sci. USA* **2002**, *99*, 10954–10957. [CrossRef]
38. Kahveci, Z.; Martínez-Tomé, M.J.; Mallavia, R.; Mateo, C.R. Fluorescent biosensor for phosphate determination based on immobilized polyfluorene-liposomal nanoparticles coupled with alkaline phosphatase. *ACS Appl. Mater. Interfaces* **2017**, *9*, 136–144. [CrossRef]
39. Ruoyu Zhan, B.L. Benzothiadiazole-Containing Conjugated Polyelectrolytes for Biological Sensing and Imaging. *Macromol. Chem. Phys.* **2014**, *216*, 131–144. [CrossRef]
40. Pina, J.; Seixas De Melo, J.S.; Eckert, A.; Scherf, U. Unusual photophysical properties of conjugated, alternating indigo-fluorene copolymers. *J. Mater. Chem. A* **2015**, *3*, 6373–6382. [CrossRef]
41. Su, H.J.; Wu, F.I.; Tseng, Y.H.; Shu, C.F. Color tuning of a light-emitting polymer: Polyfluorene-containing pendant amino-substituted distyrylarylene units. *Adv. Funct. Mater.* **2005**, *15*, 1209–1216. [CrossRef]
42. Kahveci, Z.; Martínez-Tomé, M.J.; Mallavia, R.; Mateo, C.R. Use of the Conjugated Polyelectrolyte Poly{[9,9-bis(6′-N,N,N-trimethylammonium)hexyl]fluorene-phenylene} Bromide (HTMA-PFP) as a Fluorescent Membrane Marker. *Macromolecules* **2013**, *14*, 1990–1998. [CrossRef]

43. Kahveci, Z.; Vázquez-Guilló, R.; Martínez-Tomé, M.J.; Mallavia, R.; Mateo, C.R. New Red-Emitting Conjugated Polyelectrolyte: Stabilization by Interaction with Biomolecules and Potential Use as Drug Carriers and Bioimaging Probes. *ACS Appl. Mater. Interfaces* **2016**, *8*, 1958–1969. [CrossRef] [PubMed]
44. Vázquez-Guilló, R.; Martínez-Tomé, M.J.; Kahveci, Z.; Torres, I.; Falco, A.; Mallavia, R.; Mateo, R.C. Synthesis and characterization of a novel green cationic polyfluorene and its potential use as a fluorescent membrane probe. *Polymers* **2018**, *10*, 938. [CrossRef]
45. Kahveci, Z.; Martínez-Tomé, M.J.; Esquembre, R.; Mallavia, R.; Mateo, C.R. Selective interaction of a cationic polyfluorene with model lipid membranes: Anionic versus zwitterionic lipids. *Materials* **2014**, *7*, 2120–2140. [CrossRef] [PubMed]
46. Kahveci, Z.; Vázquez-Guilló, R.; Mira, A.; Martinez, L.; Falcó, A.; Mallavia, R.; Mateo, C.R. Selective recognition and imaging of bacterial model membranes over mammalian ones by using cationic conjugated polyelectrolytes. *Analyst* **2016**, *141*, 6287–6296. [CrossRef]
47. Coutinho, A.; Prieto, M. Self-association of the polyene antibiotic nystatin in dipalmitoylphosphatidylcholine vesicles: A time-resolved fluorescence study. *Biophys. J.* **1995**, *69*, 2541–2557. [CrossRef]
48. Melo, M.N.; Castanho, M.A.R.B. Omiganan interaction with bacterial membranes and cell wall models. Assigning a biological role to saturation. *Biochim. Biophys. Acta Biomembr.* **2007**, *1768*, 1277–1290. [CrossRef]
49. Chen, J.; Dong, W.-F.; Möhwald, H.; Krastev, R. Amplified Fluorescence Quenching of Self-Assembled Polyelectrolyte–Dye Nanoparticles in Aqueous Solution. *Chem. Mater.* **2008**, *20*, 1664–1666. [CrossRef]
50. Chemburu, S.; Ji, E.; Casana, Y.; Wu, Y.; Buranda, T.; Schanze, K.S.; Lopez, G.P.; Whitten, D.G. Conjugated polyelectrolyte supported bead based assays for phospholipase A2 activity. *J. Phys. Chem. B* **2008**, *112*, 14492–14499. [CrossRef]
51. Lakowicz, J.R. *Principles of Fluorescence Spectroscopy Principles of Fluorescence Spectroscopy*, 2nd ed.; Kluwer Academic/Plenum: New York, NY, USA, 1999.
52. Sun, C.; Wang, J.; Liu, J.; Qiu, L.; Zhang, W.; Zhang, L. Liquid proliposomes of nimodipine drug delivery system: Preparation, characterization, and pharmacokinetics. *AAPS PharmSciTech* **2013**, *14*, 332–338. [CrossRef] [PubMed]
53. Yi, P.N.; MacDonald, R.C. Temperature dependence of optical properties of aqueous dispersions of phosphatidylcholine. *Chem. Phys. Lipids* **1973**, *11*, 114–134. [CrossRef]
54. Esquembre, R.; Ferrer, M.L.; Gutiérrez, M.C.; Mallavia, R.; Mateo, C.R. Fluorescence study of the fluidity and cooperativity of the phase transitions of zwitterionic and anionic liposomes confined in sol-gel glasses. *J. Phys. Chem. B* **2007**, *111*, 3665–3673. [CrossRef] [PubMed]
55. Chen, R.F.; Knutson, J.R. Mechanism of fluorescence concentration quenching of carboxyfluorescein in liposomes: Energy transfer to nonfluorescent dimers. *Anal. Biochem.* **1988**, *172*, 61–77. [CrossRef]
56. Riske, K.A.; Barroso, R.P.; Vequi-Suplicy, C.C.; Germano, R.; Henriques, V.B.; Lamy, M.T. Lipid bilayer pre-transition as the beginning of the melting process. *Biochim. Biophys. Acta Biomembr.* **2009**, *1788*, 954–963. [CrossRef]

© 2019 by the authors. Licensee MDPI, Basel, Switzerland. This article is an open access article distributed under the terms and conditions of the Creative Commons Attribution (CC BY) license (http://creativecommons.org/licenses/by/4.0/).

Article

Electrochemical Sensors Modified with Combinations of Sulfur Containing Phthalocyanines and Capped Gold Nanoparticles: A Study of the Influence of the Nature of the Interaction between Sensing Materials

Ana Isabel Ruiz-Carmuega [1], Celia Garcia-Hernandez [1,2], Javier Ortiz [3], Cristina Garcia-Cabezon [1,2], Fernando Martin-Pedrosa [1,2], Ángela Sastre-Santos [3], Miguel Angel Rodríguez-Perez [2] and Maria Luz Rodriguez-Mendez [1,2,*]

[1] Group UVASENS, Escuela de Ingenierías Industriales, Universidad de Valladolid, Paseo del Cauce, 59, 47011 Valladolid, Spain; anaisabel.ruiz@uva.es (A.I.R.-C.); celiagarciahernandez@gmail.com (C.G.-H.); crigar@eii.uva.es (C.G.-C.); fmp@eii.uva.es (F.M.-P.)
[2] BioecoUVA Research Institute, Universidad de Valladolid, 47011 Valladolid, Spain; marrod@fmc.uva.es
[3] Área de Química Orgánica, Instituto de Bioingeniería, Universidad Miguel Hernández de Elche, 03202 Elche, Spain; jortiz@umh.es (J.O.); asastre@umh.es (Á.S.-S.)
* Correspondence: mluz@eii.uva.es; Tel.: +34-983-423-540

Received: 13 September 2019; Accepted: 16 October 2019; Published: 23 October 2019

Abstract: Voltametric sensors formed by the combination of a sulfur-substituted zinc phthalocyanine (ZnPc^{RS}) and gold nanoparticles capped with tetraoctylammonium bromide (AuNP^{tOcBr}) have been developed. The influence of the nature of the interaction between both components in the response towards catechol has been evaluated. Electrodes modified with a mixture of nanoparticles and phthalocyanine (AuNP^{tOcBr}/ZnPc^{RS}) show an increase in the intensity of the peak associated with the reduction of catechol. Electrodes modified with a covalent adduct-both component are linked through a thioether bond-(AuNP^{tOcBr}-S-ZnPc^{R}), show an increase in the intensity of the oxidation peak. Voltammograms registered at increasing scan rates show that charge transfer coefficients are different in both types of electrodes confirming that the kinetics of the electrochemical reaction is influenced by the nature of the interaction between both electrocatalytic materials. The limits of detection attained are 0.9 × 10^{-6} mol·L^{-1} for the electrode modified with the mixture AuNP^{tOcBr}/ZnPc^{RS} and 1.3 × 10^{-7} mol·L^{-1} for the electrode modified with the covalent adduct AuNP^{tOcBr}-S-ZnPc^{R}. These results indicate that the establishment of covalent bonds between nanoparticles and phthalocyanines can be a good strategy to obtain sensors with enhanced performance, improving the charge transfer rate and the detection limits of voltammetric sensors.

Keywords: electrochemical sensor; phthalocyanine; gold nanoparticle; catechol

1. Introduction

Catechol is an important member of the family of phenols that can be found as an antioxidant in foods. Different types of electrodes have been described in the literature to assess the concentration of catechol in solution [1–7].

Phthalocyanines (Pcs) have attracted interest as chemical modifiers in electrochemical sensors dedicated to the detection of phenols due to their well-known electrocatalytic activity. Their electrochemical properties can be modified by introducing substituents in the aromatic ring [8–13]. Over the last decade, phthalocyanines have been linked covalently to a number of molecules, including fullerenes [14,15], perylenes [16,17], carbon-nanotubes [18,19], graphite, and nanoparticles [20–23].

The electrocatalytic properties of gold nanoparticles (AuNPs) are also well established [24–27], and a variety of uncapped and capped AuNPs have been successfully used to detect phenols [28–30].

One possible strategy to improve the performance of electrochemical sensors could be to develop composites formed by combinations of electrocatalytic materials, in order to generate synergistic effects [13,31,32]. Synergistic effects have been observed in AuNP/Pcs composites obtained by introduction weak interactions between both materials by means of mixing [32–34], self-assembling [35], the Langmuir-Blodgett (LB) [36,37], or electrodeposition techniques [38].

In spite of the interest in these combinations, the influence of the nature of the interaction between both components in the sensing properties remains largely unexplored.

The aim of this work is to develop new voltammetric sensors based on combinations of gold nanoparticles and sulfur-substituted zinc phthalocyanines and to analyze the electron transfer process, as well as the existence of synergistic effects between both components in the absence and presence of covalent links.

For this purpose, tetraoctylammonium bromide-gold nanoparticles (AuNP^{tOcBr}) and 2-{2'-[(5''-Acetylthiopentyloxo)amino]ethoxy}-9(10),16(17),23(24)-tri-tert-butylphthalocyaninate Zn(II) (ZnPc^{RS}) have been synthesized. These species have the appropriate substituents necessary to obtain a covalently linked adduct in which the nanoparticles and the phthalocyanines have been linked covalently through thiol bonds (AuNP^{tOcBr}-S-ZnPc^R).

The sensing properties towards catechol of an ITO substrate modified with the adduct, have been compared with the responses of an ITO glass covered with a mixture of both components (AuNP^{tOcBr}/ZnPc^{RS}). In addition, the response of a mixture formed by AuNP^{tOcBr} and a dimeric phthalocyanine where the sulfur groups are blocked AuNP^{tOcBr}/ZnPc^R-S-ZnPc^R have also been analyzed.

In all cases, studies at increasing scan rates have been carried out to evaluate the influence of the type of bond in the charge transfer rates. The limits of detection have also been calculated and compared.

2. Materials and Methods

Chemicals and solvents were of reagent grade (Aldrich Chemical Ltd., St. Louis, MO, USA). Reagents to prepare gold nanoparticles were: HAuCl4·xH2O (99.9%, min. 49% Au, Alfa Aesar, Haverhill, MA. USA), tetraoctylammonium bromide (98%, Aldrich Chemical. Ltd., St. Louis, MO, USA), sodium borohydride (95%, Riedel-de Haën, Seelze, Germany). Solutions were prepared in deionized water obtained using a Milli-Q system (Millipore, Direct-Q5, Madrid, Spain). The complete list of reactants can be found in Supplementary Materials S1.

2.1. Synthesis of the Sensitive Materials

Sensitive materials used in this work are depicted in Figure 1. They were synthesized as follows.

2.1.1. Tetraoctylammonium Bromide-Capped Gold Nanoparticles (AuNP^{tOcBr})

They were synthesized using the Brust method [39]. A water solution of gold tetrachloride was mixed with a toluene solution of tetraoctylammonium bromide. The mixture was stirred until the aqueous phase lost its color, and the organic phase appeared colored. Then, sodium borohydride was added drop by drop to the organic phase until a cherry color was observed. Afterwards, the mixture was stirred under nitrogen atmosphere in darkness. After decantation, gold nanoparticles capped with tetraoctylammonium bromide were obtained as a colloid in toluene.

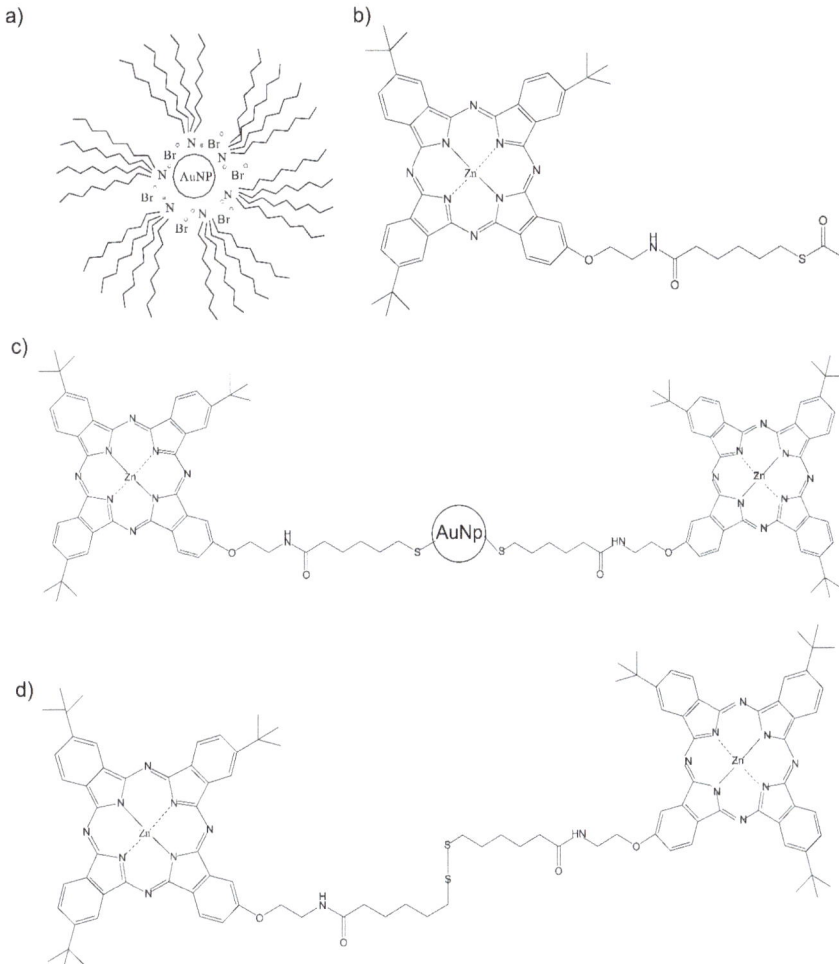

Figure 1. Scheme of the electrocatalytic materials. (**a**) Tetraoctylammonium bromide-capped gold nanoparticles (AuNP^{tOcBr}), (**b**) sulfur-substituted zinc phthalocyanine (ZnPc^{RS}), (**c**) covalent adduct (AuNP^{tOcBr}-S-ZnPc^{RS}), (**d**) dimeric sulfur-substituted zinc bisphthalocyanine: (ZnPc^{R}-S-ZnPc^{R}).

2.1.2. 6,6′-dithiodihexanoic Acid

It was obtained according to a previously published method [40]. Nine hundred eighty-eight mg (5 mmol) of 6-bromohexanoic acid, 345 mg (2.5 mmol) of K$_2$CO$_3$ and 0.5 mL of H$_2$O were heated at reflux for 20 min. A solution of 1.24 g (5 mmol) of sodium thiosulfate pentahydrate in 3 mL of H$_2$O was added and the reaction was allowed to react for 1 h at reflux. Then 1.26 g (5 mmol) of iodine was added and allowed to cool for 30 min. One hundred fifty µL (0.125 mmol) of concentrated H$_2$SO$_4$ was added, the reaction mixture was diluted in dichloromethane (DCM) and washed with H$_2$O, extracting the aqueous phase twice with DCM. The organic phases were dried with Na$_2$SO$_4$ and the solvent was removed under reduced pressure. The reaction crude was recrystallized from hot toluene to obtain 552 mg of the product (40%), mp 76.5 °C (toluene). ^{1}H-RMN (300 MHz, DMSO-d_6, 25 °C): δ = 1.43–1.48 (m, 4H), 1.64–1.73 (m, 8H), 2.2 (t, J = 7.3 Hz, 4H, CH$_2$CO), 2.5 (t, J = 7.3 Hz, 4H, CH$_2$S), 11.0 (br s, 2H,

2×CO$_2$H).^{13}C-RMN (75 MHz, DMSO-d$_6$, 25 °C): δ = 24.1, 27.3, 29.3 (3×CH$_2$), 36.1 (CH$_2$CO$_2$H), 39 (CH$_2$S) y 177 (CO$_2$H). ν$_{max}$ (KBr)/cm^{-1}: 2933, 2856, 1691, 1466, 1434, 1410, 1190 y 922.

2.1.3. Sulfur-Substituted Zinc Phthalocyanine: 2-{2′-[(5″-Acetylthiopentyloxo)amino]ethoxy}-9(10), 16(17),23(24)-tri-tert-butylphalocyaninate Zn(II) (ZnPc^{RS})

It was synthesized following a previously published procedure [21]. The corresponding dimeric structure (ZnPc^R-S-ZnPc^R) was synthetized here for the first time using the following method.

2.1.4. Dimeric Sulfur Substituted Zinc Bisphthalocyanine: (ZnPc^R-S-ZnPc^R)

As mentioned before, this compound was obtained for the first time in this work. 21 mg (0.024 mmol) of (2-aminoethoxy)-tri-tert-butylphthalocyaninate zinc (II) [16], 3.5 mg (0.012 mmol) of 6,6′-dithiodihexanoic acid and 11.2 mg (0.055 mmol) of dicyclohexylcarbodiimide (DCC) were dissolved in 700 μL of dichloromethane under argon at 0 °C. After 30 min, 1 mg (0.009 mmol) of N,N-dimethylaminopyridine (DMAP) was added and allowed to react for 3 h. The reaction mixture was diluted with dichloromethane, the organic phase was washed with NH$_4$Cl (aq.), NaHCO$_3$ (aq.) and H$_2$O, dried with Na$_2$SO$_4$ and the solvent was removed under reduced pressure. The crude was purified by column chromatography (dichloromethane: methanol/99:1) to obtain 14 mg of the compound (60%). ^{1}H-RMN (300 MHz, TFA-d$_1$, 25 °C): δ = 1.47 (m, 12H, 6×CH$_2$) 1.68 [br s, 54H, 6×(CH$_3$)$_3$C], 2.54 (br s, 4H, CH$_2$CO), 2.74 (br s, 4H, CH$_2$S), 4.16 (br s, 4H, CH$_2$N), 4.68 (br s, 4H, CH$_2$O), 7.88 (m, 3H, ArH), 8.47 (m, 6H, ArH), 8.96 (m, 2H, ArH) y 9.31–9.48 (m, 13H, ArH). ν$_{max}$ (KBr)/cm^{-1}: 3401, 2952, 2855, 1610, 1488, 1461, 1391, 1329, 1255, 1089, 1046 y 748 cm^{-1}. UV-Vis (DMF): λmax/nm (log ε): 350 (5.14), 610 (4.84), 676 (5.58). HRMS-MALDI-TOF (dithranol): m/z: for C$_{104}$H$_{108}$N$_{18}$O$_4$S$_2$Zn$_2$ calcd, 1864.682; found 1864.684 (M$^+$).

2.1.5. AuNP^{tOcBr}/ZnPc^{RS} and AuNP^{tOcBr}/ZnPc^R-S-ZnPc^R Mixtures

The non-covalent mixture of AuNP^{tOcBr}/ZnPc^{RS} was prepared from AuNP^{tOcBr} toluene colloid (Abs398 nm = 3, 5 ua) and ZnPc^{RS} (6.5 × 10^{-5} mol·L^{-1}) mixed in a proportion of 2:1 (v/v). The mixture was kept in the dark until used. A similar method was followed to obtain the mixture AuNP^{tOcBr}/ZnPc^R-S-ZnPc^R.

2.1.6. AuNP^{tOcBr}-S-ZnPc^R Covalent Adduct

The covalent adduct (AuNP^{tOcBr}-S-ZnPc^R) was obtained as follows [21]: 4 mL of the phthalocyanine toluene solution (1.3 × 10^{-3} mol·L^{-1}) was mixed with 4 mL of the nanoparticle colloid (Abs398 nm = 3.5 ua) and stirred for 24 h at room temperature, in darkness and under inert atmosphere. Next, the product was added to pentane drop by drop. The precipitate was dissolved in methane and kept overnight at −20 °C. Following centrifugation, the new precipitate of AuNP^{tOcBr}-S-ZnPc^R was re-suspended in toluene.

2.2. Preparation of the Sensors

Sensors were prepared by depositing a layer of the mixtures or of the adduct by spin coating (spin coater model 1H-D7, Micasa Co., Tokyo, Japan). Before deposition, ITO glass substrates were washed with acetone and rinsed twice with deionized water in an ultrasonic bath. Fifty μL of the corresponding material was deposited onto the substrate (1 cm^2 surface) using 120 s slope and 120 s at 1000 rpm.

The sensing materials and films were characterized by TEM microscopy (JEOL-FS2200 HRP. 200 kV emission) and UV-Vis spectroscopy with a double beam spectrophotometer (UV-2600, Shimadzu, Kyoto, Japan).

2.3. Sensing Properties

Cyclic voltammetry was used to characterize the sensing behavior of the chemically modified films. Electrochemical measurements were carried out in a Parstat 2273 (Princeton Applied Research) using a three-electrode cell. The reference electrode was Ag|AgCl/KCl sat. and the counter electrode was a platinum sheet. Modified ITO films were used as working electrodes. The electrochemical responses were analyzed towards catechol 10^{-3} mol·L^{-1} in phosphate buffer (Na$_2$HPO$_4$/NaH$_2$PO$_4$ 0.01 M pH = 7). Cyclic voltammograms were registered from −0.8 to 1.2 V at a scan rate of 0.1 V·s^{-1}. The Limits of detection (LOD) were calculated from peak current responses in voltammograms registered at concentrations from 4×10^{-6} to 1.45×10^{-4} mol·L^{-1} following the "3Sd/m" method, where "Sd" is the standard deviation (n = 5) of the signal registered in the buffer, and "m" is the slope of the calibration curve. The influence of the potential sweep rate was studied in catechol 10^{-4} mol·L^{-1} changing the scan rates from 0.01 to 1.0 V·s^{-1}.

3. Results and Discussion

The UV-Vis spectra of the individual sensing materials are shown in Figure 2a. The electronic absorption spectrum of the AuNP^{tOcBr} colloid was dominated by an intense peak at 398 nm produced by the plasmon resonance, accompanied by a small shoulder at 485 nm. The sharpness of the peak at 398 nm reflected a homogeneous distribution of the NPs size. The colloid diluted 1:10, showed the same features as the undiluted colloid, confirming the lack of aggregation. UV-Vis spectra of the ZnPc^{RS} toluene solutions showed the expected Q bands at 689 nm and at 675 nm which are usually observed in unsymmetrical phthalocyanines. The spectrum also exhibited an intense Soret band at 353 nm and a small vibronic band at 618 nm. The spectrum of the dimeric phthalocyanine was similar to the one observed in the monomeric form. The only differences were found in the intensity of the Q and Soret bands which were more intense in the dimeric compound due to the presence of two phthalocyanine rings. The UV-Vis spectra of the mixtures and of the adduct are presented in Figure 2b. The spectrum of the AuNP^{tOcBr}/ZnPc^{RS} mixture showed bands associated with each one of the components, although changes in the intensities and positions of the peaks with respect to those observed in the spectra of the individual components were observed: The Q band of the phthalocyanine appeared at 679 nm. Due to its broadness, the splitting was no longer observed. Furthermore, the Soret band increased its intensity with respect to the Q band, and appeared at 359 nm, overlapping with the plasmonic band of the nanoparticle.

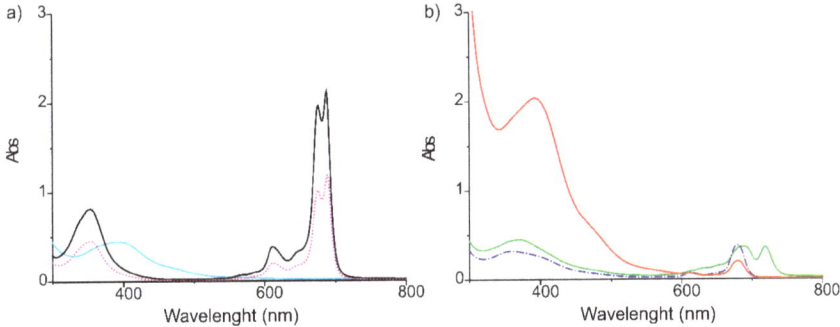

Figure 2. UV-Vis absorption spectra of (**a**) AuNP^{tOcBr} (blue —), ZnPc^{RS} (pink ······), ZnPc^{R}-S-ZnPc^{R} (black—) and of (**b**) the mixture AuNP^{tOcBr}/ZnPc^{RS} (purple ----), the mixture AuNP^{tOcBr}/ZnPc^{R}-S-ZnPc^{R} (green —) and the covalent adduct AuNP^{tOcBr}-S-ZnPc^{R} (red —), in toluene as solvent.

The UV-Vis spectrum of the covalent adduct AuNP^{tOcBr}-S-ZnPc^{R} showed the same features as shown by the mixture. However, a clear increase in the intensity of the band at 393 nm produced by the overlapping of the phthalocyanine Soret band and the band of the AuNP^{tOcBr} plasmon band was observed. This effect was consistent with a covalent interaction between the phthalocyanine and the nanoparticle that caused the modification of the π-π transition.

The mixture with the dimer AuNP^{tOcBr}/ZnPc^R-S-ZnPc^R showed two broad Q bands. The first broadband at 685 nm is produced by the substituted Pc ring similar to that observed in the monomeric species. The splitting observed in the monomer cannot be observed due to the broadness of the band. The second band at 719 nm is typical of the formation of J aggregates due to the interaction between the two Pc rings. The Soret band appears overlapped with the band corresponding to the plasmon resonance of the nanoparticles. Obviously, a covalent adduct could not be obtained by reaction of the dimer and the AuNPs because the covalent bond was not accessible.

According to TEM images (Figure 3), the estimated core diameter of the AuNP^{tOcBr} was 2–3 nm. The images of the mixtures AuNP^{tOcBr}/ZnPc^{RS} and of the adduct AuNP^{tOcBr}-S-ZnPc^R showed nanoparticles with sizes ranging from 2 to 5 nm, with an average value of 4 nm. The images also revealed the existence of a light halo surrounding the nanoparticles, which was due to the phthalocyanines located around nanoparticles. The thickness of the halo was smaller in the case of the mixtures AuNP^{tOcBr}/ZnPc^{RS} and AuNP^{tOcBr}/ZnPc^R-S-ZnPc^R films and could only be observed at higher magnifications.

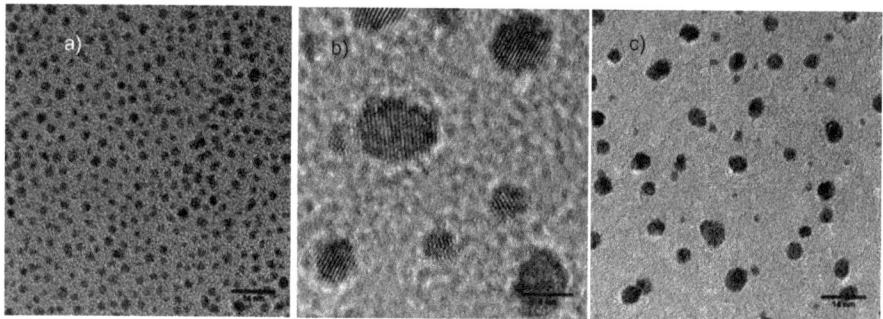

Figure 3. TEM images of (**a**) AuNP^{tOcBr}, (**b**) mixture AuNP^{tOcBr}/ZnPc^{RS}, and (**c**) adduct AuNP^{tOcBr}-S-ZnPc^R.

ITO glasses were modified with spin-coated films of the AuNP^{tOcBr}/ZnPc^{RS}, AuNP^{tOcBr}/ZnPc^R-S-ZnPc^R mixtures and of the AuNP^{tOcBr}-S-ZnPc^R adduct. Their sensing properties towards catechol were analyzed using cyclic voltammetry. The electrochemical responses of a bare ITO and films prepared from individual components AuNP^{tOcBr}, ZnPc^{RS}, and ZnPc^R-S-ZnPc^R were also analyzed for comparison purposes.

Voltammetric responses towards a 10^{-3} mol·L^{-1} catechol solution (in 0.01 M phosphate buffer as electrolyte pH = 7) are shown in Figure 4. As a general rule, responses were characterized by an anodic peak at positive potentials (produced by the oxidation of catechol to 1, 2 benzoquinone) and a cathodic peak at ca. −0.25 V produced by the corresponding reduction of the benzoquinone. However, important differences were caused by the modification of the electrode.

When a bare ITO electrode was immersed in catechol, peaks were quite weak. A small increase in the intensity of the peaks was observed when the ITO glass was coated with AuNP^{tOcBr}. In contrast, ZnPc^{RS} coated ITO glass produced an increase in the intensity of the anodic wave (from 3 μA in ITO to 30 μA in films coated with ZnPc^{RS}). The cathodic peak also increased from −7 μA to −45 μA. The observed increase proved the electrocatalytic properties of the zinc phthalocyanine derivative.

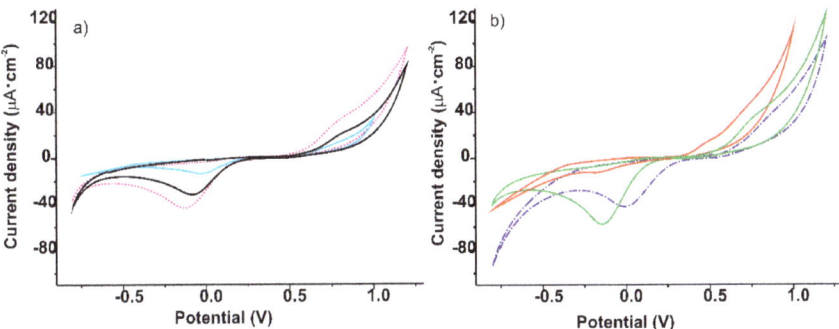

Figure 4. Cyclic voltammograms registered in catechol 10^{-3} mol·L^{-1} (0.01 M phosphate buffer as electrolyte using (**a**) AuNP^{tOcBr} (blue —), ZnPc^{RS} (pink ······), ZnPc^{R}-S-ZnPc^{R} (black —) and of (**b**) the mixture AuNP^{tOcBr}/ZnPc^{RS} (purple ·—·—·), the mixture AuNP^{tOcBr}/ZnPc^{R}-S-ZnPc^{R} (green —) and the covalent adduct AuNP^{tOcBr}-S-ZnPc^{R} (red —), Scan rate 100 mV·s^{-1}.

Voltammograms obtained using electrodes modified with the AuNP^{tOcBr}/ZnPc^{RS} mixture also showed the expected anodic and cathodic waves. The anodic peak was almost identical to that obtained with ZnPc^{RS} alone, indicating that the influence on the electrocatalytic behavior of the AuNP^{tOcBr} present in the mixture was almost negligible. In contrast, the position of the cathodic peak shifted to lower potentials and the mixture of compounds seemed to show a stronger electrocatalytic effect than the components separately. The mixture of gold nanoparticles with the dimeric species AuNP^{tOcBr}/ZnPc^{R}-S-ZnPc^{R}, produced a higher increase in the intensity of the cathodic wave than the mixture of the nanoparticle with the monomeric phthalocyanine. This could be due to the stronger interaction between the phthalocyanine rings and the gold NPs.

Responses observed using electrodes modified with the covalent adduct AuNP^{tOcBr}-S-ZnPc^{R} differed from those obtained with the AuNP^{tOcBr}/ZnPc^{RS} mixture. The main difference was observed in the anodic peak that showed an important shift to lower potentials. In contrast, the electrocatalytic effect disappeared completely in the cathodic peak.

The important differences between the mixture and the adduct confirm the importance of the nature of the interaction between the phthalocyanine and the gold nanoparticle in the electrocatalytic mechanism.

In order to further analyze the effect of the modifiers on the dynamic character of the electrochemical process, voltammograms were registered at different scan rates (from 0.01 to 1.0 V·s^{-1}). Experiments were carried out in catechol 10^{-4} mol·L^{-1} (in phosphate buffer 0.01 M, pH = 7). In all cases, the intensity of the peaks increased with the scan rate. Simultaneously, cathodic peaks shifted to more negative potentials while anodic peaks shifted to more positive potentials.

According to the literature, when the peak current varies linearly with the sweep rate (v), the transfer of the electrons from the analyte to the electrode is the limiting step of the process. If the peak current varies linearly with the square root of the scan rate ($v^{1/2}$), the electrode reaction is controlled by diffusion. Figure 5 shows the analysis of the dynamic behavior of the sensor based on the mixture AuNPtOcBr/ZnPcR-S-ZnPcR. Figure 5a shows the relationship between the current density and the sweep rate (v), according to the Laviron model (Equation (1)) and Figure 5b shows the relationship between the current density and the square root of the scan rate ($v^{1/2}$) according to the Randles–Sevcik model (Equation (2)). Slopes and correlation coefficients for all the sensors are collected in Table 1.

$$Ic = \frac{n^2F^2v\Gamma A}{4RT} \tag{1}$$

$$Ic = 0.446FA\sqrt{\frac{FDv}{RT}}[C] \tag{2}$$

where I_c is the peak current, n is the number of electrons involved in the process, F is the Faraday's constant, v is the scan rate (expressed in V·s^{-1}), Γ is the surface coverage of the electrode reaction substance (mol cm^{-2}), A is the electrode area (cm^2), R is the ideal gas constant (8.314 J·mol^{-1}·K^{-1}), T is the temperature (298 K). D is the diffusion coefficient, [C] the bulk concentration of species C in the solution.

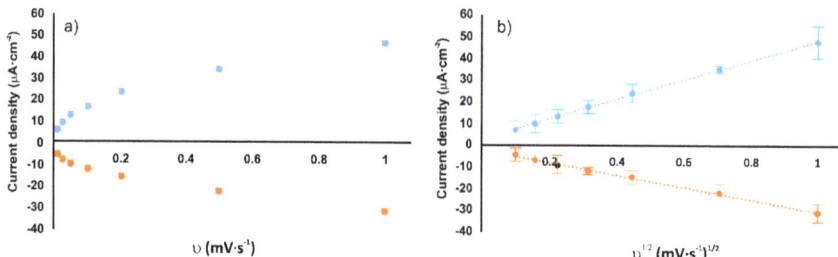

Figure 5. Analysis of the dynamic behavior of the sensor based on the mixture AuNPt^{OcBr}/ZnPc^R-S-ZnPc^R (**a**) Laviron model, graphical relationship between the current peak density and the sweep rate (v), (**b**) Randles–Sevcik model peak current density varies linearly with the square root of the scan rate ($v^{1/2}$).

Table 1. Relationship between the intensity of the peaks and the scan rate in sensors immersed in 10^{-4} mol·L^{-1} catechol. (Results shown correspond to the average values obtained from three different experiments).

	Cathodic Wave at ca. −0.15 V					
	Laviron Model: $I= f(v)$,			Randless–Sevcik Model $I = f (sqrt(v))$		
	I_c (µA·cm^{-2}) vs. v (V/s)			I_c (µA·cm^{-2}) vs. $v^{1/2}$ (V/s)$^{1/2}$		
Sensor	Slope	Intercept	R^2	Slope	Intercept	R^2
AuNPt^{OcBr}/ZnPc^{RS}	−19.32	−6.11	0.993	−21.31	−2.32	0.973
AuNPt^{OcBr}/ZnPc^R-S-ZnPc^R	−24.78	−7.59	0.953	−28.26	−2.34	0.997
AuNPt^{OcBr}-S-ZnPc^R	−54.46	−5.33	0.915	−31.65	−1.64	0.982
	Anodic wave at ca. 0.8 V					
	I_c (µA·cm^{-2}) vs. v (V/s)			I_c (µA·cm^{-2}) vs. $v^{1/2}$ (V/s)$^{1/2}$		
Sensor	Slope	Intercept	R^2	Slope	Intercept	R^2
AuNPt^{OcBr}/ZnPc^{RS}	42.12	6.34	0.980	37.20	−0.06	0.962
AuNPt^{OcBr}/ZnPc^R-S-ZnPc^R	38.85	11.71	0.935	44.76	3.30	0.998
AuNPt^{OcBr}-S-ZnPc^R	210.91	10.45	0.934	92.17	1.82	0.987

As observed in Figure 5 and Table 1, correlation coefficients R^2 show that both models, Laviron and Randless–Sevcik could explain the dynamic behavior of the sensors. This is quite common in chemically modified electrodes immersed in electroactive solutions. However, the fitting is clearly linear in the diffusion-controlled model.

In order to further analyze the nature of the limiting step of the electrode reaction, the relationship between $I/v^{1/2}$ vs. v was analyzed. If this relationship is linear, the mechanism that controls the redox process is the charge transfer of the adsorbate. On the contrary, when the current function $I/v^{1/2}$ is independent of the scan rate, the predominant mechanism is diffusion. In this case, all sensors showed a combination of both mechanisms: At low scan rates (lower than 0.20 V·s^{-1}), the charge transfer predominated. At scan rates over 0.20 V·s^{-1}, the process was limited by diffusion (Figure 6).

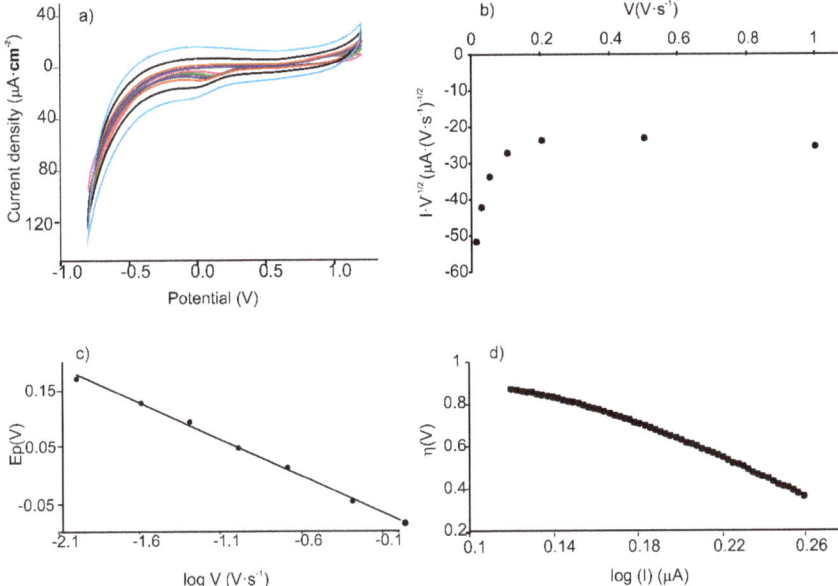

Figure 6. Analysis of the electron transfer mechanism. (**a**) Cyclic voltammograms of AuNP^{tOcBr}/ZnPc^{RS} in catechol 10^{-4} mol·L^{-1} registered at increasing scan rates from 0.01 to 1 V·s^{-1}, (**b**) representation of I·$v^{-1/2}$ vs. scan rate for the cathodic pea, (**c**) variation of peak potentials vs. the logarithm of the scan rates. (**d**) Representation of Tafel plot: overpotential η vs log (I) in cathodic peak.

At scan rates lower than 0.20 V·s^{-1}, where the charge transfer is the limiting step, the charge transfer coefficient α can be calculated from the slope of the Laviron equation (Equation (3)). This coefficient is related to the efficiency of the electron transfer between the electrode and the surface-confined redox couple [36],

$$E_c = E^0 - \frac{2.3RT}{(\alpha_c)nF} \log v \qquad (3)$$

where E_c is the cathodic peak potential, E^0 is a constant that includes the formal standard potential, R is the ideal gas constant (8.314 J·mol^{-1}·K^{-1}), T is the temperature (298 K), α_c is the charge transfer coefficient, n is the number of electrons involved in the process, F is the Faraday's constant and v is the scan rate (expressed in V·s^{-1}).

Our results showed that the slope of the E_c vs. log v gave αn values between 0.43 and 0.45 (Table 2).

Table 2. Relationship with scan rate in sensors immersed in catechol 10^{-4} mol·L^{-1}.

	Cathodic Peak								
	$I_c/v^{1/2}$ vs. v		Log I vs. η			E_c vs. log v			
Sensor	Slope	R^2	Slope	R^2	α	Slope	R^2	αn	n
AuNP^{tOcBr}/ZnPc^{AcS}	252.39	0.882	3.73	0.986	0.28	0.130	0.997	0.452	2.05
AuNP^{tOcBr}/ZnPc^R-S-ZnPc^R	177.68	0.947	4.43	0.979	0.26	−0.105	0.997	0.562	2.14
AuNP^{tOcBr}-S-ZnPc^R	102.29	0.958	2.912	0.997	0.17	0.136	0.990	0.434	2.43
	Anodic Peak								
	$I_c/v^{1/2}$ vs. v		Log I vs. η			E_c vs. log v			
Sensor	Slope	R^2	Slope	R^2	α	Slope	R^2	αn	n
AuNP^{tOcBr}/ZnPc^{AcS}	−244.14	0.872	2.34	0.998	0.76	0.189	0.953	0.313	2.28
AuNP^{tOcBr}/ZnPc^R-S-ZnPc^R	−153.15	0.888	3.88	0.999	0.77	0.113	0.985	0.522	2.28
AuNP^{tOcBr}-S-ZnPc^R	−137.50	0.964	1.579	0.999	0.90	0.331	0.992	0.165	1.76

In order to obtain information about the efficiency of the catalyst and the rate-determining step, representation of log I (µA) vs. the overpotential, η (V), Tafel plot was used to calculate values thanks to the simplified Butler–Volmer equation (Equation (4)).

$$Log I = \log I_0 - \frac{\alpha F}{2.3RT}\eta \quad (4)$$

The α values obtained can be substituted in Laviron's equation to calculate the number of electrons implicated in the redox process. All these values are listed in Table 2. Calculations indicate a two-electron redox reaction of catechol at all three electrodes.

Similar calculations were carried out using the anodic peak (Equation (5) and (6)) where E_a is the anodic peak potential.

$$E_a = E^0 - \frac{2.3RT}{(1-\alpha)nF} \log v \quad (5)$$

$$Log I = \log I_0 - \frac{(1-\alpha)F}{2.3RT}\eta \quad (6)$$

As observed in the table, the charge transfer coefficient α, showed different values in the AuNP^{tOcBr}/ZnPc^{RS} composite than in the AuNP^{tOcBr}-S-ZnPc^{R} adduct, confirming the different mechanism of the reduction process. It is noteworthy that the behavior of the mixture containing the dimer (AuNP^{tOcBr}/ZnPc^{R}-S-ZnPc^{R}) where the thiol group is protected coincided with that of the mixture AuNP^{tOcBr}/ZnPc^{RS}. This confirms that the interaction between the phthalocyanine and the nanoparticle did not occur through thiol bonds.

The limits of detection (LOD) were calculated from voltammograms registered in solutions with increasing concentrations of catechol (from 4.0×10^{-6} to 1.40×10^{-4} mol·L^{-1}). Experiments were replicated three times for each sensor. As expected, the intensity of the peaks increased with the concentration and the responses were linear in the studied range (Figure 7). Calibration curves were constructed by representing I_a (or I_c) vs. catechol concentration. Sensitivity and LODs were calculated from those plots. The results are shown in Table 3. As expected, and according to the α parameters obtained from the experiments carried out at different sweep rates, the presence or absence of a covalent bond influenced the sensitivities and the LODs.

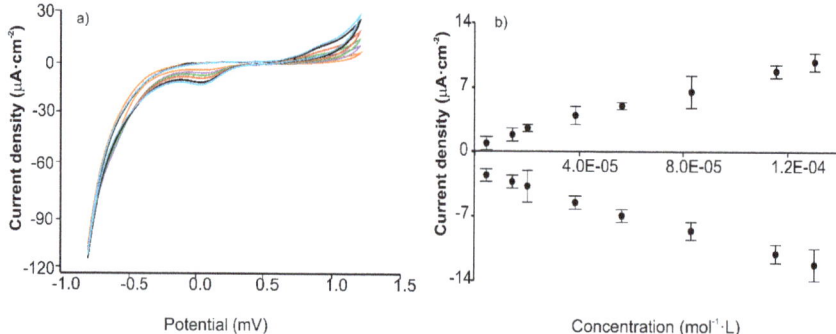

Figure 7. Analysis of limits of detection. (a) Voltammetric response of the AuNP^{tOcBr}/ZnPc^{AcS} sensor to increasing concentrations of catechol (from 4×10^{-6} to 1.40×10^{-4} mol·L^{-1} in phosphate buffer), (b) calibration curves calculated in both the anodic and the cathodic peaks.

Table 3. Sensitivity, limit of detection (LD) and correlation coefficient (R^2).

	Sensor	Sensitivity ($\mu A \cdot cm^{-2}/mol \cdot L^{-1}$)	LOD ($\times 10^{-6}$ mol·L^{-1})	R^2
Cathodic peak	AuNP^{tOcBr}	−23,747	4.0	0.992
	ZnPc^{RS}	−87,223	2.0	0.987
	AuNP^{tOcBr}/ZnPc^{RS}	−76,350	0.9	0.997
	AuNP^{tOcBr}/ZnPc^R-S-ZnPc^R	−99,039	1.2	0.989
	AuNP^{tOcBr}-S-ZnPc^R	−32,419	8.3	0.989
Anodic peak	AuNP^{tOcBr}	10,539	4.4	0.992
	ZnPc^{RS}	28,343	2.9	0.985
	AuNP^{tOcBr}/ZnPc^{RS}	68,170	2.2	0.994
	AuNP^{tOcBr}/ZnPc^R-S-ZnPc^R	44,337	2.07	0.981
	AuNP^{tOcBr}-S-ZnPc^R	45,498	0.13	0.993

According to Table 3, LODs obtained from the cathodic curves were lower in sensors modified with the mixtures (0.9 × 10^{-6} M for AuNP^{tOcBr}/ZnPc^{RS} and 1.2 × 10^{-6} M AuNP^{tOcBr}/ZnPc^R-S-ZnPc^R). These values were quite similar to those obtained with the ZnPc^R alone indicating a weak electrocatalytic effect. The AuNP^{tOcBr}-S-ZnPc^R covalent adduct did not show any electrocatalytic effect in the cathodic process. These values are similar to those obtained with other sensors modified with nanoparticles or phthalocyanines separately [2,10,36,38,41,42].

Results were completely different in the anodic wave. LODs calculated from the anodic peaks, showed that the sensor modified with the covalent adduct gave the lowest LOD values (1.38 × 10^{-7} mol·L^{-1}), confirming the strong influence of the covalent bond in the mechanism of catechol oxidation. This result indicates that the covalent interaction facilitated the electron transfer during oxidation and that the nature of the interaction between both components (weak bond in the mixture or covalent bond in the adduct) modulates the catalytic activity.

4. Conclusions

New voltammetric sensors based on combinations of gold nanoparticles and sulfur substituted zinc phthalocyanines have been developed and used as electrochemical sensors for the detection of catechol. The electron transfer process, as well as the existence of synergistic effects between both components in the absence and presence of covalent links has been analyzed.

It has been demonstrated that the electrocatalytic properties and the kinetic parameters depend on the type of interaction between both components. The AuNP^{tOcBr}/ZnPc^{RS} and the AuNP^{tOcBr}/ZnPc^R-S-ZnPc^R mixtures enhance the electron transfer rate of the catechol reduction. Both modifiers showed similar LODs of 10^{-6} mol·L^{-1}. As in the dimeric phthalocyanine the sulfur group is blocked, it can be inferred that the sulfur group does not play a role in the electrocatalytic process. In turn, the AuNP^{tOcBr}-S-ZnPc^R covalent adduct facilitates the oxidation of catechol, showing an enhanced charge transfer rate, and an LOD of 10^{-7} mol·L^{-1}.

Under the light of these results, combining covalently nanoparticles and phthalocyanines can be considered a good strategy to improve the charge transfer rate and the limits of detection of catechol. Future works should be dedicated to analyzing the effect of the interaction between electrocatalytic materials in other systems different than the nanoparticle-phthalocyanine system.

Supplementary Materials: The following are available online at http://www.mdpi.com/2079-4991/9/11/1506/s1.

Author Contributions: Conceptualization, M.L.R.-M., Á.S.-S. and C.G.-C.; methodology, M.L.R.-M.; investigation, Á.I.R.-C., J.O. and C.G.-H.; resources, Á.S.-S. and M.A.R.-P.; data curation, A.I.R.-C. and F.M.-P.; writing—original draft preparation, A.I.R.-C.; writing—review and editing, M.L.R.-M. and Á.S.-S.; funding acquisition, M.L.R.-M. and Á.S.-S.

Funding: Financial support by MINECO-FEDER (RTI2018-097990-B-I00 and CTQ2017-87102-R) and the Junta de Castilla y Leon FEDER (VA275P18) is gratefully acknowledged.

Acknowledgments: Celia Garcia-Hernandez would also like to thank Junta de Castilla y León for a grant (BOCYL-D-4112015-9).

Conflicts of Interest: The authors declare no conflict of interest.

References

1. Blasco, A.J.; González Crevillén, A.; González, M.C.; Escarpa, A. Direct Electrochemical Sensing and Detection of Natural Antioxidants and Antioxidant Capacity in Vitro Systems. *Electroanalysis* **2007**, *19*, 2275–2286. [CrossRef]
2. Medina-Plaza, C.; García-Cabezón, C.; García-Hernández, C.; Bramorski, C.; Blanco-Val, Y.; Martín-Pedrosa, F.; Kawai, T.; de Saja, J.A.; Rodríguez-Méndez, M.L. Analysis of organic acids and phenols of interest in the wine industry using Langmuir–Blodgett films based on functionalized nanoparticles. *Anal. Chim. Acta* **2015**, *853*, 572–578. [CrossRef] [PubMed]
3. Makhotkina, O.; Kilmartin, P.A. The use of cyclic voltammetry for wine analysis: Determination of polyphenols and free sulfur dioxide. *Anal. Chim. Acta* **2010**, *668*, 155–165. [CrossRef] [PubMed]
4. Yang, C.; Denno, M.E.; Pyakurel, P.; Venton, B.J. Recent trends in carbon nanomaterial-based electrochemical sensors for biomolecules: A review. *Anal. Chim. Acta* **2015**, *887*, 17–37. [CrossRef] [PubMed]
5. Gay Martín, M.; de Saja, J.A.; Muñoz, R.; Rodríguez-Méndez, M.L. Multisensor system based on bisphthalocyanine nanowires for the detection of antioxidants. *Electrochim. Acta* **2012**, *68*, 88–94. [CrossRef]
6. Liu, D.; Li, F.; Yu, D.; Yu, J.; Ding, Y.; Liu, D.; Li, F.; Yu, D.; Yu, J.; Ding, Y. Mesoporous Carbon and Ceria Nanoparticles Composite Modified Electrode for the Simultaneous Determination of Hydroquinone and Catechol. *Nanomaterials* **2019**, *9*, 54. [CrossRef]
7. Krampa, F.; Aniweh, Y.; Awandare, G.; Kanyong, P.; Krampa, F.D.; Aniweh, Y.; Awandare, G.A.; Kanyong, P. A Disposable Amperometric Sensor Based on High-Performance PEDOT:PSS/Ionic Liquid Nanocomposite Thin Film-Modified Screen-Printed Electrode for the Analysis of Catechol in Natural Water Samples. *Sensors* **2017**, *17*, 1716. [CrossRef]
8. Apetrei, C.; Casilli, S.; De Luca, M.; Valli, L.; Jiang, J.; Rodriguez-Méndez, M.L.; De Saja, J.A. Spectroelectrochemical characterisation of Langmuir-Schaefer films of heteroleptic phthalocyanine complexes. Potential applications. *Colloids Surfaces A Physicochem. Eng. Asp.* **2006**, *284–285*, 574–582. [CrossRef]
9. Rodriguez-Méndez, M.L.; Gay, M.; de Saja, J.A. New insights into sensors based on radical bisphthalocyanines. *J. Porphyr. Phthalocyanines* **2009**, *13*, 1159–1167. [CrossRef]
10. Alessio, P.; Martin, C.S.; de Saja, J.A.; Rodriguez-Mendez, M.L. Mimetic biosensors composed by layer-by-layer films of phospholipid, phthalocyanine and silver nanoparticles to polyphenol detection. *Sens. Actuators B Chem.* **2016**, *233*, 654–666. [CrossRef]
11. Matemadombo, F.; Apetrei, C.; Nyokong, T.; Rodríguez-Méndez, M.L.; de Saja, J.A. Comparison of carbon screen-printed and disk electrodes in the detection of antioxidants using CoPc derivatives. *Sens. Actuators B Chem.* **2012**, *166*, 457–466. [CrossRef]
12. Fernandes, E.G.R.; Brazaca, L.C.; Rodríguez-Mendez, M.L.; Saja, J.A.; de Zucolotto, V. Immobilization of lutetium bisphthalocyanine in nanostructured biomimetic sensors using the LbL technique for phenol detection. *Biosens. Bioelectron.* **2011**, *26*, 4715–4719. [CrossRef] [PubMed]
13. Zagal, J.H.; Griveau, S.; Silva, J.F.; Nyokong, T.; Bedioui, F. Metallophthalocyanine-based molecular materials as catalysts for electrochemical reactions. *Coord. Chem. Rev.* **2010**, *254*, 2755–2791. [CrossRef]
14. Tkachenko, N.V.; Efimov, A.; Lemmetyinen, H. Covalent phthalocyanine-fullerene dyads: Synthesis, electron transfer in solutions and molecular films. *J. Porphyr. Phthalocyanines* **2011**, *15*, 780–790. [CrossRef]
15. Martín-Gomis, L.; Peralta-Ruiz, F.; Thomas, M.B.; Fernández-Lázaro, F.; D'Souza, F.; Sastre-Santos, Á. Multichromophoric Perylenediimide-Silicon Phthalocyanine-C 60 System as an Artificial Photosynthetic Analogue. *Chem. A Eur. J.* **2017**, *23*, 3863–3874. [CrossRef]
16. Fukuzumi, S.; Ohkubo, K.; Ortiz, J.; Gutiérrez, A.M.; Fernández-Lázaro, F.; Sastre-Santos, Á. Control of Photoinduced Electron Transfer in Zinc Phthalocyanine-Perylenediimide Dyad and Triad by the Magnesium Ion. *J. Phys. Chem. A* **2008**, *112*, 10744–10752. [CrossRef]
17. Blas-Ferrando, V.M.; Ortiz, J.; Ohkubo, K.; Fukuzumi, S.; Fernández-Lázaro, F.; Sastre-Santos, Á. Submillisecond-lived photoinduced charge separation in a fully conjugated phthalocyanine–perylenebenzimidazole dyad. *Chem. Sci.* **2014**, *5*, 4785–4793. [CrossRef]

18. Aragão, J.S.; Ribeiro, F.W.P.; Portela, R.R.; Santos, V.N.; Sousa, C.P.; Becker, H.; Correia, A.N.; de Lima-Neto, P. Electrochemical determination diethylstilbestrol by a multi-walled carbon nanotube/cobalt phthalocyanine film electrode. *Sens. Actuators B Chem.* **2017**, *239*, 933–942. [CrossRef]
19. Jubete, E.; Żelechowska, K.; Loaiza, O.A.; Lamas, P.J.; Ochoteco, E.; Farmer, K.D.; Roberts, K.P.; Biernat, J.F. Derivatization of SWCNTs with cobalt phthalocyanine residues and applications in screen printed electrodes for electrochemical detection of thiocholine. *Electrochim. Acta* **2011**, *56*, 3988–3995. [CrossRef]
20. Nombona, N.; Antunes, E.; Litwinski, C.; Nyokong, T. Synthesis and photophysical studies of phthalocyanine-gold nanoparticle conjugates. *Dalt Trans.* **2011**, *40*, 11876–11884. [CrossRef]
21. Blas-Ferrando, V.M.; Ortiz, J.; Fernández-Lázaro, F.; Sastre-Santos, Á. Synthesis and characterization of a sulfur-containing phthalocyanine-gold nanoparticle hybrid. *J. Porphyr. Phthalocyanines* **2015**, *19*, 335–343. [CrossRef]
22. Baldovi, H.G.; Blas-Ferrando, V.M.; Ortiz, J.; Garcia, H.; Fernández-Lázaro, F.; Sastre-Santos, Á. Phthalocyanine-Gold Nanoparticle Hybrids: Modulating Quenching with a Silica Matrix Shell. *ChemPhysChem* **2016**, *17*, 1579–1585. [CrossRef] [PubMed]
23. Pan, X.; Liang, X.; Yao, L.; Wang, X.; Jing, Y.; Ma, J.; Fei, Y.; Chen, L.; Mi, L.; Pan, X.; et al. Study of the Photodynamic Activity of N-Doped TiO2 Nanoparticles Conjugated with Aluminum Phthalocyanine. *Nanomaterials* **2017**, *7*, 338. [CrossRef] [PubMed]
24. Pingarrón, J.M.; Yáñez-Sedeño, P.; González-Cortés, A. Gold nanoparticle-based electrochemical biosensors. *Electrochim. Acta* **2008**, *53*, 5848–5866. [CrossRef]
25. Saha, K.; Agasti, S.S.; Kim, C.; Li, X.; Rotello, V.M. Gold Nanoparticles in Chemical and Biological Sensing. *Chem. Rev.* **2012**, *112*, 2739–2779. [CrossRef]
26. Yu, C.; Wang, Q.; Qian, D.; Li, W.; Huang, Y.; Chen, F.; Bao, N.; Gu, H. An ITO electrode modified with electrodeposited graphene oxide and gold nanoclusters for detecting the release of H2O2 from bupivacaine-injured neuroblastoma cells. *Microchim. Acta* **2016**, *183*, 3167–3175. [CrossRef]
27. Singh, S.; Jain, D.S.M. Sol–gel based composite of gold nanoparticles as matix for tyrosinase for amperometric catechol biosensor. *Sens. Actuators B Chem.* **2013**, *182*, 161–169. [CrossRef]
28. Medina-Plaza, C.; Rodriguez-Mendez, M.L.; Sutter, P.; Tong, X.; Sutter, E. Nanoscale Au-In Alloy-Oxide Core-Shell Particles as Electrocatalysts for Efficient Hydroquinone Detection. *J. Phys. Chem. C* **2015**, *119*, 25100–25107. [CrossRef]
29. Lin, X.; Ni, Y.; Kokot, S. Glassy carbon electrodes modified with gold nanoparticles for the simultaneous determination of three food antioxidants. *Anal. Chim. Acta* **2013**, *765*, 54–62. [CrossRef]
30. Li, X.; Ye, X.; Li, C.; Wu, K. Substitution group effects of 2-mercaptobenzothiazole on gold nanoparticles toward electrochemical oxidation and sensing of tetrabromobisphenol A. *Electrochim. Acta* **2018**, *270*, 517–525. [CrossRef]
31. Steinebrunner, D.; Schnurpfeil, G.; Wichmann, A.; Wöhrle, D.; Wittstock, A. Synergistic Effect in Zinc Phthalocyanine—Nanoporous Gold Hybrid Materials for Enhanced Photocatalytic Oxidations. *Catalysts* **2019**, *9*, 555. [CrossRef]
32. Nyokong, T.; Antunes, E. Influence of nanoparticle materials on the photophysical behavior of phthalocyanines. *Coord. Chem. Rev.* **2013**, *257*, 2401–2418. [CrossRef]
33. Gonzalez-Anton, R.; Osipova, M.M.; Garcia-Hernandez, C.; Dubinina, T.V.; Tomilova, L.G.; Garcia-Cabezon, C.; Rodriguez-Mendez, M.L. Subphthalocyanines as electron mediators in biosensors based on phenol oxidases: Application to the analysis of red wines. *Electrochim. Acta* **2017**, *255*, 239–247. [CrossRef]
34. Muthukumar, P.; Abraham John, S. Synergistic effect of gold nanoparticles and amine functionalized cobalt porphyrin on electrochemical oxidation of hydrazine. *New J. Chem.* **2014**, *38*, 3473–3479. [CrossRef]
35. Muthukumar, P.; John, S.A. Effect of amine substituted at ortho and para positions on the electrochemical and electrocatalytic properties of cobalt porphyrins self-assembled on glassy carbon surface. *Electrochim. Acta* **2014**, *115*, 197–205. [CrossRef]
36. Medina-Plaza, C.; Furini, L.N.; Constantino, C.J.L.; De Saja, J.A.; Rodriguez-Mendez, M.L. Synergistic electrocatalytic effect of nanostructured mixed films formed by functionalised gold nanoparticles and bisphthalocyanines. *Anal. Chim. Acta* **2014**, *851*, 95–102. [CrossRef]

37. Alessio, P.; Aoki, P.H.B.; De Saja Saez, J.A.; Rodríguez-Méndez, M.L.; Constantino, C.J.L. Combining SERRS and electrochemistry to characterize sensors based on biomembrane mimetic models formed by phospholipids. *RSC Adv.* **2011**, *1*, 211–218. [CrossRef]
38. García-Hernández, C.; García-Cabezón, C.; Medina-Plaza, C.; Martín-Pedrosa, F.; Blanco, Y.; de Saja, J.A.; Rodríguez-Méndez, M.L. Electrochemical behavior of polypyrrol/AuNP composites deposited by different electrochemical methods: Sensing properties towards catechol. *Beilstein J. Nanotechnol.* **2015**, *6*, 2052–2061. [CrossRef]
39. Brust, M.; Walker, M.; Bethell, D.; Schiffrin, D.J.; Whyman, R.; Kirkland, A.I.; Logan, D.E. Synthesis of thiol-derivatised gold nanoparticles in a two-phase Liquid?Liquid system. *J. Chem. Soc. Chem. Commun.* **1994**, *7*, 801–802. [CrossRef]
40. Roth, P.J.; Theato, P. Versatile Synthesis of Functional Gold Nanoparticles: Grafting Polymers From and Onto. *Chem. Mater.* **2008**, *20*, 1614–1621. [CrossRef]
41. Alessio, P.; Pavinatto, F.J.; Oliveira, O.N.; de Saja, J.A.; Constantino, J.C.L.; Rodriguez-Mendez, M.L. Detection of catechol using mixed Langmuir–Blodgett films of a phospholipid and phthalocyanines as voltammetric sensors. *Analyst* **2010**, *135*, 2591–2599. [CrossRef] [PubMed]
42. Maximino, M.D.; Martin, C.S.; Paulovich, F.V.; Alessio, P. Layer-by-Layer Thin Film of Iron Phthalocyanine as a Simple and Fast Sensor for Polyphenol Determination in Tea Samples. *J. Food. Sci.* **2016**, *81*, C2344–C2351. [CrossRef] [PubMed]

 © 2019 by the authors. Licensee MDPI, Basel, Switzerland. This article is an open access article distributed under the terms and conditions of the Creative Commons Attribution (CC BY) license (http://creativecommons.org/licenses/by/4.0/).

Review

Bare Iron Oxide Nanoparticles: Surface Tunability for Biomedical, Sensing and Environmental Applications

Massimiliano Magro and Fabio Vianello *

Department of Comparative Biomedicine and Food Science, University of Padua, Agripolis-Viale dell'Università 16, 35020 Legnaro (PD), Italy; massimiliano.magro@unipd.it
* Correspondence: fabio.vianello@unipd.it; Tel.: +39-049-8272-638; Fax: 0039-049-8272-604

Received: 3 October 2019; Accepted: 11 November 2019; Published: 12 November 2019

Abstract: Surface modification is widely assumed as a mandatory prerequisite for the real applicability of iron oxide nanoparticles. This is aimed to endow prolonged stability, electrolyte and pH tolerance as well as a desired specific surface chemistry for further functionalization to these materials. Nevertheless, coating processes have negative consequences on the sustainability of nanomaterial production contributing to high costs, heavy environmental impact and difficult scalability. In this view, bare iron oxide nanoparticles (BIONs) are arousing an increasing interest and the properties and advantages of pristine surface chemistry of iron oxide are becoming popular among the scientific community. In the authors' knowledge, rare efforts were dedicated to the use of BIONs in biomedicine, biotechnology, food industry and environmental remediation. Furthermore, literature lacks examples highlighting the potential of BIONs as platforms for the creation of more complex nanostructured architectures, and emerging properties achievable by the direct manipulation of pristine iron oxide surfaces have been little studied. Based on authors' background on BIONs, the present review is aimed at providing hints on the future expansion of these nanomaterials emphasizing the opportunities achievable by tuning their pristine surfaces.

Keywords: iron oxide nanoparticles; surface chemistry; nanotechnology; food chemistry; biomedicine; environment

---

1. Introduction

The blossoming of nanoscience and nanotechnologies has raised increasing expectations in the fields of research and industry. By definition, nanotechnology concerns the synthesis, characterization and/or the use of materials, devices or structures displaying at least one dimension (or containing components with at least one dimension) comprised in the 1–100 nm range. When a particle size stands within these limits, its physical and chemical properties significantly differ from the corresponding bulk material. Indeed, if the latter has constant physical properties regardless its size, approaching the nanoscale the same material displays drastically variable physico-chemical properties. At the nanosize, the fraction of atoms on the material surface is increasingly relevant and properties become heavily dependent on the size, the shape, the surface morphology and on several other parameters, which are still object of intense study. Thus, the state of the art is continuously enriched by the hardly predictable and fascinating properties originated by creating novel nanomaterials, and this leads to a limitless landscape of opportunities as well as challenges.

In this view, metal oxides nanoparticles have gained particular interest as their magnetic, electronic and chemical features can be modulated by a wide choice of innovative synthetic methods [1]. As examples, the electrical properties of In_2O_3, SnO_2 and WO_3, nanoparticles were exploited for gas sensing applications [2]. TiO_2 nanoparticles were applied in photocatalysis for pollutant elimination, for the development of solar cells [3]. ZnO, thanks to its intrinsic properties as wide bandgap semiconductor, was object of extensive studies and was proposed for the development of solar cells, laser sources, gas sensors and as catalysts for various organic molecules [4]. Furthermore, one-dimensional (1-D) ZnO found application in electronics and optoelectronics [5]. GeO_2 nanoparticles demonstrated potential applications for improved optical fibers and other optoelectronic purposes [6]. Ga_2O_3 nanoparticles were employed in surface-catalyzed systems for electronic or optical applications [7] CuO nanoparticles were used as redox catalysts for several oxidation processes in photothermal and photoconductive applications [8]. MgO nanoparticles were largely applied for eliminating gaseous pollutants and as a catalysts for organic syntheses [9]. Al_2O_3 was employed as a support for immobilizing catalysts [10]. ZrO_2 nanoparticles are exploited as catalysts and gas sensing materials [11]. CeO_2 nanoparticles were applied for electrochemistry, gas sensing and material chemistry [12]. Iron oxide nanomaterials (Fe_2O_3 and Fe_3O_4) were recently employed in the development of electrochemical sensors for detecting drugs, such as paracetamol [13], chloramphenicol [14] and linagliptin [15].

Nowadays, nanotechnology can count on the evolution of novel synthetic processes aimed at the fine control of composition, size, shape, surface coating and surface charge of nanomaterials to cope with complex technological tasks [16]. Among nanomaterials, magnetic nanoparticles were widely employed and in the last decades these nanostructures have been applied for the immobilization of proteins and enzymes [17], bioseparations [18], immunoassays [19], biosensors [20] and drug delivery systems [21]. In this view, it should be mentioned that magnetic nanoparticles were classified as medical devices, and they should conform to ISO 10,993 guidelines according to the US-FDA. Accordingly, some magnetic nanoparticles have been already approved for clinical MRI applications (Feridex by AMAG Pharmaceuticals, Inc., Lexington, MA, USA; Endorem by Guerbet, Villepinte, France). Among magnetic nanoparticles, iron-oxide nanoparticles (IONPs) combine many advantageous properties, such as superparamagnetism, high values of saturation magnetization, easy control by low intensity magnetic fields, as well as non-toxicity, biodegradability and biocompatibility. IONPs can be further discriminated into various categories, such as ultra-small superparamagnetic iron oxide (USPIO) [22], cross-linked iron oxide (CLIO) [23] and mono-crystalline iron oxide nanoparticles (MIONs) [24].

Maghemite (γ-Fe_2O_3) and magnetite (Fe_3O_4) represent the most widely employed crystalline iron oxide structures, finding countless applications in many fields [25], spanning from magnetic data storage [26] and pigment production [27] to electrochemistry [28] and many others. These iron oxide nanoparticles can be synthesized by different routes, even if the production of monodisperse populations of magnetic grains and the maintenance of the colloidal nature of their dispersions is still hardly achievable [29]. Indeed, generally, magnetic nanoparticles require to be coated by organic polymers or inorganic shells for their stabilization in order to prevent particle aggregation and to preserve long-term stability, electrolyte and pH tolerance and to provide a proper surface chemistry for further functionalization [30]. As an example, lipophilic drugs can easily be loaded on iron oxide nanoparticles coated with hydrophobic polymers, which release the drug when the coating degrades into the organism [31]. Notwithstanding, many of the reported polymeric coatings used for nanoparticle stabilization suffer of physical-chemical lability and eventually desorb in the bulk solution, thus reducing the stability of the resulting dispersions. Furthermore, processes proposed to coat nanoparticles are cumbersome, time-consuming, and expensive, with low yields, limiting their large scale application. At last, nanoparticle coverage reduces the average magnetic moment by introducing a diamagnetic shell in the final nanomaterial.

A number of approaches were developed for endowing nanoparticle surfaces of proper chemical functionalities. For simplicity, they can be grouped into two main categories: covalent conjugation and physical interactions. Covalent strategies involve linkages of the moiety of interest directly to amino, carboxyl, thiol or hydroxyl functional groups on the surface of previously coated magnetic nanoparticles. Generally, methods require mild reactive conditions for binding, and hence, they are suitable for organic molecules with the tendency to degradation and denaturation. Physical interactions, such as electrostatic, hydrophilic/hydrophobic and affinity interactions, can also lead to stable conjugates [32]. In particular, the latter offers the most stable noncovalent binding [33,34].

Notwithstanding that in the last decades engineered iron oxide nanomaterials have led to exciting developments and were widely employed in interdisciplinary studies involving physics, chemistry, biology and medicine, bare iron oxide nanoparticles are (BIONs) are object of an increasing interest due to advantages as cost-effectiveness and environmental sustainability. In this view, even if scientific literature on BIONs is not very diffused yet, the proposed applications are multifaceted and range from catalysis to protein purification [35–37]. It is important to mention that a considerable body of work is emerging from the use of BIONs in microalgae harvesting for biomass exploitation purposes [38–44] and from their employment in water remediation [45,46]

BIONs couple well-known features, such as magnetism, with unique properties due to the availability of under-coordinated iron(III) sites on their surfaces as a consequence of dangling bonds derived from crystal truncation. Significantly, iron(III) sites possess multifaceted reactivity. As an example, they can display catalytic activity. Hence, the catalyst can be easily driven by an external magnet. Furthermore, surface iron(III) sites behave, in some extents, as iron(III) ions providing a competitive alternative to immobilized metal affinity chromatography (IMAC) for protein purification. The knowledge of the reactivity of iron oxide surface and the comprehension of its chemical behavior may open new avenues for challenging opportunities.

Therefore, the present review, besides being aimed to the valorization of BIONs, is aimed at highlighting their importance as building blocks for producing complex nanostructured architectures, as well as of the possible achievements obtainable by tuning pristine iron oxide surfaces without the screening of stabilizing coatings.

For simplicity, scientific publications focused on innovative applications of BIONs were grouped and reviewed into the following categories: BIONs for food industry, BIONs for biomedicine, BIONs for the environment.

2. BIONs for Food Industry

Nanotechnology based solutions are used in nearly all the segments of the food industry, ranging from agriculture (e.g., pesticides, fertilizers and vaccines delivery, plant and animal pathogen detection and targeted genetic engineering) to food processing (e.g., flavors encapsulation or odor enhancers, food textural or quality improvement, novel gelation or viscosity control agents) to food packaging (e.g., pathogen protection, gas sensors, UV-protection films) to nutrient supplements (e.g., nutraceuticals with high stability and bioavailability). In this view, iron oxide nanoparticles presenting peculiar surface chemistries are suitable for being competitive alternatives to common processes.

Innovative analytical techniques are continuously proposed to control critical steps of an increasing variety of processes and in very different environments [47]. Nano-biosensors developed from the combination of nanotechnology and biological molecules have been proposed as effective alternatives to traditional methods in terms of sensitivity and response time [48,49].

The main advantages of using BIONs for the development of biosensors are related to the immediacy of construction. Firstly, these nanoparticles possess versatile surfaces, prone to bind macromolecules by simple self-assembly and, therefore, the resulting nano-bioconjugate is enzymatically active. Conversely, common protein immobilization processes require complicated chemical reactions, which may affect the preservation of the biological activity of the macromolecules. In addition, BIONs can also count on the electro-catalytic and magnetic properties of their iron oxide core, allowing the development of multi-functional devices.

Among proposed analytical devices, electrochemical detection is a popular method for biosensors with possible applications in food industry. Electrochemical sensors arouse interest for their convenient instrumental setup, low cost, short analysis time and experimental simplicity. Compared to other analytical methods, the electrochemical approach may be useful for food samples because matrix problems from the various food components can be avoided. Among electrochemical devices, amperometric biosensors operate by applying a constant potential at the working electrode and monitoring the current associated with the oxidation or reduction of an electroactive substance involved in the recognition process. The potential at the working electrode is maintained constant with respect to a reference electrode (usually Ag/AgCl), which is at equilibrium.

Electrochemical sensor suffers from several drawbacks, such as fouling phenomena occurring upon exposure to biological matrixes, leading to difficult redox processes at the electrode surface. In this view, electrode performances can be improved by BIONs, due to their electrocatalytic properties and high surface to mass ratio [50].

In order to improve the sensing properties and the analytical performances of biosensing devices based on nanomaterials, hybrids constituted of inorganic nanostructures and biological molecules were proposed [50]. Binary hybrids have been investigated during the past few decades due to the emerging properties of nanoparticle composites. As an example, the electrocatalytic properties of electrostatically stabilized core–shell nanoclusters composed of bare iron oxide nanoparticles (BIONs) and a set of differently charged carbon nanomaterials (Gallic acid modified carbon dots, PEG modified graphene dots and quaternized carbon dots) were investigated [51]. The combination of quaternized carbon dots (Q-CDs) with an excess of iron oxide nanoparticles led to the spontaneous formation of a Q-CD@BION hybrid, which possessed peculiar electrocatalytic properties, different from the parent components, and attributable to the influence of the strong electrostatic interactions at the interface between the carbonaceous and the inorganic nanomaterials. This led to an alteration of the iron oxide surface properties (see Figure 1). Despite Q-CDs represented only a small fraction of the hybrid (about 1% w/w), they were able to orient BIONs electrocatalysis toward the selective oxidation of polyphenols at low applied potentials (+0.1 V vs. SCE). The Q-CD@IONP hybrid was used for developing a coulometric sensor constituted of a simple carbon paste electrode inside a small volume electrochemical flow cell (1 µL) and finally used on real samples for assessing the concentration of polyphenols from plant extracts. The system responded linearly to chlorogenic acid, used as reference molecule, up to 1 mM, with a sensitivity of 224.6 ± 1.5 nC mM^{-1} and a limit of detection of 26.4 mM [51].

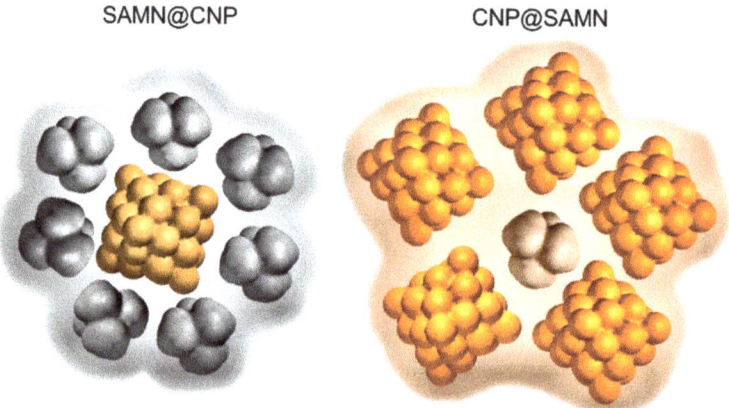

Figure 1. Graphical sketches of the hybrid nanomaterials constituted of BIONs and carbon nanoparticles. SAMNs (surface active maghemite nanoparticles) are a particular category of BIONs; CNPs are the carbon nanoparticles. CNPs constituted the shell (left) or the core (right) of the hybrids. Reproduced from [51], with permission from Royal Soc. Chem. 2019.

The preservation of the catalytic activity of enzymes is a mandatory requirement for the real applicability of hybrid nano-bioconjugates, and the control over protein–nanoparticle interactions remains a crucial task. Actually, the creation of stable and catalytically active nano-hybrids is at the core of the development of the next generation biosensing platforms. In this view, peculiar BIONs (Surface Active Maghemite Nanoparticles, SAMNs) represent an attracting option. In fact, these nanoparticles possess the ability to selectively bind proteins and, at the same time, they display peculiar electrocatalytic properties. As an example, recombinant aminoaldehyde dehydrogenase from tomato (SlAMADH1) was used and successfully bound on BIONs by self-assembly in a proof-of-concept study [52]. Significantly, the enzymatic activity of the biomolecule was preserved for more than 6 months, and the hybrid nanomaterial (BION@SlAMADH1) was applied for the development of an electrochemical biosensor for the assessment of aminoaldehydes in commercial alcoholic beverages (Figure 2). The linearity range spanned from 25 µM to 2.0 mM aminoaldehydes, the sensitivity was 5.33 µC µM^{-1} cm^{-2}, and the limit of detection (LOD) was 750 nM [52].

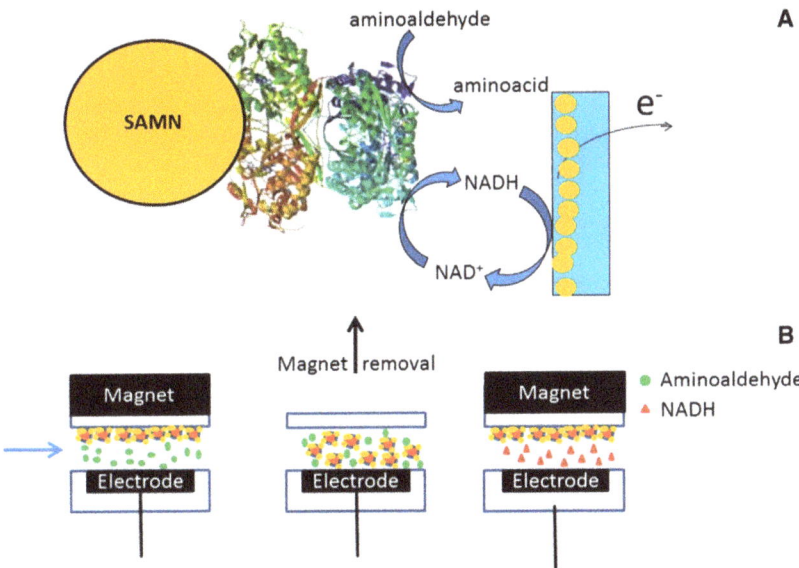

Figure 2. Biosensor for aminoaldehydes developed by immobilizing a recombinant aminoaldehyde dehydrogenase on peculiar BIONs (surface active maghemite nanoparticles, SAMNs). (**A**) Graphical sketch simplifying the working principle of the biosensor; (**B**) scheme representing the operational steps of the biosensor, exploiting the magnetic properties of the biological element (BION@SlAMADH1). Reproduced from [52], with permission from *Springer nature*, 2019.

Alternatively, BION surface can be tuned, via simple self-assembly reactions in water, to expand the spectrum of developable biosensing systems. In this view, rhodamine isothiocyanate (RITC) and tannic acid (TA) were used as bridging functionalities for the immobilization of glucose oxidase and laccase, and the as-obtained nano-bioconjugates were successfully exploited for building electrochemical biosensors for the detection of glucose in apricot juice (0–1.5 mM linearity range, 45.85 nA mM^{-1} cm^{-2} sensitivity and 0.9 µM LOD) [53] and polyphenols in blueberry extracts (100 nM-50 µM linearity range, 868.9 ± 1.9 nA µM^{-1} sensitivity and 81 nM LOD) [54].

The determination of foodborne pathogens is obviously a crucial task in food industry. Traditional diagnostic techniques for pathogens comprise colony counting, immunological-based methods and polymerase chain reaction (PCR), but all these methods present some important drawbacks [55]. Most of them need expensive, sophisticated instrumentation which requires trained personnel; hence, they are hardly suitable for developing countries. Their application use is also limited in the field, such as distribution centers.

Moreover, the short shelf-life and the high cost of some reagents, such as DNA primers and enzymes, represent a limit for the application of most conventional detection techniques for pathogens. Furthermore, notwithstanding their sensitivity and reproducibility, current technologies, such as

ELISA and PCR, require complex sample preparation procedures and count on long readout times, which delay prompt responses and disease containment.

The detection of bacteria can be obtained by mass-sensitive biosensors, in which the transduction is due to small changes of the mass of the sensing material. Mass measurements are generally performed by piezoelectric crystals, which oscillate at a specific frequency according to the application of a periodic electric signal. Therefore, when the mass increases due to the binding of an analyte (the pathogen), the frequency of oscillation of the crystal decreases and this frequency change can be used to determine the additional mass on the crystal. When a piezoelectric sensor surface coated with an antibody is placed in a solution containing the pathogen, the recognition of the bacteria by the antibody on the detecting surface results in an increase in the crystal mass, leading to a corresponding oscillation frequency shift. Magnetic nanoparticles, properly functionalized with antibodies, can be used for the fast testing of the presence of pathogens in foods by capturing and magnetically concentrating the microorganisms on the sensing device [56]. This leads to the amplification of the detectable signal. Nevertheless, even if the production of custom immuno-magnetic devices would be strongly preferred at the laboratory bench, nanoparticle engineering is still a prerogative of manufacturers. This is due to the chemical complexity of most of the common antibody immobilization protocols, as well as the need of specific competences from the user. Therefore, BIONs were proposed as versatile platform for developing tailored immuno-magnetic devices by very simple wet reactions. Significantly, this opportunity was provided by the characteristic binding properties of the pristine iron surface. Both native and biotinylated antibodies were conjugated to nanoparticles and the as-obtained multifunctional nanodevices were successfully tested for the recognition of *Campylobacter fetus* and *Listeria monocytogenes* by a piezoelectric quartz crystal microbalance (QCM) sensor (Figure 3) [57]. The system was applied for the detection of *L. monocytogenes* in milk, showing a detection limit of 3 bacterial cells per sample (200 µL).

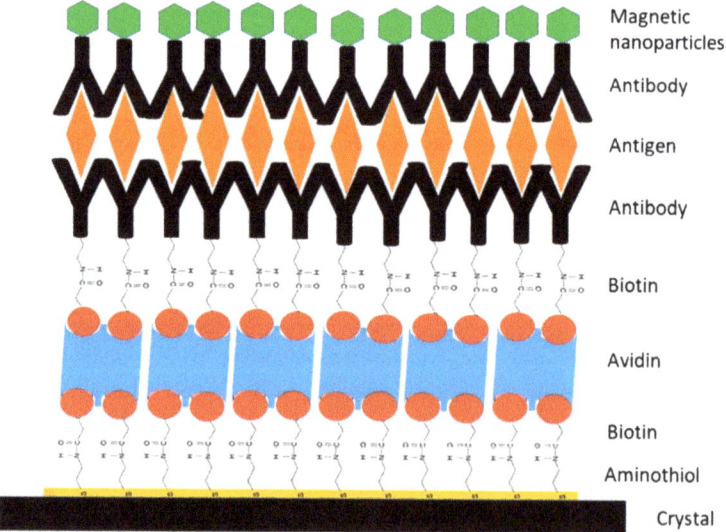

Figure 3. Scheme of a modified piezoelectric crystal for the detection of *Listeria monocytogenes* after magnetic capturing by SAMN@avidin@Anti-*Lm*. Reproduced from [57], with permission from Springer Nature, 2019.

Beyond analytical purposes, iron oxide nanomaterials were also used for the study of the metabolism of foodborne pathogens. As an example, some strains of *Pseudomonas fluorescens* produce a characteristic blue pigment that was recently investigated on cheeses for the negative economic impact on dairy industry [58–60]. The chemical nature of the pigment responsible for the blue discoloration has generated an intense debate over the last years [59,61], and it was demonstrated that blue discoloration in *Pseudomonas fluorescens* is correlated to the high resistance to oxidative stress induced by hydrogen peroxide. Indeed, this property was not displayed by white mutants [62]. Conventional colorimetric methods failed to evidence any difference between the antioxidant activity of blue and white microorganisms, and in a recent manuscript the mechanism of H_2O_2 scavenging activity by blue strains of *P. fluorescens* was studied by a nanotechnology based electrochemical approach [63].

The control of foodborne pathogens represents an important matter of concern and efforts to reduce the possibility of disease outbreaks are constantly in progress [64]. Food industries normally use antibiotics and synthetic preservatives during food production processes to prevent this problem. However, the combination of inappropriate and excessive use of chemical substances led to heavy chemical contamination of foods and to the emergence of drug resistant bacteria, increasing the difficulties of controlling the proliferation of foodborne diseases. In this view, iron oxide nanoparticles were modified with tannic acid, and the resulting core-shell hybrid nanomaterial resulted as one of the most robust tannic acid complexes to date [65]. Indeed, a drastic reorganization of the crystalline structure of the nanoparticle at the boundary with the solvent was observed, leading to the formation of a ferric tannate network that revealed antimicrobial properties. Thus, the core-shell hybrid nanomaterial was tested on *Listeria monocytogenes* showing a bacteriostatic effect. Moreover, the tannic acid modified nanostructure combined the inhibitory activity toward *L. monocytogenes* to the opportunity of being magnetically removed from the food matrix, hence leaving no residues in the final food products. The system was proposed as an effective, low-cost and environment friendly processing aid for the surface treatment in food industry [66].

The analysis of protein corona on nanomaterials represents the last frontier in proteomics, offering an alternative diagnostic strategy aimed at revealing faint proteome alterations correlated to a specific sample modification. Upon exposure to a biological matrix, nanomaterials are subjected to the spontaneous formation of a stable protein coating defined "protein corona" [67]. This common phenomenon leads to the formation of a protein shell, whose composition changes according to the proteome variations following, for example, a disease occurrence. Thus, the nanomaterial can be used as a probe for fishing specific peptides or proteins constituting a protein corona in a biological sample, simplifying the whole sample proteome [68]. This innovative approach was adopted for developing a fast analytical procedure for testing milk quality. MALDI-TOF mass spectrometry evidenced a hidden biomarker in the protein corona on nanoparticle surface, enabling the discrimination between milk coming from healthy and mastitis affected bovines (Figure 4) [69].

Another matter of concern for food industry is the large volume of residual biomass generated by the large-scale purification of natural compounds. This heavily influences processing industries, logistic sizing, magnitude of solvent consumption employed for extraction processes and the overall chemical and biological waste generation. In order to cope with this issue and, more in general, to move the applications of magnetic nanoparticles for the isolation and purification of biomolecules to an industrial level, an automatic modular pilot system was developed and applied for the continuous magnetic purification of curcuminoids, which are natural compounds of high nutraceutical value, from *Curcuma longa* roots. This novel magnetic separation device demonstrated the scalability of the purification processes by magnetic nanoparticles and embodies the fundamental principles of sustainable innovation [70].

Similarly to the presented purification of curcumin from *Curcuma longa* rhizome extracts and based as well on the specific iron binding properties of the molecule of interest, BIONs are a promising option for overcoming conventional large scale elimination of toxic substances from food matrices, such as citrinin, a nephrotoxic mycotoxin that can be synthesized by *Monascus* mold during the fermentation process in foods [71].

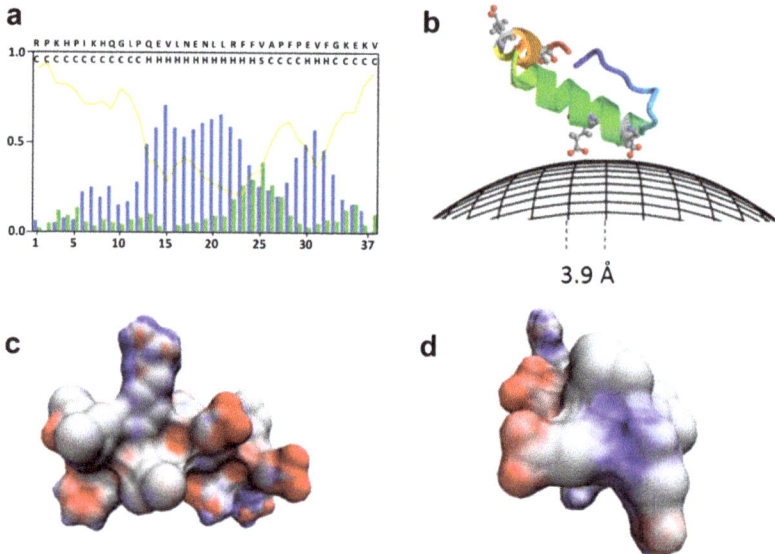

Figure 4. Binding of a milk peptide (αs1-casein 1-37)) on the surface of BIONs. (**a**) Statistical populations of ordered and disordered regions for the peptide according to the prediction of the s2D method. Coil curve describes the prediction of the probability to be unstructured and flexible. Blue and green bars represent the prediction of the statistical population for the helix and strand conformations; (**b**) graphical sketch of the peptide binding on BIONs where structured parts of the peptide are represented by helical models. The peptide structure was created by running QUARK algorithm; (**c**) and (**d**) helical model of the portion of the peptide predicted to be in helix according to s2D, two different views. The negatively charged surface is highlighted in red and computed with a Poisson–Boltzmann electrostatics calculations by 3D-HM. Reproduced from [69], with permission from Springer Nature, 2019.

3. BIONs for Biomedicine

In the last decades nanotechnology provided a wide range of new materials to biomedicine. As examples, carbon based nanomaterials, such as carbon nanotubes, fullerenes, graphene, nano-diamonds and fluorescent carbon quantum dots (CQDs) were exploited in a variety of biomedical applications [71]. CQDs, due to their advantageous chemical-physical features such as optical and fluorescence characteristics, aroused increasing interest and led to applications in biosensing and bioimaging [72]. Gold nanoparticles (AuNPs) appear among the most diffusely applied nanomaterials and are generally used after proper surface coating. The surface coating, its nature and structural organization, dominates the nanoparticle interactions with biological systems, influencing nanoparticle biodistribution and effects in biological systems. Surface coatings on AuNPs, generally by thiol containing molecules, may bear different functional groups, such as carboxylate, phosphate or sulfate. These nanosystems were extensively studied and applied in biomedicine, including for the development of biosensors, immunoassays, phototherapy, optical bioimaging, tracking of cells, targeted delivery of drugs [73]. Notably, toxicity limits the applications of AuNPs in clinical therapy. Nanomaterials based on polymeric nanoparticles were developed to improve the diagnosis and treatment of a wide range of diseases, spanning from viral infections, cardiovascular diseases and cancer to pulmonary and urinary tract infections [74]. Polymeric nanoparticles attracted the interest of the scientific community due to their structural versatility. In fact, they can be easily tuned to load and deliver their payload to the desired site of action and to respond to specific external or physiological stimuli [75]. The chemistry of polymeric nanoparticles and of their cargo influences their biocompatibility, stability, biodegradability, as well as their biodistribution, cellular and subcellular fate. Silica nanoparticles (SiO$_2$) are not endowed with particular chemical-physical properties: Nevertheless, they found applications in biomedicine as versatile tools for designing nanosized probes and carriers [76]. Indeed, silica nanoparticles possess a well-defined and easily tunable surface chemistry, which can be modified with different functionalities and linked to different biomolecules, enabling the fine control of the interplay with biological entities [77]. Mesoporous silica nanoparticles, characterized by pore sizes ranging from 2 to 50 nm, were excellent options for drug delivery and biomedical applications [78]. The mesoporous structure allows the control of drug loading and the control of release kinetics, enhancing drug therapeutic efficacy and reducing the toxicity [79]. The design of functionalized silica nanoparticles focused on the selective delivery of drugs and radionuclides to improve diagnosis and treatment were highlighted by the current progresses in the field of pharmaceutical nanotechnology [80].

The broad spectrum of favorable properties of iron oxide nanoparticles encouraged their utilization for diagnostic and therapeutic applications in biomedicine, and some were already accepted for clinical purposes or expected for new approved issues in the near future [81]. Gene therapy, cell imaging and tissue regeneration are among the most challenging biomedical applications of iron oxide nanoparticles. Indeed, BIONs are characterized by surface tunability, cost-effectiveness and biocompatibility, which make these pristine nanomaterial an attracting option in biomedicine. Moreover, BIONs can be further derivatized to be upgraded to multifunctional nano-devices for theranostic uses. Indeed, engineered BIONs were developed to meet the increasing interests for in vivo imaging, gene delivery, drug targeting and biomarker detection. As an example, a novel multifunctional nano-immunosensor coupling the magnetic properties of nanosized iron oxides and the optical features of gold nanoparticles modified with specific aptamers was developed and tested for the recognition of dengue virus [82]. The proposed complex nano-architecture allowed the rapid visual detection of the virus by a simple colorimetric assay. It should be considered that the development of this hybrid nanomaterial was extremely facilitated by the use of BIONs. Indeed, chemical approaches commonly employed for manipulating coated magnetic nanoparticles are considerably more complicated, requiring solvents, expensive chemicals, drastic conditions and trained operators. Alternatively, iron oxide nanoparticles can be considered as intelligent, universal nano-vectors, able to match drug protection (avoiding drug loss), due to the unusual specific nature of the surface interactions, with a pronounced reactivity. This, along with the extensively demonstrated colloidal stability, excellent cell uptake [83], stability in the host cells, low toxicity and great MRI contrast agent properties [84], makes iron oxide nanoparticles an elective tool for biomedical applications. These properties, in combination with the ones provided by surface ligands led to promising multifaceted nanodevices. Long-term dual imaging (fluorescence and MRI) nanoprobes were obtained by the direct immobilization of rhodamine isothycianate on BIONs and demonstrated in vitro on mesenchymal stromal cells [85]. The biological properties of curcumin were endowed to BIONs creating a magnetically drivable nanovehicle with antioxidant properties [86]. Furthermore, a plasmid (pDNA) harboring the coding gene for GFP was directly chemisorbed onto iron oxide nanoparticles, leading to a novel DNA nanovector that was successfully internalized and stored into mesenchymal stem cells. Transfection by the proposed system occurred and GFP expression was higher than that observed with lipofectamine procedure, even in the absence of external magnetic field (Figure 5) [87].

As above mentioned, the combination of different nanomaterials represents a challenge for the emerging properties and features of the resulting nano-hybrids. As an example, an electrostatically stabilized ternary hybrid, constituted of a core of BION@DNA coated by spermidine based quantum dots (Spd-CQDs), was developed, fully characterized and tested on HeLa cells in order to evaluate its cellular uptake and biocompatibility. This novel ternary nano-hybrid with multifaceted properties, ranging from superparamagnetism to fluorescence, represents a novel option for cell tracking [88]. Moreover, iron oxide nanoparticles were used for biomarker identification and detection by immobilizing different biomolecules, such as antibodies [89], aptamers [90] and enzymes [91] or peptide fragments, such as RGD [92] or polysaccharides [93].

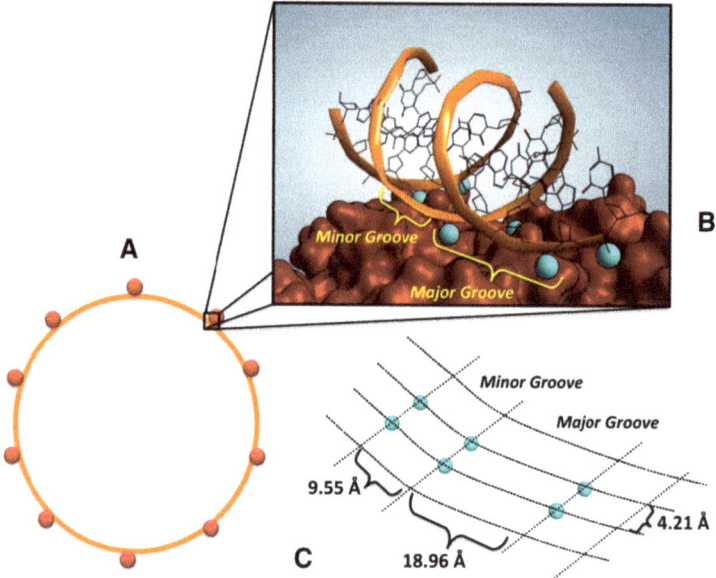

Figure 5. Schematic representation of the interactions between plasmidic DNA and BIONs. (**A**) One molecule of plasmidic DNA (in orange) binds about 10 nanoparticles (in red); (**B**) Interactions between Fe^{3+} (in blue) and DNA on the surface of BION; a possible representation of the nanoparticle surface is reported. (**C**) Projection of Fe^{3+} (in blue) on BION surface, in which the distances among Fe^{3+} are shown. Reproduced from [87], with permission from Elsevier, 2019.

The potential of bare iron oxide nanoparticles as antibiotic carriers for the treatment of specific vertically transmissible infections in fishes, such as those affecting farmed trouts caused by bacteria (such as *Flavobacterium psychrophilum* and *Bacterium salmoninarum*), was explored in zebrafish (*Danio rerio*) by providing a drug nanovehicle directly into the farming water [94]. The ability of bare iron oxide nanoparticles to circumvent the clearance by the immune system was examined in vivo using the same animal model. Iron oxide nanoparticles overcame the intestinal barrier and led to the prolonged (28 days) and specific organotropic delivery of the antibiotic to the fish ovary. Significantly, no adverse effects were observed. Interestingly, a structural analogy between high-density lipoproteins (HDL) and the complex formed between iron oxide nanoparticles and apolipoprotein A1 (Figure 6) suggests that biomimetic properties play an important role in such a complex phenomenon [95].

The presence of a specific proteome in a living organism can represent a crucial gap between in vivo and in vitro testing of nanomaterials. This aspect was recently evidenced by using fish organ cultures and testing different oxytetracycline exposure methods, including antibiotic bound on BIONs. Notably, the exposure of an isolated organ to BIONs bearing the antibiotic cannot be representative of the complexity of the interactions occurring in a whole organism [96].

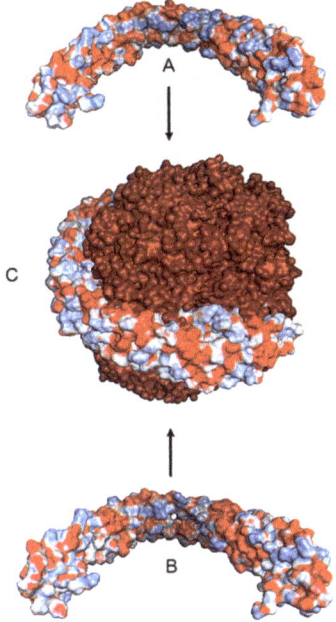

Figure 6. Pictorial representation of the binding on an iron oxide nanoparticle of a dimer of apolipoporotein A1 from gilthead bream (*Sparus aurata*). A dimer of apolipoporotein A1 (A, B) is shown as Connolly's electrostatic surface charge distribution, by AMBER99 force field. Blue color indicates positive charges and red color indicates negative charges. Adapted from [95], with permission from A.C.S., 2019.

4. BIONs for the Environment

A deep analysis of the potential environmental risks associated with the intensive extensive use of iron oxide nanoparticles cannot be further delayed, and even the factors related with epigenetic phenomena and long-term effects should be assessed. Iron oxide nanoparticles, due to their wide utilization for research purposes and perspectives in clinical practice, represent an ideal model for proposing an acceptable and comprehensive platform for the punctual classification of environmental risks of nanotechnology. In fact, the identification and definition of general protocols for the assessment of nanomaterial safety toward environmental protection and human health and the evaluation of long-term effects of nanoparticle drugs complexes and contrast agents for medical applications are of primary interest [97].

It is recognized that the effects of nanomaterials on biological systems must be faced with interdisciplinary approaches and the coordinated contributions of chemists, physicists, biologists, pharmacologists and clinical doctors. On these bases, an eco-toxicological study on iron oxide nanoparticles was carried out using *Daphnia magna* as animal model (Figure 7).

The surface reactivity of the nanomaterial emerged as a fundamental issue for determining its effects on biological systems, hence tests on acute and chronic toxicity were compared to nanoparticle distribution into the crustacean, intake/depletion rates and swimming performances. Fast depuration from the nanomaterial and absence of delayed effects indicated no retention and good tolerance within the organism, hence substantiating the safety of iron oxide nanoparticles as a tool for aquaculture purposes [98].

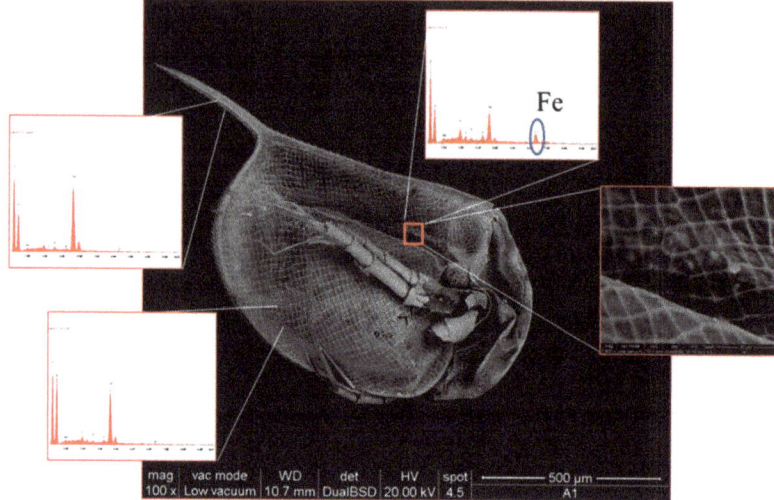

Figure 7. Micrograph by scanning electron microscopy of Daphnia Magna after incubation in a solution containing 1.25 mg L^{-1} BIONs. Insets: Elemental analysis by energy dispersive X-ray spectroscopy and higher magnification of gut detail. Reproduced from [98], with permission from Springer Nature, 2019.

In order to cope with the worldwide threat of diseases vectored by mosquitos, novel solutions aimed at the enhancement of the control over these detrimental insects are strongly desired. In this view, systems coupling low environmental impacts with high insecticidal activity represent a crucial goal nowadays, particularly in the view of their application in aquaculture. An eco-friendly photosensitizing magnetic nanocarrier with larvicidal effects on *Aedes aegypti* was proposed [99]. A core-shell nanocarrier was synthesized, combining iron oxide nanoparticles and chlorin-e6. The photosensitizing functionality was provided to the magnetic core by a simple wet reaction, exploiting the chelating groups of the organic molecule. The self-assembled hybrid possessed several advantageous properties, such as magnetic drivability, physico-chemical robustness and high colloidal stability. This last feature is a mandatory prerequisite for the application in aquatic systems. Most importantly, the proposed photosensitizing nano-device presented real chances to be a competitor of conventional insecticides, considering its high efficiency, as demonstrated on larvae of *Aedes aegypti*, and its minimal environmental impact (Figure 8) [99].

Magnetic separation strategies by nanomaterials represent competitive options or, at least, valuable complementing methods to traditional approaches for the removal of organic and inorganic pollutants in water. Magnetic nanoparticles have been demonstrated to be effective, rapid and cost-effective tools for water remediation issues [100–103]. Separation processes based on magnetic particles offer the advantage of eliminating absorbed toxic compounds by the application of an external magnetic field. In this context, iron oxide nanoparticles represent the gold standard.

Water pollution from antibiotics is a global threat due to their extensive and durable presence in the ecosystems and their bulky use [104]. Indeed, the use of antibiotics for human and veterinary medicine can be estimated to overcome 100 ktons per year worldwide [105], and most of these compounds in the environment are very stable, exerting for long time their activity, thus granting their effects for long [106]. In this context, the thermodynamically favorable complexation chemistry of different iron oxides, such as magnetite, goethite or hematite, [107,108] have the potential for representing efficient sorbents for antibiotics. In fact, many of these molecules are strong chelating agents. Nevertheless, despite the promising possibilities offered by the surface complexation chemistry of iron oxides, only rare efforts were dedicated to the use of bare iron oxide nanoparticles for water remediation from antibiotics, such as oxytetracycline [107,109]. This is not surprising as the preparation of stable colloidal suspensions of bare metal oxides remains a challenge for preparative nanotechnology [110]. A novel category of BIONs was employed for the removal of oxytetracycline in large water volumes where, in order to simulate a real in situ scenario, a population of zebrafish (*Danio rerio*) was introduced. The introduction of the nanomaterial in water led to the complete elimination of the antibiotic without any sign of toxicity for the animal model. Furthermore, according to a toxicological characterization on oxytetracycline sensitive *Escherichia coli* strains, the entrapped antibiotic resulted biologically safe. Thus, the proposed nanomaterial emerged as a competitive option for water remediation from oxytetracycline and, in particular, for in situ applications [111].

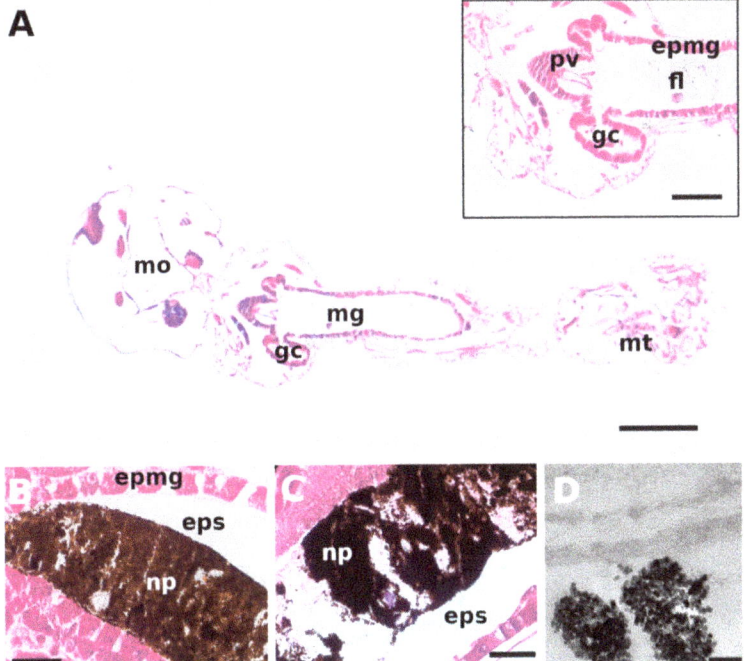

Figure 8. Microscopic characterization of *Aedes aegypti* larvae. (**A**) Optical micrograph of the sagittal section of a control sample. Inset: Optical micrograph of the section of gastric caeca, proventriculus and anterior end of mid-gut. (**B**) Optical micrograph of the midgut of a larva after exposure to 100 mg L^{-1} BIONs for 1 h, evidencing the accumulation of nanoparticles in the gut lumen. (**C**) Optical micrograph of the midgut of a larva exposed to 100 mg L^{-1} BION@chlorin for 80 min. (**D**) TEM micrograph of the gut of a *Aedes aegypti* larva exposed to 100 mg L^{-1} BION@chlorin for 1 h. (epmg), epithelium of the midgut; (eps), ectoperitrophic space; (fl), food in gut lumen; (gc), gastric caeca; (mg), midgut; (mo), mouth; (mt), malpighian tubules; (pv), proventriculus; (np), nanoparticles. Size bars: A = 250 μm, inset = 100 μm, B and C = 50 μm, D = 200 nm. Reproduced from [99], with permission from Elsevier, 12019.

Peculiar BIONs were also proposed for ex situ and in situ water remediation of Cr(VI) pollution. This nanomaterial was proposed as suitable sorbent for entrapping and recovering dissolved chromium in water. Significantly, the BION@CrVI complex showed no genotoxicity on a chromium sensitive strain of *Salmonella typhimurium TA100* [112].

5. Overall Conclusions and Perspectives

Iron oxide nanomaterials are already applied by a number of high-tech industries due to their outstanding magnetic, catalytic and electrochemical properties, as a function of their size, structure, shape and surface chemistry. Indeed, iron oxide nanoparticles, due to their structural peculiarities, were applied for developing novel platforms for biomedical, food safety and environmental issues, and efficient promising solutions to overcome the limits and drawbacks of currently available techniques were proposed. As examples, novel user-friendly, sensitive and cheap diagnostic assays to assist the control of microbial pathogens and food animal diseases were developed [113,114]. Furthermore, bare iron oxide nanoparticles offer the opportunity of covalently interacting with ligands, allowing the development of robust delivery tools. Therefore, magnetic drivable gene and drug nano-vectors can be produced [87,95,115].

In the near future, a wide scenario of innovations is promised by the novel physicochemical properties of the hybrids involving these nanomaterials, generating a constantly increasing number of innovative applications.

Among proposed nanodevices, one of the most promising achievement relies on the evidences that naked metal oxides can provide a stealth behavior within organisms with important in vivo applications. In fact, the "stealth effect" consists in the ability of avoiding the immune system clearance of the nanomaterial. It should be considered that recognition of nanoparticles by cells was recently defined as "cell vision", emphasizing the concept of what the cell "sees" when it faces nanoparticles [116]. In this context, the formation of a shell of biologically recognizable proteins could play a fundamental role for enhancing nanoparticle internalization by specific cells and tissues [117]. In fact, the protein "corona", shaping surface properties, charges, hydrodynamic size and resistance to aggregation, largely defines the biological identity of the particle [118]. Exerting control over protein-nanoparticle interactions is a problematic task, considering that the binding of enzymes and proteins on nanoparticle surfaces leads to often unavoidable unspecific interactions [119,120]. Contrary to recent findings reporting on the loss of targeting capabilities of nanoparticles upon the spontaneous formation of a protein corona [121,122], some iron oxide nanoparticles are able to select few proteins within a biological medium, forming stable nano-bioconjugates, which are easily recognizable and incorporated by cells [83]. Thus, the control of the interactions of macromolecules with nanomaterial surfaces will be a relevant aspect of future applications of iron oxide nanoparticles and possibly will be exploited for the preparation of suitably targeted nanosystems for interacting with specific tissues [123].

Finally, an important task in the future will be our improvement in the knowledge of the nature of the macromolecule-iron oxide nanoparticle interactions and their influence on the biological activity of the resulting nano-bioconjugate.

Author Contributions: M.M.: Conceptualization, Original Draft preparation, Writing – Review & Editing, Visualization. F.V.: Conceptualization, Writing – Review & Editing, Supervision, Project administration.

Funding: This research received no external funding

Conflicts of Interest: The authors declare no conflict of interest

References

1. Chavali, M.S.; Nikolova, M.P. Metal oxide nanoparticles and their applications in nanotechnology. *SN Appl. Sci.* **2019**, *1*, 607–637. [CrossRef]
2. Franke, M.E.; Koplin, T.J.; Simon, U. Metal and metal oxide nanoparticles in chemiresistors: Does the nanoscale matter? *Small* **2006**, *2*, 36–50. [CrossRef] [PubMed]
3. Wilkes, J.S.; Zaworotko, M.J. Air and water stable 1-ethyl-3-methylimidazolium based ionic liquids. *Chem. Commun.* **1992**, *13*, 965–967. [CrossRef]
4. Nuraje, N.; Su, K.; Haboosheh, A.; Samson, J.; Manning, E.P.; Yang, N.-I.; Matsui, H. Room temperature synthesis of ferroelectric barium titanate nanoparticles using peptide nanorings as templates. *Adv. Mater.* **2006**, *18*, 807–811. [CrossRef] [PubMed]
5. Yahiro, J.; Oaki, Y.; Imai, H. Biomimetic synthesis of wurtzite ZnO nanowires possessing a mosaic structure. *Small* **2006**, *2*, 1183–1187. [CrossRef] [PubMed]

6. Margaryan, A.A.; Liu, W. Prospects of using germanium dioxide-based glasses for optics. *Opt. Eng.* **1993**, *32*, 1995–1996. [CrossRef]
7. Lee, S.-Y.; Gao, X.; Matsui, H. Biomimetic and aggregation driven crystallization route for the room-temperature material synthesis: The growth of β-Ga_2O_3 nanoparticles on peptide assemblies as nanoreactors. *J. Am. Chem. Soc.* **2007**, *129*, 2954–2958. [CrossRef] [PubMed]
8. Klem, M.T.; Mosolf, J.; Young, M.; Douglas, T. Photochemical mineralization of europium, titanium, and iron oxyhydroxide nanoparticles in the ferritin protein cage. *Inorg. Chem.* **2008**, *47*, 2237–2239. [CrossRef] [PubMed]
9. Zhang, W.; Zhang, D.; Fan, T.; Ding, J.; Guo, Q.; Ogawa, H. Fabrication of ZnO microtubes with adjustable nanopores on the walls by the templating of butterfly wing scales. *Nanotechnology* **2006**, *17*, 840–844. [CrossRef]
10. Aizenberg, J.; Hanson, J.; Koetzle, T.F.; Weiner, S.; Addadi, L. Control of macromolecule distribution within synthetic and biogenic single calcite crystals. *J. Am. Chem. Soc.* **1997**, *119*, 881–886. [CrossRef]
11. Biro, L.P.; Balint, Z.; Kertesz, K.; Vertesy, Z.; Mark, G.I.; Tapaszto, L.; Vigneron, J.P.; Lousse, V. Photonic crystal structures of biologic origin: Butterfly wing scales. *Mater. Res. Soc. Symp. Proc.* **2007**, *1014*. [CrossRef]
12. Zou, D.; Xu, C.; Luo, H.; Wang, L.; Ying, T. Synthesis of Co_3O_4 nanoparticles via an ionic liquid-assisted methodology at room temperature. *Mater. Lett.* **2008**, *62*, 1976–1978. [CrossRef]
13. Vinay, M.M.; Nayaka, Y.A. Iron oxide (Fe_2O_3) nanoparticles modified carbon paste electrode as an advanced material for electrochemical investigation of paracetamol and dopamine. *J. Sci. Adv. Mater. Dev.* **2019**, *4*, 442–450. [CrossRef]
14. Jakubec, P.; Urbanová, V.; Medříková, Z.; Zbořil, R. Advanced sensing of antibiotics with magnetic gold nanocomposite: Electrochemical detection of chloramphenicol. *Chem. Eur. J.* **2016**, *22*, 14279–14284. [CrossRef] [PubMed]
15. Manal, A.; Azab, S.M.; Hendawy, H.A. A facile nano-iron oxide sensor for the electrochemical detection of the anti-diabetic drug linagliptin in the presence of glucose and metformin. *Bull. Nat. Res. Cent.* **2019**, *43*, 95.
16. Xie, J.; Jon, S. Magnetic nanoparticle-based theranostics. *Theranostics* **2012**, *2*, 122–124. [CrossRef] [PubMed]
17. Yang, H.H.; Zhang, S.Q.; Chen, X.L.; Zhuang, Z.X.; Xu, J.G.; Wang, X.R. Magnetite-containing spherical silica nanoparticles for biocatalysis and bioseparations. *Anal. Chem.* **2004**, *76*, 1316–1321. [CrossRef] [PubMed]
18. Ito, A.; Shinkai, M.; Honda, H.; Kobayashi, T. Medical application of functionalized magnetic nanoparticles. *J. Biosci. Bioeng.* **2005**, *100*, 1–11. [CrossRef] [PubMed]
19. Tanaka, T.; Matsunaga, T. Fully automated chemiluminescence immunoassay of insulin using antibody-protein A-bacterial magnetic particle complexes. *Anal. Chem.* **2000**, *72*, 3518–3522. [CrossRef] [PubMed]
20. Liu, Z.; Liu, Y.; Yang, H.; Yang, Y.; Shen, G.; Yu, R. A phenol biosensor based on immobilizing tyrosinase to modified core-shell magnetic nanoparticles supported at a carbon paste electrode. *Anal. Chim. Acta* **2005**, *533*, 3–9. [CrossRef]
21. Neuberger, T.; Schopf, B.; Hofmann, H.; Hofmann, M.; Rechenberg, B.V. Superparamagnetic nanoparticles for biomedical applications: Possibilities and limitations of a new drug delivery system. *J. Magn. Magn. Mater.* **2005**, *293*, 483–496. [CrossRef]
22. Saito, S.; Tsugeno, M.; Koto, D.; Mori, Y.; Yoshioka, Y.; Nohara, S.; Murase, K. Impact of surface coating and particle size on the uptake of small and ultrasmall superparamagnetic iron oxide nanoparticles by macrophages. *Int. J. Nanomed.* **2012**, *7*, 5415–5421.
23. Tassa, C.; Shaw, S.Y.; Weissleder, R. Dextran-coated iron oxide nanoparticles: A versatile platform for targeted molecular imaging, molecular diagnostics, and therapy. *Acc. Chem. Res.* **2011**, *44*, 842–852. [CrossRef] [PubMed]
24. Wang, J.; Huang, Y.; David, A.E.; Chertok, B.; Zhang, L.; Yu, F.; Yang, V.C. Magnetic nanoparticles for MRI of brain tumors. *Curr. Pharm. Biotechnol.* **2012**, *13*, 2403–2416. [CrossRef] [PubMed]
25. Tran, P.H.-L.; Tran, T.T.-D.; Vo, T.V.; Lee, B.-J. Promising iron oxide-based magnetic nanoparticles in biomedical engineering. *Arch. Pharm. Res.* **2012**, *35*, 2045–2061. [CrossRef] [PubMed]
26. Sasaki, Y.; Usuki, N.; Matsuo, K.; Kishimoto, M. Development of NanoCAP technology for high-density recording. *IEEE Trans. Magn.* **2005**, *41*, 3241–3243. [CrossRef]
27. Fouda, M.F.R.; El-Kholy, M.B.; Moustafa, S.A.; Hussien, A.I.; Wahba, M.A.; El-Shahat, M.F. Synthesis and characterization of nanosized Fe_2O_3 pigments. *Int. J. Inorg. Chem.* **2012**, *2012*, 989281. [CrossRef]

28. Magro, M.; Baratella, D.; Pianca, N.; Toninello, A.; Grancara, S.; Zboril, R.; Vianello, F. Electrochemical determination of hydrogen peroxide production by isolated mitochondria: A novel nanocomposite carbon–maghemite nanoparticle electrode. *Sens. Actuator B Chem.* **2013**, *176*, 315–322. [CrossRef]
29. Laurent, S.; Forge, D.; Port, M.; Roch, A.; Robic, C.; Vander Elst, L.; Muller, R.N. Magnetic iron oxide nanoparticles: Synthesis, stabilization, vectorization, physicochemical characterizations, and biological applications. *Chem. Rev.* **2008**, *108*, 2064–2110. [CrossRef] [PubMed]
30. Ahmed, N.; Fessi, H.; Elaissari, A. Theranostic applications of nanoparticles in cancer. *Drug Discov. Today* **2012**, *17*, 928–934. [CrossRef] [PubMed]
31. Soenen, S.J.; Hodenius, M.; De Cuyper, M. Magnetoliposomes: Versatile innovative nanocolloids for use in biotechnology and biomedicine. *Nanomedicine* **2009**, *4*, 177–191. [CrossRef] [PubMed]
32. Wei, H.; Insin, N.; Lee, J.; Han, H.S.; Cordero, J.M.; Liu, W.; Bawendi, M.G. Compact zwitterion-coated iron oxide nanoparticles for biological applications. *Nano Lett.* **2012**, *12*, 22–25. [CrossRef] [PubMed]
33. Wahajuddin, S.S.P.; Arora, S. Superparamagnetic iron oxide nanoparticles: Magnetic nanoplatforms as drug carriers. *Int. J. Nanomed.* **2012**, *7*, 3445–3471. [CrossRef] [PubMed]
34. Meyers, S.R.; Grinstaff, M.W. Biocompatible and bioactive surface modifications for prolonged in vivo efficacy. *Chem. Rev.* **2012**, *112*, 1615–1632. [CrossRef] [PubMed]
35. Hudson, R.; Feng, Y.; Varma, R.S.; Moores, A. Bare magnetic nanoparticles: Sustainable synthesis and applications in catalytic organic transformations. *Green Chem.* **2014**, *16*, 4493–4505. [CrossRef]
36. Schwaminger, S.P.; Fraga-García, P.; Blank-Shim, S.A.; Straub, T.; Haslbeck, M.; Muraca, F.; Dawson, K.A.; Berensmeier, S. Magnetic one-Step purification of his-tagged protein by bare iron oxide nanoparticles. *ACS Omega* **2019**, *4*, 3790–3799. [CrossRef] [PubMed]
37. Schwaminger, S.P.; Blank-Shim, S.A.; Scheifele, I.; Pipich, V.; Fraga-García, P.; Berensmeier, S. Design of interactions between nanomaterials and proteins: A highly affine peptide tag to bare iron oxide nanoparticles for magnetic protein separation. *Biotechnol. J.* **2019**, *14*, 1800055. [CrossRef] [PubMed]
38. Fraga-García, P.; Kubbutat, P.; Brammen, M.; Schwaminger, S.; Berensmeier, S. Bare iron oxide nanoparticles for magnetic harvesting of microalgae: From interaction behavior to process realization. *Nanomaterials* **2018**, *8*, 292. [CrossRef] [PubMed]
39. Wang, S.K.; Stiles, A.R.; Guo, C.; Liu, C.Z. Harvesting microalgae by magnetic separation: A review. *Algal Res.* **2015**, *9*, 178–185. [CrossRef]
40. Xu, L.; Guo, C.; Wang, F.; Zheng, S.; Liu, C.Z. A simple and rapid harvesting method for microalgae by in situ magnetic separation. *Bioresour. Technol.* **2011**, *102*, 10047–10051. [CrossRef] [PubMed]
41. Lee, Y.C.; Lee, K.; Oh, Y.K. Recent nanoparticle engineering advances in microalgal cultivation and harvesting processes of biodiesel production: A review. *Bioresour. Technol.* **2015**, *184*, 63–72. [CrossRef] [PubMed]
42. Safarik, I.; Pospiskova, K.; Baldikova, E. Magnetic particles for microalgae separation and biotechnology. In *Food Bioactives*; Puri, M., Ed.; Springer: Cham, Switzerland, 2017; pp. 153–169.
43. Prochazkova, G.; Safarik, I.; Branyik, T. Harvesting microalgae with microwave synthesized magnetic microparticles. *Bioresour. Technol.* **2013**, *130*, 472–477. [CrossRef] [PubMed]
44. Toh, P.Y.; Ng, B.W.; Ahmad, A.L.; Chieh, D.C.J.; Lim, J. The role of particle-to-cell interactions in dictating nanoparticle aided magnetophoretic separation of microalgal cells. *Nanoscale* **2014**, *6*, 12838–12848. [CrossRef] [PubMed]
45. Lin, S.; Lu, D.; Liu, Z. Removal of Arsenic Contaminants with Magnetic γ-Fe_2O_3 Nanoparticles. *Chem. Eng. J.* **2012**, *211*, 46–52. [CrossRef]
46. Liu, R.; Liu, J.F.; Zhang, L.Q.; Sun, J.F.; Jiang, G.B. Low temperature synthesized ultrathin γ-Fe_2O_3 nanosheets show similar adsorption behavior for As(III) and As(V). *J. Mater. Chem. A* **2016**, *4*, 7606–7614. [CrossRef]
47. Rana, S.; Yeh, Y.C.; Rotello, V.M. Engineering the nanoparticle protein interface: Applications and possibilities. *Curr. Opin. Chem. Biol.* **2010**, *14*, 828–834. [CrossRef] [PubMed]
48. Velusamy, V.; Arshak, K.; Korostynska, O.; Oliwa, K.; Adley, C. An overview of foodborne pathogen detection: In the perspective of biosensors. *Biotechnol. Adv.* **2010**, *28*, 232–254. [CrossRef] [PubMed]
49. Hierlemann, A.; Gutierrez-Osuna, R. Higher-order chemical sensing. *Chem. Rev.* **2008**, *108*, 563–613. [CrossRef] [PubMed]
50. Urbanova, V.; Magro, M.; Gedanken, A.; Baratella, D.; Vianello, F.; Zboril, R. Nanocrystalline Iron Oxides, Composites, and Related Materials as a Platform for Electrochemical, Magnetic, and Chemical Biosensors. *Chem. Mater.* **2014**, *26*, 6653–6673. [CrossRef]

51. Baratella, D.; Magro, M.; Jakubec, P.; Bonaiuto, E.; de Almeida Roger, J.; Gerotto, E.; Zoppellaro, G.; Tucek, J.; Safarova, K.C.; Zboril, R.; et al. Electrostatically stabilized hybrids of carbon and maghemite nanoparticles: Electrochemical study and application. *Phys. Chem. Chem. Phys.* **2017**, *19*, 11668–11677. [CrossRef] [PubMed]
52. Magro, M.; Baratella, D.; Miotto, G.; Frömmel, J.; Šebela, M.; Kopečná, M.; Agostinelli, E.; Vianello, F. Enzyme self-assembly on naked iron oxide nanoparticles for aminoaldehyde biosensing. *Amino Acids* **2019**, *51*, 679–690. [CrossRef] [PubMed]
53. Baratella, D.; Magro, M.; Sinigaglia, G.; Zboril, R.; Salviulo, G.; Vianello, F. A glucose biosensor based on surface active maghemite nanoparticles. *Biosens. Bioelectron.* **2013**, *45*, 13–18. [CrossRef] [PubMed]
54. Magro, M.; Baratella, D.; Colò, V.; Vallese, F.; Nicoletto, C.; Santagata, S.; Sambo, P.; Molinari, S.; Salviulo, G.; Venerando, A.; et al. Electrocatalytic Nanostructured Ferric Tannates as platform for enzyme conjugation: Electrochemical determination of phenolic compounds. *Bioelectrochemistry* **2019**. [CrossRef]
55. Shinde, S.B.; Fernandes, C.B.; Patravale, V.B. Recent trends in in-vitro nanodiagnostics for detection of pathogens. *J. Control. Release* **2012**, *159*, 164–180. [CrossRef] [PubMed]
56. Sanvicens, N.; Pastells, C.; Pascual, N.; Marco, M.P. Nanoparticle based biosensors for detection of pathogenic bacteria. *TrAC Trends Anal. Chem.* **2009**, *28*, 1243–1252. [CrossRef]
57. Bonaiuto, E.; Magro, M.; Fasolato, L.; Novelli, E.; Shams, S.; Piccirillo, A.; Bakhshi, B.; Moghadam, T.T.; Baratella, D.; Vianello, F. Versatile nano-platform for tailored immuno-magnetic carriers. *Anal. Bioanal. Chem.* **2018**, *410*, 7575–7589. [CrossRef] [PubMed]
58. Martin, N.H.; Murphy, S.C.; Ralyea, R.D.; Wiedmann, M.; Boor, K.J. When cheese gets the blues: Pseudomonas fluorescens as the causative agent of cheese spoilage. *J. Dairy Sci.* **2011**, *94*, 3176–3183. [CrossRef] [PubMed]
59. Andreani, N.A.; Martino, M.E.; Fasolato, L.; Carraro, L.; Montemurro, F.; Mioni, R.; Bordin, P.; Cardazzo, B. Tracking the blue: A MLST approach to characterize the Pseudomonas fluorescens group. *Food Microbiol.* **2014**, *39*, 116–126. [CrossRef] [PubMed]
60. Chierici, M.; Picozzi, C.; La Spina, M.G.; Orsi, C.; Vigentini, I.; Zambrini, V.; Foschino, R. Strain diversity of Pseudomonas fluorescens group with potential blue pigment phenotype isolated from dairy products. *J. Food Prot.* **2016**, *79*, 1430–1435. [CrossRef] [PubMed]
61. Andreani, N.A.; Carraro, L.; Martino, M.E.; Fondi, M.; Fasolato, L.; Miotto, G.; Magro, M.; Vianello, F.; Cardazzo, B. A genomic and transcriptomic approach to investigate the blue pigment phenotype in *Pseudomonas fluorescens*. *Int. J. Food Microbiol.* **2015**, *213*, 88–98. [CrossRef] [PubMed]
62. Andreani, N.A.; Carraro, L.; Zhang, L.; Vos, M.; Cardazzo, B. Transposon mutagenesis in Pseudomonas fluorescens reveals genes involved in blue pigment production and antioxidant protection. *Food Microbiol.* **2019**, *82*, 497–503. [CrossRef] [PubMed]
63. Magro, M.; Baratella, D.; Jakubec, P.; Corraducci, V.; Fasolato, L.; Cardazzo, B.; Novelli, E.; Zoppellaro, G.; Zboril, R.; Vianello, F. H_2O_2 Tolerance in Pseudomonas fluorescens: Synergy between Pyoverdine-Iron(III) Complex and a Blue Extracellular Product by a Nanotechnology based Electrochemical Approach. *ChemElectroChem.* **2019**. [CrossRef]
64. Kiran-Kumar, P.; Badarinath, V.; Halami, P. Isolation of anti-listerial bacteriocin producing Lactococcus lactis CFR-B3 from Beans (Phaseolus vulgaris). *Internet J. Microbiol.* **2008**, *6*, 1–6.
65. Magro, M.; Bonaiuto, E.; Baratella, D.; de Almeida Roger, J.; Jakubec, P.; Corraducci, V.; Tucek, J.; Malina, O.; Zboril, R.; Vianello, F. Electrocatalytic nanostructured ferric tannates: Characterization and application of a polyphenol nanosensor. *ChemPhysChem* **2016**, *17*, 3196–3203. [CrossRef] [PubMed]
66. De Almeida Roger, J.; Magro, M.; Spagnolo, S.; Bonaiuto, E.; Baratella, D.; Fasolato, L.; Vianello, F. Antimicrobial and magnetically removable tannic acid nanocarrier: A processing aid for Listeria monocytogenes treatment for food industry applications. *Food Chem.* **2018**, *267*, 430–436. [CrossRef] [PubMed]
67. Hadjidemetriou, M.; Al-Ahmady, Z.; Buggio, M.; Swift, J.; Kostarelos, K. A novel scavenging tool for cancer biomarker discovery based on the blood-circulating nanoparticle protein corona. *Biomaterials* **2019**, *188*, 118–129. [CrossRef] [PubMed]
68. Miotto, G.; Magro, M.; Terzo, M.; Zaccarin, M.; Da Dalt, L.; Bonaiuto, E.; Baratella, D.; Gabai, G.; Vianello, F. Protein corona as a proteome fingerprint: The example of hidden biomarkers for cow mastitis. *Colloids Surf. B Biointerfaces* **2016**, *140*, 40–49. [CrossRef] [PubMed]

69. Magro, M.; Zaccarin, M.; Miotto, G.; Da Dalt, L.; Baratella, D.; Fariselli, P.; Gabai, G.; Vianello, F. Analysis of hard protein corona composition on selective iron oxide nanoparticles by MALDI-TOF mass spectrometry: Identification and amplification of a hidden mastitis biomarker in milk proteome. *Anal. Bioanal. Chem.* **2018**, *410*, 2949–2959. [CrossRef] [PubMed]
70. Ferreira, M.I.; Magro, M.; Ming, L.C.; da Silva, M.B.; Rodrigues, L.F.O.S.; do Prado, D.Z.; Bonaiuto, E.; Baratella, D.; Lima, G.P.P.; de Almeida Roger, J.; et al. Sustainable production of high purity curcuminoids from Curcuma longa by magnetic nanoparticles: A case study in Brazil. *J. Clean. Prod.* **2017**, *154*, 233–241. [CrossRef]
71. Magro, M.; Esteves Moritz, D.; Bonaiuto, E.; Baratella, D.; Terzo, M.; Jakubec, P.; Malina, O.; Cépe, K.; de Falcao Aragao, G.M.; Zboril, R.; et al. Citrinin mycotoxin recognition and removal by naked magnetic nanoparticles. *Food Chem.* **2016**, *203*, 505–512. [CrossRef] [PubMed]
72. Huang, S.; Li, W.; Han, P.; Zhou, X.; Cheng, J.; Wen, H.; Xue, W. Carbon quantum dots: Synthesis, properties, and sensing applications as a potential clinical analytical method. *Anal. Methods* **2019**, *11*, 2240–2258. [CrossRef]
73. Dykman, L.A.; Khlebtsov, N.G. Gold Nanoparticles in Biology and Medicine: Recent Advances and Prospects. *Acta Naturae* **2011**, *3*, 34–55. [CrossRef] [PubMed]
74. Baetke, S.C.; Lammers, T.; Kiessling, F. Applications of nanoparticles for diagnosis and therapy of cancer. *Br. J. Radiol.* **2015**, *88*, 20150207. [CrossRef] [PubMed]
75. Elsabahy, M.; Wooley, K.L. Design of polymeric nanoparticles for biomedical delivery applications. *Chem. Soc. Rev.* **2012**, *41*, 2545–2561. [CrossRef] [PubMed]
76. Capeletti, L.B.; Loiola, L.M.D.; Picco, A.S.; da Silva Liberato, M.; Cardoso, M.B. Silica Nanoparticle Applications in the Biomedical Field. In *Smart Nanoparticles for Biomedicine*; Elsevier: Amsterdam, The Netherlands, 2018; pp. 115–129.
77. Shirshahi, V.; Soltani, M. Solid silica nanoparticles: Applications in molecular imaging. *Contrast Media Mol. Imaging* **2015**, *10*, 1–17. [CrossRef] [PubMed]
78. Wang, Y.; Zhao, Q.; Han, N.; Bai, L.; Li, J.; Liu, J.; Che, E.; Hu, L.; Zhang, Q.; Jiang, T.; et al. Mesoporous silica nanoparticles in drug delivery and biomedical applications. *Nanomed. Nanotechnol. Biol. Med.* **2015**, *11*, 313–327. [CrossRef] [PubMed]
79. Braun, K.; Stürzel, C.M.; Biskupek, J.; Kaiser, U.; Kirchhoff, F.; Lindén, M. Comparison of different cytotoxicity assays for in vitro evaluation of mesoporous silica nanoparticles. *Toxicol. In Vitro* **2018**, *52*, 214–221. [CrossRef] [PubMed]
80. Tamba, B.I.; Dondas, A.; Leon, M.; Neagu, A.N.; Dodi, G.; Stefanescu, C.; Tijani, A. Silica nanoparticles: Preparation, characterization and in vitro/in vivo biodistribution studies. *Eur. J. Pharm. Sci.* **2015**, *71*, 46–55. [CrossRef] [PubMed]
81. Colombo, M.; Carregal-Romero, S.; Casula, M.F.; Gutierrez, L.; Morales, M.P.; Böhm, I.B.; Heverhagen, J.T.; Prosperi, D.; Parak, W.J. Biological applications of magnetic nanoparticles. *Chem. Soc. Rev.* **2012**, *41*, 4306–4334. [CrossRef] [PubMed]
82. Basso, C.R.; Crulhas, B.P.; Magro, M.; Vianello, F.; Pedrosa, V.A. A new immunoassay of hybrid nanomater conjugated to aptamers for the detection of dengue virus. *Talanta* **2019**, *197*, 482–490. [CrossRef] [PubMed]
83. Venerando, R.; Miotto, G.; Magro, M.; Dallan, M.; Baratella, D.; Bonaiuto, E.; Zboril, R.; Vianello, F. Magnetic Nanoparticles with Covalently Bound Self-Assembled Protein Corona for Advanced Biomedical Applications. *J. Phys. Chem. C* **2013**, *117*, 20320–20331. [CrossRef]
84. Skopalik, J.; Polakova, K.; Havrdova, M.; Justan, I.; Magro, M.; Milde, D.; Knopfova, L.; Smarda, J.; Polakova, H.; Gabrielova, E.; et al. Mesenchymal stromal cell labeling by new uncoated superparamagnetic maghemite nanoparticles in comparison with commercial Resovist—An initial in vitro study. *Int. J. Nanomed.* **2014**, *9*, 5355–5372. [CrossRef] [PubMed]
85. Cmiel, V.; Skopalik, J.; Polakova, K.; Solar, J.; Havrdova, M.; Milde, D.; Justan, I.; Magro, M.; Starcuk, Z.; Provaznik, I. Rhodamine bound maghemite as a long-term dual imaging nanoprobe of adipose tissue-derived mesenchymal stromal cells. *Eur. Biophys.* **2017**, *46*, 433–444. [CrossRef] [PubMed]
86. Magro, M.; Campos, R.; Baratella, D.; Lima, G.P.P.; Hola, K.; Divoky, C.; Stollberger, R.; Malina, O.; Aparicio, C.; Zoppellaro, G.; et al. A Magnetically Drivable Nanovehicle for Curcumin with Antioxidant Capacity and MRI Relaxation Properties. *Chem. Eur. J.* **2014**, *20*, 11913–11920. [CrossRef] [PubMed]

87. Magro, M.; Martinello, T.; Bonaiuto, E.; Gomiero, C.; Baratella, D.; Zoppellaro, G.; Cozza, G.; Patruno, M.; Zboril, R.; Vianello, F. Covalently bound DNA on naked iron oxide nanoparticles: Intelligent colloidal nano-vector for cell transfection. *Biochim. Biophys. Acta Gen. Sub.* **2017**, *1861*, 2802–2810. [CrossRef] [PubMed]
88. Venerando, A.; Magro, M.; Baratella, D.; Ugolotti, J.; Zanin, S.; Malina, O.; Zboril, R.; Lin, H.; Vianello, F. Biotechnological applications of nanostructured hybrids of polyamine carbon quantum dots and iron oxide nanoparticles. *Amino Acids* **2019**, 1–11. [CrossRef] [PubMed]
89. Anderson, C.J.; Bulte, J.W.; Chen, K.; Chen, X.; Khaw, B.A.; Shokeen, M.; Wooley, K.L.; VanBrocklin, H.F. Design of targeted cardiovascular molecular imaging probes. *J. Nucl. Med.* **2010**, *51*, 3S–17S. [CrossRef] [PubMed]
90. Kanwar, J.R.; Roy, K.; Kanwar, R.K. Chimeric aptamers in cancer cell-targeted drug delivery. *Crit. Rev. Biochem. Mol. Biol.* **2011**, *46*, 459–477. [CrossRef] [PubMed]
91. Kubinova, S.; Sykova, E. Nanotechnologies in regenerative medicine. *Minim. Invasive Ther. Allied Technol.* **2010**, *19*, 144–156. [CrossRef] [PubMed]
92. Kiessling, F.; Huppert, J.; Zhang, C.; Jayapaul, J.; Zwick, S.; Woenne, E.C.; Mueller, M.M.; Zentgraf, H.; Eisenhut, M.; Addadi, Y.; et al. RGD-labeled USPIO inhibits adhesion and endocytotic activity of alpha v beta3-integrin-expressing glioma cells and only accumulates in the vascular tumor compartment. *Radiology* **2009**, *253*, 462–469. [CrossRef] [PubMed]
93. Dias, A.M.; Hussain, A.; Marcos, A.S.; Roque, A.C. A biotechnological perspective on the application of iron oxide magnetic colloids modified with polysaccharides. *Biotechnol. Adv.* **2011**, *29*, 142–155. [CrossRef] [PubMed]
94. Chemello, G.; Piccinetti, C.; Randazzo, B.; Carnevali, O.; Maradonna, F.; Magro, M.; Bonaiuto, E.; Vianello, F.; Pasquaroli, S.; Radaelli, G.; et al. Oxytetracycline delivery in adult female zebrafish by iron oxide nanoparticles. *Zebrafish* **2016**, *13*, 495–503. [CrossRef] [PubMed]
95. Magro, M.; Baratella, D.; Bonaiuto, E.; de Almeida Roger, J.; Chemello, G.; Pasquaroli, S.; Mancini, I.; Olivotto, I.; Zoppellaro, G.; Ugolotti, J.; et al. Stealth iron oxide nanoparticles for organotropic drug targeting. *Biomacromolecules* **2019**, *20*, 1375–1384. [CrossRef] [PubMed]
96. Chemello, G.; Randazzo, B.; Zarantoniello, M.; Fifi, A.P.; Aversa, S.; Ballarin, C.; Radaelli, G.; Magro, M.; Olivotto, I. Safety assessment of antibiotic administration by magnetic nanoparticles in in vitro zebrafish liver and intestine cultures. *Comp. Biochem. Physiol. C Toxicol. Pharmacol.* **2019**, 108559. [CrossRef] [PubMed]
97. Skjolding, L.M.; Sørensen, S.N.; Hartmann, N.B.; Hjorth, R.; Hansen, S.F.; Baun, A. A critical review of aquatic ecotoxicity testing of nanoparticles—The quest for disclosing nanoparticle effects. *Angew. Chem. Int. Ed.* **2016**, *55*, 15224–15239. [CrossRef] [PubMed]
98. Magro, M.; De Liguoro, M.; Franzago, E.; Baratella, D.; Vianello, F. The surface reactivity of iron oxide nanoparticles as a potential hazard for aquatic environments: A study on Daphnia magna adults and embryos. *Sci. Rep.* **2018**, *8*, 13017. [CrossRef] [PubMed]
99. Magro, M.; Bramuzzo, S.; Baratella, D.; Ugolotti, J.; Zoppellaro, G.; Chemello, G.; Olivotto, I.; Ballarin, C.; Radaelli, G.; Arcaro, B.; et al. Self-assembly of chlorin-e6 on γ-Fe2O3 nanoparticles: Application for larvicidal activity against Aedes aegypti. *J. Photochem. Photobiol. B Biol.* **2019**, *194*, 21–31. [CrossRef] [PubMed]
100. Tang, S.C.N.; Lo, I.M.C. Magnetic nanoparticles: Essential factors for sustainable environmental applications. *Water Res.* **2013**, *47*, 2613–2632. [CrossRef] [PubMed]
101. Plachtová, P.; Medříková, Z.; Zbořil, R.; Tuček, J.; Varma, R.S.; Maršálek, B. Iron and iron oxide nanoparticles synthesized with green tea extract: Differences in ecotoxicological profile and ability to degrade malachite green. *ACS Sustain. Chem. Eng.* **2018**, *6*, 8679–8687. [CrossRef] [PubMed]
102. Markova, Z.; Novak, P.; Kaslik, J.; Plachtova, P.; Brazdova, M.; Jancula, D.; Siskova, K.M.; Machala, L.; Marsalek, B.; Zboril, R.; et al. Iron(II,III)-polyphenol complex nanoparticles derived from green tea with remarkable ecotoxicological impact. *ACS Sustain. Chem. Eng.* **2014**, *2*, 1674–1680. [CrossRef]
103. Mwilu, S.K.; Siska, E.; Baig, R.B.N.; Varma, R.S.; Heithmar, E.; Rogers, K.R. Separation and measurement of silver nanoparticles and silver ions using magnetic particles. *Sci. Total Environ.* **2014**, *472*, 316–323. [CrossRef] [PubMed]
104. Coyne, R.; Smith, P.; Moriarty, C. The fate of oxytetracycline in the marine environment of a salmon cage farm, Marine Environment and Health Series No. 3, Marine Institute 2001. *Mar. Environ. Health Ser.* **2001**, *3*, 1–24.

105. Wise, R. Antimicrobial resistance: Priorities for action. *J. Antimicrob. Chemother.* **2002**, *49*, 585–586. [CrossRef] [PubMed]
106. De La Torre, A.; Iglesias, I.; Carballo, M.; Ramírez, P.; Muñoz, M.J. An approach for mapping the vulnerability of European Union soils to antibiotic contamination. *Sci. Total Environ.* **2012**, *414*, 672–679. [CrossRef] [PubMed]
107. Rakshit, S.; Sarkar, D.; Elzinga, E.; Punamiya, P.; Datta, R. Surface complexation of oxytetracycline by magnetite: Effect of solution properties. *Vadose Zone J.* **2014**, *13*. [CrossRef]
108. Figueroa, R.A.; Mackay, A.A. Sorption of oxytetracycline to iron oxides and iron oxide-rich soils. *Environ. Sci. Technol.* **2005**, *39*, 6664–6671. [CrossRef] [PubMed]
109. Ihara, I.; Toyoda, K.; Beneragama, N.; Umetsu, K. Magnetic separation of antibiotics by electrochemical magnetic seeding. *J. Phys. Conf. Ser.* **2009**, *156*, 012034. [CrossRef]
110. Xiao, L.; Li, J.; Brougham, D.F.; Fox, E.K.; Feliu, N.; Bushmelev, A.; Schmidt, A.; Mertens, N.; Kiessling, F.; Valldor, M.; et al. Water-soluble superparamagnetic magnetite nanoparticles with biocompatible coating for enhanced magnetic resonance imaging. *ACS Nano* **2011**, *5*, 6315–6324. [CrossRef] [PubMed]
111. Magro, M.; Baratella, D.; Molinari, S.; Venerando, A.; Salviulo, G.; Chemello, G.; Olivotto, I.; Zoppellaro, G.; Ugolotti, J.; Aparicio, C.; et al. Biologically safe colloidal suspensions of naked iron oxide nanoparticles for in situ antibiotic suppression. *Colloids Surf. B Biointerfaces* **2019**, *181*, 102–111. [CrossRef] [PubMed]
112. Magro, M.; Domeneghetti, S.; Baratella, D.; Jakubec, P.; Salviulo, G.; Bonaiuto, E.; Venier, P.; Malina, O.; Tuček, J.; Ranc, V.; et al. Colloidal Surface Active Maghemite Nanoparticles for biologically safe CrVI remediation: From core-shell nanostructures to pilot plant development. *Chem. Eur. J.* **2016**, *22*, 14219–14226. [CrossRef] [PubMed]
113. Colino, C.; Millán, C.; Lanao, J. Nanoparticles for signaling in biodiagnosis and treatment of infectious diseases. *Int. J. Mol. Sci.* **2018**, *19*, 1627. [CrossRef] [PubMed]
114. Cho, I.H.; Ku, S. Current technical approaches for the early detection of foodborne pathogens: Challenges and opportunities. *Int. J. Mol. Sci.* **2017**, *18*, 2078. [CrossRef] [PubMed]
115. Magro, M.; Baratella, D.; Bonaiuto, E.; de Almeida Roger, J.; Vianello, F. New Perspectives on Biomedical Applications of Iron Oxide Nanoparticles. *Curr. Med. Chem.* **2017**, *25*, 540–555. [CrossRef] [PubMed]
116. Mahmoudi, M.; Lynch, I.; Ejtehadi, M.R.; Monopoli, M.P.; Baldelli Bombelli, F.; Laurent, S. Protein–Nanoparticle Interactions: Opportunities and Challenges. *Chem. Rev.* **2011**, *111*, 5610–5637. [CrossRef] [PubMed]
117. Jersmann, H.P.; Dransfield, I.; Hart, S.P. Fetuin/Alpha2-HS Glycoprotein Enhances Phagocytosis of Apoptotic Cells and Macropinocytosis by Human Macrophages. *Clin. Sci.* **2003**, *105*, 273–278. [CrossRef] [PubMed]
118. Cedervall, T.; Lynch, I.; Lindman, S.; Berggård, T.; Thulin, E.; Nilsson, H.; Dawson, K.A.; Linse, S. Understanding the nanoparticle–protein corona using methods to quantify exchange rates and affinities of proteins for nanoparticles. *Proc. Natl. Acad. Sci. USA* **2007**, *104*, 2050–2055. [CrossRef] [PubMed]
119. Garcia, J.; Zhang, Y.; Taylor, H.; Cespedes, O.; Webb, M.E.; Zhou, D. Multilayer Enzyme-Conjugated Magnetic Nanoparticles as Efficient, Reusable Biocatalysts and Biosensors. *Nanoscale* **2001**, *3*, 3721–3730. [CrossRef] [PubMed]
120. Niemirowicz, K.; Markiewicz, K.H.; Wilczewska, A.Z.; Car, H. Magnetic nanoparticles as new diagnostic tools in medicine. *Adv. Med. Sci.* **2012**, *57*, 196–207. [CrossRef] [PubMed]
121. Mirshafiee, V.; Mahmoudi, M.; Lou, K.; Cheng, J.; Kraft, M.L. Protein corona significantly reduces active targeting yield. *Chem. Commun.* **2013**, *49*, 2557–2559. [CrossRef] [PubMed]
122. Salvati, A.; Pitek, A.S.; Monopoli, M.P.; Prapainop, K.; Bandelli Bombelli, F.; Hristov, D.R.; Kelly, P.M.; Åberg, C.; Mahon, E.; Dawson, K.A. Transferrin-functionalized nanoparticles lose their targeting capabilities when a biomolecule corona adsorbs on the surface. *Nat. Nanotechnol.* **2013**, *8*, 137–143. [CrossRef] [PubMed]
123. Laurent, S.; Burtea, C.; Thirifays, C.; Rezaee, F.; Mahmoudi, M. Significance of cell "observer" and protein source in nanobiosciences. *J. Colloid Interface Sci.* **2013**, *392*, 431–445. [CrossRef] [PubMed]

© 2019 by the authors. Licensee MDPI, Basel, Switzerland. This article is an open access article distributed under the terms and conditions of the Creative Commons Attribution (CC BY) license (http://creativecommons.org/licenses/by/4.0/).

Review

Microbial Nanotechnology: Challenges and Prospects for Green Biocatalytic Synthesis of Nanoscale Materials for Sensoristic and Biomedical Applications

Gerardo Grasso *, Daniela Zane and Roberto Dragone

Consiglio Nazionale delle Ricerche—Istituto per lo Studio dei Materiali Nanostrutturati c/o Dipartimento di Chimica, 'Sapienza' Università di Roma, P. le Aldo Moro 5, 00185 Roma, Italy; daniela.zane@cnr.it (D.Z.); roberto.dragone@cnr.it (R.D.)
* Correspondence: gerardo.grasso@cnr.it; Tel.: +39-064-991-3380

Received: 19 November 2019; Accepted: 13 December 2019; Published: 18 December 2019

Abstract: Nanomaterials are increasingly being used in new products and devices with a great impact on different fields from sensoristics to biomedicine. Biosynthesis of nanomaterials by microorganisms is recently attracting interest as a new, exciting approach towards the development of 'greener' nanomanufacturing compared to traditional chemical and physical approaches. This review provides an insight about microbial biosynthesis of nanomaterials by bacteria, yeast, molds, and microalgae for the manufacturing of sensoristic devices and therapeutic/diagnostic applications. The last ten-year literature was selected, focusing on scientific works where aspects like biosynthesis features, characterization, and applications have been described. The knowledge, challenges, and potentiality of microbial-mediated biosynthesis was also described. Bacteria and microalgae are the main microorganism used for nanobiosynthesis, principally for biomedical applications. Some bacteria and microalgae have showed the ability to synthetize unique nanostructures: bacterial nanocellulose, exopolysaccharides, bacterial nanowires, and biomineralized nanoscale materials (magnetosomes, frustules, and coccoliths). Yeasts and molds are characterized by extracellular synthesis, advantageous for possible reuse of cell cultures and reduced purification processes of nanomaterials. The intrinsic variability of the microbiological systems requires a greater protocols standardization to obtain nanomaterials with increasingly uniform and reproducible chemical-physical characteristics. A deeper knowledge about biosynthetic pathways and the opportunities from genetic engineering are stimulating the research towards a breakthrough development of microbial-based nanosynthesis for the future scaling-up and possible industrial exploitation of these promising 'nanofactories'.

Keywords: applied microbiology; white biotechnology; green chemistry; nanostructured materials; diatom nanotechnology; sensoristic devices; drug delivery; theranostics

---

1. Introduction

During the period of 2016–2022 the global nanomaterials market is expected to grow with a compound annual growth rate of about 20% or more [1]. One of the major challenges for the global advancement of nanomaterials market is the environmental sustainability of nanomanufacturing processes. Indeed, traditional top-down or bottom-up chemical and physical nanomanufacturing approaches have a greater energy-intensity compared to manufacturing processes of bulk materials. Further, they are often characterized by low process yields (using acidic/basic chemicals and organic solvents), generation of greenhouse gases, and they require specific facilities, operative conditions (e.g., from moderate to high vacuum), and high purity levels of starting materials [2–4]. The principles of green chemistry ("the invention, design and application of chemical products and processes to reduce or to eliminate the use and generation of hazardous substances") combined with white biotechnology ("biotechnology that uses living cells—yeasts, molds, bacteria, plants, and enzymes to synthesize products at industrial scale") can really contribute to the development of more sustainable industrial processes [5], also for nanomanufacturing. The microbial-mediated biosynthesis of nanomaterials is a promising biotechnological-based nanomanufacturing process that represents a 'green' alternative approach to physical and chemical strategies of nanosynthesis [6,7]. The microbial-mediated biosynthesis of metallic (also as alloys), non-metallic, or metal oxides nanoparticles have been reported for many microbial strains of bacteria, yeast, molds, and microalgae [8] (Figure 1).

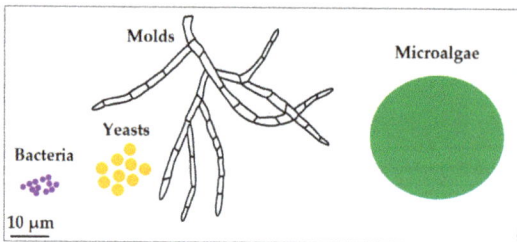

Figure 1. Schematic comparing average sizes of the microorganisms described in this review.

In addition, some microorganisms have shown the capability to biosynthesize unique nanostructured materials, i.e., biomineralized nanostructures like silicified frustules [9], calcified coccoliths [10], magnetosomes [11], and organic nanomaterials like bacterial nanocellulose [12] exopolysaccharide nanoparticles [13] and bacterial nanowires [14]. The microbial-mediated biosynthesis of nanomaterials has been extensively explored showing different advantages and features including the following: (i) synthetized nanomaterials have defined chemical composition, size and morphology, (ii) biosynthesis is performed at mild physico-chemical conditions, (iii) easily handling and cultivation of microbial cells and possibility of cell culture scale-up, (iv) possibility of in vivo tuning nanomaterial characteristics by changing key parameters of cell culture operational set up or through genetically engineering tools [15]. In order to enable a broad applicability of microbial-mediated biosynthesis of nanomaterials as a real alternative to 'traditional' synthetic approaches to nanomanufacturing, many hurdles still need to be overcome: a reduction of polidispersity of nanoparticles, a more complete characterization of biocapping layer agents, effectiveness of removal procedures of biocapping layer and nanomaterials purifications, standardization of microbial cell culture protocols for reproducibility of nanosynthesis processess, as well as production costs and yields. Overeaching the challenge for the development of reliable eco-friendly nanotechnologies for nanomaterial synthesis is of utmost importance for future exploitations of broad-impact nanostructured-based technologies and applications, like innovative optical and electrochemical (bio) sensoristic devices [16] and therapeutic and diagnostic applications of nanostructured materials e.g., for drug delivery, in vivo/in vitro imaging and development of antimicrobial and antitumoral drugs [17,18]. In the first part of this review, we reported an overview of scientific literature (mainly from the last ten years) about in vivo microbial biosynthesis of nanomaterials that have been used for (bio) sensoristic and biomedical purposes. We focused on works that have covered several key aspects of nanomaterials: (i) type of biosynthesis (in some cases post-biosynthesis functionalization), (ii) biosynthetic pathways (presumptive or demonstrated), (iii) characterization, (iv) applications. In the second part, main acquired knowledge, challenges, and potentiality of microbial-mediated biosynthesis has been described.

2. Microbial-Mediated Biosynthesis of Nanomaterials for Sensoristic and Biomedical Applications

2.1. Bacteria

In the last ten years, bacteria have been used to synthesize inorganic nanomaterials (mainly selenium, gold, and silver nanoparticles) with interesting properties for the development of voltammetric sensoristic devices [19], and third-generation biosensors [20], for possible diagnostic applications [21] like cell imaging and biolabeling [22] and for applications where no surface coat is required, like annealing and thin film formation [23] (Table 1). Bacterial-biosynthesized nanoparticles have mainly shown in vitro antimicrobial activity against some pathogenic bacterial strains [24–28] and properties i.e., antioxidant [29], anti-proliferative, anti-migration [30], anticoagulant [31], and anticancer [26–33]. Biochemical mechanisms which mediate the bacterial biosynthesis of nanoparticles have been proposed or they are currently under investigation. Many of these biochemical mechanisms have been described as part of microbial resistance mechanisms for cellular detoxification which involves changes in solubility of inorganic ions by enzymatic reduction and/or precipitation of soluble toxic to insoluble non-toxic nanostructures. Both extracellular and intracellular biocatalytic synthesis (with and possible excretion) mainly involves oxidoreductase enzymes (e.g., NADH-dependent nitrate reductase, NADPH-dependent sulphite reductase flavoprotein subunit α, and cysteine desulfhydrase) and cellular transporters. Physicochemical processes like biosorption, complexation, nucleation, growth, and stabilization mediated by biomolecules (e.g., proteins and carbohydrates) have also been described. In addition to inorganic nanomaterials, some bacteria genera have shown the ability to biosynthesize very peculiar organic nanostructures. Bacterial nanocellulose is a 3-D network of cellulose nanofibrils produced by aerobic acetic bacteria like those belonging to the genus *Gluconacetobacter*, the most efficient bacteria for nanocellulose biosynthesis. Compared to the nanocrystalline cellulose and nanofibrillated cellulose, bacterial nanocellulose shows higher purity, crystallinity and mechanical stability [34]. Therefore, bacterial nanocellulose is a nanomaterial which has attracted great attention for use in biomedical applications (e.g., as antimicrobial agent, for drug delivery systems and scaffolds for tissue engineering) and on biosensoristic platforms (as nanocomposite and as support for the immobilization of biological recognition elements) [35–37]. Exopolysaccharides are microbial extracellular biopolymers with different roles in adhesion of bacterial biofilms and as protection agents. In a recent work a novel self-assembled and spherical nanosized non-glucan exopolysaccharide has been described for bacteria *Lactobacillus plantarum*-605. Results have showed its reducing actions for rapid (30 min.) biosynthesis of good monodispersed gold and silver nanoparticles without any pretreatment or modification [38]. Bacterial nanowires are conductive proteinaceous pilus-like nanostructures involved in extracellular electron transport processes of anaerobic dissimilatory metal-reducing bacteria like *Geobacter* and *Shewanella* genera [39], aerobic bacteria like *Pseudomonas aeruginosa* [40] and aerobic photosynthetic cyanobacteria like *Microcystis* and *Synechocystis* genera [41]. Metallic-like conductivity (due to aromatic amino acids-richness in PilA proteic fibers) and a redox-based conductivity (mediated by cytochrome OmcS present on fibers surface) have been hypothesized for bacterial nanowires in *G. sulfurreducens* [39]. Studies on nanowires *Shewanella oneidensis* MR-1 strain have showed a p-type, tunable electronic behavior with electrical conductivities comparable to moderately doped inorganic semiconductors used in synthetic organic semiconductor-based devices like field-effect transistors [42]. The bacterium *S. oneidensis* have been also described for biosynthesis of gold and silver nanomaterials [23,24]. Bacterial nanowires are also very promising nanostructures in the bioelectronic field for the development of new biomaterial for microbial fuel cells and electrochemical (bio) sensoristic devices i.e., as direct electron transfer mediator between bacteria biofilm and the solid-state electrode surfaces. Different silicon-based electrodes for rapid biochemical oxygen demand (BOD) determination and water integral toxicity monitoring have been described in recent literature [43–45]. Bacterial magnetosomes are organic-coated intracellular nanocrystals of Fe_3O_4 and/or Fe_3S_4, biosynthesized by both magnetotactic and non-magnetotactic bacteria. The composition of magnetic inorganic part is species-specific, and the external organic coating layer is derived from bacterial phospholipid bilayer membrane. The putative functions of protein component of the external organic coating layer in the magnetosome biomineralization process have been hypothesized [11]. Bacterial Fe_3O_4 magnetosomes are stable single-magnetic domains permanently magnetic at ambient temperature, possessing peculiar characteristics of high chemical purity, a narrow size range and consistent crystal morphology [46]. Some recent applications include molecular imaging [47], cancer therapy [48], and the development of a chip-based whole-cell biosensor for toxicity assessment [49].

2.2. Yeasts and Molds

The research focused on biosynthesis of nanomaterials by fungi, like yeasts and molds, have brought to the coinage of the term 'myconanotechnology', in order to refer to a newly emerging domain of nanotechnology. Yeasts are unicellular fungi mainly known in nanosynthesis for their ability to produce semiconductor nanoparticles [8]. Biosynthesis of high water-soluble and biocompatible cadmium telluride quantum dots by model organism yeast *Saccharomyces cerevisiae* have been reported in literature. These cadmium telluride quantum dots have showed interesting characteristics of size-tunable (changing culture time and temperature) emission and photoluminescence quantum yield as good candidate for bio-imaging and bio-labelling applications [50]. *S. cerevisiae* have been also used for biosynthesis of Au–Ag alloy nanoparticles for electrochemical sensors fabrication [51,52], aimed to the determination of paracetamol in tablet samples and vanillin in vanilla bean and vanilla tea sample, respectively (see Table 2). Possible biosynthesis mechanisms of nanoparticles by *S. cerevisiae* could involve membrane bound and cytosolic oxidoreductases as well as extracellular 1,3-β-glucan synthase-mediated formation and growth of nanoparticles [50–53]. Molds are a large group of microscopic filamentous fungi that include many genera like to *Penicillum*, *Aspergillus*, and *Fusarium*. Compared to bacteria, molds possess many distinctive advantages for biosynthesis of nanomaterials: (i) higher metal tolerance, (ii) higher metal binding and uptake capabilities, (iii) easy culturing and fast growing; (iv) higher extracellular nanosynthesis (mediated by extracellular enzyme, reductive proteins, and secondary secreted metabolites). Extracellular biosynthesis of nanomaterials poses advantages in terms of a possible reuse of cell cultures for new biosynthesis (cell lysis not required) and reduced nanoparticle downstream purification processes [54]. Proposed mechanisms behind fungal synthesis of nanoparticles hypothesized a possible involvement of biomolecules secreted in formation and stabilization of nanoparticles [55], secreted reductases [56,57] and possible trapping of metal ions by electrostatic interaction with positively charged groups in enzymes present in cell wall of the mycelia [58]. In the last ten years, several works have described mold-based biosynthesis of nanoparticles (silver, gold, and tellurium) and quantum dots (zinc sulfide, zinc sulfide with gadolinium, and lead sulfide). These nanoparticles have shown both antibacterial activity [54,57,59] and antitumoral activity [55,56,58] beside possible employment in optical detection of heavy metals and arsenic in water [60,61] (Table 2).

Table 1. Nanomaterials synthesized by bacteria.

Microorganism	Culture Conditions (Synthesis Time)	Nanomaterial	Characterization	Biosynthetic Pathway	Application	Ref.
Bacillus subtilis	Enrichment medium, 35 °C, stirred at 170 rpm + 4 mM Na_2SeO_3 (48 h)	Se NPs	50–400 nm; spherical regular morphology; 100 nm uniform single-crystalline; nanowires	Reduction mechanism of SeO_3^{2-} ions to Se^0 is yet to be elucidated	H_2O_2 sensoristic device	[19]
Streptomyces minutiscleroticus M10A62	5 g of wet bacterial biomass from 120 h cell culture + 1 mM Na_2SeO_3, stirred at 200 rpm (72 h)	Se NPs	10–250 nm; spherical shape; crystalline; ζ-potential −19.1 mV	Extracellular synthesis not described	Anti-biofilm, antioxidant activity, antiviral activity against Dengue virus; anti-proliferative activity against HeLa and HepG2 cell lines	[21]
Pantoea agglomerans strain UC-32	1% (v/v) of an overnight cell culture in tryptic soy broth + 1 mM Na_2SeO_3, 25 °C (24 h)	Se NPs	<100 nm; spherical shape; amorphous form size vary with culture time (10–24 h);	Intracellular reduction of Se (IV) to Se (0) and subsequent excretion	High antioxidant activity (when stabilized with L-cysteine)	[29]
Streptomyces bikiniensis strain Ess. amA-1	1 mL fresh bacteria inoculums (OD_{600} = 0.5 a.u.) in international *Streptomyces* Project 2 medium + 1 mM SeO_2, 30 °C, stirred at 150 rpm (48 h)	Se NPs	600 nm length, 17 nm diameter	Possible involvement of proteins/enzymes in SeO_2 reduction nucleation, growth, stabilization of nanorods	In vitro anticancer activity against human breast adenocarcinoma cell line and human liver carcinoma cell line	[32]
Escherichia coli DH5α	10 h culture, resuspended in sterile distilled water + 1 mM $HAuCl_4$, room temperature (120 h)	Au NPs	25 ± 8 nm; spherical shape; crystalline form (face centered cubic phase)	Extracellular synthesis possibly modulated by sugars or enzymes present onto bacteria surface	Direct electro-chemistry of hemoglobin	[20]
Shewanella oneidensis MR-1	Washed cell pellet from a 24 h cell culture + 1 mM $HAuCl_4$, 30 °C, stirred at 200 rpm (48 h)	Au NPs	12 ± 5 nm; spherical shape, capping proteins easily removable but not identified	Extracellular synthesis possible enzymatic shuttle-based enzymatic reduction of ionic Au^{3+} to Au^0	No antibacterial properties/annealing and thin film formation	[23]
Nocardiopsis sp. MBRC-48	Cell-free supernatant (from a 96 h cell culture) + 0.9 mM $HAuCl_4$, incubated in the dark, 35 °C, stirred at 180 rpm (48 h)	Au NPs	11.57 ± 1.24 nm; spherical shape; face centered cubic; polydispersed without significant structure	Extracellular synthesis using the cell free supernatant, proteins, enzymes and metabolites	High antimicrobial activity against *Staphylococcus aureus* and *Candida albicans*, antioxidant activity and cytotoxic activities	[25]
Brevibacterium casei	1 g of wet bacterial biomass + 1 × 10^{-3} M $AgNO_3$ + 1 × 10^{-3} M $HAuCl_4$, 37 °C, stirred at 200 rpm (24 h)	Au and Ag NPs	Ag 10–50 nm, Au, 0–50 nm; spherical shape, crystalline form (face centered cubic phase)	Intracellular synthesis, possible roles of NADH-dependent nitrate reductase (for Ag NPs) and α-NADPH-dependent sulfite reductase (for Au NPs)	Anti-coagulant properties	[31]
Shewanella oneidensis MR-1	~3–5 g of wet bacterial biomass from 24 h cell culture + 1 mM $AgNO_3$, 30 °C stirred at 200 rpm (48 h)	Ag NPs	~2–11 nm spherical shape; crystalline form; ζ-potential −16.5 mV	Extracellular synthesis by secreted factors (e.g., NADH-dependent reductases, quinines, soluble electron-shuttles)	Antibacterial activity against *Escherichia coli* and *Bacillus subtilis*	[24]
Lyngbya majuscula (CUH/Al/MW-150)	100 mg of fresh weight biomass + 9 mM Ag(I) solution (pH 4) incubated in the dark, room temperature (72 h)	Ag NPs	~5–50 nm; spherical shape, crystalline form (face-centered cubic), smooth surface morphology, both (sonication) ζ-potential = −35.2 mV	Extracellular and intracellular synthesis not described	Effective antibacterial activity against *Pseudomonas aeruginosa*; appreciable anti-proliferative effect on leukemic cells, especially on the REH cell line	[26]
Streptomyces s. Al-Dhabi-87	Broth-free cell pellets (14-days cell culture) in sterile distilled water for 1 h; cell removed from the suspension + 1–5 mM $AgNO_3$, 37 °C (48 h)	Ag NPs	20–50 nm; spherical shape	Extracellular synthesis possibly via hydrophilic and hydrophobic small metabolites attached on the bacteria cell wall	In vitro antimicrobial activity against *Bacillus subtilis*, *Enterococcus faecalis*, *Staphylococcus epi-dermidis*, and multidrug resistant *Staphylococcus aureus* strain	[27]
Bacillus licheniformis	2 g of wet bacterial biomass + 1 mM $AgNO_3$, 37 °C, stirred at 200 rpm (24 h)	Ag NPs	40 nm to 50 nm	N/A	Possible application as anti-proliferative and anti-migration agent e.g., against diabetic retinopathy, neoplasia and rheumatoid arthritis	[30]

Table 1. Cont.

Microorganism	Culture Conditions (Synthesis Time)	Nanomaterial	Characterization	Biosynthetic Pathway	Application	Ref.
Escherichia coli K12 (ATCC 29181)	Bacterial culture (OD$_{600}$ = 0.6 a.u.), Luria Bertani medium + 3 mM CdCl$_2$ + 6 mM Na$_2$C$_6$H$_5$O$_7$ + 0.8 mM Na$_2$TeO$_3$, 8 mM C$_4$H$_6$O$_4$S + 26 mM NaBH$_4$, 37 °C, stirred at 200 rpm (24 h)	CdTe QDs	~2–3 nm; uniform size, cubic crystals; strong fluorescence emission shift with increasing quantum dots size, capping proteins were not identified but enhance QDs biocompatibility; ζ-potential = −19.1 mV	Extracellular synthesis possibly via protein-assisted nucleation biosynthesis	Possible application in vitro cell imaging (demonstrated on HeLa cells) and bio-labeling	[2]
Acetobacter xylinum GIM1.327	Static culture in polysaccharides enriched medium, 30 °C (120 h)	Bacterial nanocellulose nanofibrils impregnated with Ag-NPs	Nanoporous three-dimensional network structure with a random arrangement of ribbon-shaped microfibrils without any preferential orientation; 2 to 100 nm (Ag NPs)	Intracellular-extracellular synthesis via enzymes glucokinase, phosphoglucomutase, UDPG, pyro-phospho-rylase and cellulose synthase	In vitro pH-responsive antimicrobial activity against *Escherichia coli* ATCC 25922, *Staphylococcus aureus* ATCC 6538, *Bacillus subtilis* ATCC 9372 and *Candida albicans* CMCC(F) 98001	[28,35]
Acetobacter xylinum	N/A	Ag NPs and bacterial nano-paper composite	AgNPs 10–50 nm	Intracellular-extracellular synthesis of bacterial nanocellulose via enzymes glucokinase, phosphoglucomutase, UDPG, pyro-phospho-rylase and cellulose synthase AgNPs synthesis via direct chemical reduction of Ag$^+$ mediated by baring hydroxyl groups of bacterial nanocellulose	Optical detection of cyanide ion and 2-mercaptobenzo-thiazole in water samples	[35,36]
Acetobacter xylinum	Static culture containing 50 g/L glucose, 5 g/L yeast extract, 5 g/L (NH$_4$)$_2$SO$_4$, 4 g/L KH$_2$PO$_4$ and 0.1 g/L MgSO$_4$·7H$_2$O, 28 °C (366 h)	Nanocomposites of bacterial nanocellulose with AgNP, Au-NPs CdSe@ZnS quantum dots functionalized with biotinylated antibodies, aminosilica-coated lanthanide-doped up-conversion NPs	(bacterial nanocellulose) 45 ± 10 nm (fiber mean diameter); estimated length > 10 µm	Intracellular-extracellular synthesis via enzymes glucokinase, phosphoglucomutase, UDPG, pyro-phospho-rylase and cellulose synthase	Optical detection of methimazole, thiourea, cyanide, and iodide and *Escherichia coli*; possible uses in analytes pre-concentration platform	[35,37]
Bacillus marisflavi CS3	200 mg biomass + 2.4 × 10^{-5} M graphene oxide dispersion mixture, 37 °C (72 h)	Reduced graphene oxide nanosheets	~4.3 nm (average thickness), significant reduction of GO (assessed by XRD analysis); several layers stacked on top of one another like silky sheets of paper (SEM image)	Extracellular synthesis not described	Inhibition of cell viability, reactive oxygen species (ROS) generation, and membrane integrity alteration in MCF-7 cell line	[33]

Table 1. Cont.

Microorganism	Culture Conditions (Synthesis Time)	Nanomaterial	Characterization	Biosynthetic Pathway	Application	Ref.
Magnetospirillum magneticum AMB-1 (Genetically modified)	Anaerobically grown in 5 ml/L of Wolfe's mineral solution (without iron), + 5 mM KH_2PO_4 + 10 mM $NaNO_3$ + 0.85 mM $C_2H_3NaO_2$ + 0.2 mM $C_6H_8O_6$ + 2.5 mM $C_4H_4O_6$ + 0.6 mM $Na_2S_2O_3$, pH 6.9; cell pellets were resuspended in 20 mM HEPES + 1 mM EDTA + 8% glycerol + 0.9% NaCl, pH 7.5	Magnetosome (bio-mineralized iron-oxide nanoparticles coated by a biological membrane)	Magnetosome membrane modified with Venus-RGD protein as specific and sensitive molecular imaging probe	Natural mechanism of magneto-somes formation (biomineralization) + genetic modification for Venus protein-RGD peptide expression	Contrast agent for in vivo magnetic resonance-based molecular imaging	[47]
Magnetospirillum magneticum strain AMB-1	Micro-anaerobically grown in a similar culture medium of [47]	Whole inactive magnetotactic bacteria γ-Fe_2O_3 magnetosomes chains individual γ-Fe_2O_3 magnetosomes	Magnetosomes chains (length) ~150 or ~300 nm; individual magnetosomes mean size ~45 nm; well-crystallized monodomain with a ferromagnetic behavior at physiological temperature	Natural mechanism of magneto-somes formation + genetic modification for Venus protein- RGD peptide expression	Antitumoral activity against MDA-MB-231 breast cancer cells under alternating magnetic field stimulation	[48]
Magnetospirillum gryphiswaldense strain MSR-1	Micro-anaerobically grown in a similar culture medium of [47] and [48] + 50μM Fe(III) citrate	Chains of magnetosomes	Magnetosome membrane modified with Red-emitting Click Beetle luciferase (CBR)	Natural mechanism of magneto-somes formation + genetic modification for red-emitting click beetle luciferase expression	Toxicity assay on microfluidic chip for the detection of toxicity effect on membrane by DMSO and TCDCA	[49]

Table 2. Nanomaterials synthesized by yeasts and molds.

Microorganism	Culture Conditions (Synthesis Time)	Nanomaterial	Characteristics (Average Size, Morphology, etc.)	Biosynthetic Pathway	Application	Ref.
Saccharomyces cerevisiae	Aerobic two days growth in a modified Czapek's medium, 5 °C; aliquot of cell suspension (OD_{600} = 0.6) + 3 mM $CdCl_2$ + 0.8 mM Na_2TeO_3 + 1.5 mM CH_3SO_3H + 2.6 mM NaB^-I_4, stirred at 500 rpm (N/A)	CdTe QDs	2.0–3.6 nm; cubic zinc blende crystals	Extracellular synthesis not described	Good candidate for bio-imaging and bio-labelling applications	[50]
Aspergillus welwitschiae KY766958	Growth in Czapek's medium; pH 7.3 ± 0.2, 30 °C for 7 days shaken at 150 rpm + 2 mmol K_2TeO_3 (48 h)	Te NPs	60.80 nm; oval to spherical shape	Mechanism not described	Antibacterial activity against E. coli and methicillin resistant Staphylococcus aureus (MRSA)	[59]
Commercially available instant dry yeast (Angel Yeast Co.—Yichang, China)	Sucrose solution (5 g/L) + instant dry yeast (600 mg), 30 °C for 24 h; cells pellet in sterile water (10^6 cells/mL) + $AgNO_3$ solution + $HAuCl_4$ solution (final concentrations N/A), 30 °C. (24 h)	Au-Ag alloy NPs	Reduced metallic form (XPS analysis); large superficial area and desirable conductivity (electrochemical impedance spectroscopy)	Extracellular synthesis not described	Electrochemical sensor for paracetamol	[51]
		Au-Ag alloy NPs	9–25 nm	Extracellular synthesis not described	Electrochemical sensor for vanillin	[52]
Humicola sp.	MGYP medium, pH 9, shaken at 200 rpm, 50 °C; harvested mycelial mass + 1 mM $AgNO_3$, shaken at 200 rpm, 50 °C (96 h)	Ag NPs	5–25 nm; spherical shape; face centered cubic crystals	Extracellular synthesis through a possible involvement of biomolecules secreted by the fungus	In vitro cytotoxicity against NIH3T3 mouse embryonic fibroblast cell line and MDA-MB-231 human breast carcinoma cell line	[55]
Fusarium oxysporum f. sp. lycopersici	5 days growth, potato dextrose broth, 28 °C; filtered biomass + 1 mM $AgNO_3$, 28 °C, dark condition (120 h)	Ag NPs	5 to 13 nm; spherical shape; face centered cubic crystals	Extracellular synthesis, possible involvement of a secreted reductase	Antibacterial activity against pathogenic bacteria Escherichia coli and Staphylococcus aureus; antitumoral activity against human breast carcinoma cell line MCF-7	[56]
Penicillium brevicompactum KCCM 60390	72 h growth, potato dextrose broth, 30 °C, shaken at 200 rpm; filtered biomass (5 g) in Milli-Q sterile deionized water and agitated, 72 h at 200 rpm, 30 °C; supernatant from filtered biomass + 1 mM $HAuCl_4$, shaken at 200 rpm, dark condition 30 °C (N/A)	Au NPs	(live cell filtrate) 25–60 nm; spherical shape; 20–80 nm (potato dextrose broth), spherical and triangular and hexagonal shape (culture supernatant broth) 20 to 50 nm; well dispersed and uniform in shape and size; good stability against aggregation after 3 months	Extracellular synthesis; possible ion trapping on the fungal cells surface via electrostatic interaction; possible involvement of organic reagents used for the microbial cultivations as potential reducing agents	Inhibitory effect and cytotoxicity against mouse cancer C_2C_{12} cell lines	[58]
Trichoderma harzianum (SKCGW008)	72 h cultured spores in wheat bran broth media, 28 °C shaken at 180 rpm; supernatant + 0.5% (wp) of low molecular weight chitosan in agitation (30 min)	Chitosan NPs	90.8 nm; spherical shape; amorphous structure	Extracellular synthesis via enzyme secreted (not identified)	Antioxidant activity; bactericidal activity against Staphylococcus aureus and Salmonella enterica; biocompatibility (no cytotoxic effect on murine fibroblast NIH–3T3 cells)	[57]
Aspergillus flavus	Growth in potato dextrose broth, 28 °C, 115 rpm; harvested fungal biomass + 3 mM $ZnSO_4$, 27 °C, 200 rpm; for ZnS:Gd nanoparticle 0.3 M $Cd(NO_3)_3$ (96 h)	ZnS and ZnS:Gd NPs	Nanocrystalline and a narrow size distribution: 12–24 nm (ZnS); for and 10–18 nm (ZnS:Gd)	Extracellular synthesis not described	Optical detection of Pb (II), Cd (II), Hg (II), Cu (II), and Ni (II) in water	[60]
Aspergillus flavus	Growth in potato dextrose broth + 0.5 mM $Pb(CH_3COO)_2$ + 6.4 mM Na_2S, 30 °C, 115 rpm (120 h)	PbS NPs	35–100 nm; cubic crystal	Extracellular synthesis not described	Optical detection of As (III) in water	[61]

2.3. Microalgae

Microalgae are unicellular photosynthetic microorganisms that have attracted significant interest in the field of nanomanufacturing [62] (see Table 3). Microalgae like *Tetraselmis kochinensis*, *Scenedesmus* and *Desmodesmus* have been used for the biosynthesis of noble metal nanoparticles with good antimicrobial activity, useful for applications in biomedical tool designing but also in drug delivery, catalysis and electronics [63–65]. For these microalgae, mechanisms described for nanoparticles biosynthesis include phenomena of nucleation, control of dimension, and stabilization of nanoparticle structure, mediated by reducing agents [64], enzymes present in the cell wall cytoplasmic membrane [63], biomolecules like polysaccharides, proteins, polyphenols and phenolic compounds [65]. Mechanisms behind biological mineralization (or biomineralization), i.e., the in vivo inorganic minerals formation, have been extensively studied for possible development of new nanomaterials. Diatoms are unicellular microalgae with very peculiar biomineralized silica cell wall called frustules. Diatom frustules possess a highly periodic and hierarchical 3D-porous micro-nanostructure of different morphology (pennate and centric). Hypotheses about their natural functions include mechanical protection, biological protections, filtration, DNA protection from UV and optimization of light harvesting [66–68]. Compared to analogous synthetic mesoporous silica materials, e.g., MCM-4, diatom frustules possess different advantages, including higher biocompatibility, reduced toxicity and easily purification. Diatom frustules also exhibit interesting optical and optoelectronic properties [66,69–74]. The abundance of silanol groups (Si-OH) make the diatom frustules surface easily functionalizable (also in vivo), thus allowing to fully exploit the potentiality of structural nanopatterning of diatom frustules [75]. Functionalization with antibodies [72–77] and gold nanoparticles [78,79] has been described for different sensitive materials in optical or electrochemical immunosensors. Diatom frustules have showed potential application in drug delivery systems [80–82]. Diatomaceous earth (or diatomite) is a large available microfossil material from diatom frustules with extensive commercial use in abrasives or filters. Recently, diatomite—gold nanoparticle nanocomposites have been described for on-chip chromatography and surface-enhanced Raman scattering-based sensoristic devices for the detection of cocaine in biological samples [83] and of histamine in salmon and tuna samples [84]. The "diatom nanotechnology" is a rapidly evolving research field which aims to fully exploit the unique properties of diatom frustules and the great potential of silica biomineralization cellular pathway for the development of new functionalized nanomaterials for emerging applications in sensing, photonic and drug delivery [85]. Another characteristic of diatom frustules is the presence of xanthophyll pigment fucoxanthin. Recent studies have highlighted the active role of fucoxanthin as photo-reducing agent of metal ions to stabilize silver nanoparticles. These silver nanoparticles have showed a significant in vitro antimicrobial activity against *Escherichia coli*, *Bacillus stearothermophilus*, and *Streptococcus mutans* [86] and possible application in optical chemosensing of dissolved ammonia in water samples [87]. Compared to diatoms, microalgae coccolithophores have received less attention. Coccolithophores are calcifying nanoplankton that produces $CaCO_3$ microparticles (coccoliths), in form of arrays of nanoscaled substructures. Interesting optical properties (e.g., light scattering) of coccoliths have been described but also some drawbacks including low electrical conductivity, dissolution at low pH values and scarcity of surface functional groups. Despite these, coccoliths morphologies have showed a great applicative potential for nanodevices fabrications, especially following appropriate in vivo or in vitro modification and functionalizations [10,78]. In very recent work the fabrication of an electrochemical aptamer-based sandwich-type biosensor for the detection of type 2 diabetes biomarker Vaspin has been described [88].

Table 3. Nanomaterials synthesized by microalgae.

Microorganism	Culture Conditions (Synthesis Time)	Nanomaterial	Characteristics (Average Size, Morphology, Modification)	Biosynthetic Pathway	Application	Ref.
Tetraselmis kochinensis	Guillard's Marine Enrichment medium at 28 °C, 200 rpm, 15 days, light condition. 10 g of washed harvested cells + 1 mM $HAuCl_4$, 200 rpm, 28–29 °C (48 h)	Au NPs	5–35 nm; spherical and triangular shape	Intracellular synthesis; possible reduction via enzymes present in the cell wall and in the cytoplasmic membrane	Various applications including catalysis, electronics and coatings	[63]
Scenedesmus sp. (IMMTCC-25)	Growth in Modified Bold Basal medium, 28 ± 2 °C, 16:8 h light: dark cycle, 126 rpm; washed pelleted biomass (harvested in the logarithmic growth phase) + 5 mM $AgNO_3$, 28 °C in the same growth conditions (72 h)	Ag NPs	(living cells) 3–35 nm; spherical shape, highly crystalline cluster; (raw algal extract) (5–10 nm), spherical shape; (boiled algal extract) >50 nm; less stable; colloidal stability >3 months (assessed UV-Vis measures at 420 nm)	Intracellular synthesis not described. Extracellular synthesis (raw algal extracts); reducing and stabilizing agents involved in nucleation points and size control	Good antimicrobial activity against *Streptococcus mutans* and *Escherichia coli* (boiled cell extract)	[64]
Desmodesmus sp. (KR 261937)	Growth in BG-11 medium for 15–20 days, 12:12 h light: dark cycle, 28 ± 2 °C, 120 rpm; centrifuged harvested biomass + 5 mM $AgNO_3$, 28 °C in the same growth condition (72 h)	Ag NPs	(whole cells); 10–30 nm; ζ-potential = −20.2 mV; (raw algal extract) 4–8 nm; ζ-potential = −19.9 mV; (boiled algal extract) 3–6 nm; ζ-potential = −14.2 mV	Intracellular synthesis not described Extracellular synthesis: biocomponents (e.g., polysaccharides, proteins, polyphenols and phenolic compounds) possibly involved in control of dimension and stabilization	Antibacterial effect against *Salmonella* sp. and *Listeria monocytogenes*; antifungal activity against *Candida parapsilosis*	[65]
Cosinodiscus concinnus Wm.	One-week growth (cell density 10^6 cells mL^{-1}) in silicate-enriched seawater media, 18–20 °C, 12:12 h light: dark cycle	Biogenic silica (frustules) modified with murine monoclonal antibody UN1	Green photoluminescence (peaked between 520 and 560 nm) of silanized frustules	Natural silicification process (bio-mineralization)	Using the biogenic silica photo-luminescence for immunosensors development	[72]
Cyclotella sp.	Growth in Harrison's Artificial Seawater Medium enriched with f/2 nutrients + 0.7 mM Na_2SiO_3, 22 °C 14:10 h light: dark cycle. The cell suspension was subcultured at 10% v/v every 14 days (336 h)	Biogenic silica (frustules) functionalized with IgG	~200-nm (perimetrical pores) ~100-nm (linear arrays of pores from the center to the rim) at the base of each ~100-nm pore, a thin layer of silica containing four to five nanopores of ~20-nm diameter	Natural silicification process (bio-mineralization)	Label-free photoluminescence-based immunosensor	[73]
Cosinodiscus wailesii	Growth in F/2 seawater medium, 20 °C, continuous photoperiod	Functionalized biogenic silica (frustules)	100–200 μm	Natural silicification process (bio-mineralization)	Electrochemical immunosensor for the detection of C-reactive protein and myelo-peroxidase in buffer and human serum samples	[75]
Cosinodiscus argus and *Nitzschia soratensis*	Growth in F/2 medium, 20 °C, 12:12 h light: dark cycle. The culture media volume was doubled every week to keep high the diatom reproduction rate. About 4000 cells/ml and about 5.5×10^5 cells/ml for *C. argus* and *N. soratensis* respectively); (about 1000 h)	Multi-layered package array of biogenic silica (frustules) functionalized with purified primary rabbit IgG	*C. argus* 80–100 μm uniformly distributed sub-micron elliptical holes (~170–300 nm) and nanopores (~90–100 nm); *N. soratensis* ~10–15 μm (long axis) and ~5μm (short axis) with nanopores (60–80 nm)	Natural silicification process (bio-mineralization)	Optical immunochip for fluorophore-labeled donkey anti-rabbit IgG detection	[76]
Pseudostaurosira trainorii	Growth in F/2 medium + silica 7 mg mL^{-1}, under aeration : 2:12 h light: dark cycle	Biogenic silica (frustules) integrated with Au NPs functionalized with 5,5′-dithiobis (2-nitrobenzoic acid) + anti-interleukin-8 antibodies	4–5 μm; 98% silica. Perpendicular oriented rows of 4–5 pores (100–200 nm) decreasing in size towards the central axis; neighboring rows separated by ~450 nm; neighboring pores in a row separated by ~100 nm	Natural silicification process (bio-mineralization)	Surface-enhanced Raman scattering immunosensor for the detection of interleukin-8 in blood plasma	[77]

Table 3. Cont.

Microorganism	Culture Conditions (Synthesis Time)	Nanomaterial	Characteristics (Average Size, Morphology, Modification)	Biosynthetic Pathway	Application	Ref.
Pinnularia sp. (UTEX #B679)	Growth in Harrison's artificial seawater medium + 0.5 mM Na_2SiO_3, 22 °C, 14:10 h light: dark cycle for 21 days. (336 h)	Biogenic silica (frustules) functionalized with anti-2,4,6-TNT single chain variable fragment derived from the monoclonal antibody 2C5B5	Ellipsoidal shape with major axe ~20 μm, minor axe ~6 μm; pores in rectangular array (~200 nm diameter) spaced 300-400 nm apart. 4-5 nanopores (~50 nm diameter) at the base of each pore	Natural silicification process (bio-mineralization)	Label-free photo-luminescence quenching -based sensor for 2,4,6-trinitro-toluene detection	[77]
Aulacoseira sp.	N/A	Biogenic silica (frustules) coated with gold (multiple layers of Au particles)	5-10 μm cylindrical-shaped frustules	Natural silicification process (bio-mineralization)	Functional support for surface-enhanced Raman scattering sensor	[78]
Melosira preicelanica	N/A	biogenic silica (frustules) tailored with Au NPs	~20 nm cylindrical-shaped frustules	Natural silicification process (bio-mineralization)	Detection of bovine serum albumin and mineral oil by surface-enhanced Raman spectroscopy	[79]
Coscinodiscus concinnus	Same conditions reported in [70]	Biogenic silica (frustules) loaded with streptomycin	Homogeneous size distribution with a radius of 220 ± 15 μm	Natural silicification process (biomineralization)	Drug delivery	[80]
Thalassiosira weissflogii CCAP strain 1085/10	Growth in silicate-enriched seawater media, 18-20 °C, 12:12 h light: dark cycle; final cell density 10^6 cells mL^{-1} (168 h)	Biogenic silica (frustules)	Mainly composed of separated valves, porosity and hierarchically ordered nanostructure; luminescent and nanostructured silica shells, combining the dye photoluminescence with the photonic silica nanostructure	Natural silicification process (bio-mineralization)	Loading and delivery of fluoro-quinolone ciprofloxacin	[81]
Fossil diatoms	N/A	Biogenic silica (frustules) integrated with 50-60 nm gold nanoparticles	~400 μm (width of the diatomite channels porous); disk-shaped; extremely high confinement of the analyte and increase the concentration of target molecules at the sensor surface; photonic crystals (substrate for surface-enhanced Raman scattering) with 50-60 nm Au NPs	N/A	On-chip chromatography-surface-enhanced Raman scattering -based microfluidic label-free device for cocaine detection in biological samples	[84]
Fossil diatoms	N/A	Biogenic silica (frustules) integrated with 50-60 nm Au nanoparticles	10 to 30 μm; disk-shaped with two-dimensional periodic pores; thickness of the diatomite layer on the glass ~20 μm, (one-third of that of a commercial Thin Layer Chromatography chip) photonic crystals (substrate for surface-enhanced Raman scattering	N/A	On-chip chromatography-surface-enhanced Raman scattering -based microfluidic label-free device for histamine in salmon and tuna	[85]
Amphora-46	Growth in F/2 medium made with filter sterile brackish water (salinity 3%, pH 8.2), 30 °C, 16:8 h light: dark cycle, 130 rpm; Aqueous cell extract + 2 mM $AgNO_3$, 35-40 °C (30 h)	polycrystalline Ag NPs	20-25 nm	Extracellular synthesis; photosynthetic pigment fucoxanthin acts as a reducing agent	Antimicrobial activity against *Escherichia coli*, *Bacillus stearothermophilus*, and *Streptococcus mutans*	[86]
Emiliania huxleyi strain CCMP371	Growth in Artificial seawater (ASW) + f/2 nutrients (without added Si), 20 °C, 12:12 h light: dark cycle, 130 rpm. Cells were harvested at late exponential phase	Aptamer-modified coccolith electrodeposited on the screen-printed Au electrode	N/A	Natural calcification process (coccolith-genesis)	Aptamer-based sandwich-type electrochemical biosensor for Vaspin (type 2 diabetes biomarker)	[88]

3. Towards a Large-Scale Applicability: Knowledge, Issues, and Potentiality

The in vivo microbial nanobiosynthesis and possible control and tuning of nanomaterial properties represent a concrete opportunity for future development and promising uses in biosensoristics and biomedical fields. Despite all the advantages, microbial nanotechnology still has very limited uses [89]. Bacteria have showed the ability to synthetize nanomaterials either by extracellular or intracellular mechanisms. These mechanisms generally produce opposite advantages and disvantages in terms of metal nanoparticles dispersity and purification. Extracellularly produced nanoparticles are generally more polydispersed (i.e., with a great variability in size) than intracellularly produced nanoparticles. By contrast, in extracellular nanomaterial productions less downstream extraction/purification steps (e.g., ultrasound treatment and detergent uses) are required. Thus, the extracellular mediated synthesis described for yeast and molds can greatly simplify the purification steps, besides being an advantage for a possible reuse of microorganisms for more biosynthesis cycles. However, the characterization and identification of the enzymes responsible for nanobiosynthesis in molds is still uncomplete. The photoautotrophic metabolism of microalgae and cyanobacteria is based on carbon dioxide (as carbon source), light (as energy source), inorganic nutrients and water. This condition generally reduce the costs of culture media (compared to culture media used for the growth of bacteria, yeasts, and molds) and it can strongly spur the future scaling-up from the laboratory to the industrial scale, also through the design and the development of solar photobioreactors for the fixation (and reduction) of atmospheric carbon dioxide.

3.1. Nanoparticles Dispersion and Capping Layers

One of the main challenges in microbial nanobiosynthesis is the control of dispersity of nanostructure materials, which heavily influence electronic and optical properties, and the isolation and purification of plural form. Dispersity, i.e., the size distribution of the nanoparticle population, is a key property that strongly influences the particle's behavior in fluids. Improvement and optimization of extraction and purification protocols are required, both for intracellular and extracellular biosynthesis: methods like freeze-thawing, osmotic shock and centrifugation could lead to changes in nanoparticle structures as well as aggregation and precipitation phenomena. Through the adoption of suitable strategies, microbial biosynthesis of nanoparticles could be improved. Selection of appropriate microbial strains (in terms of growth rate and biocatalytic activities), optimization of culturing conditions and uses of genetic engineering tools could help to overcome drawbacks linked to slower producing rate and polidispersity (compared to chemical-based nanomanufacturing) [90]. Microbial biosynthetic nanoparticles are characterized by the presence of a capping layer of biomolecules adsorbed on the surface that act as stabilizing agent and biological active layer of nanoparticles [21]. A deep knowledge of capping characteristics, a clear identification of capping agents (mainly peptides like glutathione, metallothioneins, membrane associated proteins etc.), and possible purification of nanoparticles [23] are fundamental for future in vivo medical applications [15,91].

3.2. Cell Culture Conditions

For future large-scale productions, costs of culture media for microbial growth should be seriously considered to not limit the applications of microbial biosynthetic nanomaterials. One current example is bacterial nanocellulose whose applications are still limited to a few biomedical devices, mainly because of costs of culture medium [92]. In addition, optimization and standardization of microbial cell culture growth protocols and modifications culture conditions are pivotal for control, tune, and to improve characteristics of microbial biosynthesized nanomaterials. The influence of physico-chemical parameters of cell culture operational set up on nanomaterials biosynthesis have been previously highlighted. These factors include (i) microbial cell concentration; (ii) precursor concentration; (iii) pH; and (iv) temperature. The optimum conditions of pH, temperature, and NaCl concentration have been studied to achieve high purity and high synthesis rate of cadmium selenide nanoparticles by bacterium

Pseudomonas aeruginosa strain RB. Interestingly, the results of this work have showed that optimum conditions for nanoparticles synthesis did not match with optimum growth conditions for *Pseudomonas aeruginosa* strain RB [93]. Effects of precursor concentration, temperature, and pH on silver nanoparticle synthesis and particle sizes have been described for the bacterium *E. coli* strain DH5α [94]. Other recent examples include: (i) the temperature-dependence of size and monodispersity of silver nanoparticles biosynthesis by mold *Trichoderma viride* [95], (ii) the influence on the type of gold nanostructures synthetized (nanoparticles or nanoplates) and them relative size in yeast *Yarrowia lipolytica* strain NCIM 3589 by changing the proportion of cell concentration and precursor gold salt concentration, (iii) the effect of temperature on gold nanostructures release from cell wall into the aqueous phase [96] and control of bacterial growth kinetics of the bacterium *Morganella psychrotolerans* to achieve shape anisotropy of silver nanoparticles [97]. Elsoud et al. (2018) observed an improvement in tellurium nanoparticles production by a 1 kGy of gamma irradiation of mold *Aspergillus welwitschiae* KY766958 broth culture (compared to the non-irradiated broth culture control). These results have been ascribed to the activation of enzyme (s) involved in biosynthetic pathway [59]. Although the controlling of biomineralization process still remains a challenge, the optimization of frustules morphological properties (e.g., pore sizes and pore density) has been explored by changing operational parameters of experimental setup (e.g., pH, salinity, temperature, nutrient concentration, precursor $Si(OH)_4$ concentration, and light regime) [98]. Interestingly Townley et al. (2007) reported alteration in pore sizes of *Coscinodiscus wailesii* frustules when exposed to sublethal concentration of nickel [99]. The possibility of in vivo chemical modification of frustules or other biomineralized structures has been recently described. These in vivo chemical modifications lead to the inorganic elements/compounds-doping of biomineralized structures through the addition to the culture medium of given precursors at sublethal concentrations. Several works described the doping of diatom *Pinnularia* sp. frustules or diatoms *Thalassiosira weissflogii* frustules with titania (TiO_2) [100,101] nanobiosynthesis containing Si-Ge oxides nanocomb in diatom *Nitzschia* and *Pinnularia* sp. by adding $Ge(OH)_4$ or GeO_2 in the diatom culture medium at photobioreactor scale-productions [102–105]. Compared to diatoms, possible nanotechnological applications of the calcareous-based shell of marine protozoa foraminifera have not been so widely explored. A recent work described the in vivo preparation of a bionic material through the inclusion of fluorescent magnetite nanoparticles within calcite skeletal structure of the unicellular organism foraminifer *Amphistrigina lesson*. Such in vitro synthetic approach exploited the natural biomineralization process of growth in the presence of magnetic nanoparticles functionalized with a hydroxylated dextran shell [106].

3.3. Biochemistry, Molecular Biology, and Genetics

A deeper knowledge of molecular biology and genetic aspects behind microbial nanobiosynthetic pathways is strongly required. For instance, the characterization of not fully understood biochemical mechanisms and a complete identification of extracellular enzymes secreted by filamentous fungi could lead to an improved control in chemical compositions, shapes, and sizes of nanoparticles [55]. The availability of microorganism genome sequences could considerably increase the range of possibilities in genetic manipulation of microorganism to implement the nanoparticles biosynthesis and the in vivo tuning of nanoparticle characteristics. A recent example has come from the study of Zhang et al. (2017), which showed how CdSe quantum dots biosynthesis can be improved through genetic modification of the ATP metabolism pathway in yeast *S. cerevisiae* [107]. Biotechnological approaches based on genetic engineering and recombinant technologies could allow the identification of sequences of gene involved in nanoparticle synthesis and a possible heterologous expression (i.e., controlled expression of one or more gene sequences in a host organism) to enhance nanomaterial production efficiency [91]. The bacterium *E. coli* is a highly efficient model host microorganism that has been exploited as heterologous expression system for phytochelatin synthase and/or metallothionein for the in vivo synthesis of various metal nanoparticles (e.g., CdSeZn, PrGd, CdCs, and FeCo) never synthesized before by chemical methods [108,109] or cadmium selenide quantum dots [110]. Thanks to recent advances in the characterization of biochemical mechanisms involved in bacterial nanocellulose biosynthesis, the future development of new genetically engineered bacterial nanocellulose-producing strains could be achieved. This could eventually lead to a reduction of production costs, an improvement of production yield, and biosynthesis of nanocellulose with new properties, suitable for broader range of technological applications [92,111]. Concerning genetic manipulation of nanostructure-producing microorganisms, Tan et al. (2016) have reported a 2000-fold increase in electrical conductivity and diameter of nanowires filaments produced by model microorganism *Geobacter sulfurreducens*. In this case, genetic modification has been concerned the modification of aminoacidic composition of the carboxyl end of PilA protein, the structural component of bacterial nanowires [112]. The biomineralization process behind magnetosome biogenesis in bacterial species (*Magnetospirillum* species are the most studied) is very complex and not completely elucidated. The mechanism of magnetosome formation has been shown to be under tight genetic control and induced by growth conditions. To date, six different models have been proposed to elucidate magnetosome formation, but they are still not completed [11]. Delalat et al. (2015) have reported a genetically engineered diatom *Thalassiosira pseudonana* (whose genome has been completely sequenced) to incorporate immunoglubulin G-binding domain of protein G (GB1) into the frustules surface. Such antibody-labelled genetically modified diatom enables an in vitro selectively cell targeting and selectively killing of neuroblastoma and B-lymphoma cells [113]. Furthermore, the role of gene Silicanin-1 in the control of biosilica morphology has been recently highlighted in *Thalassiosira pseudonana*, opening new possibilities for future genetic engineering of frustules architectures [114]. A deeper knowledge of biosilicification process as well as the role of organic components of diatom frustules (i.e., proteins silaffins and long-chain polyamines) in biogenesis and formation of nanopatterns in diatom frustules are still a challenge. To date, biocalcification system of coccolithophores remains unclear, even though a very recent genetic and proteomic study about expression of transcripts and proteins in coccolithophore *Emiliania huxleyi* will help future identification and more detailed characterization of molecular mechanisms and metabolic pathways underlying calcification in coccolithophores [115].

4. Conclusions

In the light of recent literature herein reported, microbial nanotechnology is a fascinating and booming field for future breakthrough nanomaterial synthesis. Through a 'green' and sustainable approach, microbial nanotechnology can really spur innovation in nanomanufacturing with a potential strong impact in several fields, including sensoristics and biomedicine.

Author Contributions: Conceptualization, G.G.; writing—original draft preparation, G.G., D.Z., R.D.; writing—review and editing, G.G., D.Z., R.D.; visualization, G.G., D.Z., R.D.; supervision, R.D. All authors have read and agreed to the published version of the manuscript.

Funding: This research received no external funding.

Conflicts of Interest: The authors declare no conflict of interest.

References

1. Inshakova, E., Inshakov, O. World market for nanomaterials: Structure and trends. In Proceedings of the MATEC Web of Conferences, Sevastopol, Russia, 11–15 September 2017; EDP Sciences: Les Ulis, France, 2017; Volume 129, p. 02013. [CrossRef]
2. Şengül, H.; Theis, T.L.; Ghosh, S. Toward sustainable nanoproducts: An overview of nanomanufacturing methods. *J. Ind. Ecol.* **2008**, *12*, 329–359. [CrossRef]
3. Fang, F.Z.; Zhang, X.D.; Gao, W.; Guo, Y.B.; Byrne, G.; Hansen, H.N. Nanomanufacturing—Perspective and applications. *CIRP Ann.* **2017**, *66*, 683–705. [CrossRef]
4. Yuan, C.; Zhang, T. Environmental Implications of Nano-manufacturing. In *Green Manufacturing*; Dornfeld, D., Ed.; Springer: Boston, MA, USA, 2013. [CrossRef]

5. Ribeiro, B.D.; Coelho, M.A.Z.; de Castro, A.M. Principles of Green Chemistry and White Biotechnology. In *White Biotechnology for Sustainable Chemistry*; Coelho, M.A.Z., Ribeiro, B.D., Eds.; The Royal Society of Chemistry: London, UK, 2016; pp. 1–8. [CrossRef]
6. Li, X.; Xu, H.; Chen, Z.S.; Chen, G. Biosynthesis of nanoparticles by microorganisms and their applications. *J. Nanomater.* **2011**, *2011*. [CrossRef]
7. Khan, T.; Abbas, S.; Fariq, A.; Yasmin, A. Microbes: Nature's cell factories of nanoparticles synthesis. In *Exploring the Realms of Nature for Nanosynthesis*; Springer: Cham, Switzerland, 2018; pp. 25–50. [CrossRef]
8. Hulkoti, N.I.; Taranath, T.C. Biosynthesis of nanoparticles using microbes—A review. *Colloids Surf. B Biointerfaces* **2014**, *121*, 474–483. [CrossRef] [PubMed]
9. Kröger, N.; Poulsen, N. Diatoms—From cell wall biogenesis to nanotechnology. *Annu. Rev. Genet.* **2008**, *42*, 83–107. [CrossRef] [PubMed]
10. Skeffington, A.W.; Scheffel, A. Exploiting algal mineralization for nanotechnology: Bringing coccoliths to the fore. *Curr. Opin. Biotechnol.* **2018**, *49*, 57–63. [CrossRef] [PubMed]
11. Yan, L.; Da, H.; Zhang, S.; Lopez, V.M.; Wang, W. Bacterial magnetosome and its potential application. *Microbiol. Res.* **2017**, *203*, 19–28. [CrossRef]
12. Gama, M.; Gatenholm, P.; Klemm, D. (Eds.) *Bacterial Nanocellulose: A sophisticated Multifunctional Material*; CRC Press: Boca Raton, FL, USA, 2012.
13. Nwodo, U.; Green, E.; Okoh, A. Bacterial exopolysaccharides: Functionality and prospects. *Int. J. Mol. Sci.* **2012**, *13*, 14002–14015. [CrossRef]
14. Malvankar, N.S.; Lovley, D.R. Microbial nanowires: A new paradigm for biological electron transfer and bioelectronics. *ChemSusChem* **2012**, *5*, 1039–1046. [CrossRef]
15. Prasad, R.; Pandey, R.; Barman, I. Engineering tailored nanoparticles with microbes: Quo vadis? *Wiley Interdiscip. Rev. Nanomed. Nanobiotechnol.* **2016**, *8*, 316–330. [CrossRef]
16. Dragone, R.; Grasso, G.; Muccini, M.; Toffanin, S. Portable bio/chemosensoristic devices: Innovative systems for environmental health and food safety diagnostics. *Front. Public Health* **2017**, *5*, 80. [CrossRef] [PubMed]
17. Petros, R.A.; DeSimone, J.M. Strategies in the design of nanoparticles for therapeutic applications. *Nature Rev. Drug Discov.* **2010**, *9*, 615. [CrossRef] [PubMed]
18. Kiessling, F.; Mertens, M.E.; Grimm, J.; Lammers, T. Nanoparticles for imaging: Top or flop? *Radiology* **2014**, *273*, 10–28. [CrossRef] [PubMed]
19. Wang, T.; Yang, L.; Zhang, B.; Liu, J. Extracellular biosynthesis and transformation of selenium nanoparticles and application in H_2O_2 biosensor. *Colloids Surf. B Biointerfaces* **2010**, *80*, 94–102. [CrossRef] [PubMed]
20. Du, L.; Jiang, H.; Liu, X.; Wang, E. Biosynthesis of gold nanoparticles assisted by *Escherichia coli* DH5α and its application on direct electrochemistry of hemoglobin. *Electrochem. Commun.* **2007**, *9*, 1165–1170. [CrossRef]
21. Ramya, S.; Shanmugasundaram, T.; Balagurunathan, R. Biomedical potential of actinobacterially synthesized selenium nanoparticles with special reference to anti-biofilm, anti-oxidant, wound healing, cytotoxic and anti-viral activities. *J. Trace Elem. Med. Biol.* **2015**, *32*, 30–39. [CrossRef]
22. Bao, H.; Lu, Z.; Cui, X.; Qiao, Y.; Guo, J.; Anderson, J.M.; Li, C.M. Extracellular microbial synthesis of biocompatible CdTe quantum dots. *Acta Biomater.* **2010**, *6*, 3534–3541. [CrossRef]
23. Suresh, A.K.; Pelletier, D.A.; Wang, W.; Broich, M.L.; Moon, J.W.; Gu, B.; Allison, D.P.; Joy, D.C.; Phelps, T.J.; Doktycz, M.J. Biofabrication of discrete spherical gold nanoparticles using the metal-reducing bacterium *Shewanella oneidensis*. *Acta Biomater.* **2011**, *7*, 2148–2152. [CrossRef]
24. Suresh, A.K.; Pelletier, D.A.; Wang, W.; Moon, J.W.; Gu, B.; Mortensen, N.P.; David, P.; Allison, D.C.; Joy, C.; Phelps, T.J.; et al. Silver nanocrystallites: Biofabrication using *Shewanella oneidensis*, and an evaluation of their comparative toxicity on gram-negative and gram-positive bacteria. *Environ. Sci. Technol.* **2010**, *44*, 5210–5215. [CrossRef]
25. Manivasagan, P.; Alam, M.S.; Kang, K.H.; Kwak, M.; Kim, S.K. Extracellular synthesis of gold bionanoparticles by *Nocardiopsis* sp. and evaluation of its antimicrobial, antioxidant and cytotoxic activities. *Bioprocess Biosyst. Eng.* **2015**, *38*, 1167–1177. [CrossRef]
26. Roychoudhury, P.; Gopal, P.K.; Paul, S.; Pal, R. Cyanobacteria assisted biosynthesis of silver nanoparticles—A potential antileukemic agent. *J. Appl. Phycol.* **2016**, *28*, 3387–3394. [CrossRef]
27. Al-Dhabi, N.; Mohammed Ghilan, A.K.; Arasu, M. Characterization of silver nanomaterials derived from marine *Streptomyces* sp. al-dhabi-87 and its in vitro application against multidrug resistant and extended-spectrum beta-lactamase clinical pathogens. *Nanomaterials* **2018**, *8*, 279. [CrossRef] [PubMed]

28. Shao, W.; Liu, H.; Liu, X.; Sun, H.; Wang, S.; Zhang, R. pH-responsive release behavior and anti-bacterial activity of bacterial cellulose-silver nanocomposites. *Int. J. Biol. Macromol.* **2015**, *76*, 209–217. [CrossRef] [PubMed]
29. Torres, S.K.; Campos, V.L.; León, C.G.; Rodríguez-Llamazares, S.M.; Rojas, S.M.; Gonzalez, M.; Mondaca, M.A. Biosynthesis of selenium nanoparticles by *Pantoea agglomerans* and their antioxidant activity. *J. Nanopart. Res.* **2012**, *14*, 1236. [CrossRef]
30. Kalishwaralal, K.; Banumathi, E.; Pandian, S.R.K.; Deepak, V.; Muniyandi, J.; Eom, S.H.; Gurunathan, S. Silver nanoparticles inhibit VEGF induced cell proliferation and migration in bovine retinal endothelial cells. *Colloids Surf. B Biointerfaces* **2009**, *73*, 51–57. [CrossRef]
31. Kalishwaralal, K.; Deepak, V.; Pandian, S.R.K.; Kottaisamy, M.; BarathManiKanth, S.; Kartikeyan, B.; Gurunathan, S. Biosynthesis of silver and gold nanoparticles using *Brevibacterium casei*. *Colloids Surf. B Biointerfaces* **2010**, *77*, 257–262. [CrossRef]
32. Ahmad, M.S.; Yasser, M.M.; Sholkamy, E.N.; Ali, A.M.; Mehanni, M.M. Anticancer activity of biostabilized selenium nanorods synthesized by Streptomyces bikiniensis strain Ess_amA-1. *Int. J. Nanomed.* **2015**, *10*, 3389. [CrossRef]
33. Gurunathan, S.; Han, J.W.; Eppakayala, V.; Kim, J.H. Green synthesis of graphene and its cytotoxic effects in human breast cancer cells. *Int. J. Nanomed.* **2013**, *8*, 1015. [CrossRef]
34. Golmohammadi, H.; Morales-Narváez, E.; Naghdi, T.; Merkoçi, A. Nanocellulose in sensing and biosensing. *Chem. Mater.* **2017**, *29*, 5426–5446. [CrossRef]
35. Saxena, I.M.; Dandekar, T.; Brown, R.M. Mechanisms in cellulose biosynthesis. *Mol. Biol.* **2004**. Available online: http://www.esf.edu/outreach/pd/2000/cellulose/saxena.pdf (accessed on 10 November 2019).
36. Pourreza, N.; Golmohammadi, H.; Naghdi, T.; Yousefi, H. Green in-Situ Synthesized Silver Nanoparticles Embedded in Bacterial Cellulose Nanopaper as a Bionanocomposite Plasmonic Sensor. *Biosens. Bioelectron.* **2015**, *74*, 353–359. [CrossRef] [PubMed]
37. Morales-Narváez, E.; Golmohammadi, H.; Naghdi, T.; Yousefi, H.; Kostiv, U.; Horák, D.; Pourreza, N.; Merkoçi, A. Nanopaper as an Optical Sensing Platform. *ACS Nano* **2015**, *9*, 7296–7305. [CrossRef]
38. Li, C.; Zhou, L.; Yang, H.; Lv, R.; Tian, P.; Li, X.; Zhang, Y.; Chen, Z.; Lin, F. Self-assembled exopolysaccharide nanoparticles for bioremediation and green synthesis of noble metal nanoparticles. *ACS Appl. Mater. Interfaces* **2017**, *9*, 22808–22818. [CrossRef] [PubMed]
39. Simonte, F.; Sturm, G.; Gescher, J.; Sturm-Richter, K. Extracellular electron transfer and biosensors. In *Bioelectrosynthesis*; Springer: Cham, Switzerland, 2017; pp. 15–38. [CrossRef]
40. Liu, X.; Wang, S.; Xu, A.; Zhang, L.; Liu, H.; Ma, L.Z. Biological synthesis of high-conductive pili in aerobic bacterium *Pseudomonas aeruginosa*. *Appl. Microbiol. Biotechnol.* **2019**, *103*, 1535–1544. [CrossRef] [PubMed]
41. Sure, S.; Torriero, A.A.; Gaur, A.; Li, L.H.; Chen, Y.; Tripathi, C.; Adholeya, A.; Ackland, M.L.; Kochar, M. Inquisition of *Microcystis aeruginosa* and *Synechocystis* nanowires: Characterization and modelling. *Antonie Van Leeuwenhoek* **2015**, *108*, 1213–1225. [CrossRef] [PubMed]
42. Leung, K.M.; Wanger, G.; El-Naggar, M.Y.; Gorby, Y.; Southam, G.; Lau, W.M.; Yang, J. *Shewanella oneidensis* MR-1 bacterial nanowires exhibit p-type, tunable electronic behavior. *Nano Lett.* **2013**, *13*, 2407–2411. [CrossRef]
43. Davila, D.; Esquivel, J.P.; Sabate, N.; Mas, J. Silicon-based microfabricated microbial fuel cell toxicity sensor. *Biosens. Bioelectron.* **2011**, *26*, 2426–2430. [CrossRef]
44. Wang, X.; Gao, N.; Zhou, Q. Concentration responses of toxicity sensor with *Shewanella oneidensis* MR-1 growing in bioelectrochemical systems. *Biosens. Bioelectron.* **2013**, *43*, 264–267. [CrossRef]
45. Webster, D.P.; TerAvest, M.A.; Doud, D.F.; Chakravorty, A.; Holmes, E.C.; Radens, C.M.; Sureka, S.; Gralnick, J.A.; Angenent, L.T. An arsenic-specific biosensor with genetically engineered *Shewanella oneidensis* in a bioelectrochemical system. *Biosens. Bioelectron.* **2014**, *62*, 320–324. [CrossRef]
46. Vargas, G.; Cypriano, J.; Correa, T.; Leão, P.; Bazylinski, D.A.; Abreu, F. Applications of magnetotactic bacteria, magnetosomes and magnetosome crystals in biotechnology and nanotechnology: Mini-review. *Molecules* **2018**, *23*, 2438. [CrossRef]
47. Boucher, M.; Geffroy, F.; Prévéral, S.; Bellanger, L.; Selingue, E.; Adryanczyk-Perrier, G.; Péan, M.; Lefèvre, C.T.; Pignol, D.; Ginet, N.; et al. Genetically tailored magnetosomes used as MRI probe for molecular imaging of brain tumor. *Biomaterials* **2017**, *121*, 167–178. [CrossRef] [PubMed]

48. Alphandery, E.; Faure, S.; Seksek, O.; Guyot, F.; Chebbi, I. Chains of magnetosomes extracted from AMB-1 magnetotactic bacteria for application in alternative magnetic field cancer therapy. *ACS Nano* **2011**, *5*, 6279–6296. [CrossRef] [PubMed]
49. Roda, A.; Cevenini, L.; Borg, S.; Michelini, E.; Calabretta, M.M.; Schüler, D. Bioengineered bioluminescent magnetotactic bacteria as a powerful tool for chip-based whole-cell biosensors. *Lab Chip* **2013**, *13*, 4881–4889. [CrossRef]
50. Luo, Q.Y.; Lin, Y.; Li, Y.; Xiong, L.H.; Cui, R.; Xie, Z.X.; Pang, D.W. Nanomechanical analysis of yeast cells in CdSe quantum dot biosynthesis. *Small* **2014**, *10*, 699–704. [CrossRef] [PubMed]
51. Wei, R. Biosynthesis of Au–Ag alloy nanoparticles for sensitive electrochemical determination of paracetamol. *Int. J. Electrochem. Sci.* **2017**, *12*, 9131–9140. [CrossRef]
52. Zheng, D.; Hu, C.; Gan, T.; Dang, X.; Hu, S. Preparation and application of a novel vanillin sensor based on biosynthesis of Au–Ag alloy nanoparticles. *Sens. Actuators B Chem.* **2010**, *148*, 247–252. [CrossRef]
53. Korbekandi, H.; Mohseni, S.; Mardani Jouneghani, R.; Pourhossein, M.; Iravani, S. Biosynthesis of silver nanoparticles using *Saccharomyces cerevisiae*. *Artif. Cells Nanomed. Biotechnol.* **2016**, *44*, 235–239. [CrossRef]
54. Dhillon, G.S.; Brar, S.K.; Kaur, S.; Verma, M. Green approach for nanoparticle biosynthesis by fungi: Current trends and applications. *Crit. Rev. Biotechnol.* **2012**, *32*, 49–73. [CrossRef]
55. Syed, A.; Saraswati, S.; Kundu, G.C.; Ahmad, A. Biological synthesis of silver nanoparticles using the fungus *Humicola* sp. and evaluation of their cytoxicity using normal and cancer cell lines. *Spectrochim. Acta Part A Mol. Biomol. Spectrosc.* **2013**, *114*, 144–147. [CrossRef]
56. Husseiny, S.M.; Salah, T.A.; Anter, H.A. Biosynthesis of size-controlled silver nanoparticles by *Fusarium oxysporum*, their antibacterial and antitumor activities. *Beni Suef Univ. J. Basic Appl. Sci.* **2015**, *4*, 225–231. [CrossRef]
57. Saravanakumar, K.; Chelliah, R.; MubarakAli, D.; Jeevithan, E.; Oh, D.H.; Kathiresan, K.; Wang, M.H. Fungal enzyme-mediated synthesis of chitosan nanoparticles and its biocompatibility, antioxidant and bactericidal properties. *Int. J. Biol. Macromol.* **2018**, *118*, 1542–1549. [CrossRef] [PubMed]
58. Mishra, A.; Tripathy, S.K.; Wahab, R.; Jeong, S.H.; Hwang, I.; Yang, Y.B.; Kim, Y.S.; Shin, H.S.; Yun, S.I. Microbial synthesis of gold nanoparticles using the fungus *Penicillium brevicompactum* and their cytotoxic effects against mouse mayo blast cancer C 2 C 12 cells. *Appl. Microbiol. Biotechnol.* **2011**, *92*, 617–630. [CrossRef] [PubMed]
59. Elsoud, M.M.A.; Al-Hagar, O.E.; Abdelkhalek, E.S.; Sidkey, N.M. Synthesis and investigations on tellurium myconanoparticles. *Biotechnol. Rep.* **2018**, *18*, e00247. [CrossRef] [PubMed]
60. Uddandarao, P.; Balakrishnan, R.M.; Ashok, A.; Swarup, S.; Sinha, P. Bioinspired ZnS: Gd Nanoparticles Synthesized from an Endophytic Fungi *Aspergillus flavus* for Fluorescence-Based Metal Detection. *Biomimetics* **2019**, *4*, 11. [CrossRef]
61. Priyanka, U.; Akshay Gowda, K.M.; Elisha, M.G.; Nitish, N. Biologically synthesized PbS nanoparticles for the detection of arsenic in water. *Int. Biodeterior. Biodegrad.* **2017**, *119*, 78–86. [CrossRef]
62. Dahoumane, S.A.; Mechouet, M.; Alvarez, F.J.; Agathos, S.N.; Jeffryes, C. Microalgae: An outstanding tool in nanotechnology. *Bionatura* **2016**, *1*, 196–201. [CrossRef]
63. Senapati, S.; Syed, A.; Moeez, S.; Kumar, A.; Ahmad, A. Intracellular synthesis of gold nanoparticles using alga Tetraselmis kochinensis. *Mater. Lett.* **2012**, *79*, 116–118. [CrossRef]
64. Jena, J.; Pradhan, N.; Nayak, R.R.; Dash, B.P.; Sukla, L.B.; Panda, P.K.; Mishra, B.K. Microalga *Scenedesmus* sp.: A potential low-cost green machine for silver nanoparticle synthesis. *J. Microbiol. Biotechnol.* **2014**, *24*, 522–533. [CrossRef]
65. Öztürk, B.Y. Intracellular and extracellular green synthesis of silver nanoparticles using *Desmodesmus* sp.: Their Antibacterial and antifungal effects. *Caryol. Int. J. Cytol. Cytosystem. Cytogenet.* **2019**, *72*, 29–43. [CrossRef]
66. Fuhrmann, T.; Landwehr, S.; El Rharbi-Kucki, M.; Sumper, M. Diatoms as living photonic crystals. *Appl. Phys. B* **2004**, *78*, 257–260. [CrossRef]
67. Losic, D.; Mitchell, J.G.; Voelcker, N.H. Diatomaceous lessons in nanotechnology and advanced materials. *Adv. Mater.* **2009**, *21*, 2947–2958. [CrossRef]
68. Aguirre, L.E.; Ouyang, L.; Elfwing, A.; Hedblom, M.; Wulff, A.; Inganäs, O. Diatom frustules protect DNA from ultraviolet light. *Sci. Rep.* **2018**, *8*, 5138. [CrossRef] [PubMed]

69. Mazumder, N.; Gogoi, A.; Kalita, R.D.; Ahmed, G.A.; Buragohain, A.K.; Choudhury, A. Luminescence studies of fresh water diatom frustules. *Indian J. Phys.* **2010**, *84*, 665–669. [CrossRef]
70. LeDuff, P.; Roesijadi, G.; Rorrer, G.L. Micro-photoluminescence of single living diatom cells. *Luminescence* **2016**, *31*, 1379–1383. [CrossRef]
71. Ragni, R.; Cicco, S.R.; Vona, D.; Farinola, G.M. Multiple routes to smart nanostructured materials from diatom microalgae: A chemical perspective. *Adv. Mater.* **2018**, *30*, 1704289. [CrossRef]
72. De Stefano, L.; Rotiroti, L.; De Stefano, M.; Lamberti, A.; Lettieri, S.; Setaro, A.; Maddalena, P. Marine diatoms as optical biosensors. *Biosens. Bioelectron.* **2009**, *24*, 1580–1584. [CrossRef]
73. Gale, D.K.; Gutu, T.; Jiao, J.; Chang, C.H.; Rorrer, G.L. Photoluminescence detection of biomolecules by antibody-functionalized diatom biosilica. *Adv. Funct. Mater.* **2009**, *19*, 926–933. [CrossRef]
74. Lin, K.C.; Kunduru, V.; Bothara, M.; Rege, K.; Prasad, S.; Ramakrishna, B.L. Biogenic nanoporous silica-based sensor for enhanced electrochemical detection of cardiovascular biomarkers proteins. *Biosens. Bioelectron.* **2010**, *25*, 2336–2342. [CrossRef]
75. Li, A.; Cai, J.; Pan, J.; Wang, Y.; Yue, Y.; Zhang, D. Multi-layer hierarchical array fabricated with diatom frustules for highly sensitive bio-detection applications. *J. Micromech. Microeng.* **2014**, *24*, 025014. [CrossRef]
76. Kamińska, A.; Sprynskyy, M.; Winkler, K.; Szymborski, T. Ultrasensitive SERS immunoassay based on diatom biosilica for detection of interleukins in blood plasma. *Anal. Bioanal. Chem.* **2017**, *409*, 6337–6347. [CrossRef]
77. Zhen, L.; Ford, N.; Gale, D.K.; Roesijadi, G.; Rorrer, G.L. Photoluminescence detection of 2, 4, 6-trinitrotoluene (TNT) binding on diatom frustule biosilica functionalized with an anti-TNT monoclonal antibody fragment. *Biosens. Bioelectron.* **2016**, *79*, 742–748. [CrossRef] [PubMed]
78. Pannico, M.; Rea, I.; Chandrasekaran, S.; Musto, P.; Voelcker, N.H.; De Stefano, L. Electroless gold-modified diatoms as surface-enhanced Raman scattering supports. *Nanoscale Res. Lett.* **2016**, *11*, 315. [CrossRef] [PubMed]
79. Onesto, V.; Villani, M.; Coluccio, M.L.; Majewska, R.; Alabastri, A.; Battista, E.; Schirato, A.; Calestani, D.; Coppedè, N.; Cesarelli, M.; et al. Silica diatom shells tailored with Au nanoparticles enable sensitive analysis of molecules for biological, safety and environment applications. *Nanoscale Res. Lett.* **2018**, *13*, 94. [CrossRef] [PubMed]
80. Gnanamoorthy, P.; Anandhan, S.; Prabu, V.A. Natural nanoporous silica frustules from marine diatom as a biocarrier for drug delivery. *J. Porous Mater.* **2014**, *21*, 789–796. [CrossRef]
81. Lo Presti, M.; Ragni, R.; Vona, D.; Leone, G. In vivo doped biosilica from living *Thalassiosira weissflogii* diatoms with a triethoxysilyl functionalized red emitting fluorophore. *Biomater. Soft Mater.* **2018**, *3*, 1509–1517. [CrossRef]
82. Terracciano, M.; De Stefano, L.; Rea, I. Diatoms Green Nanotechnology for Biosilica-Based Drug Delivery Systems. *Pharmaceutics* **2018**, *10*, 242. [CrossRef] [PubMed]
83. Kong, X.; Chong, X.; Squire, K.; Wang, A.X. Microfluidic diatomite analytical devices for illicit drug sensing with ppb-Level sensitivity. *Sens. Actuators B Chem.* **2018**, *259*, 587–595. [CrossRef]
84. Kong, X.; Yu, Q.; Li, E.; Wang, R.; Liu, Q.; Wang, A. Diatomite photonic crystals for facile on-chip chromatography and sensing of harmful ingredients from food. *Materials* **2018**, *11*, 539. [CrossRef]
85. Panwar, V.; Dutta, T. Diatom Biogenic Silica as a Felicitous Platform for Biochemical Engineering: Expanding Frontiers. *ACS Appl. Bio Mater.* **2019**, *2*, 2295–2316. [CrossRef]
86. Jena, J.; Pradhan, N.; Dash, B.P.; Panda, P.K.; Mishra, B.K. Pigment mediated biogenic synthesis of silver nanoparticles using diatom *Amphora* sp. and its antimicrobial activity. *J. Saudi Chem. Soc.* **2015**, *19*, 661–666. [CrossRef]
87. Chetia, L.; Kalita, D.; Ahmed, G.A. Synthesis of Ag nanoparticles using diatom cells for ammonia sensing. *Sens. Bio Sens. Res.* **2017**, *16*, 55–61. [CrossRef]
88. Kim, S.H.; Nam, O.; Jin, E.; Gu, M.B. A new coccolith modified electrode-based biosensor using a cognate pair of aptamers with sandwich-type binding. *Biosens. Bioelectron.* **2019**, *123*, 160–166. [CrossRef] [PubMed]
89. Luo, C.H.; Shanmugam, V.; Yeh, C.S. Nanoparticle biosynthesis using unicellular and subcellular supports. *NPG Asia Mater.* **2015**, *7*, e209. [CrossRef]
90. Narayanan, K.B.; Sakthivel, N. Biological synthesis of metal nanoparticles by microbes. *Adv. Colloid Interface Sci.* **2010**, *156*, 1–13. [CrossRef]

91. Voeikova, T.A.; Zhuravliova, O.A.; Bulushova, N.V.; Veiko, V.P.; Ismagulova, T.T.; Lupanova, T.N.; Shaitan, K.V.; Debabov, V.G. The "Protein Corona" of Silver-Sulfide Nanoparticles Obtained Using Gram-Negative and-Positive Bacteria. *Mol. Genet. Microbiol. Virol.* **2017**, *32*, 204–211. [CrossRef]
92. Jacek, P.; Dourado, F.; Gama, M.; Bielecki, S. Molecular aspects of bacterial nanocellulose biosynthesis. *Microb. Biotechnol.* **2019**, *12*, 633–649. [CrossRef]
93. Ayano, H.; Kuroda, M.; Soda, S.; Ike, M. Effects of culture conditions of Pseudomonas aeruginosa strain RB on the synthesis of CdSe nanoparticles. *J. Biosci. Bioeng.* **2015**, *119*, 440–445. [CrossRef]
94. Gurunathan, S.; Kalishwaralal, K.; Vaidyanathan, R.; Venkataraman, D.; Pandian, S.R.K.; Muniyandi, J.; Hariharana, N.; Eom, S.H. Biosynthesis, purification and characterization of silver nanoparticles using Escherichia coli. *Colloids Surf. B Biointerfaces* **2009**, *74*, 328–335. [CrossRef]
95. Fayaz, A.M.; Balaji, K.; Kalaichelvan, P.T.; Venkatesan, R. Fungal based synthesis of silver nanoparticles—An effect of temperature on the size of particles. *Colloids Surf. B Biointerfaces* **2009**, *74*, 123–126. [CrossRef]
96. Pimprikar, P.S.; Joshi, S.S.; Kumar, A.R.; Zinjarde, S.S.; Kulkarni, S.K. Influence of biomass and gold salt concentration on nanoparticle synthesis by the tropical marine yeast *Yarrowia lipolytica* NCIM 3589. *Colloids Surf. B Biointerfaces* **2009**, *74*, 309–316. [CrossRef]
97. Ramanathan, R.; O'Mullane, A.P.; Parikh, R.Y.; Smooker, P.M.; Bhargava, S.K.; Bansal, V. Bacterial kinetics-controlled shape-directed biosynthesis of silver nanoplates using *Morganella psychrotolerans*. *Langmuir* **2010**, *27*, 714–719. [CrossRef] [PubMed]
98. Su, Y.; Lundholm, N.; Ellegaard, M. Effects of abiotic factors on the nanostructure of diatom frustules—Ranges and variability. *Appl. Microbiol. Biotechnol.* **2018**, *102*, 5889–5899. [CrossRef] [PubMed]
99. Townley, H.E.; Woon, K.L.; Payne, F.P.; White-Cooper, H.; Parker, A.R. Modification of the physical and optical properties of the frustule of the diatom *Coscinodiscus wailesii* by nickel sulfate. *Nanotechnology* **2007**, *18*, 295101. [CrossRef]
100. Jeffryes, C.; Gutu, T.; Jiao, J.; Rorrer, G.L. Metabolic insertion of nanostructured TiO_2 into the patterned biosilica of the diatom *Pinnularia* sp. by a two-stage bioreactor cultivation process. *ACS Nano* **2008**, *2*, 2103–2112. [CrossRef]
101. Lang, Y.; Del Monte, F.; Rodriguez, B.J.; Dockery, P.; Finn, D.P.; Pandit, A. Integration of TiO_2 into the diatom *Thalassiosira weissflogii* during frustule synthesis. *Sci. Rep.* **2013**, *3*, 3205. [CrossRef]
102. Rorrer, G.L.; Chang, C.H.; Liu, S.H.; Jeffryes, C.; Jiao, J.; Hedberg, J.A. Biosynthesis of silicon–germanium oxide nanocomposites by the marine diatom Nitzschia frustulum. *J. Nanosci. Nanotechnol.* **2005**, *5*, 41–49. [CrossRef]
103. Qin, T.; Gutu, T.; Jiao, J.; Chang, C.H.; Rorrer, G.L. Biological fabrication of photoluminescent nanocomb structures by metabolic incorporation of germanium into the biosilica of the diatom *Nitzschia* frustulum. *ACS Nano* **2008**, *2*, 1296–1304. [CrossRef]
104. Jeffryes, C.; Gutu, T.; Jiao, J.; Rorrer, G.L. Two-stage photobioreactor process for the metabolic insertion of nanostructured germanium into the silica microstructure of the diatom *Pinnularia* sp. *Mater. Sci. Eng. C* **2008**, *28*, 10–118. [CrossRef]
105. Jeffryes, C.; Solanki, R.; Rangineni, Y.; Wang, W.; Chang, C.H.; Rorrer, G.L. Electroluminescence and photoluminescence from nanostructured diatom frustules containing metabolically inserted germanium. *Adv. Mater.* **2008**, *20*, 2633–2637. [CrossRef]
106. Magnabosco, G.; Hauzer, H.; Fermani, S.; Calvaresi, M.; Corticelli, F.; Christian, M.; Albonetti, C.; Morandi, V.; Ere, J.; Falini, G. Bionics synthesis of magnetic calcite skeletal structure through living foraminifera. *Mater. Horiz.* **2019**, *6*, 1862–1867. [CrossRef]
107. Zhang, R.; Shao, M.; Han, X.; Wang, C.; Li, Y.; Hu, B.; Pang, D.; Xie, Z. ATP synthesis in the energy metabolism pathway: A new perspective for manipulating CdSe quantum dots biosynthesized in *Saccharomyces cerevisiae*. *Int. J. Nanomed.* **2017**, *12*, 3865. [CrossRef] [PubMed]
108. Park, T.J.; Lee, S.Y.; Heo, N.S.; Seo, T.S. In vivo synthesis of diverse metal nanoparticles by recombinant Escherichia coli. *Angew. Chem. Int. Ed.* **2010**, *49*, 7019–7024. [CrossRef] [PubMed]
109. Choi, Y.; Park, T.J.; Lee, D.C.; Lee, S.Y. Recombinant Escherichia coli as a biofactory for various single-and multi-element nanomaterials. *Proc. Natl. Acad. Sci. USA* **2018**, *115*, 5944–5949. [CrossRef] [PubMed]
110. Mi, C.; Wang, Y.; Zhang, J.; Huang, H.; Xu, L.; Wang, S.; Fang, X.; Fang, J.; Mao, C.; Xu, S. Biosynthesis and characterization of CdS quantum dots in genetically engineered Escherichia coli. *J. Biotechnol.* **2011**, *153*, 125–132. [CrossRef]

111. Taweecheep, P.; Naloka, K.; Matsutani, M.; Yakushi, T.; Matsushita, K.; Theeragool, G. Superfine bacterial nanocellulose produced by reverse mutations in the bcsC gene during adaptive breeding of *Komagataeibacter oboediens*. *Carbohyd. Polym.* **2019**, 115243. [CrossRef]
112. Tan, Y.; Adhikari, R.Y.; Malvankar, N.S.; Pi, S.; Ward, J.E.; Woodard, T.L.; Nevin, K.P.; Xia, Q.; Tuominen, M.T.; Lovley, D.R. Synthetic biological protein nanowires with high conductivity. *Small* **2016**, *12*, 4481–4485. [CrossRef]
113. Delalat, B.; Sheppard, V.C.; Ghaemi, S.R.; Rao, S.; Prestidge, C.A.; McPhee, G.; Rogers, M.-L.; Donoghue, J.F.; Pillay, V.; Johns, T.G.; et al. Targeted drug delivery using genetically engineered diatom biosilica. *Nat. Commun.* **2015**, *6*, 8791. [CrossRef]
114. Görlich, S.; Pawolski, D.; Zlotnikov, I.; Kröger, N. Control of biosilica morphology and mechanical performance by the conserved diatom gene Silicanin-1. *Commun. Biol.* **2019**, *2*, 245. [CrossRef]
115. Nam, O.; Park, J.M.; Lee, H.; Jin, E. De novo transcriptome profile of coccolithophorid alga *Emiliania huxleyi* CCMP371 at different calcium concentrations with proteome analysis. *PLoS ONE* **2019**, *14*, e0221938. [CrossRef]

© 2019 by the authors. Licensee MDPI, Basel, Switzerland. This article is an open access article distributed under the terms and conditions of the Creative Commons Attribution (CC BY) license (http://creativecommons.org/licenses/by/4.0/).

Article

Influence of Magnetic Micelles on Assembly and Deposition of Porphyrin J-Aggregates

Maria Angela Castriciano [1,*], Mariachiara Trapani [1], Andrea Romeo [1,2], Nicoletta Depalo [3], Federica Rizzi [3,4], Elisabetta Fanizza [3,4], Salvatore Patanè [5] and Luigi Monsù Scolaro [1,2,*]

1. CNR-ISMN, Istituto per lo Studio dei Materiali Nanostrutturati c/o Dipartimento di Scienze Chimiche, Biologiche, Farmaceutiche ed Ambientali, University of Messina V.le F. Stagno D'Alcontres, 31 98166 Messina, Italy; mariachiara.trapani@cnr.it (M.T.); anromeo@unime.it (A.R.)
2. Dipartimento di Scienze Chimiche, Biologiche, Farmaceutiche ed Ambientali and C.I.R.C.M.S.B., University of Messina V.le F. Stagno D'Alcontres, 31 98166 Messina, Italy
3. CNR-IPCF, Istituto Per i Processi Chimico-Fisici, 70124 Bari, Italy; n.depalo@ba.ipcf.cnr.it (N.D.); f.rizzi@ba.ipcf.cnr.it (F.R.); elisabetta.fanizza@uniba.it (E.F.)
4. Dipartimento di Chimica, Università degli Studi di Bari Aldo Moro, Via Orabona 4, 70125 Bari, Italy
5. Dipartimento di Scienze Matematiche e Informatiche, Scienze Fisiche e Scienze della Terra, University of Messina V.le F. Stagno D'Alcontres, 31 98166 Messina, Italy; salvatore.patane@unime.it
* Correspondence: maria.castriciano@cnr.it (M.A.C.); lmonsu@unime.it (L.M.S.)

Received: 23 December 2019; Accepted: 16 January 2020; Published: 21 January 2020

Abstract: Clusters of superparamagnetic iron oxide nanoparticles (SPIONs) have been incorporated into the hydrophobic core of polyethylene glycol (PEG)-modified phospholipid micelles. Two different PEG-phospholipids have been selected to guarantee water solubility and provide an external corona, bearing neutral (SPIONs@PEG-micelles) or positively charged amino groups (SPIONs@NH$_2$-PEG-micelles). Under acidic conditions and with specific mixing protocols (porphyrin first, PF, or porphyrin last, PL), the water-soluble 5,10,15,20-tetrakis-(4-sulfonatophenyl)-porphyrin (TPPS) forms chiral J-aggregates, and in the presence of the two different types of magnetic micelles, an increase of the aggregation rates has been generally observed. In the case of the neutral SPIONs@PEG-micelles, PL protocol affords a stable nanosystem, whereas PF protocol is effective with the charged SPIONs@NH$_2$-PEG-micelles. In both cases, chiral J-aggregates embedded into the magnetic micelles (TPPS@SPIONs@micelles) have been characterized in solution through UV/vis absorption and circular/linear dichroism. An external magnetic field allows depositing films of the TPPS@SPIONs@micelles that retain their chiroptical properties and exhibit a high degree of alignment, which is also confirmed by atomic force microscopy.

Keywords: SPION; porphyrin; chirality; self-assembly; hybrid nanosystems

1. Introduction

Spintronics open a new scenario for the fabrication of efficient devices and refer to electronics based on the electron spin, which is typically controlled by magnetic fields or by ferro-/paramagnetic materials. An important prerequisite for realizing operative spintronic devices is to achieve high spin injection coefficient materials. With this aim, spin filters characterized by the coexistence of one spin conducting and one insulating channel are required. Organic molecules were proposed as a spin filter for "molecular spintronics" due to their relatively longer coherent spin lifetimes and spin transport distances, resulting from weak spin–orbit coupling and weak hyperfine interactions [1,2]. The latter may allow the development of nanoscale devices with improved performance or new functionalities. Porphyrin derivatives, which have already been investigated for molecular electronic devices due to their peculiar electronic and optical properties, have been recently proposed as a class of molecules that is particularly promising as a building block for spintronic technology [3,4]. One-dimensional chromium porphyrin arrays showing half-metallic behavior [5], manganese porphyrin molecules connected with a p-phenylene-ethynylene group [6], and porphyrin/graphene hybrid materials [7] have been reported so far. Theoretical and experimental investigations allowed demonstrating that metallic substrates are able to induce magnetic ordering and the switching of paramagnetic porphyrins, which is due to a superexchange interaction between Fe atoms in the chromophores and Co or Ni atoms in the substrate [8]. Spin-dependent transport properties in an iron-porphyrin such as a carbon nanotube have been investigated, reporting a magnetoresistance ratio that is strongly dependent on the magnetic configuration of the system [9]. Recently, a new promising effective approach for spintronics has emerged using spin selectivity in electron transport through chiral molecules [10–14]. This effect, defined as chiral-induced spin selectivity (CISS), is due to the special property of chiral symmetry that couples the electron spin and its linear momentum acting as a spin filter depending on the handedness of the molecules [15]. The spin-polarized electron current due to the CISS effect can be used to magnetize ferromagnets, potentially allowing the fabrication of less expensive and high-density devices. A DNA double helix [16,17], as well as chiral molecules with (i.e., oligopeptides or chiral polymers) [12,18–21] or without (helicenes) [22] stereogenic carbon centers have been reported so far. The driving of electrons through chiral layers has been reported for α helix L-polyalanine and CdSe nanocrystals by local light-induced magnetization [23] and for self-assembled monolayers of polyalanine to magnetize a Ni layer [24]. However, the development of this technology is hindered by the fact that solid thin films and nanostructures need a precise control of homogeneity, morphology, and chirality. Spatially uniform chiral films [25,26], supramolecular structures with a programmed helicity [27,28], artificial assemblies [29], and the patterning of chiral nanostructures [30,31] have been developed up to now. These latter showed significant improvements toward technological applications, even if the uniformity and spatial control of the local handedness of chiral self-assembled systems on a surface is still a fundamental open challenge. In this framework, we already investigated the self-assembly of the achiral water soluble 5,10,15,20-tetrakis-(4-sulfonatophenyl)-porphyrin (TPPS) into chiral aggregates on a substrate by combining a wet lithographic method with the local induction of specific chirality imprinted by a chiral templating agent [32]. It is well known that the diacid form of this porphyrin in solution under the opportune experimental conditions is able to self-organize into highly ordered chiral J-aggregates with or without the assistance of chiral species [33–35]. Indeed, in the absence of a chiral bias, TPPS could be assembled into chiral aggregates showing a dichroic signal characterized by a positive bisignate Cotton effect [34,36–41] whose shape and magnitude is strictly related to the experimental conditions used for the preparation of the aggregates [34,37,42–46]. The formation of chiral supramolecular assemblies from achiral building blocks is a particularly interesting phenomenon [47–51] for the possible correlation with homochirality observed in nature. In many bioprocesses, the interactions between molecules induce a redistribution of the electronic charge accompanied in a chiral system by spin polarization. It has been experimentally demonstrated that this spin polarization adds an enantioselective term to the forces, thus leading to homochiral interaction energies different from heterochiral ones [12]. Herein, we report on the role of organic capped superparamagnetic iron oxide nanoparticles (SPIONs), after their incorporation in the hydrophobic core of polyethylene glycol (PEG)-modified phospholipid micelles (SPIONs@micelles) on the kinetic and spectroscopic behavior of porphyrin J-aggregates, pointing to the formation of a hybrid system. We anticipate that the SPION-loaded micelles are able to efficiently trigger the growth of TPPS J-aggregates under proper experimental conditions and reagent mixing order protocol. Moreover, this assembling strategy allows conjugating the magnetic properties of the SPIONs@micelles with the chiral optical properties of the J-aggregates, thus obtaining hybrid architectures in solution and on solid state by means of an applied magnetic field. By the application of an external magnetic field, SPIONs show magnetization values comparable and magnetic susceptibility much higher than bulk paramagnetic materials [52]. However, they exhibit high coercivity or residual magnetization value with zero magnetic field [53], anisotropy, and large magnetic saturation values [54]. For these magnetic properties and their good biocompatibility, iron oxide nanoparticles (NPs) have been largely reported for biomedical applications [55,56] and in particular as contrast agents for magnetic resonance imaging (MRI) [52,57], in drug delivery [58,59], magnetic hyperthermia [60], magnetically assisted genetic transfection [61], and/or in combined therapeutic and diagnostic use (theranostics) [62]. To the best of our knowledge, no examples of SPIONs involved in the self-assembly of porphyrins in solution and on the solid state of chiral porphyrin aggregates have been reported so far. Besides, iron oxides thin films with an Fe^{3+}/Fe^{2+} ratio of 2 have been reported as a promising candidates for spintronic applications, showing high saturation magnetization and magnetic susceptibility [63]. The merging of the NPs magnetic properties and chiroptical properties of the porphyrin aggregates makes the hybrid SPIONs@micelle J-agg assemblies interesting candidates for a variety of potential applications, ranging from optics to electronics and spintronics.

2. Materials and Methods

Chemicals. Oleic acid (90%), dodecan-1,2-diol (90%), iron pentacarbonyl (98%), oleyl amine (70%), and 1-octadecene (90%) were purchased from Sigma-Aldrich. 1,2-dipalmitoyl-sn-glycero-3-phosphoethanolamine-N-[methoxy (poly(ethylene glycol))-2000] (16:0 PEG-2-PE, ammonium salt) and amine 1,2-distearoyl-sn-glycero-3-phosphoethanolamine-N-[amino(polyethylene glycol)-2000] (DSPE-PEG-amine, ammonium salt) were from Avanti Polar Lipids. The 5,10,15,20-tetrakis-(4 sulfonatophenyl)porphyrin (TPPS) was purchased from Aldrich Chemicals, and its solutions of known concentration were prepared using the extinction coefficient at the Soret maximum ($\varepsilon = 5.33 \times 10^5$ $M^{-1}cm^{-1}$ at $\lambda = 414$ nm). All the reagents were used without further purification, and the solutions were prepared in dust-free Milli-Q water.

Preparation of PEG-modified phospholipid micelles loaded with SPION. Organic capped SPIONs were synthesized according to the experimental protocols reported in the literature [64]. For the preparation of the micelles loaded with SPIONs and composed only of PEG-2-PE (SPIONs@PEG-micelles), 150 µL of PEG-2-PE in chloroform (3.5×10^{-2} M) were mixed with 240 µL of a SPION stock chloroform dispersion (0.08 M); while for the preparation of amine functionalized superparamagnetic micelles (SPIONs@NH_2-PEG-micelles), 120 µL of PEG-2-PE (3.5×10^{-2} M) and 30 µL of DSPE-PEG-amine (3.5×10^{-2} M) were used. After the complete evaporation of organic solvent, the SPION/PEG-modified lipid film was treated with 2 mL of phosphate buffer (PBS, 10 mM, pH 7.4). Three consecutive cycles of heating–cooling, at 80 °C and room temperature respectively, were carried out in order to obtain SPION-loaded micelles. The excess of organic capped SPIONs, not eventually encapsulated in the micelles, was removed by mild centrifugation at 5000× g for 1 min; the empty micelles were subsequently removed by ultracentrifugation (200,000× g) for 16 h. The SPIONs@micelles that were recovered as pellets were resuspended in water, filtered by using 0.2 µm filters (Anotop, Whatman, Merck, Italy), and lyophilized. Water or PBS was used to reconstitute the dried micelles before their characterization or application [65,66].

Aggregation and deposition procedure. TPPS@SPIONs@micelle J-aggregates were prepared in 0.3 M HCl following two different mixing protocol procedures: (i) porphyrin-first protocol (PF), consisting of the addition of a proper volume of acid stock solution to a diluted solution of porphyrin and SPION-loaded micelles, and (ii) porphyrin-last protocol (PL), in which a known amount of porphyrin stock solution is added to diluted magnetic micelles in acidic solution. The concentration of magnetic micelles is reported in terms of the phospholipid monomer concentration. This has been calculated taking into account a weight percentage of about 74% for the phospholipid into the micelles [66]. For the amine-terminated phospholipid, the PEG-PE/DSPE-PEG-amine mixture is in a 4:1 ratio. Glass slides, carefully cleaned with Piranha acidic solution, were immersed into 3 mL of solution containing TPPS@SPIONs@micelle J-aggregates. We used a neodymium magnet cube (1 cm × 1 cm × 1 cm) placed below the glass slides to deposit aggregates onto glass surface from the solution. After an overnight aging time at room temperature, the slides were washed by quick immersion in aqueous acidic solution and dried under a gentle nitrogen flow.

Spectroscopic and morphological characterization. UV-Vis spectra were collected on a diode-array spectrophotometer Agilent model 8452. Kinetic experiments were carried out in the thermostated compartment of the spectrophotometer, with a temperature accuracy of ± 0.1 K. The analysis of the kinetic profiles has been performed by a non-linear fit of the absorption data according to Equation (1):

$$E_{xt} = E_{xt\infty} + (E_{xt0} - E_{xt\infty})(1 + (m-1)\{k_0 t + (n+1)^{-1}(k_c t)^{n+1}\})^{-1/(m-1)} \tag{1}$$

with E_{xt0}, $E_{xt\infty}$, k_0, k_c, m and n as the parameters to be optimized or Equation (2):

$$E_{xt} = E_{xt0} + (E_{xt\infty} - E_{xt0})(\exp(-(kt)^n)) \tag{2}$$

with E_{xt0}, $E_{xt\infty}$, k and n as the parameters to be optimized (E_{xt}, E_{xt0} and E_{xt} are the extinction at time t, at starting time, and at the end of aggregation, respectively). The circular (CD) and linear (LD) dichroism spectra were recorded on a JASCO J-720 spectropolarimeter equipped with a 450 W xenon lamp. The LD spectra under an applied magnetic field have been recorded by setting a couple of neodymium magnets (1 cm × 1 cm × 1 cm) close to the cuvette walls in a perpendicular direction with respect to the light beam. CD and LD spectra were corrected both for the cell and solvent contributions.

Fluorescence emission and resonance light scattering (RLS) experiments were performed on a Jasco mod. FP-750 spectrofluorimeter. A synchronous scan protocol with a right angle geometry was adopted for collecting RLS spectra [67], which were not corrected for the absorption of the samples. Time-resolved fluorescence emission measurements were performed on a Jobin Yvon-Spex Fluoromax 4 spectrofluorimeter using the time-correlated single-photon counting technique. A NanoLED (λ = 390 nm) has been used as the excitation source.

For the transmission electron microscopy (TEM) investigation, a Jeol JEM-1011 microscope, working at an accelerating voltage of 100 kV, was used. TEM micrographs were acquired by an Olympus Quemesa Camera (11 Mpx). A total of 400 mesh amorphous carbon-coated Cu grids were dipped in SPION chloroform dispersion to achieve the sample deposition.

Hydrodynamic particle sizes and size distributions were measured by Dynamic Light Scattering (DLS)and carried out at 25 °C by a Zetasizer Nano-ZS (Malvern Instruments) equipped with a 633 nm He−Ne laser using backscattering detection. Each DLS sample was measured several times, and the results were averaged.

Atomic force microscopy (AFM) and magnetic force microscopy (MFM) measurements were performed using a NT-MDT Smena head working in tapping mode and equipped with a CoCr coated tip (mod. MFM01). To collect the MFM data, the instrument was configured to work in the "double pass AC Magnetic Force" mode. In this working mode, the system produces two images; the first consists of the bare morphology collected line by line in a standard non-contact scanning mode. In the second step, the tip is uplifted at some tenths of a nanometer from the surface, and the phase shift due to the magnetic interaction between the tip and the sample is recorded. During the second step, the profile data collected in the first step are used to assure that the distance between the tip and the sample is constant. To improve the sensitivity, before the measurement, the tip was exposed to the field of a neodymium magnet for a few hours.

3. Results

The TPPS porphyrin is present in neutral aqueous solution as free base characterized by the presence of a Soret band centered at 414 nm and four Q-bands between 500 and 700 nm in the UV/Vis spectrum. Under acid conditions, the protonation of the pyrrole nitrogen atoms of the central core occurs (pKa: approximately 4.9), with the formation of a protonated species showing a Soret band at 434 nm and two Q-bands. This diacid specie is able to self-arrange in J-aggregates characterized by a linear arrangement of the chromophores and stabilized by electrostatic interactions between the negatively charged benzenesulfonate groups and the positively charged nitrogen atoms of the pyrrole rings, as well as by hydrogen bonds and stacking interactions [41,68–74].

The formation in solution of these aggregates leads in the absorption spectrum to the formation of a new band, bathochromically shifted (Δλ: approximately 50 nm) with respect to the monomeric species. Their aggregation kinetics can be influenced by different experimental parameters such as the nature of the acid [45], the ionic strength [41,74], or the reagent mixing order protocol [34,72,73]. In particular, this latter influences the dynamics of the growth and eventually both the morphology and size of the assemblies [73]. Here, we investigated on the ability of SPION-loaded micelles to efficiently promote the formation of hybrid TPPS@SPIONs@micelle J-aggregates. In particular, two different magnetic micelles were prepared to investigate the self-assembly process. The aqueous dispersibility of the hydrophobic SPIONs was guaranteed by the formation of SPIONs encapsulating micelles composed of 1,2-dipalmitoyl-sn-glycero-3-phosphoethanolamine-N-[methoxy (poly(ethylene glycol))-2000] (PEG-2-PE), or a mixture of PEG-2-PE and amine 1,2-distearoyl-sn-glycero-3-phosphoethanolamine-N-[amino(polyethylene glycol)-2000] (DSPE-PEG-amine], bearing PEG chains and amine terminated PEG chains, respectively (Figure 1) [66,75]. Indeed, magnetic micelles composed only of PEG-2-PE (SPIONs@PEG-micelles, Figure 1A) and amine-functionalized magnetic micelles based on a PEG-2-PE/DSPE-PEG-amine mixture (SPIONs@NH$_2$-PEG-micelles, Figure 1B) were obtained by the encapsulation of a certain number of organic-capped SPIONs having an average diameter of about 9 nm (Figure 1C) clustered in the hydrophobic core into the single micelle.

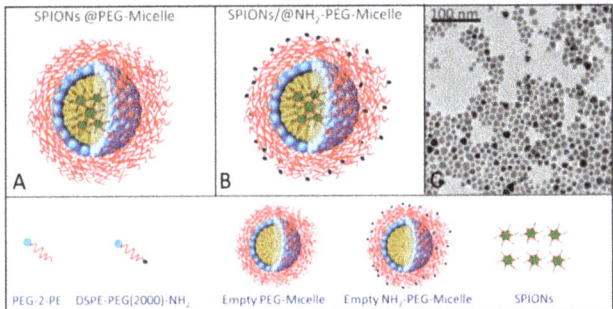

Figure 1. Schematic representation of superparamagnetic iron oxide nanoparticles, after their incorporation in the hydrophobic core of polyethylene glycol (PEG)-modified phospholipid micelles (SPIONs@PEG-micelles) (**A**)- and amine functionalized superparamagnetic micelles (SPIONs@NH$_2$-PEG-micelles) (**B**), along with the corresponding legend. TEM micrograph of organic capped SPIONs (**C**).

DLS measurement was carried out on SPIONs@PEG-micelles to test their stability. From the analysis of the data, the micelles both in neutral and acidic aqueous solution show an average R_H of 150 (±10) nm. When SPIONs@PEG@micelles (15 µM) were added to a solution of TPPS at neutral or mild acidic pH, no evidence of modification in the spectroscopic behavior of the chromophore, in terms of absorption (inset of Figure 2), fluorescence emission, fluorescence lifetimes, and time-resolved fluorescence anisotropy has been observed (data not shown). The experimental evidence excludes a preinteraction among porphyrins and magnetic micelles before aggregation. In order to promote the formation of supramolecular assemblies among porphyrin aggregates and magnetic micelles, we lowered the pH (0.3 M HCl) following the two previously mentioned mixing order protocols. Using a PF protocol, although the J-aggregates electronic spectrum remains unchanged (Figure 2), their formation kinetics are conversely deeply influenced by the presence of SPIONs@PEG-micelles. A kinetic analysis of the extinction/time traces on the aggregate band (491 nm) has been performed by using an autocatalytic model already reported in the literature and largely reported for kinetic investigations on porphyrin aggregation [34,39,44,45,76].

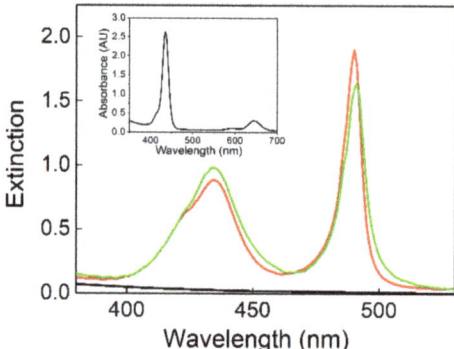

Figure 2. UV-vis spectra of SPIONs@PEG-micelles (black line) and chiral J-aggregates formed by 5,10,15,20-tetrakis-(4-sulfonatophenyl)-porphyrin (TPPS) embedded into the magnetic micelles (TPPS@SPIONs@PEG-micelles) in aqueous solution, porphyrin first (PF) (red line) and porphyrin last (PL) mixing protocol (green line). Experimental conditions: [SPIONs@PEG-micelles] = 15 µM, [TPPS] = 5 µM, [HCl] = 0.3 M. For comparison, TPPS@SPIONs@PEG-micelles at pH 3 is reported in the inset.

The kinetic profiles exhibit a sigmoidal behavior characterized by the presence of an initial induction period that was shortened in the presence of magnetic micelles, (Figure 3A) with an observed rate constant k_c that is about an order of magnitude larger than that observed for the pure system. All the kinetic parameters are summarized in Table 1.

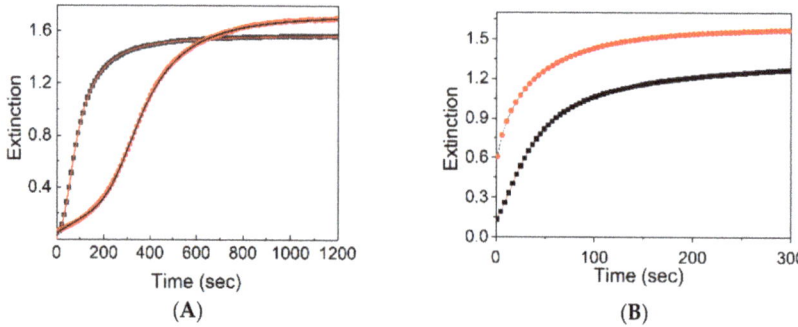

Figure 3. Extinction kinetic traces at 491 nm with (black line) and without (red line) SPIONs@PEG-micelles. The solid lines are the best-fitting of the experimental data ($\lambda_{491\ nm}$) to Equation (1) (**A**) and Equation (2) (**B**). The best-fitting parameters are collected in Table 1. Experimental conditions: [SPIONs@PEG-micelles] = 15 µM, [TPPS] = 5 µM, [HCl] = 0.3 M PF, T = 298 K mixing order protocol (**A**), PL mixing order protocol (**B**).

An increase of the aggregation kinetic rate has been observed also for the PL mixing order protocol. According to the literature, the PL protocol induces a dramatic difference in the kinetic profiles, which now obey a stretched exponential form (Equation (2)). The presence of magnetic micelles further increases the rate constants of the aggregation process (k), whose value goes from 1.8×10^{-2} s^{-1} to a 2.7×10^{-2} s^{-1} (Figure 3B). It is noteworthy that unlike the previous experimental protocol, in this case, the electronic spectrum of the final aggregates shows a slight bathochromic shifted and widened J-absorption (Figure 2). DLS experiments point to the presence in solution of particles having a R_H value of approximately 1 µm. Since it is known that TPPS aggregates in solution in analogous experimental conditions are nanorods sizing hundreds of nanometers [41], our experimental findings suggest the formation of a hybrid system formed by porphyrin aggregates and magnetic micelles. As TPPS J-aggregates can undergo spontaneously to symmetry breaking, circular dichroism spectra have also been collected. At the end of aggregation kinetics for the hybrid system obtained by PL protocol, the CD spectrum shows a profile with a positive bisignate Cotton effect centered at the aggregate absorption band that is much higher with respect to the neat sample (Figure 4).

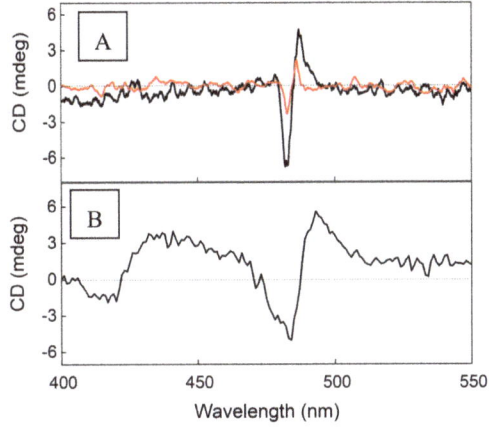

Figure 4. Circular dichorism (CD) spectra of J-aggregates in solution (**A**) and on solid state (**B**) in the presence (black line) or in the absence (red line) of SPIONs@PEG-micelles. Experimental conditions: [SPIONs@PEG-micelles] = 15 µM, [TPPS] = 5 µM, [HCl] = 0.3 M, PL mixing order protocol.

It is interesting to note that according to the literature, for "pure" TPPS J-aggregates, whatever the pretreatment of the samples, the dissymmetry g-factor generally decreases on increasing the value of the kinetic rate constant [44]. On the contrary, for the hybrid system here reported, we observe an increase on the CD intensity signal, despite the increase of the aggregation rate constant. Taking advantage of the magnetic properties of the SPIONs@PEG-micelles, we used a magnet field below the sample to try to deposit aggregates obtained by PF and PL protocols, respectively, onto glass surface from the solutions. After aging overnight at room temperature, the formation of a green film from the PL samples and of a dark orange precipitate from the PF protocol samples respectively, is evident. As a blank experiment, the magnet field has been applied also for the aggregate solutions in the absence of magnetic micelles and, as expected, no deposition has been observed. The extinction spectrum of the solid sample shows unequivocally the presence of the characteristic J-aggregate band (Figure 5), whereas in solution, only the presence of porphyrin in its diacid monomeric form has been observed (Figure 5, inset). Interestingly, the CD spectrum of the film shows the chiroptical properties observed in solution with an induced bisignate Cotton effect in the aggregate absorption region (Figure 4B). This dichroic signal cannot be ascribed to linear dichroism, as no variation due to the sample orientation with respect to incident light has been observed.

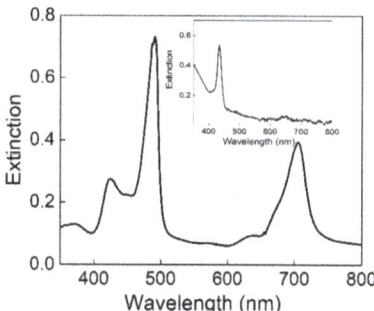

Figure 5. Extinction spectrum of TPPS@SPIONs@PEG-Micelle J-aggregates deposited on the glass surface by applying a magnetic field through a neodymium magnet cube 1 cm × 1 cm × 1 cm and residual specie in solution (inset). Experimental conditions: [SPIONs@PEG-Micelles] = 15 µM, [TPPS] = 5 µM, [HCl] = 0.3 M, PL mixing order protocol.

These experimental findings prove the formation of a hybrid system only for the PL protocol. So, in our case, the reagent mixing order affects not only the kinetic rates of the self-assembly processes but also the formation of the hybrid supramolecular system. This sort of "YES/NO" effect has already been reported for TPPS J-aggregates obtained in the presence of polyamines containing less than three protonable nitrogen atoms [73]. In particular, the nucleation step is the critical parameter affecting both the kinetic and mesoscopic structure of the resulting aggregates [34,77]. In the present case, we hypothesize that the PL protocol, due to a concentration effect, induces a very quick formation of a larger number of porphyrin seeds that can be entrapped in peripheral hydrophilic chains of the SPIONs@PEG-micelles, leading to hybrid aggregates. On the contrary, in the PF protocol, a longer nucleation period allows the organization of porphyrins in much larger aggregates, thus preventing their entrapment in the magnetic micelles.

In order to characterize the morphology of the deposited sample on a glass surface, we performed atomic force microscopy (AFM). Figure 6 shows the sample topography and the line profile acquired along the blue line, as depicted on the picture. The morphology consists of an almost homogeneous layer of compactly arranged nanostructures, whose dimension ranges between 100 and 150 nm. It is worth noting that a preferential direction is present; this alignment is probably due to the magnetic field used during the deposition method. Even if no clear detection of the TPPS aggregates is possible, their embedding into the film is confirmed by the absorption of the sample (Figure 5).

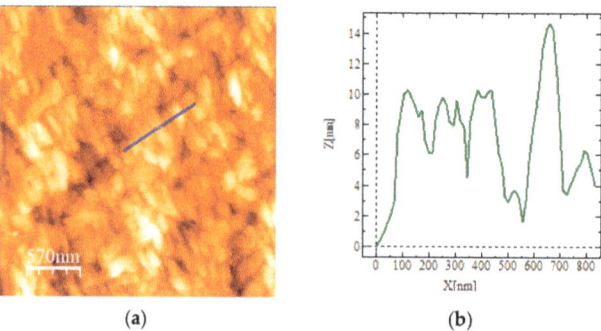

Figure 6. Atomic force microscopy (AFM) topography images (**a**) and relative profile (**b**) of the TPPS@SPIONs@PEG-micelle J-aggregates deposit on a glass substrate after applying a magnetic field.

As the interaction among TPPS porphyrins and functional molecules bearing protonable nitrogen atoms such as polyamines or PAMAM dendrimers has been already reported in aqueous solution [72,73,78–80], we exploited amine functionalized magnetic micelles based on a PEG-2-PE/DSPE-PEG-amine mixture (SPIONs@NH_2-PEG-micelles) to trigger the aggregation process. The presence of protonable nitrogen atoms in the periphery of the micelles should be useful for interaction with the negatively charged groups present on the porphyrin ring. Accordingly, we tested the preinteraction between the chromophoric unit in its monomeric neutral and diacid forms with SPIONs@NH_2-PEG-micelles at different stoichiometric ratios. In all the investigated samples, no changes on the spectroscopic behavior of the porphyrin have been observed so excluding the presence of the chromophoric units embedded in the micelles or entangled in their peripheral chains, even if some interaction with the peripheral counter-cations cannot be ruled out. As for the previous system, the stability of the SPIONs@NH_2-PEG-micelles under the investigated experimental conditions has been tested by UV/Vis and DLS measurements. This latter technique evidences the presence in solutions of objects with R_H of about 140 (±10) nm. The aggregation process in the presence of TPPS has been induced by employing the two different mixing order protocols, PF and PL, respectively. When PL protocol is adopted on SPIONs@NH_2-PEG-micelles under acidic conditions (HCl 0.3 M), J-aggregates form almost instantaneously. The resulting solutions are highly unstable, and precipitation occurs in a very short time. DLS investigations confirm the presence of micrometric objects, thus justifying the experimental observations. Due to the scarce reproducibility of these samples, we decided to not further investigate them. On the contrary, when PF protocol is used, the aggregation is fostered by the addition of acidic solution of SPIONs@NH_2-PEG-micelles at different concentrations; the extinction and kinetic features are reported in Figure 7. In particular, the inset shows a set of kinetic traces obtained from the extinction increase at 491 nm (Figure 7). These curves display a sigmoidal profile characterized by a nucleation early stage, which shortens upon increasing the concentration of magnetic micelles in solution (inset Figure 7). The values of the parameter m (~3), that is the critical size for the nuclei, and n (~3–4), the time exponent, are comparable to those reported in the literature for similar systems in aqueous solutions [34]. Differently, the values of the rate constants for the uncatalyzed pathway, k_0, and those for the catalyzed pathway, k_c, in the presence of magnetic micelles are higher with respect to the pure self-assembled system and increase, exhibiting an exponential dependence on SPIONs@NH_2-PEG-micelles concentration. All the kinetic parameters are collected in Table 1. After equilibration, the UV–Vis spectra provide evidence that the amount of the final aggregate increases with the concentration of SPIONs@NH_2-PEG-micelles.

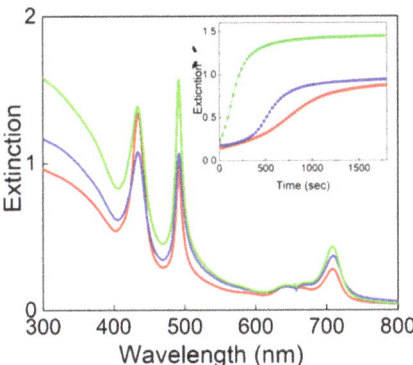

Figure 7. Extinction spectra for TPPS@SPIONs@NH_2-PEG-micelles J-aggregates formed at different SPIONs@NH_2-PEG-micelles concentrations. [SPIONs@NH_2-PEG-micelles] = 15 µM (red line), [SPIONs@NH_2-PEG-micelles] = 22 µM (blue line), [SPIONs@NH_2-PEG-micelles] = 45 µM (green line). In the inset, the extinction kinetic traces at 491 nm and the best fitting experimental data according to Equation (1). Experimental conditions: [TPPS] = 5 µM, [HCl] = 0.3 M PF mixing order protocol, T = 298 K.

The aggregated samples are stable in solution, and DLS measurements show the presence of objects with R_H values of about 700 (±50) nm. The CD spectra recorded at the end of the aggregation process show in the absence of magnetic nanoparticles the typical positive bisignate spectrum, whereas unusual profiles can be observed in the presence of SPIONs@NH$_2$-PEG-micelles at different porphyrin/magnetic micelles concentration ratios (Figure 8a). A generally inverted and consistently bathochromically shifted band is present at the absorption of the J-component, while the H-band shows the usual positive bisignate feature, even if it is slightly red shifted. Since linear dichroism (LD) could be responsible for the observed effects in the case of alignment of the aggregates, we performed LD measurements on the same TPPS@SPIONs@NH$_2$-PEG-micelles J-aggregates (Figure 8b). The LD spectra show for all the samples the presence of a positive band at 491 nm and a less intense and negative one at 420 nm, similarly to those reported for analogous systems, where the J-aggregates were aligned by flow. The presence of different signs for the two LD bands has been attributed to the different polarization within the porphyrin aggregate of the J and the H electronic transitions [81].

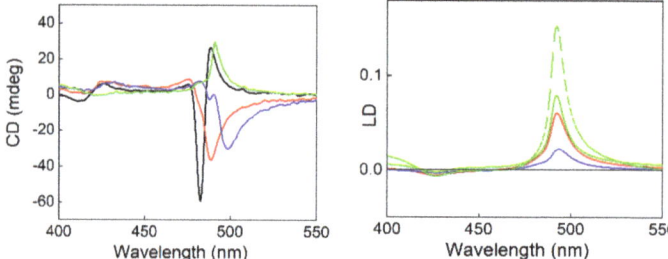

Figure 8. CD (**a**) and linear dichroism (LD) (**b**) spectra for TPPS@SPIONs@NH$_2$-PEG-micelles J-aggregates at different SPIONs@NH$_2$-PEG-Micelles concentrations. In the absence of SPION-NH$_2$ (black line) [SPIONs@NH$_2$-PEG-Micelles] = 15 μM (red line), [SPIONs@NH$_2$-PEG-Micelles] = 22 μM (blue line), [SPIONs@NH$_2$-PEG-Micelles] = 45 μM (green line), and under an applied magnetic field perpendicularly to the light path (green dashed line). Experimental conditions: [TPPS] = 5 μM, [HCl] = 0.3 M, PF mixing order protocol.

Table 1. Kinetic parameters k_0, k_c, m, and n for the aggregation of TPPS@SPIONs@PEG-micelles and TPPS@SPIONs@NH$_2$-PEG-micelles as a function of the mixing order protocol and/or magnetic micelles concentration (T = 298 K). [a] Data obtained according to Equation (2); [b] Data obtained according to Equation (1).

J-Aggregates	K		N	
[SPIONs@PEG-micelle]/μM [a]				
—	1.8 × 10⁻² ± 1 × 10⁻⁴		0.8 ± 0.05	
15	2.7 × 10⁻² ± 1 × 10⁻³		0.7 ± 0.02	
	k_0	k_c	m	n
[SPIONs@PEG-micelle]/μM [b]				
0	2.7 × 10⁻⁴ ± 7 × 10⁻⁶	1.9 × 10⁻³ ± 8 × 10⁻⁶	3.0 ± 0.1	4.0 ± 0.1
15	—	1.6 × 10⁻² ± 4 × 10⁻⁵	2.3 ± 0.1	1.3 ± 0.1
[SPIONs@NH$_2$-PEG-micelle]/μM [b]				
0	2.7 × 10⁻⁴ ± 7 × 10⁻⁶	1.9 × 10⁻³ ± 8 × 10⁻⁶	3.0 ± 0.1	4.0 ± 0.1
15	4.1 × 10⁻⁴ ± 2 × 10⁻⁷	1.7 × 10⁻³ ± 9 × 10⁻⁶	2.2 ± 0.1	3.6 ± 0.3
22	1.2 × 10⁻⁴ ± 5 × 10⁻⁶	2.6 × 10⁻³ ± 5 × 10⁻⁶	3.0 ± 0.3	4.3 ± 0.1
45	2.5 × 10⁻³ ± 7 × 10⁻⁷	9.1 × 10⁻³ ± 8 × 10⁻⁶	3.5 ± 0.2	3.0 ± 0.2

This experimental evidence suggests that the obtained aggregates being large enough are somehow aligned, probably by effect of barodiffusion. Taking advantage of the magnetic property of

TPPS@SPIONs@NH$_2$-PEG-micelles J-aggregates, a further LD spectrum was taken on the solution containing the larger amount of NPs by applying an external magnetic field perpendicularly to the direction of the optical beam during the measurement. In comparison with the LD spectrum recorded at zero field (Figure 8b, full green line), the profile in the presence of the magnetic field exhibits a twofold increase in the intensity of the linear contributions with no alteration of the shape (Figure 8b, dashed green line), thus suggesting an enhanced alignment effect. All our experimental findings point to the formation of a hybrid system, evidencing the role of the magnetic NPs on their kinetics of growth and final structure. In this case, also deposition on a glass surface has been achieved by applying a magnetic field, and the electronic spectrum of the film shows the presence of the J-aggregates band (Figure 9).

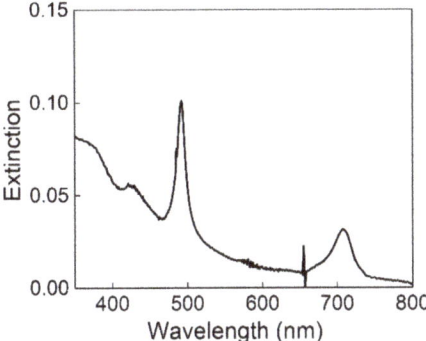

Figure 9. Extinction spectrum of TPPS@SPIONs@NH$_2$-PEG-micelles J-aggregates deposited on a glass surface by applying a magnetic field under the solid substrate. Experimental conditions: [SPIONs@NH$_2$-PEG-micelles] = 45 µM, [TPPS] = 5 µM, [HCl] = 0.3 M, PF mixing order protocol.

Figure 10 displays the morphology (a) and the magnetic response (b) of this sample obtained through MFM. The morphology consists of an almost homogeneous compact layer of nanoparticles. The surface is also populated by a small number of aggregates with larger sizes. The MFM map reveals a general magnetic activity, which is highlighted by the signal modulation distributed over almost all the sample surface, as the intrinsically limited resolution of the technique is not able to map the magnetic response of the smallest nanoparticles. The bigger aggregates clearly show a much stronger magnetic signal. Interestingly, some particles show a bright contrast, while others highlight a dark contrast. This behavior is due to the different interaction with the magnetic tip that operates in tapping mode at its resonance frequency: a repulsive magnetic force gradient will cause the resonance curve to shift to a higher frequency, an increase in phase shift, and a bright contrast. Conversely, an attractive magnetic force gradient results in the resonance curve shifting to a lower frequency, and a decrease in phase shift resulting in a dark contrast. Some particles with no magnetic activity are also observed. Therefore, all the spectroscopic and microscopy evidence suggests that the deposited film is constituted by the hybrid TPPS@SPIONs@NH$_2$-PEG-micelles J-aggregates.

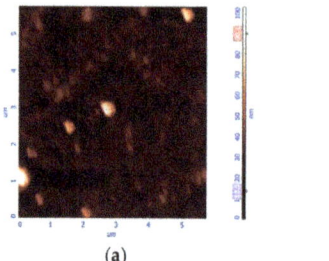 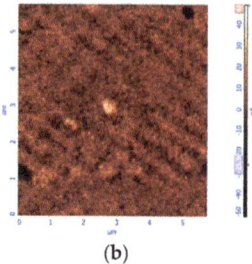

(a) (b)

Figure 10. AFM images for TPPS@SPIONs@NH$_2$-PEG-micelles J-aggregates deposited on glass by applying a magnetic field under the solid substrate. Topographic (**a**) and magnetic (**b**) images. Experimental conditions: [SPIONs@NH$_2$-PEG-Micelles] = 45 µM, [TPPS] = 5 µM, [HCl] = 0.3 M, PF mixing order protocol.

4. Conclusions

J-aggregates of TPPS porphyrin are interesting nanomaterials, since depending on the experimental conditions and templating agents, they can be prepared in a variety of different morphologies. The use of magnetic micelles offers a way to obtain hybrid organic/inorganic nanosystems that conjugate the properties of each constituent component. The porphyrin nanoaggregates exhibit peculiar extinction bands, with enhanced resonant light-scattering effects [34], and chiroptical properties that can spontaneously appear or be induced by doping them with a proper chiral reagent [40]. The SPIONs micelles, with the two different capping phospholipids, while allowing for the water solubility of the inorganic cluster core, afford a hydrophilic surface where eventually TPPS can nucleate, leading to the final J-aggregates. The presence of an amino-terminated chain in the case of SPIONs@NH$_2$-PEG-micelles determines a distinct change in the supramolecular assembling kinetics of TPPS. Furthermore, the nature of the terminal PEG chain on the phospholipid influences the way in which the two different mixing protocols control the formation of stable nanosystems. In both cases, taking advantage of the magnetic properties of SPIONs, it is possible to drive the deposition of TPPS J-aggregates embedded into the micelles onto surface, thus achieving a high level of alignment into the film. Considering the optical and chiroptical response, still conserved in the solid phase, and their magnetic properties, we expect that these nanosystems could find potential applications in a variety of fields, including spintronics, optoelectronics and theranostics.

Author Contributions: Conceptualization, L.M.S. and M.A.C.; investigation, M.T., N.D., E.F., F.R., and S.P.; data curation, M.T., A.R., E.F., and M.A.C.; writing—original draft preparation, M.A.C. and N.D.; writing—review and editing, L.M.S., M.A.C., A.R., N.D., and M.T.; visualization, A.R., E.F., and S.P.; Authorship must be limited to those who have contributed substantially to the work reported. All authors have read and agree to the published version of the manuscript.

Funding: This research received no external funding.

Conflicts of Interest: The authors declare no conflict of interest.

References

1. Cornia, A.; Seneor, P. The molecular way. *Nat. Mater.* **2017**, *16*, 505–506. [CrossRef] [PubMed]
2. Sanvito, S. Molecular spintronics. *Chem. Soc. Rev.* **2011**, *40*, 3336–3355. [CrossRef] [PubMed]
3. Sun, Q.L.; Dai, Y.; Ma, Y.D.; Li, X.R.; Wei, W.; Huang, B.B. Two-dimensional metalloporphyrin monolayers with intriguing electronic and spintronic properties. *J. Mater. Chem. C* **2015**, *3*, 6901–6907. [CrossRef]
4. Le Roy, J.J.; Cremers, J.; Thomlinson, I.A.; Slota, M.; Myers, W.K.; Horton, P.H.; Coles, S.J.; Anderson, H.L.; Bogani, L. Tailored homo- and hetero-lanthanide porphyrin dimers: A synthetic strategy for integrating multiple spintronic functionalities into a single molecule. *Chem. Sci.* **2018**, *9*, 8474–8481. [CrossRef]
5. Cho, W.J.; Cho, Y.; Min, S.K.; Kim, W.Y.; Kim, K.S. Chromium Porphyrin Arrays As Spintronic Devices. *J. Am. Chem. Soc.* **2011**, *133*, 9364–9369. [CrossRef]

6. Zeng, J.; Chen, K.Q. Spin filtering, magnetic and electronic switching behaviors in manganese porphyrin-based spintronic devices. *J. Mater. Chem. C* **2013**, *1*, 4014–4019. [CrossRef]
7. Zeng, J.; Chen, K.Q. A nearly perfect spin filter and a spin logic gate based on a porphyrin/graphene hybrid material. *Phys. Chem. Chem. Phys.* **2018**, *20*, 3997–4004. [CrossRef]
8. Wende, H.; Bernien, M.; Luo, J.; Sorg, C.; Ponpandian, N.; Kurde, J.; Miguel, J.; Piantek, M.; Xu, X.; Eckhold, P.; et al. Substrate-induced magnetic ordering and switching of iron porphyrin molecules. *Nat. Mater.* **2007**, *6*, 516–520. [CrossRef]
9. Zeng, J.; Chen, K.Q. Magnetic configuration dependence of magnetoresistance in a Fe-porphyrin-like carbon nanotube spintronic device. *Appl. Phys. Lett.* **2014**, *104*. [CrossRef]
10. Naaman, R.; Fontanesi, C.; Waldeck, D.H. Chirality and its role in the electronic properties of peptides: Spin filtering and spin polarization. *Curr. Opin. Electrochem.* **2019**, *14*, 138–142. [CrossRef]
11. Abendroth, J.M.; Stemer, D.M.; Bloom, B.P.; Roy, P.; Naaman, R.; Waldeck, D.H.; Weiss, P.S.; Mondal, P.C. Spin Selectivity in Photoinduced Charge-Transfer Mediated by Chiral Molecules. *ACS Nano* **2019**, *13*, 4928–4946. [CrossRef] [PubMed]
12. Kumar, A.; Capua, E.; Kesharwani, M.K.; Martin, J.M.L.; Sitbon, E.; Waldeck, D.H.; Naaman, R. Chirality-induced spin polarization places symmetry constraints on biomolecular interactions. *Proc. Natl. Acad. Sci. USA* **2017**, *114*, 2474–2478. [CrossRef] [PubMed]
13. Michaeli, K.; Kantor-Uriel, N.; Naaman, R.; Waldeck, D.H. The electron's spin and molecular chirality—how are they related and how do they affect life processes? *Chem. Soc. Rev.* **2016**, *45*, 6478–6487. [CrossRef] [PubMed]
14. Mondal, P.C.; Fontanesi, C.; Waldeck, D.H.; Naaman, R. Field and Chirality Effects on Electrochemical Charge Transfer Rates: Spin Dependent Electrochemistry. *ACS Nano* **2015**, *9*, 3377–3384. [CrossRef]
15. Naaman, R.; Waldeck, D.H. Spintronics and Chirality: Spin Selectivity in Electron Transport Through Chiral Molecules. *Annu. Rev. Phys. Chem.* **2015**, *66*, 263–281. [CrossRef]
16. Ray, S.G.; Daube, S.S.; Leitus, G.; Vager, Z.; Naaman, R. Chirality-induced spin-selective properties of self-assembled monolayers of DNA on gold. *Phys. Rev. Lett.* **2006**, *96*. [CrossRef]
17. Xie, Z.; Markus, T.Z.; Cohen, S.R.; Vager, Z.; Gutierrez, R.; Naaman, R. Spin Specific Electron Conduction through DNA Oligomers. *Nano Lett.* **2011**, *11*, 4652–4655. [CrossRef]
18. Di Nuzzo, D.; Kulkarni, C.; Zhao, B.; Smolinsky, E.; Tassinar, F.; Meskers, S.C.J.; Naaman, R.; Meijer, E.W.; Friend, R.H. High Circular Polarization of Electroluminescence Achieved via Self-Assembly of a Light-Emitting Chiral Conjugated Polymer into Multidomain Cholesteric Films. *Acs Nano* **2017**, *11*, 12713–12722. [CrossRef]
19. Kettner, M.; Goehler, B.; Zacharias, H.; Mishra, D.; Kiran, V.; Naaman, R.; Fontanesi, C.; Waldeck, D.H.; Sek, S.; Pawlowski, J.; et al. Spin Filtering in Electron Transport Through Chiral Oligopeptides. *J. Phys. Chem. C* **2015**, *119*, 14542–14547. [CrossRef]
20. Mondal, P.C.; Kantor-Uriel, N.; Mathew, S.P.; Tassinari, F.; Fontanesi, C.; Naaman, R. Chiral Conductive Polymers as Spin Filters. *Adv. Mater.* **2015**, *27*, 1924–1927. [CrossRef]
21. Koplovitz, G.; Leitus, G.; Ghosh, S.; Bloom, B.P.; Yochelis, S.; Rotem, D.; Vischio, F.; Striccoli, M.; Fanizza, E.; Naaman, R.; et al. Single Domain 10 nm Ferromagnetism Imprinted on Superparamagnetic Nanoparticles Using Chiral Molecules. *Small* **2019**, *15*. [CrossRef] [PubMed]
22. Kiran, V.; Mathew, S.P.; Cohen, S.R.; Delgado, I.H.; Lacour, J.; Naaman, R. Helicenes-A New Class of Organic Spin Filter. *Adv. Mater.* **2016**, *28*, 1957–1962. [CrossRef] [PubMed]
23. Ben Dor, O.; Morali, N.; Yochelis, S.; Baczewski, L.T.; Paltiel, Y. Local Light-Induced Magnetization Using Nanodots and Chiral Molecules. *Nano Lett.* **2014**, *14*, 6042–6049. [CrossRef] [PubMed]
24. Ben Dor, O.; Yochelis, S.; Mathew, S.P.; Naaman, R.; Paltiel, Y. A chiral-based magnetic memory device without a permanent magnet. *Nat. Commun.* **2013**, *4*. [CrossRef]
25. Verbiest, T.; Elshocht, S.V.; Kauranen, M.; Hellemans, L.; Snauwaert, J.; Nuckolls, C.; Katz, T.J.; Persoons, A. Strong Enhancement of Nonlinear Optical Properties Through Supramolecular Chirality. *Science* **1998**, *282*, 913–915. [CrossRef]
26. Kim, C.-J.; Sánchez Castillo, A.; Ziegler, Z.; Ogawa, Y.; Noguez, C.; Park, J. Chiral atomically thin films. *Nat. Nanotechnol.* **2016**, *11*, 520–524. [CrossRef]

27. Leclere, P.; Surin, M.; Lazzaroni, R.; Kilbinger, A.F.M.; Henze, O.; Jonkheijm, P.; Biscarini, F.; Cavallini, M.; Feast, W.J.; Meijer, E.W.; et al. Surface-controlled self-assembly of chiral sexithiophenes. *J. Mater. Chem.* **2004**, *14*, 1959–1963. [CrossRef]
28. Tahara, K.; Yamaga, H.; Ghijsens, E.; Inukai, K.; Adisoejoso, J.; Blunt, M.O.; De Feyter, S.; Tobe, Y. Control and induction of surface-confined homochiral porous molecular networks. *Nat. Chem.* **2011**, *3*, 714–719. [CrossRef]
29. Jack, C.; Karimullah, A.S.; Tullius, R.; Khorashad, L.K.; Rodier, M.; Fitzpatrick, B.; Barron, L.D.; Gadegaard, N.; Lapthorn, A.J.; Rotello, V.M.; et al. Spatial control of chemical processes on nanostructures through nano-localized water heating. *Nat. Commun.* **2016**, *7*. [CrossRef]
30. Bystrenova, E.; Facchini, M.; Cavallini, M.; Cacace, M.G.; Biscarini, F. Multiple length-scale patterning of DNA by stamp-assisted deposition. *Angew. Chem. Int. Ed.* **2006**, *45*, 4779–4782. [CrossRef]
31. van Hameren, R.; Schön, P.; van Buul, A.M.; Hoogboom, J.; Lazarenko, S.V.; Gerritsen, J.W.; Engelkamp, H.; Christianen, P.C.M.; Heus, H.A.; Maan, J.C.; et al. Macroscopic Hierarchical Surface Patterning of Porphyrin Trimers via Self-Assembly and Dewetting. *Science* **2006**, *314*, 1433–1436. [CrossRef] [PubMed]
32. Castriciano, M.A.; Gentili, D.; Romeo, A.; Cavallini, M.; Scolaro, L.M. Spatial control of chirality in supramolecular aggregates. *Sci. Rep.* **2017**, *7*. [CrossRef] [PubMed]
33. Short, J.M.; Berriman, J.A.; Kübel, C.; El-Hachemi, Z.; Naubron, J.-V.; Balaban, T.S. Electron Cryo-Microscopy of TPPS4·2HCl Tubes Reveals a Helical Organisation Explaining the Origin of their Chirality. *ChemPhysChem* **2013**, *14*, 3209–3214. [CrossRef] [PubMed]
34. Romeo, A.; Castriciano, M.A.; Occhiuto, I.; Zagami, R.; Pasternack, R.F.; Scolaro, L.M. Kinetic Control of Chirality in Porphyrin J-Aggregates. *J. Am. Chem. Soc.* **2014**, *136*, 40–43. [CrossRef]
35. Castriciano, M.A.; Donato, M.G.; Villari, V.; Micali, N.; Romeo, A.; Scolaro, L.M. Surfactant-like behavior of short-chain alcohols in porphyrin aggregation. *J. Phys. Chem. B* **2009**, *113*, 11173–11178. [CrossRef]
36. Rubires, R.; Farrera, J.A.; Ribo, J.M. Stirring effects on the spontaneous formation of chirality in the homoassociation of diprotonated meso-tetraphenylsulfonato porphyrins. *Chem. A Eur. J.* **2001**, *7*, 436–446. [CrossRef]
37. Zagami, R.; Castriciano, M.A.; Romeo, A.; Trapani, M.; Pedicini, R.; Monsù Scolaro, L. Tuning supramolecular chirality in nano and mesoscopic porphyrin J-aggregates. *Dye. Pigment.* **2017**, *142*, 255–261. [CrossRef]
38. Zagami, R.; Castriciano, M.A.; Romeo, A.; Scolaro, L.M. Spectroscopic investigations on chiral J-aggregates induced by tartaric acid in alcoholic solution. *J. Porphyr. Phthalocyanines* **2017**, *21*, 327–333. [CrossRef]
39. Castriciano, M.A.; Romeo, A.; Zagami, R.; Micali, N.; Scolaro, L.M. Kinetic effects of tartaric acid on the growth of chiral J-aggregates of tetrakis(4-sulfonatophenyl)porphyrin. *Chem. Commun.* **2012**, *48*, 4872–4874. [CrossRef]
40. Castriciano, M.A.; Romeo, A.; De Luca, G.; Villari, V.; Scolaro, L.M.; Micali, N. Scaling the chirality in porphyrin J-nanoaggregates. *J. Am. Chem. Soc.* **2011**, *133*, 765–767. [CrossRef]
41. Micali, N.; Villari, V.; Castriciano, M.A.; Romeo, A.; Scolaro, L.M. From fractal to nanorod porphyrin J-aggregates. concentration-induced tuning of the aggregate size. *J. Phys. Chem. B* **2006**, *110*, 8289–8295. [CrossRef] [PubMed]
42. El-Hachemi, Z.; Escudero, C.; Acosta-Reyes, F.; Casas, M.T.; Altoe, V.; Aloni, S.; Oncins, G.; Sorrenti, A.; Crusats, J.; Campos, J.L.; et al. Structure vs. properties—Chirality, optics and shapes—In amphiphilic porphyrin J-aggregates. *J. Mater. Chem. C* **2013**, *1*, 3337–3346. [CrossRef]
43. Zagami, R.; Romeo, A.; Castriciano, M.A.; Monsù Scolaro, L. Inverse Kinetic and Equilibrium Isotope Effects on Self-Assembly and Supramolecular Chirality of Porphyrin J-Aggregates. *Chem. A Eur. J.* **2017**, *23*, 70–74. [CrossRef] [PubMed]
44. Romeo, A.; Castriciano, M.A.; Zagami, R.; Pollicino, G.; Monsù Scolaro, L.; Pasternack, R.F. Effect of zinc cations on the kinetics of supramolecular assembly and the chirality of porphyrin J-aggregates. *Chem. Sci.* **2017**, *8*, 961–967. [CrossRef] [PubMed]
45. Occhiuto, I.G.; Zagami, R.; Trapani, M.; Bolzonello, L.; Romeo, A.; Castriciano, M.A.; Collini, E.; Monsù Scolaro, L. The role of counter-anions in the kinetics and chirality of porphyrin J-aggregates. *Chem. Commun.* **2016**, *52*, 11520–11523. [CrossRef] [PubMed]
46. Trapani, M.; Castriciano, M.A.; Romeo, A.; De Luca, G.; Machado, N.; Howes, B.D.; Smulevich, G.; Scolaro, L.M. Nanohybrid Assemblies of Porphyrin and Au-10 Cluster Nanoparticles. *Nanomaterials* **2019**, *9*, 1026. [CrossRef]

47. Dressel, C.; Reppe, T.; Prehm, M.; Brautzsch, M.; Tschierske, C. Chiral self-sorting and amplification in isotropic liquids of achiral molecules. *Nat. Chem.* **2014**, *6*, 971–977. [CrossRef]
48. Micali, N.; Engelkamp, H.; van Rhee, P.G.; Christianen, P.C.M.; Scolaro, L.M.; Maan, J.C. Selection of supramolecular chirality by application of rotational and magnetic forces. *Nat. Chem.* **2012**, *4*, 201–207. [CrossRef]
49. Abdulrahman, N.A.; Fan, Z.; Tonooka, T.; Kelly, S.M.; Gadegaard, N.; Hendry, E.; Govorov, A.O.; Kadodwala, M. Induced Chirality through Electromagnetic Coupling between Chiral Molecular Layers and Plasmonic Nanostructures. *Nano Lett.* **2012**, *12*, 977–983. [CrossRef]
50. Lei, S.; Surin, M.; Tahara, K.; Adisoejoso, J.; Lazzaroni, R.; Tobe, Y.; Feyter, S.D. Programmable Hierarchical Three-Component 2D Assembly at a Liquid–Solid Interface: Recognition, Selection, and Transformation. *Nano Lett.* **2008**, *8*, 2541–2546. [CrossRef]
51. Gaeta, M.; Oliveri, I.P.; Fragala, M.E.; Failla, S.; D'Urso, A.; Di Bella, S.; Purrello, R. Chirality of self-assembled achiral porphyrins induced by chiral Zn(ii) Schiff-base complexes and maintained after spontaneous dissociation of the templates: A new case of chiral memory. *Chem. Commun.* **2016**. [CrossRef] [PubMed]
52. Mahmoudi, M.; Sant, S.; Wang, B.; Laurent, S.; Sen, T. Superparamagnetic iron oxide nanoparticles (SPIONs): Development, surface modification and applications in chemotherapy. *Adv. Drug Deliv. Rev.* **2011**, *63*, 24–46. [CrossRef] [PubMed]
53. Perkas, N.; Palchik, O.; Brukental, I.; Nowik, I.; Gofer, Y. A Mesoporous Iron–Titanium Oxide Composite Prepared Sonochemically. *J. Phys. Chem. B* **2003**, *107*, 8772–8778. [CrossRef]
54. Sun, S.; Zeng, H.; Robinson, D.B.; Raoux, S.; Rice, P.M.; Wang, S.X.; Li, G. Monodisperse MFe2O4 (M = Fe, Co, Mn) Nanoparticles. *J. Am. Chem. Soc.* **2004**, *126*, 273–279. [CrossRef] [PubMed]
55. Dulinska-Litewka, J.; Lazarczyk, A.; Halubiec, P.; Szafranski, O.; Karnas, K.; Karewicz, A. Superparamagnetic Iron Oxide NanoparticlesCurrent and Prospective Medical Applications. *Materials* **2019**, *12*, 617. [CrossRef]
56. Dadfar, S.M.; Roemhild, K.; Drude, N.I.; von Stillfried, S.; Knuechel, R.; Kiessling, F.; Lammers, T. Iron oxide nanoparticles: Diagnostic, therapeutic and theranostic applications. *Adv. Drug Deliv. Rev.* **2019**, *138*, 302–325. [CrossRef]
57. Mao, X.; Xu, J.; Cui, H. Functional nanoparticles for magnetic resonance imaging. *Wiley Interdiscip. Rev. Nanomed. Nanobiotechnol.* **2016**, *8*, 814–841. [CrossRef]
58. Ulbrich, K.; Hola, K.; Subr, V.; Bakandritsos, A.; Tucek, J.; Zboril, R. Targeted Drug Delivery with Polymers and Magnetic Nanoparticles: Covalent and Noncovalent Approaches, Release Control, and Clinical Studies. *Chem. Rev.* **2016**, *116*, 5338–5431. [CrossRef]
59. Wahajuddin, S.A.; Arora, S. Superparamagnetic iron oxide nanoparticles: Magnetic nanoplatforms as drug carriers. *Int. J. Nanomed.* **2012**, *7*, 3445–3471. [CrossRef]
60. Kumar, C.S.S.R.; Mohammad, F. Magnetic nanomaterials for hyperthermia-based therapy and controlled drug delivery. *Adv. Drug Deliv. Rev.* **2011**, *63*, 789–808. [CrossRef]
61. Dobson, J. Gene therapy progress and prospects: Magnetic nanoparticle-based gene delivery. *Gene Ther.* **2006**, *13*, 283–287. [CrossRef]
62. Thandu, M.; Rapozzi, V.; Xodo, L.; Albericio, F.; Comuzzi, C.; Cavalli, S. "Clicking" Porphyrins to Magnetic Nanoparticles for Photodynamic Therapy. *Chempluschem* **2014**, *79*, 90–98. [CrossRef]
63. Akbar, A.; Riaz, S.; Bashir, M.; Naseem, S. Effect of Fe3+/Fe2+ Ratio on Superparamagnetic Behavior of Spin Coated Iron Oxide Thin Films. *IEEE Trans. Magn.* **2014**, *50*, 4. [CrossRef]
64. Depalo, N.; Carrieri, P.; Comparelli, R.; Striccoli, M.; Agostiano, A.; Bertinetti, L.; Innocenti, C.; Sangregorio, C.; Curri, M.L. Biofunctionalization of Anisotropic Nanocrystalline Semiconductor–Magnetic Heterostructures. *Langmuir* **2011**, *27*, 6962–6970. [CrossRef] [PubMed]
65. Valente, G.; Depalo, N.; de Paola, I.; Iacobazzi, R.M.; Denora, N.; Laquintana, V.; Comparelli, R.; Altamura, E.; Latronico, T.; Altomare, M.; et al. Integrin-targeting with peptide-bioconjugated semiconductor-magnetic nanocrystalline heterostructures. *Nano Res.* **2016**, *9*, 644–662. [CrossRef]
66. Depalo, N.; Iacobazzi, R.M.; Valente, G.; Arduino, I.; Villa, S.; Canepa, F.; Laquintana, V.; Fanizza, E.; Striccoli, M.; Cutrignelli, A.; et al. Sorafenib delivery nanoplatform based on superparamagnetic iron oxide nanoparticles magnetically targets hepatocellular carcinoma. *Nano Res.* **2017**, *10*, 2431–2448. [CrossRef]
67. Pasternack, R.F.; Collings, P.J. Resonance Light-Scattering—A New Technique for Studying Chromophore Aggregation. *Science* **1995**, *269*, 935–939. [CrossRef]

68. Akins, D.L.; Zhu, H.R.; Guo, C. Absorption and raman-scattering by aggregated meso-tetrakis(p-sulfonatophenyl)porphine. *J. Phys. Chem.* **1994**, *98*, 3612–3618. [CrossRef]
69. Ribo, J.M.; Crusats, J.; Farrera, J.A.; Valero, M.L. Aggregation in water solutions of tetrasodium diprotonated meso-tetrakis(4-sulfonatophenyl)porphyrin. *J. Chem. Soc. Chem. Commun.* **1994**, 681–682. [CrossRef]
70. Gandini, S.C.M.; Gelamo, E.L.; Itri, R.; Tabak, M. Small angle X-ray scattering study of meso-tetrakis(4-sulfonatophenyl) porphyrin in aqueous solution: A self-aggregation model. *Biophys. J.* **2003**, *85*, 1259–1268. [CrossRef]
71. Schwab, A.D.; Smith, D.E.; Rich, C.S.; Young, E.R.; Smith, W.F.; de Paula, J.C. Porphyrin nanorods. *J. Phys. Chem. B* **2003**, *107*, 11339–11345. [CrossRef]
72. Scolaro, L.M.; Romeo, A.; Castriciano, M.A.; Micali, N. Unusual optical properties of porphyrin fractal J-aggregates. *Chem. Commun.* **2005**, *24*, 3018–3020. [CrossRef] [PubMed]
73. Romeo, A.; Castriciano, M.A.; Scolaro, L.M. Spectroscopic and kinetic investigations on porphyrin J-aggregates induced by polyamines. *J. Porphyr. Phthalocyanines* **2010**, *14*, 713–721. [CrossRef]
74. Castriciano, M.A.; Romeo, A.; Villari, V.; Micali, N.; Scolaro, L.M. Structural rearrangements in 5,10,15,20-tetrakis(4-sulfonatophenyl)porphyrin J-aggregates under strongly acidic conditions. *J. Phys. Chem. B* **2003**, *107*, 8765–8771. [CrossRef]
75. Depalo, N.; Fanizza, E.; Vischio, F.; Denora, N.; Laquintana, V.; Cutrignelli, A.; Striccoli, M.; Giannelli, G.; Agostiano, A.; Curri, M.L.; et al. Imaging modification of colon carcinoma cells exposed to lipid based nanovectors for drug delivery: A scanning electron microscopy investigation. *RSC Adv.* **2019**, *9*, 21810–21825. [CrossRef]
76. Pasternack, R.F.; Gibbs, E.J.; Collings, P.J.; de Paula, J.C.; Turzo, L.C.; Terracina, A. A nonconventional approach to supramolecular formation dynamics. The kinetics of assembly of DNA-bound porphyrins. *J. Am. Chem. Soc.* **1998**, *120*, 5873–5878. [CrossRef]
77. Monsù Scolaro, L.; Castriciano, M.; Romeo, A.; Mazzaglia, A.; Mallamace, F.; Micali, N. Nucleation effects in the aggregation of water-soluble porphyrin aqueous solutions. *Phys. A Stat. Mech. Its Appl.* **2002**, *304*, 158–169. [CrossRef]
78. Paulo, P.M.R.; Costa, S.M.B. Non-covalent dendrimer–porphyrin interactions: The intermediacy of H-aggregates? *Photochem. Photobiol. Sci.* **2003**, *2*, 597–604. [CrossRef]
79. Castriciano, M.A.; Romeo, A.; Angelini, N.; Micali, N.; Longo, A.; Mazzaglia, A.; Scolaro, L.M. Structural features of meso-tetrakis(4-carboxyphenyl)porphyrin interacting with amino-terminated poly(propylene oxide). *Macromolecules* **2006**, *39*, 5489–5496. [CrossRef]
80. Castriciano, M.A.; Romeo, A.; Scolaro, L.M. Aggregation of meso-tetrakis(4-sulfonatophenyl)porphyrin on polyethyleneimine in aqueous solutions and on a glass surface. *J. Porphyr. Phthalocyanines* **2002**, *6*, 431–438. [CrossRef]
81. Vlaming, S.M.; Augulis, R.; Stuart, M.C.A.; Knoester, J.; van Loosdrecht, P.H.M. Exciton Spectra and the Microscopic Structure of Self-Assembled Porphyrin Nanotubes. *J. Phys. Chem. B* **2009**, *113*, 2273–2283. [CrossRef]

© 2020 by the authors. Licensee MDPI, Basel, Switzerland. This article is an open access article distributed under the terms and conditions of the Creative Commons Attribution (CC BY) license (http://creativecommons.org/licenses/by/4.0/).